T0231333

Computational Photography

Methods and Applications

Digital Imaging and Computer Vision Series

Series Editor

Rastislav Lukac

Foveon, Inc./Sigma Corporation
San Jose, California, U.S.A.

Computational Photography: Methods and Applications, *edited by Rastislav Lukac*
Super-Resolution Imaging, *edited by Peyman Milanfar*

Computational Photography

Methods and Applications

Edited by
Rastislav Lukac

CRC Press
Taylor & Francis Group
Boca Raton London New York

CRC Press is an imprint of the
Taylor & Francis Group, an **informa** business

CRC Press
Taylor & Francis Group
6000 Broken Sound Parkway NW, Suite 300
Boca Raton, FL 33487-2742

© 2011 by Taylor and Francis Group, LLC
CRC Press is an imprint of Taylor & Francis Group, an Informa business

No claim to original U.S. Government works

International Standard Book Number: 978-1-4398-1749-0 (Hardback)

Library of Congress Cataloging-in-Publication Data

Computational photography : methods and applications / editor, Rastislav Lukac.
    p. cm. -- (Digital imaging and computer vision)
    Includes bibliographical references and index.
    ISBN 978-1-4398-1749-0 (hardcover : alk. paper)
    1. Computational photography. I. Lukac, Rastislav.

TR267.3.C65 2011
775--dc22   2010034106

Visit the Taylor & Francis Web site at
http://www.taylorandfrancis.com

and the CRC Press Web site at
http://www.crcpress.com

Photography suits the temper of this age — of active bodies and minds. It is a perfect medium for one whose mind is teeming with ideas, imagery, for a prolific worker who would be slowed down by painting or sculpting, for one who sees quickly and acts decisively, accurately.

—Edward Henry Weston, photographer

Dedication

To my lovely daughter, Sofia

Contents

Preface

Computational photography is a new and rapidly developing research field. It has evolved from computer vision, image processing, computer graphics, and applied optics, and refers broadly to computational imaging techniques that enhance or extend the capabilities of digital photography. The output of these techniques is an image which cannot be produced by today's common imaging solutions and devices. Despite the recent establishment of computational photography as a recognized research area, numerous commercial products capitalizing on its principles have already appeared in diverse market applications due to the gradual migration of computational algorithms from computers to image-enabled consumer electronic devices and imaging software.

Image processing methods for computational photography are of paramount importance in the research and development community specializing in computational imaging due to the urgent needs and challenges of emerging digital camera applications. There exist consumer digital cameras which use face detection to better focus and expose the image, while others perform preliminary panorama stitching directly in the camera and use local tone mapping to manage difficult lighting situations. There are also successful attempts to use the information from a set of images, for instance, to reduce or eliminate image blur, suppress noise, increase image resolution, and remove objects from or add them to a captured image.

Thus it is not difficult to see that many imaging devices and applications already rely on research advances in the field of computational photography. The commercial proliferation of digital still and video cameras, image-enabled mobile phones and personal digital assistants, surveillance and automotive apparatuses, machine vision systems, and computer graphic systems has increased the demand for technical developments in the area. It is expected that the growing interest in image processing methods for computational photography and their use in emerging applications such as digital photography and art, visual communication, online sharing in social networks, digital entertainment, surveillance, and multimedia will continue.

The purpose of this book is to fill the existing gap in the literature and comprehensively cover the system design, implementation, and application aspects of image processing-driven computational photography. Due to the rapid developments in specialized areas of computational photography, the book is a contributed volume in which well-known experts deal with specific research and application problems. It presents the state-of-the-art as well as the most recent trends in image processing methods and applications for computational photography. It serves the needs of different readers at different levels. It can be used as textbook in support of graduate courses in computer vision, digital imaging, visual data processing and computer graphics or as stand-alone reference for graduate students, researchers, and practitioners. For example, a researcher can use it as an up-to-date reference since it will offer a broad survey of the relevant literature. Development engineers, technical managers, and executives may also find it useful in the design and implementation of various digital image and video processing tasks.

This book provides a strong, fundamental understanding of theory and methods, and a foundation upon which solutions for many of today's most interesting and challenging computational imaging problems can be built. It details recent advances in digital imaging, camera image processing, and computational photography methods and explores their applications. The book begins by focusing on single capture image fusion technology for consumer digital cameras. This is followed by the discussion of various steps in a camera image processing pipeline, such as data compression, color correction and enhancement, denoising, demosaicking, super-resolution reconstruction, deblurring, and high-dynamic range imaging. Then, the reader's attention is turned to bilateral filtering and its applications, painterly rendering of digital images, shadow detection for surveillance applications, and camera-driven document rectification. The next part of the book presents machine learning methods for automatic image colorization and digital face beautification. The remaining chapters explore light field acquisition and processing, space-time light field rendering, and dynamic view synthesis with an array of cameras.

Chapters 1 and 2 discuss concepts and technologies that allow effective design and high performance of single-sensor digital cameras. Using a four-channel color filter array, an image capture system can produce images with high color fidelity and improved signal-to-noise performance relative to traditional three-channel systems. This is accomplished by adding a panchromatic or spectrally nonselective channel to the digital camera sensor to decouple sensing luminance information from chrominance information. To create a full-color image on output, as typically required for storage and display purposes, *single capture image fusion* techniques and methodology are used as the means for reducing the original four-channel image data down to three channels in a way that makes the best use of the additional fourth channel. *Single capture image fusion with motion consideration* enhances these concepts to provide a capture system that can additionally address the issue of motion occurring during a capture. By allowing different integration times for the panchromatic and color pixels, an imaging system produces images with reduced motion blur.

Chapters 3 and 4 address important issues of data compression and color manipulation in the compressed domain of captured camera images. *Lossless compression of Bayer color filter array images* has become *de facto* a standard solution of image storage in single-lens reflex digital cameras, since stored raw images can be completely processed on a personal computer to achieve higher quality compared to resource-limited in-camera processing. This approach poses a unique challenge of spectral decorrelation of spatially interleaved samples of three or more sampling colors. Among a number of reversible lossless transforms, algorithms that rely on predictive and entropy coding seem to be very effective in removing statistical redundancies in both spectral and spatial domains using the spatial correlation in the raw image and the statistical distribution of the prediction residue.

Color restoration and enhancement in the compressed domain address the problem of adjusting a camera image represented in the block discrete cosine transform space. The goal is to compensate for shifts from perceived color in the scene due to the ambient illumination and a poor dynamic range of brightness values due to the presence of strong background illumination. The objective of restoring colors from varying illumination is to derive an illumination-independent representation of an image to allow its faithful rendering, whereas the enhancement process aims at improving the color reproduction capability of a display device depending upon its displayable range of color gamut and the brightness range it can handle. Both color correction and enhancement for the display of color images may simultaneously be required when the scene suffers from both widely varying spectral components and brightness of illuminants.

Chapters 5 to 8 are intended to cover the basics of and review recent advances in camera image processing based on the concept of data estimation. Since digital camera images usually suffer from the presence of noise, the denoising step is one of the crucial components of the imaging pipeline to meet certain image quality requirements. *Principal component analysis-based denoising of color filter array images* addresses the problem of noise removal in raw image data captured using a sensor equipped with a color filter array. Denoising such raw image data avoids color artifacts that are introduced in the color restoration process through the combination of noisy sensor readings corresponding to different color channels. Principal component analysis can be used as the underlying concept of a spatially adaptive denoising algorithm to analyze the local image statistics inside a supporting window. By exploiting the spatial and spectral correlation characteristics of the color filter array image, the denoising algorithm can effectively suppress noise while preserving color edges and details. The denoised color filter array image is convenient for subsequent demosaicking, which is an image processing operation used to restore the color image from the raw sensor data acquired by a digital camera equipped with a color filter array.

Regularization-based color image demosaicking constitutes an effective strategy which consists of considering demosaicking as an inverse problem that can be solved by making use of some prior knowledge about natural color images. Taking advantage of assumptions based on the smoothness of the color components and the high-frequency correlation between the color channels in the regularization process allows the design of efficient demosaicking algorithms that are suitable for any color filter array and that can be coupled with other frequent problems in image reconstruction and restoration.

Super-resolution imaging is another image restoration operation that has become more and more important in modern imaging systems and applications. It is used to produce a high-resolution image or a sequence of high-resolution images from a set of low-resolution images. The process requires an image acquisition model that relates a high-resolution image to multiple low-resolution images and involves solving the resulting inverse problem. The acquisition model includes aliasing, blurring, and noise as the main sources of information loss. A super-resolution algorithm increases the spatial detail in an image, recovering the high-frequency information that is lost during the imaging process.

Focusing on removing the blurring effect, which is mainly caused by camera motion during exposure or a lens that is out-of-focus, the conventional approach is to construct an image degradation model and then solve the inverse problem of the given model. *Image deblurring using multi-exposed images* constitutes a new approach that takes advantage of recent advances in image sensing technology that enable splitting or controlling the exposure time. This approach exploits the mutually different pieces of information from multi-exposed images of the same scene to produce a deblurred image that faithfully represents a real scene.

Chapters 9 and 10 deal with the enhancement of the dynamic range of an image using multiple captures of the scene. *Color high-dynamic range imaging* enables access to a wider range of color values than traditional digital photography. Methods for capture, composition, and display of high-dynamic range images have become quite common in modern imaging systems. In particular, luminance-chrominance space-driven composition techniques seem to be suitable in various real-life situations where the source images are corrupted by noise and/or misalignment and the faithful treatment of color is essential. As the objects in the scene often move during the capture process, *high-dynamic range imaging for dynamic scenes* is needed to enhance the performance of an imaging system and extend the range of its applications by integrating motion and dynamic scenes in underlying technology targeting both photographs and movies.

Chapter 11 focuses on *shadow detection in digital images and videos*, with application to video surveillance. Addressing the problem of color modeling of cast shadows in real-life situations requires a robust adaptive model for shadow segmentation without strong restrictions on *a priori* probabilities, image quality, objects' shapes, and processing speed. Such a modeling framework can be generalized for and used to compare different color spaces, as the appropriate color space selection is a key in reliable shadow detection and classification, for example, using color-based pixel clustering and Bayesian foreground/background shadow segmentation.

Chapter 12 presents another way of using information from more than one image. *Document image rectification using single-view or two-view camera input* in digital camera-driven systems for document image acquisition, analysis, and processing represents an alternative to flatbed scanners. A stereo-based method can be employed to complete the rectification task using explicit three-dimensional reconstruction. Since the method works irrespective of document contents and removes specular reflections, it can be used as a pre-processing tool for optical character recognition and digitization of figures and pictures. In situations when a user-provided bounding box is available, a single-view method allows rectifying a figure inside this bounding box in an efficient, robust, and easy-to-use manner.

Chapter 13 discusses both the *theory and applications of the bilateral filter*. This filter is widely used in various image processing and computer vision applications due to its ability to preserve edges while performing spatial smoothing. The filter is shown to relate to popular approaches based on robust estimation, weighted least squares estimation, and partial differential equations. It has a number of extensions and variations that make the bilateral filter an indispensable tool in modern image and video processing systems, although a fast implementation is usually critical for practical applications.

Chapter 14 focuses on *painterly rendering* methods. These methods convert an input image into an artistic image in a given style. Artistic images can be generated by simulating the process of putting paint on paper or canvas. A synthetic painting is represented as a list of brush strokes that are rendered on a white or canvas textured background. Brush strokes can be mathematically modeled or their attributes can be extracted from the source image. Another approach is to abstract from the classical tools that have been used by artists and focus on the visual properties, such as sharp edges or absence of natural texture, which distinguish painting from photographic images.

Chapters 15 and 16 deal with two training-based image analysis and processing steps. *Machine learning methods for automatic image colorization* focus on adding colors to a grayscale image without any user intervention. This can be done by formally stating the color prediction task as an optimization problem with respect to an energy function. Different machine learning methods, in particular nonparametric methods such as Parzen window estimators and support vector machines, provide a natural and efficient way of incorporating information from various sources. In order to cope with the multimodal nature of the problem, the solution can be found directly at the global level with the help of graph cuts, which makes the approach more robust to noise and local prediction errors and allows resolving large-scale ambiguities and handling cases with more texture noise. The approach provides a way of learning local color predictors along with spatial coherence criteria and permits a large number of possible colors.

In another application of training-based methods, *machine learning for digital face beautification* constitutes a powerful tool for automatically enhancing the attractiveness of a face in a given portrait. It aims at introducing only subtle modifications to the original image by manipulating the geometry of the face, such that the resulting beautified face maintains a strong, unmistakable similarity to the original. Using a variety of facial locations to calculate a feature vector of a given face, a feature space is searched for a vector that corresponds to a more attractive face. This can be done by employing an automatic facial beauty rating machine which has the form of two support vector regressors trained separately on a database of female and male faces with accompanying facial attractiveness ratings collected from a group of human raters. The feature vector output by the regressor serves as a target to define a two-dimensional warp field which maps the original facial features to their beautified locations. The method augments image enhancement and retouching tools available in existing digital image editing packages.

Finally, Chapters 17 and 18 discuss various light field-related issues. *High-quality light field acquisition and processing* methods rely on various hardware and software approaches to overcome the lack of the spatial resolution and avoid photometric distortion and aliasing in output images. Programmable aperture is an example of a device for high-resolution light field acquisition. It exploits the fast multiple-exposure feature of digital sensors without trading off sensor resolution to capture the light field sequentially, which, in turn, enables the multiplexing of light rays. The quality of the captured light field can be further improved by employing a calibration algorithm to remove the photometric distortion unique to the light field without using any reference object by estimating this distortion directly from the captured light field and a depth estimation algorithm utilizing the multi-view property of light field and visibility reasoning to generate view-dependent depth maps for view interpolation. The device and algorithms constitute a complete system for high-quality light field acquisition.

Light field-style rendering techniques have an important position among image-based modeling methods for *dynamic view synthesis with an array of cameras*. These techniques can be extended for dynamic scenes, constituting an approach termed as space-time light field rendering. Instead of capturing the dynamic scene in strict synchronization and treating each image set as an independent static light field, the notion of a space-time light field assumes a collection of video sequences that may or may not be synchronized and can have different capture rates. In order to be able to synthesize novel views from any viewpoint at any instant in time, feature correspondences are robustly identified across frames and used as land markers to digitally synchronize the input frames and improve view synthesis quality. This concept is further elaborated in reconfigurable light field rendering where both the scene content and the camera configurations can be dynamic. Automatically adjusting the cameras' placement allows achieving optimal view synthesis results for different scene contents.

The bibliographic links included in all chapters of the book provide a good basis for further exploration of the presented topics. The volume includes numerous examples and illustrations of computational photography results, as well as tables summarizing the results of quantitative analysis studies. Complementary material is available online at *http://www.colorimageprocessing.org*.

I would like to thank the contributors for their effort, valuable time, and motivation to enhance the profession by providing material for a wide audience while still offering their individual research insights and opinions. I am very grateful for their enthusiastic support, timely response, and willingness to incorporate suggestions from me to improve the quality of contributions. I also thank Rudy Guttosch, my colleague at Foveon, Inc., for his help with proofreading some of the chapters. Finally, a word of appreciation for CRC Press / Taylor & Francis for giving me the opportunity to edit a book on computational photography. In particular, I would like to thank Nora Konopka for supporting this project, Jennifer Ahringer for coordinating the manuscript preparation, Shashi Kumar for his LaTeX assistance, Karen Simon for handling the final production, Phoebe Roth for proofreading the book, and James Miller for designing the book cover.

Rastislav Lukac
Foveon, Inc. / Sigma Corp., San Jose, CA, USA
E-mail: lukacr@colorimageprocessing.com
Web: www.colorimageprocessing.com

The Editor

Rastislav Lukac (www.colorimageprocessing.com) received M.S. (Ing.) and Ph.D. degrees in telecommunications from the Technical University of Kosice, Slovak Republic, in 1998 and 2001, respectively. From February 2001 to August 2002, he was an assistant professor with the Department of Electronics and Multimedia Communications at the Technical University of Kosice. From August 2002 to July 2003, he was a researcher with the Slovak Image Processing Center in Dobsina, Slovak Republic. From January 2003 to March 2003, he was a postdoctoral fellow with the Artificial Intelligence and Information Analysis Labora-
tory, Aristotle University of Thessaloniki, Thessaloniki, Greece. From May 2003 to August 2006, he was a postdoctoral fellow with the Edward S. Rogers Sr. Department of Electrical and Computer Engineering, University of Toronto, Toronto, Ontario, Canada. From September 2006 to May 2009, he was a senior image processing scientist at Epson Canada Ltd., Toronto, Ontario, Canada. In June 2009, he was a visiting researcher with the Intelligent Systems Laboratory, University of Amsterdam, Amsterdam, the Netherlands. Since August 2009, he has been a senior digital imaging scientist – image processing manager at Foveon, Inc. / Sigma Corp., San Jose, California, USA. Dr. Lukac is the author of four books and four textbooks, a contributor to nine books and three textbooks, and he has published over 200 scholarly research papers in the areas of digital camera image processing, color image and video processing, multimedia security, and microarray image processing. He is the author of over 25 patent-pending inventions in the areas of digital color imaging and pattern recognition. He has been cited more than 650 times in peer-review journals covered by the *Science Citation Index* (SCI).

Dr. Lukac is a senior member of the Institute of Electrical and Electronics Engineers (IEEE), where he belongs to the Circuits and Systems, Consumer Electronics, and Signal Processing societies. He is an editor of the books *Single-Sensor Imaging: Methods and Applications for Digital Cameras* (Boca Raton, FL: CRC Press / Taylor & Francis, September 2008) and *Color Image Processing: Methods and Applications* (Boca Raton, FL: CRC Press / Taylor & Francis, October 2006). He is a guest editor of *Real-Time Imaging*, Special Issue on Multi-Dimensional Image Processing, *Computer Vision and Image Understanding*, Special Issue on Color Image Processing, *International Journal of Imaging Systems and Technology*, Special Issue on Applied Color Image Processing, and *International Journal of Pattern Recognition and Artificial Intelligence*, Special Issue on Facial Image Processing and Analysis. He is an associate editor for the *IEEE Transactions on Circuits and Systems for Video Technology* and the *Journal of Real-Time Image Processing*. He is an editorial board member for *Encyclopedia of Multimedia* (2nd Edition, Springer, September 2008). He is a *Digital Imaging and Computer Vision* book series founder and editor for CRC Press / Taylor & Francis. He serves as a technical reviewer for various scientific journals, and participates as a member of numerous international conference committees. He is the recipient of the 2003 North Atlantic Treaty Organization / National Sciences and Engineering Research Council of Canada (NATO/NSERC) Science Award, the Most Cited Paper Award for the *Journal of Visual Communication and Image Representation* for the years 2005–2007, and the author of the #1 article in the ScienceDirect Top 25 Hottest Articles in *Signal Processing* for April–June 2008.

Contributors

James E. Adams, Jr. Eastman Kodak Company, Rochester, New York, USA

Yasemin Altun Max Planck Institute for Biological Cybernetics, Tübingen, Germany

Csaba Benedek Computer and Automation Research Institute, Budapest, Hungary

Ilja Bezrukov Max Planck Institute for Biological Cybernetics, Tübingen, Germany

Giancarlo Calvagno University of Padova, Padova, Italy

Yuk-Hee Chan The Hong Kong Polytechnic University, Hong Kong SAR

Guillaume Charpiat INRIA, Sophia-Antipolis, France

Homer H. Chen National Taiwan University, Taipei, Taiwan R.O.C.

Nam Ik Cho Seoul National University, Seoul, Korea

King-Hong Chung The Hong Kong Polytechnic University, Hong Kong SAR

Aaron Deever Eastman Kodak Company, Rochester, New York, USA

Gideon Dror The Academic College of Tel-Aviv-Yaffo, Tel Aviv, Israel

Alessandro Foi Tampere University of Technology, Tampere, Finland

Atanas Gotchev Tampere University of Technology, Tampere, Finland

Bahadir K. Gunturk Louisiana State University, Baton Rouge, Louisiana, USA

John F. Hamilton, Jr. Rochester Institute of Technology, Rochester, New York, USA

Matthias Hofmann Institute for Biological Cybernetics, Tübingen, Germany

Katrien Jacobs University College London, London, UK

Seung-Won Jung Korea University, Seoul, Korea

Sung-Jea Ko Korea University, Seoul, Korea

Hyung Il Koo Seoul National University, Seoul, Korea

Mrityunjay Kumar Eastman Kodak Company, Rochester, New York, USA

Chia-Kai Liang National Taiwan University, Taipei, Taiwan R.O.C.

Celine Loscos Universitat de Girona, Girona, Spain

Rastislav Lukac   Foveon, Inc. / Sigma Corp., San Jose, California, USA

Daniele Menon   University of Padova, Padova, Italy

Sanjit K. Mitra   University of Southern California, Los Angeles, California, USA

Efraín O. Morales   Eastman Kodak Company, Rochester, New York, USA

Jayanta Mukherjee   Indian Institute of Technology, Kharagpur, India

Russell Palum   Eastman Kodak Company, Rochester, New York, USA

Giuseppe Papari   University of Groningen, Groningen, The Netherlands

Nicolai Petkov   University of Groningen, Groningen, The Netherlands

Bruce H. Pillman   Eastman Kodak Company, Rochester, New York, USA

Ossi Pirinen   OptoFidelity Ltd., Tampere, Finland

Bernhard Schölkopf   Institute for Biological Cybernetics, Tübingen, Germany

Tamás Szirányi   Péter Pázmány Catholic University, Budapest, Hungary

Huaming Wang   University of California, Berkeley, Berkeley, California, USA

Ruigang Yang   University of Kentucky, Lexington, Kentucky, USA

Cha Zhang   Microsoft Research, Redmond, Washington, USA

Lei Zhang   The Hong Kong Polytechnic University, Hong Kong

1

Single Capture Image Fusion

James E. Adams, Jr., John F. Hamilton, Jr., Mrityunjay Kumar, Efraín O. Morales, Russell Palum, and Bruce H. Pillman

1.1 Introduction

A persistent challenge in the design and manufacture of digital cameras is how to improve the signal-to-noise performance of these devices while simultaneously maintaining high color fidelity captures. The present industry standard three-color channel system is constrained in that the fewest possible color channels are employed for the purposes of both luminance and chrominance image information detection. Without additional degrees of freedom, for instance, additional color channels, digital camera designs are generally limited to solutions based on improving sensor hardware (larger pixels, lower readout noise, etc.) or better image processing (improved denoising, system-wide image processing chain optimization, etc.) Due to being constrained to three channels, the requirements for improved signal-to-noise and high color fidelity are frequently in opposition to each other, thereby providing a limiting constraint on how much either can be improved. For example, to improve the light sensitivity of the sensor system, one might wish to make the color channels broader spectrally. While this results in lower image noise in the raw capture, the color correction required to restore the color fidelity amplifies the noise so much that there can be a net *loss* in overall signal-to-noise performance.

This chapter explores the approach of adding a fourth, *panchromatic* or spectrally non-selective, channel to the digital camera sensor in order to decouple sensing luminance (spatial) information from chrominance (color) information [1]. Now the task of improving signal-to-noise can be largely confined to just the panchromatic channel while leaving the requirement for high color fidelity captures to the three color channels. As any such system must eventually create a full-color image on output, some means is needed for reducing the original four-channel image data down to three channels in a way that makes the best use of the additional fourth channel. To this end, *image fusion* techniques and methodology are selected as the means for accomplishing this task. Therefore, the remainder of this introduction first reviews the concept of image fusion and then sets the stage for how this body of work can be applied to the problem of capturing and processing a single four-channel digital camera capture to produce a full-color image with improved signal-to-noise and high color fidelity.

1.1.1 Image Fusion

It is a well-known fact that despite tremendous advances in sensor technologies, no single sensor can acquire all the required information of the target reliably at all times. This naturally leads to deployment of multiple sensors, having complementary properties, to inspect the target, thus capturing as much information as possible. Data fusion provides a framework to integrate redundant as well as complementary information provided by multiple sensors in such a manner that the fused information describes the target better than any of the individual sensors. The exploitation of redundant information improves accuracy and reliability, whereas integration of complementary information improves the interpretability of the target [2], [3], [4], [5], [6]. In a typical data fusion application, multiple sensors observe a common target and a decision is taken based on the collective information [7].

The concept of data fusion is not new. It is naturally performed by living beings to achieve a more accurate assessment of the surrounding environment and identification of threats, thereby improving their chances of survival. For example, human beings and animals use a combination of sight, touch, smell, and taste to perceive the quality of an edible object. Tremendous advances in sensor and hardware technology and signal processing techniques have provided the ability to design software and hardware modules to mimic the natural data fusion capabilities of humans and animals [3].

Data fusion applied to image-based applications is commonly referred to as image fusion. The goal of image fusion is to extract information from input images such that the fused image provides better information for human or machine perception as compared to any of the input images [8], [9], [10], [11]. Image fusion has been used extensively in various areas of image processing such as digital camera imaging, remote sensing, and biomedical imaging [12], [13], [14].

From the perspective of fusion, information present in the observed images that are to be fused can be broadly categorized in the following three classes: i) common information – these are features that are present in all the observed images, ii) complementary information – features that are present only in one of the observed images, and iii) noise – features that are random in nature and do not contain any relevant information. Note that this categorization of the information could be global or local in nature. A fusion algorithm should be able to select the feature type automatically and then fuse the information appropriately. For example, if the features are similar, then the algorithm should perform an operation similar to averaging, but in the case of complementary information, should select the feature that contains relevant information.

Due to the large number of applications as well as the diversity of fusion techniques, considerable efforts have been made in developing standards for data fusion. Several models for data fusion have been proposed in the recent past [15], [16]. One of the models commonly used in signal processing applications is the three-level fusion model that is based on the levels at which information is represented [17]. This model classifies data fusion into three levels depending on the way information present in the data/image is represented and combined. If the raw images are directly used for fusion then it is called low level fusion. In the case when the features of the raw images such as edge, texture, etc., are used for fusion then it is called feature or intermediate level fusion. The third level of fusion is known as decision or high level fusion in which decisions made by several experts are combined. A detailed description of these three levels of fusion is given below.

- *Low level fusion*: At this level, several raw images are combined to produce a new "raw" image that is expected to be more informative than the inputs [18], [19], [20], [21], [22]. The main advantage of low level fusion is that the original measured quantities are directly involved in the fusion process. Furthermore, algorithms are computationally efficient and easy to implement. Low level fusion requires a precise registration of the available images.

- *Feature or intermediate level fusion*: Feature level fusion combines various features such as edges, corners, lines, and texture parameters [23], [24], [25]. In this model several feature extraction methods are used to extract the features of input images and a set of relevant features is selected from the available features. Methods of feature fusion include, for example, Principal Component Analysis (PCA) and Multi-layer Perceptron (MLP) [26]. The fused features are typically used for computer vision applications such as segmentation and object detection [26]. In feature level fusion, all the pixels of the input images do not contribute in fusion. Only the salient features of the images are extracted and fused.

- *Decision or high level fusion*: This stage combines decisions from several experts [27], [28], [29], [30]. Methods of decision fusion include, for instance, statistical methods [27], voting methods [28], fuzzy logic-based methods [29], and Dempster-Schafer's method [30].

Typically, in image fusion applications, input images for fusion are captured using multiple sensors. For example, in a typical remote sensing system, multispectral sensors are used to obtain information about the Earth's surface and image fusion techniques are used to fuse the outputs of multispectral sensors to generate a thematic map [4], [6]. As another example, recent developments in medical imaging have resulted in many imaging sensors to capture various aspects of the patient's anatomy and metabolism [31], [32]. For example, magnetic resonance imaging (MRI) is very useful for defining anatomical structure whereas metabolic activity can be captured very reliably using positron emission tomography (PET). The concept of image fusion is used to combine the output of MRI and PET sensors to obtain a single image that describes anatomical as well as metabolic activity of the patient effectively [33].

Image fusion techniques can also be applied to fuse multiple images obtained from a single sensor [34], [35], [36], [37], [38], [39], [40]. An excellent example of generating complementary information from a single sensor is the Bayer color filter array (CFA) pattern [41] extensively used in digital cameras. To reduce cost and complexity, most digital cameras are designed using a single CCD or CMOS sensor that has the panchromatic responsivity of silicon. As shown in Figure 1.1 panchromatic responsivity, which passes all visible wavelengths, is higher than color (red, green, and blue) responsivities [42].

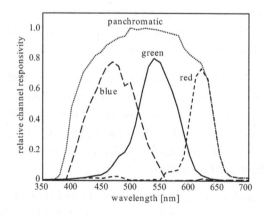

FIGURE 1.1

Relative responsivity of panchromatic, red, green, and blue channels.

G	R	G	R
B	G	B	G
G	R	G	R
B	G	B	G

FIGURE 1.2

Bayer CFA pattern.

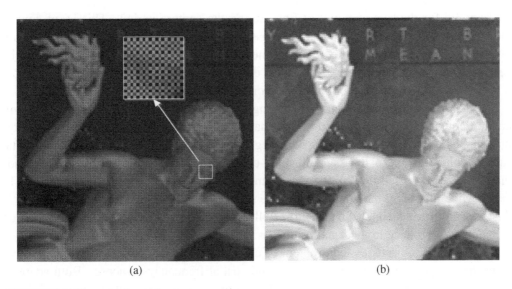

(a)　　　　　　　　　　　　　　　　(b)

FIGURE 1.3 (See color insert.)

CFA-based digital imaging: (a) Bayer CFA image, and (b) full-color reconstructed image.

Note that the pixels with panchromatic responsivity are spectrally nonselective in nature. Therefore, digital cameras use a color filter array (CFA) to capture color images, an example of which is the Bayer CFA pattern as shown in Figure 1.2. The CFA pattern provides only a single color sample at each pixel location and the missing color samples at each pixel location are estimated using a CFA interpolation or demosaicking algorithm [43], [44], [45]. An example of a Bayer CFA image is shown in Figure 1.3a. The inset shows the individual red, green, and blue pixels in the captured image. The corresponding full color image generated by applying CFA interpolation to the Bayer CFA image is shown in Figure 1.3b.

1.1.2 Chapter Overview

With the basics of image fusion stated, the remainder of the chapter discusses the design and analysis of a single sensor / single capture image fusion system. This system will be based on a four-channel CFA consisting of one panchromatic and three color channels. As a somewhat arbitrary means of simplifying the discussion, it will be assumed that all pixels

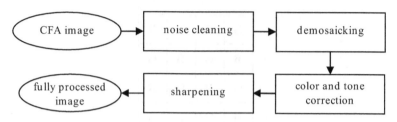

FIGURE 1.4

Example image processing chain for a four-channel system.

are exposed for the same amount of time during the capture period. This is, of course, the standard situation. In Chapter 2, this assumption will be dropped and systems with different, although concurrent, exposure times for panchromatic and color pixels will be described. As such systems are ideal for the detection and compensation of image motion, it is convenient to delay all motion-related considerations until the next chapter.

Figure 1.4 is an example image processing chain for the four-channel system that will be the reference for the discussion in this chapter. Section 1.2 focuses on CFA image formation, and color and tone correction. It begins by reviewing the fundamentals of digital camera color imaging using standard three-channel systems. From this basis it extends the discussion into the design of four-channel CFA patterns that produce high color fidelity while providing the additional degree of freedom of a panchromatic channel. It is this panchromatic channel that, in turn, will be used to enable the use of image fusion techniques to produce higher quality images not possible with typical three-channel systems. Section 1.3 discusses the problem of demosaicking four-channel CFA images both from the perspective of algorithm design and spatial frequency response. Both adaptive and nonadaptive approaches are presented and comparisons are made to standard Bayer CFA processing methods and results. Section 1.4 focuses on noise cleaning and sharpening. This includes an analytical investigation into the effects of the relative photometric gain differences between the panchromatic and color channels and how, through the use of image fusion, these gain differences can result in a fundamentally higher signal-to-noise image capture system compared to three-channel systems. Explicit investigations of how image fusion techniques are applied during both the demosaicking and sharpening operations to achieve these advantages are discussed. Section 1.5 brings the preceding material in the chapter together to illustrate with an example how the entire system shown in Figure 1.4 uses image fusion techniques to produce the final image. The performance of this system compared to a standard Bayer system is assessed both numerically and qualitatively through example images. Finally, the chapter is summarized in Section 1.6.

1.2   Color Camera Design

This chapter considers a camera that captures images for reproduction by a display or print. Color reproduction is complex because the goal is a human perception, not a simple matching of measured phenomena [46], [47]. Surround effects and viewer adaptation to different illuminants make the reproduction of a color image a challenge well beyond the technological one of producing different color stimuli. While color reproduction is very complex, the problem of camera design is slightly simpler. The camera design objective is to capture scene information to support the best possible reproduction and to do this under a wide range of imaging conditions. Because the reproduction is judged by a human observer, information about the human visual system is used in determining whether image information is common, complementary, or noise.

To set the context for this image capture problem, some history of color imaging will be reviewed. One of the first photographic color reproduction systems was demonstrated in Reference [48], later described in Reference [46]. This system captured black-and-white photographs of a scene through red, green, and blue filters, then projects them through the same filters. This technique was used as a demonstration of the trichromatic theory of color vision, although the film he used was not sensitive to red light and limited the quality of the demonstration [49]. Reference [50] reproduced color images by a very different technique. This technique captured the spectrum of light from a scene in an analog fashion and reproduced the actual spectrum when viewed by reflected light under the correct conditions. This allowed good reproduction of color, yet the process was extremely slow — the emulsion used very fine grains (10 to 40 nm in diameter) and the process required minutes of exposure even in strong daylight.

Fortunately, human color sensitivity is essentially a trichromatic system and capture of detailed spectral information is not necessary for good color reproduction. That is, human visual color response to different spectral stimuli is essentially based on three integrals over wavelength, most commonly represented as follows:

$$X = k \int_{\lambda_{min}}^{\lambda_{max}} S(\lambda) R(\lambda) \bar{x}(\lambda) \, d\lambda,$$

$$Y = k \int_{\lambda_{min}}^{\lambda_{max}} S(\lambda) R(\lambda) \bar{y}(\lambda) \, d\lambda,$$

$$Z = k \int_{\lambda_{min}}^{\lambda_{max}} S(\lambda) R(\lambda) \bar{z}(\lambda) \, d\lambda,$$

where $S(\lambda)$ is an illuminant spectral power distribution varying with wavelength λ, and $R(\lambda)$ is a spectral reflectance curve. The visual response functions $\bar{x}(\lambda)$, $\bar{y}(\lambda)$, and $\bar{z}(\lambda)$ are standardized color matching functions, defined over the wavelength range 380 nm to 780 nm and zero outside this range. The constant k is a normalization factor, normally computed to produce a value of 100 for Y with a spectrally flat 100% reflector under a chosen illumination, and X, Y, and Z are standard tristimulus values.

Different stimuli are perceived as matching colors if the different stimuli produce the same three tristimulus values. Color matching studies in the 1920s and 1930s provided the basis for the CIE standard color matching functions, providing a quantitative reference for how different colors can be matched by a three primary system. While spectral cameras continue to be developed and serve research needs, they are more complex than a trichromatic camera and usually require more time, light, or both, to capture a scene. For most color reproduction purposes, the trichromatic camera is a better match to the human visual system.

FIGURE 1.5

Color and tone correction block details.

Many different approaches have been developed to acquire trichromatic color images. Some cameras scan a scene with a trilinear sensor incorporating color filters to capture three channels of color information for every pixel in the scene [51], [52]. These cameras can achieve very high spatial resolution and excellent color quality. Because the exposure of each scan line of the scene is sequential, the exposure time for each line must be a small fraction of the capture time for the whole scene. These cameras often take seconds or even minutes to acquire a full scene, so they work poorly for recording scenes that have any motion or variation in lighting. Some cameras use a single panchromatic area sensor and filters in the lens system to sequentially capture three color channels for every pixel in the scene. These are faster than the linear scanning cameras, since all pixels in a color channel are exposed simultaneously, although the channels are exposed sequentially. These are commonly used and particularly effective in astronomy and other scientific applications where motion is not a factor. Some cameras use dichroic beam splitters and three area sensors to simultaneously capture three channels of color for each pixel in the scene [53], [54]. This is particularly common and successful in high-quality video cameras, where a high pixel rate makes the processing for demosaicking difficult to implement well. These are the fastest cameras, since all color channels are exposed simultaneously. Each of these approaches can perform well, but they increase the cost and complexity of the camera and restrict the range of camera operation.

Cost and complexity issues drive consumer digital color cameras in the direction of a single sensor that captures all color information simultaneously. Two approaches based on single array sensors are presently used. One approach fabricates a sensor with three layers of photodiodes and uses the wavelength-dependent depth of photon absorption to provide spectral sensitivity [55], [56]. This system allows sampling three channels of color information at every pixel, although the spectral sensitivity poses several challenges for image processing. The approach more commonly used is the fusion of multiple color channels from a sensor with a color filter array into a full-color image. This fusion approach, specifically with a four-channel color filter array, is the focus of this chapter.

A camera embodying this approach includes a single lens, a single area array sensor with a color filter array, and a processing path to convert pixel values read from the sensor to a suitable image for color reproduction. The image processing chain used is shown as a block diagram in Figure 1.4. The block labeled as color and tone processing is examined in more detail here, as shown in Figure 1.5. This figure breaks the overall color and tone processing down into three distinct steps. The first, white balance, applies gain factors to each color channel of camera pixel values to provide an image with equal mean code values in each color channel for neutral scene content. The second, color correction, converts the white balanced image to a known set of color primaries, such as the primaries used for sRGB. The final step applies a tone correction to convert the image to a rendered image

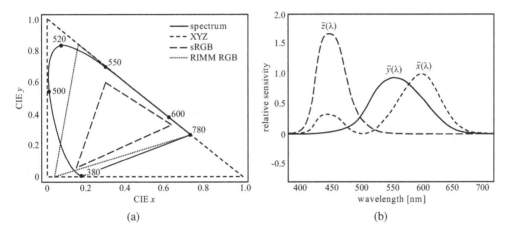

FIGURE 1.6

(a) CIE xy chromaticity diagram, and (b) CIE 1931 color matching functions.

suitable for viewing. This is often referred to as gamma correction, although optimal tone correction is rarely as simple as correcting for a standard display nonlinearity. More details behind these operations are discussed in Reference [47].

1.2.1 Three-Channel Arrays

This section discusses spectral sensitivity and color correction for three-channel systems to illustrate their limitations and show the motivation for a camera with four color channels. Color correction from capture device spectral sensitivity to output device color will be illustrated using an additive three-primary system, such as a video display.

The colors that can be reproduced with a three-color additive display are defined by the tristimulus values of its three primaries [47]. The range of colors, or gamut, of the display is a triangle in chromaticity space, such as shown in Figure 1.6a. This figure shows the spectral locus and several sets of primaries. The spectral locus traces the chromaticity of monochromatic (narrow band) light of each wavelength in the visible spectrum, with several wavelengths marked for illustration. All visible colors are contained in this horseshoe region. The XYZ set of primaries is a hypothetical set of primaries that bound a triangle including the entire spectral locus. The sRGB primaries are standard video primaries similar to most television and computer displays, specifically ones based on CRT technology. The Reference Input Medium Metric (RIMM) RGB [57] primaries are used in applications where a gamut larger that the usual video gamut is desired, since many real world colors extend beyond the gamut of sRGB. Two of the RIMM primaries are also hypothetical, lying outside the spectral locus.

Spectral sensitivities that are linear combinations of those shown in Figure 1.6b support the lowest possible color errors in reproduction. The curves shown are the standard CIE XYZ color matching functions (CMFs) corresponding to the XYZ primaries shown in Figure 1.6a. Linear combinations of these curves are also color matching functions, corresponding to other sets of primaries. Data captured with any set of color matching functions can be converted to another set of color matching functions using a 3×3 matrix

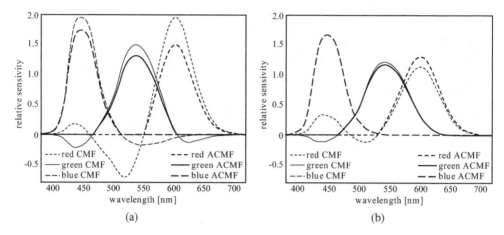

FIGURE 1.7

Color matching functions and approximations: (a) sRGB, and (b) RIMM.

as $\mathbf{P}_C = \mathbf{MP}_O$, where $\mathbf{P}_C$ and $\mathbf{P}_O$ are 3×1 vectors of converted pixel values and original color pixel values, respectively. This matrix operation is also referred to as color correction.

Each set of primaries has a corresponding set of color matching functions. An example shown in Figure 1.7a presents the color matching functions for the sRGB primaries. Because cameras cannot provide negative spectral sensitivity, cameras use all-positive approximations to color matching functions (ACMF) instead. The figure also shows one simple approximation to the sRGB color matching functions, formed by clipping at zero and eliminating the red channel sensitivity to blue light. A second example, for the RIMM set of primaries, is shown in Figure 1.7b, along with a set of simple all-positive approximations to these color matching functions. Note the RIMM color matching functions have smaller negative lobes than the sRGB color matching functions. The size of the negative excursions in the color matching functions correspond to how far the spectral locus lies outside the color gamut triangle, as can be seen by comparing the curves in Figures 1.7 and 1.6b with the gamut triangles in Figure 1.6a. Cameras with spectral sensitivities that are not color matching functions produce color errors because the camera integration of the spectrum is different from the human integration of the spectrum. In a successful color camera, the spectral sensitivities must be chosen so these color errors are acceptable for the intended application.

Digital camera images are usually corrected to one of several standardized RGB color spaces, such as sRGB [58], [59], RIMM RGB [57], [60], and Adobe RGB (1998) [61], each with somewhat different characteristics. Some of these color spaces and others are compared in Reference [62].

The deviation of a set of spectral sensitivities from color matching functions was considered in Reference [63], which proposed a q factor for measuring how well a single spectral sensitivity curve compared with its nearest projection onto color matching functions. This concept was extended in Reference [64], to the v factor, which considers the projection of a set of spectral sensitivities (referred to as scanning filters) onto the human visual sensitivities. Because q and v are computed on spectral sensitivities, the factors are not well correlated to color errors calculated in a visually uniform space, such as CIE Lab.

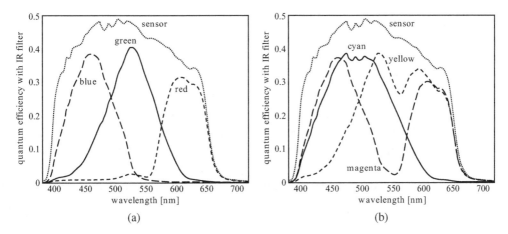

FIGURE 1.8

Example quantum efficiencies: (a) typical RGB, and (b) CMY from typical RGB.

Several three-channel systems are used to illustrate the impact of spectral sensitivity on image noise. These examples use sample spectral sensitivity curves for a typical RGB camera from Reference [56] converted to quantum efficiencies and cascaded with a typical infrared cut filter. The resulting overall quantum efficiency curves are shown, together with the quantum efficiency of the underlying sensor, in Figure 1.8a. One way to improve the signal-to-noise ratio of this camera would be to increase the quantum efficiency of the sensor itself. This is difficult and begs the question of selecting the optimal quantum efficiencies for the three color channels. Given the sensor quantum efficiency as a limit for peak quantum efficiency for any color, widening the spectral response for one or more color channels is the available option to significantly improve camera sensitivity. The effects of widening the spectral sensitivity are illustrated in this chapter by considering a camera with red, panchromatic, and blue channels and a camera with cyan, magenta, and yellow channels, shown in Figure 1.8. The CMY quantum efficiencies were created by summing pairs of the RGB quantum efficiency curves and thus are not precisely what would normally be found on a CMY sensor. In particular, the yellow channel has a dip in sensitivity near a wavelength of 560 nm, which is not typical of yellow filters. The primary effect of this dip is to reduce color errors rather than change the color correction matrix or sensitivity significantly.

Reference [65] considers the trade-off of noise and color error by examining the sensitivity and noise in sensors with both RGB and CMYG filters. It is concluded that the CMYG system has more noise in a color-corrected image than the RGB system. Reference [66] proposes optimal spectral sensitivity curves for both RGB and CMY systems considering Poisson noise, minimizing a weighted sum of color errors and noise. Fundamentally, the overlap between color matching functions drives use of substantial color correction to provide good color reproduction. All three systems in the current illustration produce reasonable color errors, so the illustration will compare the noise in the three systems.

This chapter focuses on random noise from two sources. The first is Poisson-distributed noise associated with the random process of photons being absorbed and converted to photo-electrons within a pixel, also called *shot noise*. The second is electronic amplifier read noise, which is modeled with a Gaussian distribution. These two processes are independent, so the resulting pixel values are the sum of the two processes. A pixel value Q may be modeled as $Q = k_Q(q+g)$, where k_Q is the amplifier gain, q is a Poisson random variable with mean m_q and variance σ_q^2, and g is a Gaussian random variable with mean m_g and variance σ_g^2. Note that $\sigma_q^2 = m_q$ since q is a Poisson variable, and it is entirely defined by the spectral power distribution impinging upon the sensor and the channel spectral responsivities. Also note that for a given sensor m_g and σ_g^2 are independent from m_q. For this discussion, the original pixel values are assumed to be independent, so the covariance matrix of the original pixel values, $\mathbf{K}_O$, is a diagonal matrix. Because the two random processes are independent, the variance of the pixel values is the sum of the two variances:

$$\mathbf{K}_O = \text{diag}\left(k_Q^2(m_{q,1}+\sigma_g^2), k_Q^2(m_{q,2}+\sigma_g^2), k_Q^2(m_{q,3}+\sigma_g^2)\right), \tag{1.1}$$

where $m_{q,i}$ is the mean original signal level (captured photo-electrons) for channel $i \in \{1,2,3\}$ and σ_g^2 is the read noise. In the processing path of Figure 1.5, the white balance gain factors scale camera pixel values to equalize the channel responses for neutral scene content. The gain factors are represented here with a diagonal matrix, $\mathbf{G}_B = \text{diag}(G_1, G_2, G_3)$. Accordingly, the covariance matrix for white balanced pixels, $\mathbf{K}_B$, is

$$\mathbf{K}_B = \mathbf{G}_B \mathbf{K}_O \mathbf{G}_B^T, \tag{1.2}$$

where the superscript T denotes a transposed matrix or vector. Color correction is also a 3×3 matrix; the covariance matrix for color corrected pixels is

$$\mathbf{K}_C = \mathbf{M}\mathbf{K}_B\mathbf{M}^T. \tag{1.3}$$

Photometric sensitivity and noise amplification will be compared by examining the diagonal elements of $\mathbf{K}_C$ and $\mathbf{K}_B$. The elements on the diagonal of the covariance matrix are the variance of each color channel. Since the visual impression of noise is affected by all three color channels, the sum of the variance terms can be used to compare noise levels. This sum is referred to as $\text{Tr}(\mathbf{a})$, the trace of matrix $\mathbf{a}$. More precise noise measurements convert the color image to provide a luminance channel and consider the variance in the luminance channel [67]. The luminance coefficients recommended in the ISO standard are $\mathbf{L} = [0.2125, 0.7154, 0.0721]$, so the appropriate estimate for the luminance variance is

$$\sigma_L^2 = \mathbf{L}\mathbf{K}_C\mathbf{L}^T, \tag{1.4}$$

where σ_L^2 is the variance observed in a luminance channel. The weighting values shown are specified in the ISO standard and come from ITU-R BT.709, which specifies primaries that sRGB also uses.

The following equation shows the calculation for the number of photo-electrons captured by each set of spectral quantum efficiencies:

$$P_{O,i} = \frac{l^2}{I_{EI}}\frac{55.6}{683\int I_0(\lambda)V(\lambda)d\lambda}\int\frac{I_0(\lambda)R(\lambda)}{hc/\lambda}Q_i(\lambda)d\lambda, \tag{1.5}$$

where $P_{O,i}$ is the mean number of photo-electrons captured in a square pixel with size l, the term I_0 denotes the illuminant relative spectral power distribution, R is the scene spectral

TABLE 1.1

Summary of channel sensitivity and color correction matrices. The balance gains and the sensitivity gain are respectively denoted by $\{G_1, G_2, G_3\}$ and G_E.

QE Set	Channel Response	$\{G_1, G_2, G_3\}$	G_E	**M**
RGB	261.6 397.2 315.9	1.518 1.000 1.257	2.616	1.558 −0.531 −0.027 −0.078 1.477 −0.399 0.039 −0.508 1.469
RPB	261.6 1039.0 315.9	3.972 1.000 3.289	1.000	2.000 −1.373 0.373 −1.062 3.384 −1.322 0.412 −1.248 1.836
CMY	513.4 548.6 592.9	1.155 1.081 1.000	1.752	−2.554 2.021 1.533 0.941 −1.512 1.571 1.201 1.783 −1.984

reflectance, Q_i is the quantum efficiency, and I_{EI} is the exposure index. The additional values are Planck's constant h, the speed of light c, the spectral luminous efficiency function V, and normalization constants arising from the definition of exposure index. Using a relative spectral power distribution of D65 for the illuminant, a pixel size of $l = 2.2\,\mu m$, and a spectrally flat 100% diffuse reflector, the mean number of photo-electrons captured in each pixel at an exposure index of ISO 1000 are shown under "Channel Response" in Table 1.1.

The balance gains listed are factors to equalize the color channel responses. The sensitivity gain shown is calculated to equalize the white balanced pixel values for all sets of quantum efficiencies. The color correction matrix shown for each set of quantum efficiencies was computed by calculating Equation 1.5 for 64 different color patch spectra, then finding a color correction matrix that minimized errors between color corrected camera data and scene colorimetry, as described in Reference [68].

The illustration compares the noise level in images captured at the same exposure index and corrected to pixel value P_C. For a neutral, the mean of the balanced pixel values is the same as the color corrected pixel values. Since the raw signals are related to the balanced signal by the gains shown in Table 1.1, the original signal levels can be expressed as follows:

$$\mathbf{P}_O = \begin{bmatrix} \frac{1}{G_E G_1} & 0 & 0 \\ 0 & \frac{1}{G_E G_2} & 0 \\ 0 & 0 & \frac{1}{G_E G_3} \end{bmatrix} \begin{bmatrix} P_C \\ P_C \\ P_C \end{bmatrix}. \tag{1.6}$$

Defining a modified balance matrix including the sensitivity equation gain along with the white balance gains, $\mathbf{G}_B = G_E \text{diag}(G_1, G_2, G_3)$ and substituting Equation 1.6 into Equation 1.1 produces the following covariance matrix for the white balanced and gain corrected pixel values:

$$\mathbf{K}_B = \begin{bmatrix} G_E^2 G_1^2 \left(\frac{P_C}{G_E G_1} + \sigma_g^2 \right) & 0 & 0 \\ 0 & G_E^2 G_2^2 \left(\frac{P_C}{G_E G_2} + \sigma_g^2 \right) & 0 \\ 0 & 0 & G_E^2 G_3^2 \left(\frac{P_C}{G_E G_3} + \sigma_g^2 \right) \end{bmatrix}. \tag{1.7}$$

TABLE 1.2

Summary of Relative Noise in White Balanced and Color Corrected Signals.

QE Set	S_B	Tr(S_B)	$\sigma_{L,B}$	S_C	Tr(S_C)	$\sigma_{L,C}$
RGB	3.97 0.00 0.00 0.00 2.62 0.00 0.00 0.00 3.29	9.88	1.24	10.38 −2.50 0.82 −2.50 6.25 −3.90 0.82 −3.90 7.78	24.41	1.60
RPB	3.97 0.00 0.00 0.00 1.00 0.00 0.00 0.00 3.29	8.26	0.84	18.22 −14.70 7.24 −14.70 21.68 −13.94 7.24 −13.94 13.32	53.22	2.51
CMY	2.02 0.00 0.00 0.00 1.89 0.00 0.00 0.00 1.75	5.67	1.03	25.05 −6.43 −4.71 −6.43 10.45 −8.28 −4.71 −8.28 15.84	51.34	1.90

In the case where $\sigma_g^2 << P_C/(G_E G_i)$, where $i \in [1,2,3]$, this simplifies to

$$\mathbf{K}_B = \begin{bmatrix} G_1 & 0 & 0 \\ 0 & G_2 & 0 \\ 0 & 0 & G_3 \end{bmatrix} G_E P_C. \tag{1.8}$$

To focus on the relative sensitivity, the matrix $\mathbf{S}_B$ is defined by leaving out the factor of P_C:

$$\mathbf{S}_B = \begin{bmatrix} G_1 & 0 & 0 \\ 0 & G_2 & 0 \\ 0 & 0 & G_3 \end{bmatrix} G_E. \tag{1.9}$$

The values on the diagonal of $\mathbf{S}_B$ show the relative noise levels in white balanced images before color correction, accounting for the differences in photometric sensitivity. To finish the comparison, the matrix $\mathbf{S}_C$ is defined as $\mathbf{M}\mathbf{S}_B\mathbf{M}^T$. The values on the diagonal of $\mathbf{S}_C$ indicate the relative noise levels in color corrected images. The values $\sigma_{L,B}$ and $\sigma_{L,C}$ indicate the estimated relative standard deviation for a luminance channel based on Equation 1.4.

As shown in Table 1.2, the Tr$(\mathbf{S}_B)$ and $\sigma_{L,B}$ are smaller for CMY and for RPB than for RGB, reflecting the sensitivity advantage of the broader spectral sensitivities. However, Tr$(\mathbf{S}_C)$ and $\sigma_{L,C}$ are greater for RPB and CMY than for RGB, reflecting the noise amplification from the color correction matrix. In summary, while optimal selection of spectral sensitivity is important for limiting noise, a well-selected relatively narrow set of RGB spectral sensitivies is close to optimum, as found in References [65] and [66]. Given these results, it is tempting to consider narrower spectral bands for each color channel, reducing the need for color correction. This would help to a limited extent, but eventually the signal loss from narrower bands would take over. Further, narrower spectral sensitivities would produce substantially larger color errors, leading to lower overall image quality. The fundamental problem is that providing acceptable color reproduction constrains the three channel system, precluding substantial improvement in sensitivity.

Reference [65] considers the possibility of reducing the color saturation of the image, lowering the noise level at the expense of larger color errors. However, the concept of lowering the color saturation can be applied with RGB quantum efficiencies as well. Reference [66] shows that by allowing larger color errors at higher exposure index values, the optimum set of quantum efficiencies changes with exposure index. In particular, at a high exposure index, the optimum red quantum efficiency peaks at a longer wavelength and has less overlap with the green channel. This is another way to accept larger color errors to reduce the noise in the color corrected image.

1.2.2 Four-Channel Arrays

Previous work considering optimal spectral sensitivities has repeatedly concluded that adding a fourth channel can significantly reduce color errors without increasing noise substantially. References [69] and [66] both conclude that a four-channel system satisfies the multiple objectives of a color camera better than a three-channel system. Reference [69] considers the spatial sampling of the color filter array and reconstruction errors arising from demosaicking. Reference [66] focuses on noise in the optimization, and recommends using two different red channels in the CFA pattern, allowing a degree of adaptation as a function of exposure index.

1.2.3 Color Fidelity versus Spatial Resolution

The family of CFA patterns selected in this chapter is motivated by recalling contrast sensitivity research showing human sensitivity to luminance contrast is very different from human sensitivity to chrominance contrast. Reference [70] examines the dependence of chrominance contrast sensitivity on spatial frequency and on illuminance level; it was found that contrast sensitivity degrades at lower luminance levels, for both chrominance and luminance. Despite limited comparison with luminance contrast sensitivity, the results suggest that chrominance sensitivity degrades past one cycle/degree, while luminance sensitivity peaks near two cycles/degree. Reference [71] provides a more in depth comparison of chrominance and luminance contrast sensitivity. This work finds that red-green and blue-yellow contrast sensitivity functions have similar spatial bandwidth, which is roughly 1/3 of the bandwidth of luminance contrast sensitivity. It was also found that luminance contrast sensitivity degrades below about 1 to 2 cycles/degree, while chrominance sensitivity is constant below about 1/3 to 2/3 cycles/degree. The similarity of red-green and blue-yellow contrast sensitivities and their substantial different from the luminance contrast sensitivity suggests the decoupling of spatial detail and luminance sensitivity from color sensitivity. The clearest way to accomplish this is to provide a highly sensitive luminance channel in addition to three channels for chrominance data. The spectral response of the panchromatic channel is *colorimetrically* inaccurate for luminance, but it provides the best possible signal-to-noise for a given sensor.

Introducing panchromatic pixels into a three-channel CFA pattern and allowing the color sampling to drop off provide color resolution that is roughly 1/3 to 1/4 the panchromatic sampling. Assuming that color artifacts and chromatic aliasing are successfully limited in demosaicking, this approach, fusing a panchromatic image with a lower resolution color image, provides a capture system most closely mimicking the capabilities of the human visual system under most imaging conditions.

1.3   Demosaicking

Demosaicking, or color filter array interpolation, is the process of producing a full-color image from the sparsely sampled digital camera capture. It generally involves some sort of interpolation of neighboring pixel values within a given support region. This process may be based on strictly linear, shift-invariant systems theory, or may be conducted in a more heuristic nonlinear, adaptive manner. Both approaches will be described below. Because of the large breadth of knowledge now available on demosaicking in general, the following discussion will be restricted to a particular body of research conducted in the area of four-channel color filter array image processing [72], [73], [74], [75].

1.3.1   Special Functions and Transforms

The following notation, conventions, and special functions used in the rest of this section can be found in Reference [76]. In the following, b and d are positive values. Given a function $f(x)$, the Fourier transform and its inverse are:

$$F(\xi) = \int_{-\infty}^{\infty} f(\alpha) e^{-i2\pi\xi\alpha} d\alpha,$$

$$f(x) = \int_{-\infty}^{\infty} F(\beta) e^{i2\pi\beta x} d\beta.$$

The delta function, $\delta(x)$, is the function that has the following properties:

$$\delta(x - x_0) = 0, \; x \neq x_0,$$

$$\int_{x_1}^{x_2} f(\alpha) \delta(\alpha - x_0) d\alpha = f(x_0), \; x_1 < x_0 < x_2,$$

$$\delta\left(\frac{x - x_0}{b}\right) = |b| \, \delta(x - x_0).$$

For convenience, pairs of delta functions can be defined as follows:

$$\delta\delta\left(\frac{x - x_0}{b}\right) = |b| \left[\delta(x - x_0 + b) + \delta(x - x_0 - b)\right].$$

The comb function

$$\text{comb}\left(\frac{x - x_0}{b}\right) = |b| \sum_{n=-\infty}^{\infty} \delta(x - x_0 - nb)$$

is used for describing sampling arrays within the CFA pattern.

Linear interpolation is modeled as a convolution with the tri function

$$\text{tri}\left(\frac{x - x_0}{b}\right) = \begin{cases} 0 & \text{if } |(x - x_0)/b| \geq 1, \\ 1 - |(x - x_0)/b| & \text{if } |(x - x_0)/b| < 1. \end{cases}$$

Fourier analysis of the tri function is expressed in terms of the sinc function:

$$\text{sinc}\left(\frac{x-x_0}{b}\right) = \frac{\sin\left[\pi\left(\frac{x-x_0}{b}\right)\right]}{\pi\left(\frac{x-x_0}{b}\right)}.$$

The forward Fourier transform pairs of the aforementioned special functions are defined as follows:

$$\delta\left(\frac{x-x_0}{b}\right) \xrightarrow{\mathscr{F}} |b|\,e^{-i2\pi x_0\xi},$$

$$\delta\delta\left(\frac{x-x_0}{b}\right) \xrightarrow{\mathscr{F}} 2be^{-i2\pi x_0\xi}\cos\left(2\pi b\xi\right),$$

$$\text{comb}\left(\frac{x-x_0}{b}\right) \xrightarrow{\mathscr{F}} |b|\,e^{-i2\pi x_0\xi}\text{comb}\left(b\xi\right),$$

$$\text{tri}\left(\frac{x-x_0}{b}\right) \xrightarrow{\mathscr{F}} |b|\,e^{-i2\pi x_0\xi}\text{sinc}^2\left(b\xi\right).$$

Two-dimensional versions of these special functions as well as their Fourier transforms can be constructed by multiplying together one-dimensional versions, resulting in the following (note that the results are separable):

$$\delta\left(\frac{x-x_0}{b},\frac{y-y_0}{d}\right) = \delta\left(\frac{x-x_0}{b}\right)\delta\left(\frac{y-y_0}{d}\right),$$

$$\delta\delta\left(\frac{x-x_0}{b},\frac{y-y_0}{d}\right) = \delta\delta\left(\frac{x-x_0}{b}\right)\delta\delta\left(\frac{y-y_0}{d}\right),$$

$$\text{comb}\left(\frac{x-x_0}{b},\frac{y-y_0}{d}\right) = \text{comb}\left(\frac{x-x_0}{b}\right)\text{comb}\left(\frac{y-y_0}{d}\right),$$

$$\text{tri}\left(\frac{x-x_0}{b},\frac{y-y_0}{d}\right) = \text{tri}\left(\frac{x-x_0}{b}\right)\text{tri}\left(\frac{y-y_0}{d}\right),$$

$$\text{sinc}\left(\frac{x-x_0}{b},\frac{y-y_0}{d}\right) = \text{sinc}\left(\frac{x-x_0}{b}\right)\text{sinc}\left(\frac{y-y_0}{d}\right),$$

$$\cos\left(2\pi bx,2\pi dy\right) = \cos\left(2\pi bx\right)\cos\left(2\pi dy\right),$$

$$\delta\left(\frac{x-x_0}{b},\frac{y-y_0}{d}\right) \xrightarrow{\mathscr{F}} |bd|\,e^{-i2\pi(x_0\xi+y_0\eta)},$$

$$\delta\delta\left(\frac{x-x_0}{b},\frac{y-y_0}{d}\right) \xrightarrow{\mathscr{F}} 4|bd|\,e^{-i2\pi(x_0\xi+y_0\eta)}\cos\left(2\pi b\xi,2\pi d\eta\right),$$

$$\text{comb}\left(\frac{x-x_0}{b},\frac{y-y_0}{d}\right) \xrightarrow{\mathscr{F}} |bd|\,e^{-i2\pi(x_0\xi+y_0\eta)}\text{comb}\left(b\xi,d\eta\right),$$

$$\text{tri}\left(\frac{x-x_0}{b},\frac{y-y_0}{d}\right) \xrightarrow{\mathscr{F}} |bd|\,e^{-i2\pi(x_0\xi+y_0\eta)}\text{sinc}^2\left(b\xi,d\eta\right).$$

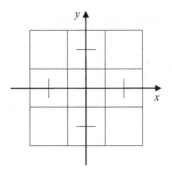

FIGURE 1.9

Panchromatic pixel neighborhood.

Finally, it is necessary to look at functions that have been rotated and skewed and their corresponding Fourier transforms. The general rule to be used can be written as follows:

$$f\left(\frac{x-x_0}{b}-\frac{y-y_0}{d},\frac{x-x_0}{b}+\frac{y-y_0}{d}\right)\xrightarrow{\mathcal{F}}\frac{|bd|}{2}e^{-i2\pi(x_0\xi+y_0\eta)}F\left(\frac{b\xi-d\eta}{2},\frac{b\xi+d\eta}{2}\right),$$

where

$$f(x,y)\xrightarrow{\mathcal{F}}F(\xi,\eta).$$

1.3.2   The Panchromatic Sensor

As a preliminary step, an all-panchromatic pixel sensor is examined first. Each pixel in the corresponding CFA has a broadband, spectrally nonselective response. Casually, these pixels maybe thought of as being "white" or "clear." Figure 1.9 shows a small pixel neighborhood with a coordinate system superimposed. Many of the subsequent algorithm analyses will deal with one-dimensional pixel neighborhoods. Therefore, restricting attention to the pixels lying along the x-axis, the row of panchromatic pixels f_P' can be expressed (ignoring pixel size[1]) by the continuous panchromatic image f_P and a comb sampling function s_P as follows:

$$s_P=\mathrm{comb}(x),$$

$$f_P'=f_Ps_P. \tag{1.10}$$

The Fourier transform of Equation 1.10 can be used for evaluation of both signal fidelity as well as sample aliasing. These Fourier transform terms can be written as

$$S_P=\mathrm{comb}(\xi),$$

$$F_P'=F_P*S_P=\int_{-\infty}^{\infty}F_P(\alpha)\sum_{n=-\infty}^{\infty}\delta(\xi-\alpha-n)d\alpha=\sum_{n=-\infty}^{\infty}F_P(\xi-n). \tag{1.11}$$

---

[1]In this analysis the pixels are considered to be point entities modeled by delta functions. These delta functions could be convolved with a finite area mask such as rect function of Reference [77] to more accurately simulate their physical dimensions. However, as doing so would not significantly impact the results of this analysis, this is omitted for the sake of simplicity. The interested reader is referred to Reference [77] for a more detailed discussion of this topic.

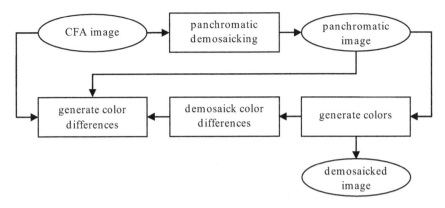

FIGURE 1.10

Demosaicking Algorithm Flowchart.

It can be seen from Equation 1.11 that if the initial panchromatic image is appropriately bandlimited to be zero beyond $|\xi| \geq 1/2$, then the fundamental component $(n = 0)$ is not aliased by any of the sidebands $(n \neq 0)$. In practice, this bandlimiting is usually imposed by an optical antialiasing filter [78]. Restricting attention to the portion of the resulting panchromatic spectrum $0 \leq \xi < 1/2$ and considering this to be the rendered portion of the reconstructed image, this idealized case can be seen to produce perfect image reconstruction, that is, $F'_P = F_P, 0 \leq \xi < 1/2$.

1.3.3 Demosaicking Algorithm Overview

Figure 1.10 shows a flowchart of the demosaicking algorithm discussed in this section. Although its terminology reflects a four-channel CFA pattern (i.e., CFA image including panchromatic pixels), it works equally well for three-channel systems with one of the color channels, usually green, taking the place of the panchromatic channel. First, the panchromatic pixel data is demosaicked to produce a full-resolution panchromatic image. Color differences are formed next by subtracting the panchromatic value from the color value at each color pixel location; that is, $D_{ij} = C_{ij} - P_{ij}$, where C is a color value (usually red, green, or blue), P is a panchromatic value, and D is the resulting color difference. These color differences are then demosaicked to produce full-resolution color difference image channels. Finally, panchromatic values are added back to the color difference values to produce the final full-resolution color image; that is, $C_{ij} = D_{ij} + P_{ij}$.

The demosaicking of the panchromatic and color difference CFA data can be done in either an adaptive or nonadaptive manner. Nonadaptive demosaicking generally employs the standard methods of pixel replication, bilinear interpolation, or bicubic interpolation. Frequently, these interpolation operations can be executed as simple convolution operations with appropriately chosen kernels and initializing missing pixel values to zero. Adaptive demosaicking also uses standard interpolation methods, but in a one-dimensional, directional manner. In the CFA patterns analyzed below, each pixel location to be demosaicked can be done in at least two possible directions, usually horizontal and vertical. Sometimes diagonal direction interpolation is also possible. To select the preferred direction for in-

D_{x_0, y_0+N}	O	D_{x_0+M, y_0+N}
O	O	O
D_{x_0, y_0}	O	D_{x_0+M, y_0}

FIGURE 1.11

Color Difference Interpolation Rectilinear Neighborhood.

terpolation, primitive edge detection computations are performed and the direction of least edge activity chosen. One set of terminology exists for describing this process. *Classification* is the selection of a preferred interpolation direction through edge detection. In this context, the edge detectors become *classifiers*. *Prediction* is the estimation of the missing pixel value. The expressions used for computing these missing values are then called *predictors*.

The simplest demosaicking algorithms will use nonadaptive methods for both panchromatic and color difference interpolation. Since nonadaptive methods are not able to respond to or take advantage of any feature (edge) information in the image, the algorithmic simplicity comes as the cost of reconstruction image fidelity. Note that this is a liability for the panchromatic channel, as the color differences are predominantly low spatial frequency records, similar to the chrominance channels in a YCC color space. Color differences, being largely devoid of edge information, are well suited to nonadaptive demosaicking methods. For improved reconstruction image fidelity, adaptive methods can be used for the demosaicking of the panchromatic channel. This, of course, comes at the price of increasing the interpolation algorithm complexity.

1.3.3.1   Rectilinear Grid-Based Nonadaptive Interpolation

The nonadaptive method used most frequently for demosaicking is standard bilinear interpolation. For the CFA patterns discussed in this chapter, the original pixels used in the interpolation process are arranged either in a rectilinear pattern or a diamond (rotated or skewed rectilinear) pattern. The former case is considered here and the latter case in the next section. This method applies equally well to panchromatic, color, or color difference values, although color difference values will be assumed below for convenience.

The rectilinear CFA interpolation neighborhood is shown in Figure 1.11. The term D_{ij} is a color difference and the $\bigcirc$ entries are luminance[2] (generally panchromatic or green) values initially devoid of color difference values. The interpolation of a color difference value at each luminance value location is accomplished by a weighted average of each of the four corner color difference values. This weighting function for bilinear interpolation is a discrete version of the two-dimensional tri function defined as follows:

$$b = \mathrm{tri}\left(\frac{x}{M}, \frac{y}{N}\right)\mathrm{comb}\,(x, y).$$

---

[2]In this analysis the term *luminance* is used to refer to the CFA channel that is the primary source for the spatial detail of the image.

The interpolation operation can be treated as a convolution operation. Therefore, the interpolation color difference plane, f'_D, can be expressed in terms of the original color difference plane, f_D, a sampling function, s_D, and the convolution kernel, b, as follows:

$$f'_D = (f_D s_D) * b. \tag{1.12}$$

The sampling function for the color differences in Figure 1.11 is given by

$$s_D = \frac{1}{MN} \text{comb} \left(\frac{x - x_0}{M}, \frac{y - y_0}{N} \right).$$

Standard Fourier analysis produces the spatial frequency response for Equation 1.12, as follows:

$$F'_D = (F_D * S_D) B.$$

This translates into the general frequency response for bilinear interpolation on a rectilinear grid:

$$F'_D = \sum_{m=-\infty}^{\infty} \sum_{n=-\infty}^{\infty} A_{mn}(\xi, \eta) F_D \left(\xi - \frac{m}{M}, \eta - \frac{n}{N} \right), \tag{1.13}$$

where

$$A_{mn} = \frac{e^{-i2\pi \left(x_0 \frac{m}{M} + y_0 \frac{n}{N} \right)}}{MN} B(\xi, \eta) \tag{1.14}$$

denotes the transfer function and B is defined as follows (see Appendix for more details):

$$B = MN \sum_{p=-\infty}^{\infty} \sum_{q=-\infty}^{\infty} \text{sinc}^2 [M(\xi - p), N(\eta - q)],$$

$$= \left[1 + 2 \sum_{j=1}^{M-1} \text{tri} \left(\frac{j}{M} \right) \cos(2\pi j \xi) \right] \left[1 + 2 \sum_{k=1}^{N-1} \text{tri} \left(\frac{k}{N} \right) \cos(2\pi k \eta) \right]. \tag{1.15}$$

Equations of the form of Equation 1.13 occur several times in the subsequent analysis. These equations can be viewed as consisting of two components: repeated spectral components, for instance, $F_D \left(\xi - \frac{m}{M}, \eta - \frac{n}{N} \right)$, which describe the aliasing behavior, and the transfer functions, $A_{mn}(\xi, \eta)$, which describe the spectral component fidelity. As a rule of thumb, the larger the values of M and N, the more likely the CFA is prone to aliasing artifacts. Similarly, the greater the departure of the transfer functions from a unity response over all spatial frequencies of interest, the more distorted the demosaicked image appears, usually as a lack of sharpness or definition.

1.3.3.2 Diamond Grid-Based Nonadaptive Interpolation

The CFA interpolation neighborhood for color differences arranged in a diamond pattern is shown in Figure 1.12. As before, D_{ij} is a color difference and the $\bigcirc$ entries are luminance values initially devoid of color difference values. The interpolation of a color difference value at each luminance value location is accomplished by a weighted average of each of the four corner color difference values. This weighting function for bilinear interpolation is a rotated and skewed version of the tri function defined as follows:

$$b = \text{tri} \left(\frac{x}{2M} - \frac{y}{2N}, \frac{x}{2M} + \frac{y}{2N} \right) \text{comb}(x, y).$$

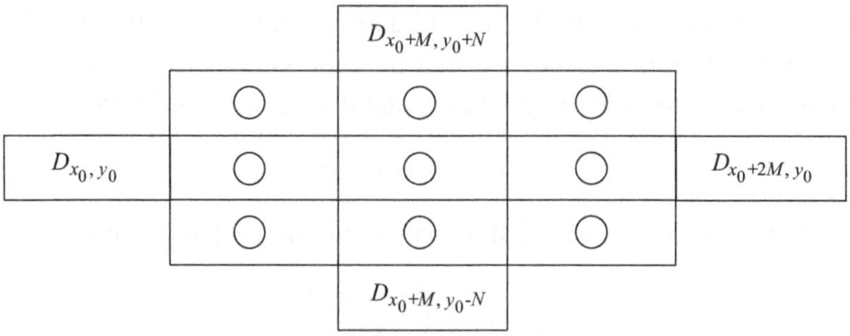

FIGURE 1.12

Color Difference Interpolation Diamond Neighborhood.

The interpolation operation is formally the same as in the rectilinear case (Equation 1.12) with only a change in the sampling function, s_D, and the convolution kernel, b. The sampling function for the color differences in Figure 1.12 is given by

$$s_D = \frac{1}{2MN}\text{comb}\left(\frac{x-x_0}{2M} - \frac{y-y_0}{2N}, \frac{x-x_0}{2M} + \frac{y-y_0}{2N}\right).$$

Standard Fourier analysis produces the equivalent spatial frequency response for Equation 1.12 using the new values for s_D and b as follows (see Appendix for more details):

$$S_D = e^{-i2\pi(x_0\xi+y_0\eta)}\text{comb}\left(M\xi - N\eta, M\xi + N\eta\right)$$

$$B = 2MN\sum_{p=-\infty}^{\infty}\sum_{q=-\infty}^{\infty}\text{sinc}^2\left[M(\xi-p) - N(\eta-q), M(\xi-p) + N(\eta-q)\right]$$

$$= 1 + 2\sum_{j=1}^{2M-1}\text{tri}^2\left(\frac{j}{2M}\right)\cos(2\pi j\xi) + 2\sum_{k=1}^{2N-1}\text{tri}^2\left(\frac{k}{2N}\right)\cos(2\pi k\eta)$$

$$+ 4\sum_{j=1}^{2M-1}\sum_{k=1}^{2N-1}\text{tri}\left(\frac{j}{2M} - \frac{k}{2N}, \frac{j}{2M} + \frac{k}{2N}\right)\cos(2\pi j\xi, 2\pi k\eta) \qquad (1.16)$$

The final result can be expressed as:

$$F_D' = \sum_{m=-\infty}^{\infty}\sum_{n=-\infty}^{\infty}A_{mn}(\xi,\eta)F_D\left(\xi - \frac{m+n}{2M}, \eta - \frac{-m+n}{2N}\right), \qquad (1.17)$$

with the transfer function

$$A_{mn} = \frac{e^{-i\pi\left(x_0\frac{m+n}{M} + y_0\frac{-m+n}{N}\right)}}{2MN}B(\xi,\eta). \qquad (1.18)$$

1.3.4 The Bayer Color Filter Array

Any four-channel CFA will ultimately be compared to the accepted industry standard Bayer three-channel CFA. Figure 1.13a shows the CFA with an imposed coordinate system.

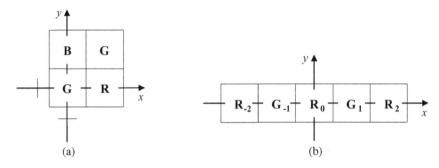

FIGURE 1.13

Bayer pattern: (a) CFA, and (b) row neighborhood.

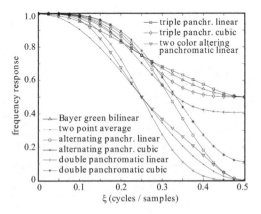

FIGURE 1.14

Fundamental transfer function frequency responses.

For the purposes of analysis, in the Bayer pattern the green channel will be taken to be the luminance channel and the color differences will be red minus green and blue minus green.[3]

1.3.4.1   Bilinear Interpolation

In the case of demosaicking using solely bilinear interpolation, the green channel reconstruction can be accomplished by using the method of Section 1.3.3.2 with $M = 1$, $N = 1$, $x_0 = 0$, and $y_0 = 0$. The bilinear interpolating function, b, can be explicitly written as

$$b = \frac{1}{4}\left[4\delta(x,y) + \delta\delta(x)\delta(y) + \delta(x)\delta\delta(y)\right].$$

The equivalent convolution kernel can be expressed as

$$\mathbf{b} = \frac{1}{4}\begin{pmatrix} 0 & 1 & 0 \\ 1 & 4 & 1 \\ 0 & 1 & 0 \end{pmatrix}.$$

---

[3]While formally one justifies interpolating color differences by performing the computations in a logarithmic space [79], for all but the most extreme pixel differences computing color differences in video gamma or even linear space is usually visually acceptable.

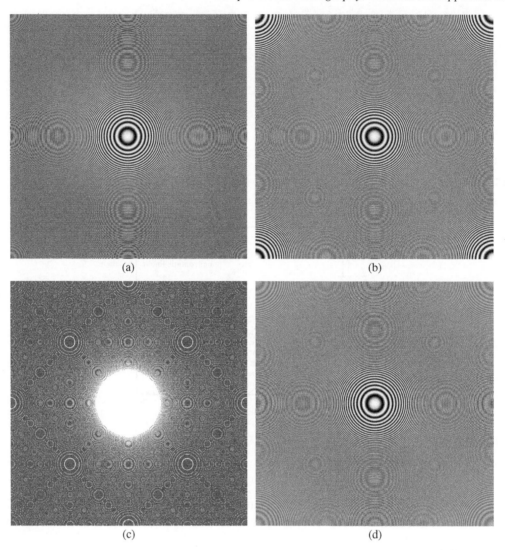

FIGURE 1.15

Bayer green bilinear interpolation results: (a) original image, (b) bilinear green interpolation, (c) interpolation error map, and (d) bilinear interpolation full color result.

The corresponding frequency response of the reconstructed green channel is then

$$F'_G = \sum_{m=-\infty}^{\infty} \sum_{n=-\infty}^{\infty} A_{mn}(\xi, \eta) F_G\left(\xi - \frac{m+n}{2}, \eta - \frac{-m+n}{2}\right),$$

where

$$A_{mn} = \frac{2 + \cos(2\pi\xi) + \cos(2\pi\eta)}{4}.$$

The ξ-axis response of A_{mn} is plotted in Figure 1.14 as "Bayer green bilinear." Another way to analyze the performance of the Bayer bilinear algorithm is to test the algorithm on a chirp circle test chart. Figure 1.15a is a chirp circle target in which the spatial frequency of the circles increases linearly from the center out. Figure 1.15b is the equivalent demosaicked green channel produced by Bayer bilinear interpolation. Since it is difficult to see differences between Figures 1.15a and 1.15b an interpolation error map of the two is presented in Figure 1.15c. This map is created as follows:

$$\varepsilon = \begin{cases} 1 & \text{if } |f_G - f'_G| \leq t, \\ 0 & \text{otherwise,} \end{cases} \tag{1.19}$$

where t is the threshold set to a value of 22 for Figure 1.15c as well as all subsequent interpolation error maps. Note that the original image code value range of Figure 1.15a is 0 to 255. The central circular region in Figure 1.15c represents an area of low interpolation error whereas the rest of the error map is dominated by aliasing and transfer function distortions. A qualitative assessment of the resulting aliasing can be made from the full-color results of the bilinear interpolation in Figure 1.15d. In this figure, the green-magenta aliasing patterns in the corners of the image represent the aliasing due to bilinear interpolation of the green channel.

Once the green channel has been fully populated by the interpolation process, red and blue color differences, $D_R = R - G$ and $D_B = B - G$, can be formed at each red and blue pixel location and the method of Section 1.3.3.1 can be used with $M = 2$, $N = 2$, $x_0 = 1$, and $y_0 = 0$ for the red channel and $x_0 = 0$ and $y_0 = 1$ for the blue channel. The bilinear interpolation function

$$b = \frac{1}{4} [2\delta(x) + \delta\delta(x)][2\delta(y) + \delta\delta(y)]$$

is the same for both color difference channels. The equivalent convolution kernel would be

$$\mathbf{b} = \frac{1}{4} \begin{pmatrix} 1 & 2 & 1 \\ 2 & 4 & 2 \\ 1 & 2 & 1 \end{pmatrix}.$$

The frequency response of the interpolated color difference channels becomes

$$F'_D = \sum_{m=-\infty}^{\infty} \sum_{n=-\infty}^{\infty} A_{mn}(\xi, \eta) F_D\left(\xi - \frac{m}{2}, \eta - \frac{n}{2}\right),$$

with the red and blue transfer functions, respectively, defined as follows:

$$A_{mn,R} = \frac{(-1)^m}{4} B(\xi, \eta),$$

$$A_{mn,B} = \frac{(-1)^n}{4} B(\xi, \eta),$$

and the frequency response of the bilinear interpolating function defined as

$$B = [1 + \cos(2\pi\xi)][1 + \cos(2\pi\eta)].$$

The aliasing consequences of the final image can be seen in Figure 1.15d with the addition of blue-orange aliasing patterns in the centers of the image sides.

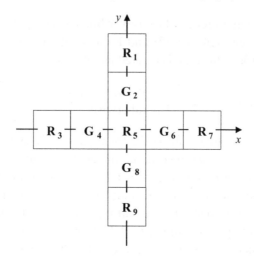

FIGURE 1.16

Bayer adaptive interpolation neighborhood.

1.3.4.2   Adaptive Interpolation

Adaptive demosaicking algorithms that respond to local edge activity are well-known in the literature [43], [80], [81], [82]. When the green pixel value in the Bayer pattern is being interpolated at a red or blue pixel location, there are several choices of pixel neighborhoods that can be used to account for local edges. Figure 1.16 shows a typical neighborhood. Interpolation can occur either horizontally or vertically. One strategy is to blend the results based on the relative strengths of the classifiers u and v:

$$u = 2\,|\delta\,(x+1) - \delta\,(x-1)| + |\delta\,(x+2) - 2\delta\,(x) + \delta\,(x-2)|,$$
$$= 2\,|G_4 - G_6| + |R_3 - 2R_5 + R_7|,$$

$$v = 2\,|\delta\,(y+1) - \delta\,(y-1)| + |\delta\,(y+2) - 2\delta\,(y) + \delta\,(y-2)|,$$
$$= 2\,|G_2 - G_8| + |R_1 - 2R_5 + R_9|,$$

as follows:

$$f_G' = G_5' = \frac{u}{u+v}V + \frac{v}{u+v}U, \tag{1.20}$$

where U and V are the horizontal and vertical predictors to be derived below. It can be seen in Equation 1.20 that the direction of the smaller classifier gives the greater weight to the corresponding predictor; for example, a smaller value of u will produce a dominant weighting of U. It can also be seen that the classifiers freely combine color Laplacians and green gradients. This is an image fusion technique that will be discussed shortly.

In this adaptive algorithm the derivation of a suitable predictor becomes a one-dimensional interpolation problem. Figure 1.13b shows an example of a five-point horizontal neighborhood. The corresponding predictor is defined as follows (these results will be stated more broadly in Section 1.3.5.1):

$$f_G' = f_G s_G + (f_G * b)\,s_R + (f_R * h)\,s_R \approx f_G s_G + (f_G * b)\,s_R + (f_G * h)\,s_R, \tag{1.21}$$

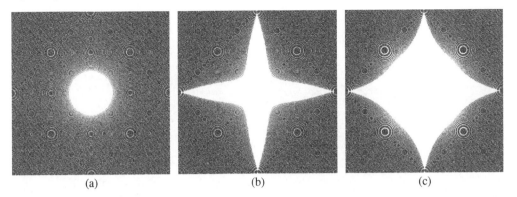

FIGURE 1.17

Bayer green interpolation error maps: (a) bilinear interpolation error map, (b) adaptive linear interpolation with $\alpha = 0$, and (c) adaptive linear interpolation with $\alpha = 1/2$.

where b and h denote, respectively, a low-pass filter and a high-pass filter, defined as:

$$b = \frac{1}{2}\delta\delta(x),$$

$$h = \frac{\alpha}{4}\left[2\delta(x) - \frac{1}{2}\delta\delta\left(\frac{x}{2}\right)\right],$$

where α is a design parameter. The terms s_G and s_R are defined as follows:

$$s_G = \frac{1}{2}\text{comb}\left(\frac{x-1}{2}\right),$$

$$s_R = \frac{1}{2}\text{comb}\left(\frac{x}{2}\right).$$

Image fusion occurs in the substitution of the high-pass image component $(f_R * h)\, s_R$ for the unavailable high-pass image component $(f_G * h)\, s_R$. This is justified on the assumption that $G = R + \text{constant}$ over the pixel neighborhood [83]. The corresponding frequency response is given by

$$F'_G = \sum_{n=-\infty}^{\infty} A_n(\xi) F_G\left(\xi - \frac{n}{2}\right),$$

$$A_n = \frac{(-1)^n\left[1 + \cos(2\pi\xi)\right] + \alpha\sin^2(2\pi\xi)}{2}.$$

The design parameter α can be set to satisfy a number of different constraints. Here, α will be set to make the slope of the fundamental transfer function as follows:

$$A_0 = \frac{1 + \cos(2\pi\xi) + \alpha\sin^2(2\pi\xi)}{2},$$

$$\left.\frac{dA_0}{d\xi}\right|_{\xi=0} \Rightarrow \alpha = \frac{1}{2},$$

with zero at the origin. Therefore, h can be restated with this value of α and the predictors written in terms of the pixel values in Figure 1.16 as follows:

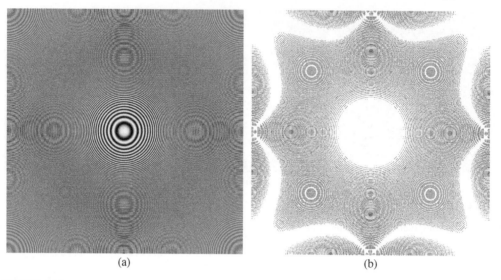

(a)   (b)

FIGURE 1.18

Bayer adaptive interpolation results: (a) adaptive interpolation full color result, and (b) bilinear-adaptive interpolation difference green channel.

$$h = \frac{1}{8}\left[2\delta\left(x\right) - \frac{1}{2}\delta\delta\left(\frac{x}{2}\right)\right],$$

$$U = \frac{G_4 + G_6}{2} + \frac{-R_3 + 2R_5 - R_7}{8},$$

$$V = \frac{G_2 + G_8}{2} + \frac{-R_1 + 2R_5 - R_9}{8}.$$

The response of A_0 is equivalent to one of the four-channel situations analyzed below and is therefore the same as the plot in Figure 1.14 labeled "alternating panchromatic linear." The interpolation error map is shown in Figure 1.17c. If α is set to zero and just the linear interpolation of green values is used in the adaptive interpolation, the fundamental transfer function becomes

$$A_0 = \frac{1 + \cos\left(2\pi\xi\right)}{2}.$$

This response of A_0 is labeled as "two-point average" in Figure 1.14. The interpolation error map is shown in Figure 1.17b. Comparing the bilinear interpolation error map shown in Figure 1.17a with Figures 1.17b and 1.17c reveals that interpolation error is greatest with bilinear interpolation and least with adaptive interpolation and $\alpha = 1/2$. The adaptive interpolation error with $\alpha = 0$ is clearly between these two extremes.

Bilinear interpolation of color differences is still used for demosaicking the red and blue channels. The resulting full color image from using adaptive interpolation for green and bilinear interpolation for red and blue is shown in Figure 1.18a. A difference map of the green channel between the all-bilinear interpolation case of Figures 1.15d and 1.18a is shown in Figure 1.18b, indicating that the largest region of improvement realized in the adaptive interpolation case is in the middle spatial frequency range of the green channel.

G	P	R	P
P	P	P	P
B	P	G	P
P	P	P	P

(a)

G	P	P
P	R	P
P	P	B

(b)

G	P	P	P
P	R	P	P
P	P	G	P
P	P	P	B

(c)

G	P	G	P
P	R	P	B
G	P	G	P
P	R	P	B

(d)

P	G	P	R
G	P	R	P
P	B	P	G
B	P	G	P

(e)

FIGURE 1.19

Four-channel CFA patterns.

1.3.5 Four-Channel CFA Demosaicking

When considering four-channel CFA patterns, such as illustrated in Figure 1.19, a number of pixel neighborhoods for adaptive panchromatic interpolation soon suggest themselves. These are illustrated in Figure 1.20. Generalizing from the Bayer adaptive interpolation case, there are two main components to panchromatic predictors. The panchromatic pixel values can be used to compute linear, cubic, or possibly even higher-order interpolation values. To these interpolated panchromatic values an image fusion component computed from the color pixel values can be added. These options are analyzed in turn below.

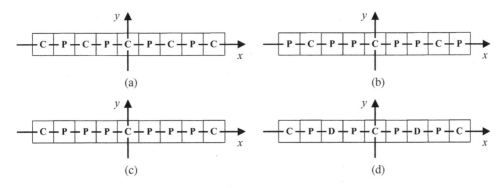

FIGURE 1.20

Four-channel row neighborhoods: (a) alternating panchromatic, (b) double panchromatic, (c) triple panchromatic, and (d) two-color alternating panchromatic.

1.3.5.1 Adaptive Linear Interpolation

For linear interpolation of panchromatic values plus a color Laplacian image fusion component, the first three cases shown in Figure 1.20 can be generalized into one expression. If there are N panchromatic pixels between subsequent color pixels of the same color, then the following predictor can be written:

$$f'_P = f_P s_P + (f_P * b + f_C * h) s_C, \tag{1.22}$$

where

$$s_C = \tfrac{1}{N+1} \mathrm{comb}\left(\tfrac{x}{N+1}\right), \quad s_P = \mathrm{comb}(x) - s_C,$$
$$b = \tfrac{1}{2}\delta\delta(x), \quad h = \tfrac{1}{2(N+1)^2}\left[2\delta(x) - \tfrac{1}{N+1}\delta\delta\left(\tfrac{x}{N+1}\right)\right]. \tag{1.23}$$

The resulting frequency response is given by

$$F'_P = \sum_{n=-\infty}^{\infty} A_n(\xi) F_P\left(\xi - \frac{n}{N+1}\right), \tag{1.24}$$

where

$$A_n = \frac{c_n + B\left(\xi - \frac{n}{N+1}\right) + H\left(\xi - \frac{n}{N+1}\right)}{N+1},$$

$$c_n = \begin{cases} N & \text{for } \frac{n}{N+1} \in \mathbb{Z}, \\ -1 & \text{otherwise}, \end{cases}$$

$$B = \cos(2\pi\xi) \; H = \frac{2}{(N+1)^2}\sin^2\left[(N+1)\pi\xi\right]. \tag{1.25}$$

The classifier is given below. The scale factor in front of the gradient term is to balance the contributions of the gradient and Laplacian terms:

$$u = \frac{(N+1)^2}{2}|\delta(x+1) - \delta(x-1)| + |-\delta(x+N+1) + 2\delta(x) - \delta(x-N-1)|.$$

In the case of Figure 1.20d only h in Equation 1.23 (and H in Equation 1.25) and the classifier need be modified as follows:

$$h = \frac{1}{2(2N+2)^2}\left[2\delta(x) - \frac{1}{2N+2}\delta\delta\left(\frac{x}{2N+2}\right)\right], \quad H = \frac{2}{(2N+2)^2}\sin^2\left[(2N+2)\pi\xi\right],$$

$$u = 2(N+1)^2|\delta(x+1) - \delta(x-1)| + |-\delta(x+N+1) + 2\delta(x) - \delta(x-N-1)|$$

As in Section 1.3.4.2, h has been defined so that $dA_n/d\xi = 0$ at $\xi = 0$.

1.3.5.2 Adaptive Cubic Interpolation

For cubic interpolation of panchromatic values plus a color Laplacian image fusion component, one need only change the expression for b in the linear interpolation case and adjust the results accordingly. As it turns out, the design parameter α now becomes zero and the image fusion term h drops out of the results.

$$b = \frac{1}{6}\left[4\delta\delta(x) - \frac{1}{2}\delta\delta\left(\frac{x}{2}\right)\right], \quad B = (4\cos(2\pi\xi) - \cos(4\pi\xi))/3$$

$$F_P' = \sum_{n=-\infty}^{\infty} A_n(\xi) F_P\left(\xi - \frac{n}{N+1}\right), \tag{1.26}$$

where

$$A_n = \frac{c_n + B\left(\xi - \frac{n}{N+1}\right)}{N+1},$$

$$c_n = \begin{cases} N & \text{for } \frac{n}{N+1} \in \mathbb{Z}, \\ -1 & \text{otherwise.} \end{cases}$$

Whereas the color Laplacian is not a part of the predictor, it can still be used as part of the classifier.

$$u = \frac{(N+1)^2}{2} |\delta(x+1) - \delta(x-1)| + |-\delta(x+N+1) + 2\delta(x) - \delta(x-N-1)|.$$

These expressions works for $N > 1$, but the $N = 1$ case needs a slightly different expression for b:

$$b = \frac{1}{16}\left[9\delta\delta(x) - \frac{1}{3}\delta\delta\left(\frac{x}{3}\right)\right], \quad B = (9\cos(2\pi\xi) - \cos(6\pi\xi))/8 \tag{1.27}$$

The design parameter α is still zero so Equation 1.26 is still applicable. Note that $dA_n/d\xi$ is still zero at the origin.

1.3.5.3 Alternating Panchromatic

The alternating panchromatic neighborhood of Figure 1.20a can be treated in the same manner as the Bayer pattern. In this figure, C is one of the color channels, either red, green, or blue. This occurs, for example, in two directions with the green channel, and in one direction with the red and blue channels in the CFA pattern of Figure 1.19d. Applying the general solution with linear interpolation with $N = 1$ produces the following:

$$F_P' = \sum_{n=-\infty}^{\infty} A_n(\xi) F_P\left(\xi - \frac{n}{2}\right), \tag{1.28}$$

$$A_n = \frac{2(-1)^n[1 + \cos(2\pi\xi)] + \sin^2(2\pi\xi)}{4}, \tag{1.29}$$

$$u = 2|\delta(x+1) - \delta(x-1)| + |-\delta(x+2) + 2\delta(x) - \delta(x-2)|.$$

The cubic interpolation solution with $N = 1$ is the special case of the general solution with the cubic interpolation section. The only change to Equation 1.28 is in the following transfer function:

$$A_n = \frac{(-1)^n[8 + 9\cos(2\pi\xi) - \cos(6\pi\xi)]}{16}. \tag{1.30}$$

The fundamental components of Equations 1.29 and 1.30 are plotted in Figure 1.14 as "alternating panchromatic linear" and "alternating panchromatic cubic," respectively. Interpolation error maps of these algorithms assuming the pattern of Figure 1.19d are shown in Figure 1.21. Due to the horizontal versus vertical asymmetry of Figure 1.19d, linear interpolation can only be applied vertically at red and blue pixel locations. Horizontally,

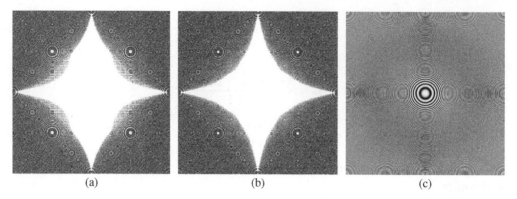

FIGURE 1.21

Alternating panchromatic interpolation: (a) linear interpolation error map, (b) cubic interpolation error map, and (c) fully processed image.

cubic interpolation is used. This is why Figure 1.21a appears to be a blend of Figures 1.17c (linear in both directions) and 1.21b (cubic in both directions). As a result, the linear interpolation method appears to have marginally lower error overall than the cubic interpolation method, as least along the vertical axis.

Color difference interpolation is done in the standard nonadaptive manner. Again referring to the CFA pattern of Figure 1.19d, the green color difference interpolation can be cast as a convolution with the following kernel:

$$\mathbf{b}_G = \frac{1}{4} \begin{pmatrix} 1\ 2\ 1 \\ 2\ 4\ 2 \\ 1\ 2\ 1 \end{pmatrix}.$$

Using the results of Section 1.3.3.1 with $M = 2$, $N = 2$ and $x_0 = 0$, $y_0 = 0$ the corresponding frequency response can be written as follows

$$F_D' = \sum_{m=-\infty}^{\infty} \sum_{n=-\infty}^{\infty} A_{mn}(\xi, \eta) F_D\left(\xi - \frac{m}{2}, \eta - \frac{n}{2}\right),$$

$$A_{mn} = \frac{1}{4}[1 + \cos(2\pi\xi)][1 + \cos(2\pi\eta)].$$

In the case of the red channel, $M = 4$, $N = 2$, $x_0 = 1$, and $y_0 = 1$. The corresponding convolution kernel and frequency response are as follows:

$$\mathbf{b}_{RB} = \frac{1}{8} \begin{pmatrix} 1\ 2\ 3\ 4\ 3\ 2\ 1 \\ 2\ 4\ 6\ 8\ 6\ 4\ 2 \\ 1\ 2\ 3\ 4\ 3\ 2\ 1 \end{pmatrix},$$

$$F_D' = \sum_{m=-\infty}^{\infty} \sum_{n=-\infty}^{\infty} A_{mn}(\xi, \eta) F_D\left(\xi - \frac{m}{4}, \eta - \frac{n}{2}\right),$$

$$A_{mn} = \frac{e^{-i\pi\left(\frac{m}{2}+n\right)}}{8}\left[1 + \frac{3}{2}\cos(2\pi\xi) + \cos(4\pi\xi) + \frac{1}{2}\cos(6\pi\xi)\right][1 + \cos(2\pi\eta)].$$

Finally, in the case of the blue channel, $M = 4$, $N = 2$, and $x_0 = -1$, $y_0 = -1$. The convolution kernel $\mathbf{b}_{RB}$ is used for both the red and blue channels, providing

$$F_D' = \sum_{m=-\infty}^{\infty} \sum_{n=-\infty}^{\infty} A_{mn}(\xi, \eta) F_D\left(\xi - \frac{m}{4}, \eta - \frac{n}{2}\right),$$

$$A_{mn} = \frac{e^{i\pi\left(\frac{m}{2}+n\right)}}{8}\left[1 + \frac{3}{2}\cos(2\pi\xi) + \cos(4\pi\xi) + \frac{1}{2}\cos(6\pi\xi)\right][1 + \cos(2\pi\eta)].$$

The aliasing characteristics of Figure 1.19d can be observed in Figure 1.21c. The aliasing patterns along the edge of the image are different from the Bayer case, and some new faint bands have appeared along the horizontal axis halfway out from the center.

1.3.5.4 Double Panchromatic

The CFA pattern of Figure 1.19b can be viewed as consisting of double panchromatic neighborhoods of Figure 1.20b. These patterns, therefore, can be demosaicked with the general solution with linear interpolation and $N = 2$, for which

$$F_P' = \sum_{n=-\infty}^{\infty} A_n(\xi) F_P\left(\xi - \frac{n}{3}\right), \tag{1.31}$$

$$A_n = \frac{9\cos\left(\frac{2}{3}n\pi\right)[2 + \cos(2\pi\xi)] + 9\sin\left(\frac{2}{3}n\pi\right)\sin(2\pi\xi) + 2\sin^2(3\pi\xi)}{27}, \tag{1.32}$$

$$u = \frac{9}{2}|\delta(x+1) - \delta(x-1)| + |-\delta(x+3) + 2\delta(x) - \delta(x-3)|$$
$$\Rightarrow 9|\delta(x+1) - \delta(x-1)| + 2|-\delta(x+3) + 2\delta(x) - \delta(x-3)|.$$

The general solution with cubic interpolation and $N = 2$ has the same functional form as Equation 1.31 with a different transfer function:

$$A_n = \frac{\cos\left(\frac{2}{3}n\pi\right)[6 + 4\cos(2\pi\xi) - \cos(4\pi\xi)] + \sin\left(\frac{2}{3}n\pi\right)[4\sin(2\pi\xi) + \sin(4\pi\xi)]}{9}.$$
$$\tag{1.33}$$

The fundamental components of Equations 1.32 and 1.33 are plotted in Figure 1.14 as "double panchromatic linear" and "double panchromatic cubic," respectively. Interpolation error maps of these algorithms assuming the pattern of Figure 1.19b are shown in Figure 1.22. Using the aliasing patterns as a visual guide, no more than subtle differences can be seen between the two error maps. It would appear that both interpolation methods are comparable.

A benefit of the CFA pattern Figure 1.19b is that all three color difference channels can be interpolated in the same manner. The corresponding convolution kernel is expressed as

$$\mathbf{b}_{RGB} = \frac{1}{9}\begin{pmatrix} 1 & 2 & 3 & 2 & 1 \\ 2 & 4 & 6 & 4 & 2 \\ 3 & 6 & 9 & 6 & 3 \\ 2 & 4 & 6 & 4 & 2 \\ 1 & 2 & 3 & 2 & 1 \end{pmatrix}.$$

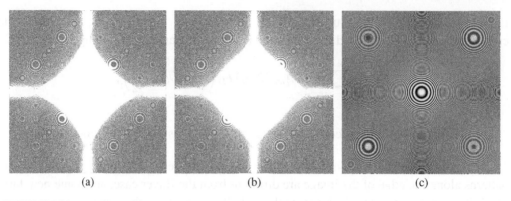

FIGURE 1.22

Double Panchromatic Interpolation: (a) Linear Interpolation Error Map, (b) Cubic Interpolation Error Map, and (c) Fully Processed.

Using the results of Section 1.3.3.1 with $M = 3$, $N = 3$, $x_0 = 0$, and $y_0 = 0$, the corresponding green color difference frequency response can be written as follows:

$$F'_D = \sum_{m=-\infty}^{\infty} \sum_{n=-\infty}^{\infty} A_{mn}(\xi, \eta) F_D\left(\xi - \frac{m}{3}, \eta - \frac{n}{3}\right),$$

$$A_{mn} = \frac{1}{9}\left[1 + \frac{4}{3}\cos(2\pi\xi) + \frac{2}{3}\cos(4\pi\xi)\right]\left[1 + \frac{4}{3}\cos(2\pi\eta) + \frac{2}{3}\cos(4\pi\eta)\right].$$

In the case of the red channel, $M = 3$, $N = 3$, $x_0 = 1$, AND $y_0 = -1$. The corresponding frequency response is as follows:

$$F'_D = \sum_{m=-\infty}^{\infty} \sum_{n=-\infty}^{\infty} A_{mn}(\xi, \eta) F_D\left(\xi - \frac{m}{3}, \eta - \frac{n}{3}\right),$$

$$A_{mn} = \frac{e^{-i\frac{2\pi}{3}(m-n)}}{9}\left[1 + \frac{4}{3}\cos(2\pi\xi) + \frac{2}{3}\cos(4\pi\xi)\right]\left[1 + \frac{4}{3}\cos(2\pi\eta) + \frac{2}{3}\cos(4\pi\eta)\right].$$

Finally, in the case of the blue channel, $M = 3$, $N = 3$, $x_0 = -1$, and $y_0 = 1$, for which

$$F'_D = \sum_{m=-\infty}^{\infty} \sum_{n=-\infty}^{\infty} A_{mn}(\xi, \eta) F_D\left(\xi - \frac{m}{3}, \eta - \frac{n}{3}\right),$$

$$A_{mn} = \frac{e^{i\frac{2\pi}{3}(m-n)}}{9}\left[1 + \frac{4}{3}\cos(2\pi\xi) + \frac{2}{3}\cos(4\pi\xi)\right]\left[1 + \frac{4}{3}\cos(2\pi\eta) + \frac{2}{3}\cos(4\pi\eta)\right].$$

The aliasing characteristics of Figure 1.19b can be observed in Figure 1.22c. The aliasing patterns along the edge of the image have been largely eliminated at the expense of four significant aliasing patterns two-thirds of the distance away from the center in both horizontal and vertical directions. The two patterns along the color pixel diagonal of Figure 1.19b are colored, whereas the other two patterns are neutral (i.e., luminance patterns).

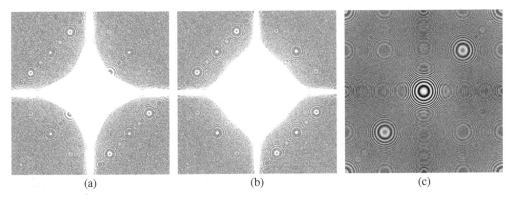

FIGURE 1.23

Triple panchromatic interpolation: (a) linear interpolation error map, (b) cubic interpolation error map, and (c) fully processed image.

1.3.5.5 Triple Panchromatic

The triple panchromatic neighborhood of Figure 1.20c can be found in the CFA pattern of Figure 1.19c. Therefore, this pattern can be demosaicked with the general solution with linear interpolation and $N = 3$, characterized as follows:

$$F'_P = \sum_{n=-\infty}^{\infty} A_n(\xi) F_P\left(\xi - \frac{n}{4}\right), \tag{1.34}$$

$$A_n = \frac{8(-1)^n + 8\cos\left(\frac{\pi}{2}n\right)\left[2 + \cos(2\pi\xi)\right] + 8\sin\left(\frac{\pi}{2}n\right)\sin(2\pi\xi) + \sin^2(4\pi\xi)}{32}, \tag{1.35}$$

$$u = 8\left|\delta(x+1) - \delta(x-1)\right| + \left|-\delta(x+4) + 2\delta(x) - \delta(x-4)\right|.$$

The general solution with cubic interpolation and $N = 3$ has the same functional form as Equation 1.34 with a different transfer function:

$$A_n = \frac{8(-1)^n + \cos\left(\frac{n}{2}\pi\right)\left[16 + 9\cos(2\pi\xi) - \cos(6\pi\xi)\right]}{32}$$
$$+ \frac{\sin\left(\frac{n}{2}\pi\right)\left[9\sin(2\pi\xi) + \sin(6\pi\xi)\right]}{32}. \tag{1.36}$$

The fundamental components of Equations 1.35 and 1.36 are plotted in Figure 1.14 as "triple panchromatic linear" and "triple panchromatic cubic," respectively. Interpolation error maps of these algorithms assuming the pattern of Figure 1.19c are shown in Figure 1.23. A crossover has clearly occurred with the cubic interpolation method starting to clearly produce less error overall than the linear interpolation method.

Color difference interpolation is once again done in the standard nonadaptive manner. Referring to the CFA pattern of Figure 1.19c, the green color difference interpolation can be cast as a convolution with the following kernel:

$$\mathbf{b}_G = \frac{1}{16} \begin{pmatrix} 0\,0\,0\,1\,0\,0\,0 \\ 0\,0\,3\,4\,3\,0\,0 \\ 0\,3\,8\,9\,8\,3\,0 \\ 1\,4\,9\,16\,9\,4\,1 \\ 0\,3\,8\,9\,8\,3\,0 \\ 0\,0\,3\,4\,3\,0\,0 \\ 0\,0\,0\,1\,0\,0\,0 \end{pmatrix}.$$

Using the results of Section 1.3.3.2 with $M = 2$, $N = 2$, $x_0 = 0$, and $y_0 = 0$, the corresponding frequency response can be written as follows:

$$F_D' = \sum_{m=-\infty}^{\infty} \sum_{n=-\infty}^{\infty} A_{mn}(\xi, \eta) F_D\left(\xi - \frac{m+n}{4}, \eta - \frac{-m+n}{4}\right),$$

$$\begin{aligned} A_{mn} = \frac{1}{8} + \frac{1}{4}\left[\frac{9}{16}\cos(2\pi\xi) + \frac{1}{4}\cos(4\pi\xi) + \frac{1}{16}\cos(6\pi\xi)\right] \\ + \frac{1}{4}\left[\frac{9}{16}\cos(2\pi\eta) + \frac{1}{4}\cos(4\pi\eta) + \frac{1}{16}\cos(6\pi\eta)\right] \\ + \frac{1}{2}\left[\frac{1}{2}\cos(2\pi\xi, 2\pi\eta) + \frac{3}{16}\cos(2\pi\xi, 4\pi\eta) + \frac{3}{16}\cos(4\pi\xi, 2\pi\eta)\right]. \end{aligned}$$

In the case of the red channel, the results of Section 1.3.3.1, $M = 4$, $N = 4$, $x_0 = -1$, and $y_0 = 1$. The corresponding convolution kernel and frequency response are as follows:

$$\mathbf{b}_{RB} = \frac{1}{16} \begin{pmatrix} 1\,2\,3\,\ \ 4\,\ \ 3\,2\,1 \\ 2\,4\,6\,\ \ 8\,\ \ 6\,4\,2 \\ 3\,6\,9\,\ 12\,\ 9\,6\,3 \\ 4\,8\,12\,16\,12\,8\,4 \\ 3\,6\,9\,\ 12\,\ 9\,6\,3 \\ 2\,4\,6\,\ \ 8\,\ \ 6\,4\,2 \\ 1\,2\,3\,\ \ 4\,\ \ 3\,2\,1 \end{pmatrix},$$

$$F_D' = \sum_{m=-\infty}^{\infty} \sum_{n=-\infty}^{\infty} A_{mn}(\xi, \eta) F_D\left(\xi - \frac{m}{4}, \eta - \frac{n}{4}\right),$$

$$\begin{aligned} A_{mn} = \frac{e^{i\pi\left(\frac{m-n}{2}\right)}}{16}\left\{1 + 2\left[\frac{3}{4}\cos(2\pi\xi) + \frac{1}{2}\cos(4\pi\xi) + \frac{1}{4}\cos(6\pi\xi)\right]\right\} \\ \times \left\{1 + 2\left[\frac{3}{4}\cos(2\pi\eta) + \frac{1}{2}\cos(4\pi\eta) + \frac{1}{4}\cos(6\pi\eta)\right]\right\}. \end{aligned}$$

Finally, in the case of the blue channel, $M = 4$, $N = 4$, $x_0 = 1$, and $y_0 = -1$. The convolution kernel $\mathbf{b}_{RB}$ is used for both the red and blue channels. The corresponding frequency response is as follows:

$$F_D' = \sum_{m=-\infty}^{\infty} \sum_{n=-\infty}^{\infty} A_{mn}(\xi, \eta) F_D\left(\xi - \frac{m}{4}, \eta - \frac{n}{4}\right),$$

$$A_{mn} = \frac{e^{-i\pi\left(\frac{m-n}{2}\right)}}{16} \left\{ 1 + 2 \left[\frac{3}{4}\cos(2\pi\xi) + \frac{1}{2}\cos(4\pi\xi) + \frac{1}{4}\cos(6\pi\xi) \right] \right\}$$
$$\times \left\{ 1 + 2 \left[\frac{3}{4}\cos(2\pi\eta) + \frac{1}{2}\cos(4\pi\eta) + \frac{1}{4}\cos(6\pi\eta) \right] \right\}.$$

The aliasing characteristics of Figure 1.19c can be observed in Figure 1.23c. Colored aliasing patterns are evident along the edges of the image half-way of the distance to the corners from both the horizontal and vertical axes. There are also four strong aliasing patterns half-way out from the center in both the horizontal and vertical directions. The two of these patterns along the color pixel diagonal of Figure 1.19c are colored, whereas the other two patterns are neutral (that is, luminance patterns). There are also strong luminance aliasing patterns in the corners of the image itself.

1.3.5.6 Two-Color Alternating Panchromatic

The two-color alternating panchromatic neighborhood of Figure 1.20d occurs in the CFA patterns of Figures 1.19a and 1.19e. These patterns can be demosaicked in the same way as the alternating panchromatic neighborhood with a change in h. As with the alternating panchromatic neighborhood, $N = 1$, which gives

$$h = \frac{1}{16} \left[2\delta(x) - \frac{1}{4}\delta\delta\left(\frac{x}{4}\right) \right],$$

$$F_P' = \sum_{n=-\infty}^{\infty} A_n(\xi) F_P\left(\xi - \frac{n}{2}\right),$$

$$A_n = \frac{8(-1)^n[1 + \cos(2\pi\xi)] + \sin^2(4\pi\xi)}{16}, \qquad (1.37)$$

$$u = 8|\delta(x+1) - \delta(x-1)| + |-\delta(x+4) + 2\delta(x) - \delta(x-4)|.$$

Since there is no contribution due to h in the cubic interpolation solution, the cubic solution for the two-color alternating panchromatic neighborhood is the same as the alternating panchromatic neighborhood. The fundamental component of Equation 1.37 is plotted in Figure 1.14 as "two-color alternating panchromatic linear." The "alternating panchromatic cubic" plot would apply in this two-color alternating panchromatic cubic interpolation case as well. Interpolation error maps of these algorithms assuming the pattern of Figure 1.19a are shown in Figure 1.24. Note that the error maps corresponding to Figure 1.19e are similar. The interpolation error maps are almost identical, although minute differences can be detected upon close inspection. Clearly the cubic interpolation terms are dominant in the predictors for these CFA patterns.

The color difference interpolation for the CFA pattern of Figure 1.19a is identical in execution to the previously described method used for the CFA pattern of Figure 1.19c. The only difference is a small change to the phase term in the transfer functions. Using the results of Section 1.3.3.2 with $M = 2$, $N = 2$, $x_0 = 0$, and $y_0 = 0$, the corresponding frequency response can be written as follows:

$$F_D' = \sum_{m=-\infty}^{\infty} \sum_{n=-\infty}^{\infty} A_{mn}(\xi,\eta) F_D\left(\xi - \frac{m+n}{4}, \eta - \frac{-m+n}{4}\right),$$

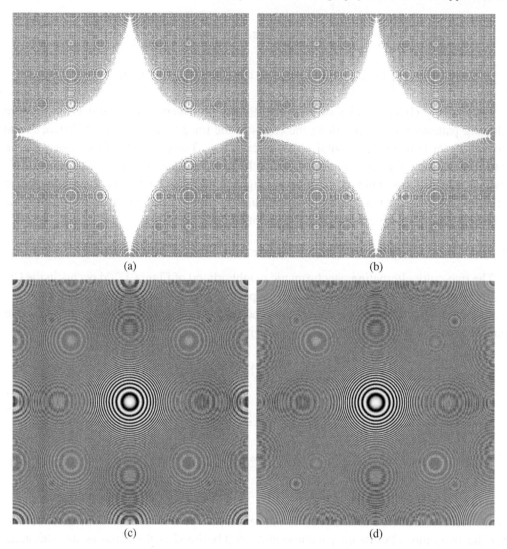

FIGURE 1.24

Two-color alternating panchromatic interpolation: (a) linear interpolation error map, (b) cubic interpolation error map, (c) fully processed image using Figure 1.19a, and (d) fully processed image using Figure 1.19e.

$$A_{mn} = \frac{1}{8} + \frac{1}{4}\left[\frac{9}{16}\cos(2\pi\xi) + \frac{1}{4}\cos(4\pi\xi) + \frac{1}{16}\cos(6\pi\xi)\right]$$
$$+ \frac{1}{4}\left[\frac{9}{16}\cos(2\pi\eta) + \frac{1}{4}\cos(4\pi\eta) + \frac{1}{16}\cos(6\pi\eta)\right]$$
$$+ \frac{1}{2}\left[\frac{1}{2}\cos(2\pi\xi, 2\pi\eta) + \frac{3}{16}\cos(2\pi\xi, 4\pi\eta) + \frac{3}{16}\cos(4\pi\xi, 2\pi\eta)\right].$$

Adjustments are made to x_0 and y_0 in the case of the red and blue channels. For the red channel the results of Section 1.3.3.1, $M = 4$, $N = 4$, $x_0 = 2$, and $y_0 = 0$ are used. The corresponding convolution kernel $\mathbf{b}_{RB}$ is as described in the previous section whereas the frequency response is expressed as follows:

$$F'_D = \sum_{m=-\infty}^{\infty} \sum_{n=-\infty}^{\infty} A_{mn}(\xi, \eta) F_D\left(\xi - \frac{m}{4}, \eta - \frac{n}{4}\right),$$

$$A_{mn} = \frac{(-1)^m}{16}\left\{1 + 2\left[\frac{3}{4}\cos(2\pi\xi) + \frac{1}{2}\cos(4\pi\xi) + \frac{1}{4}\cos(6\pi\xi)\right]\right\}$$
$$\times \left\{1 + 2\left[\frac{3}{4}\cos(2\pi\eta) + \frac{1}{2}\cos(4\pi\eta) + \frac{1}{4}\cos(6\pi\eta)\right]\right\}.$$

For the blue channel, $M = 4$, $N = 4$, $x_0 = 0$, and $y_0 = 2$. The convolution kernel $\mathbf{b}_{RB}$ is used for both the red and blue channels. The corresponding frequency response is as follows:

$$F'_D = \sum_{m=-\infty}^{\infty} \sum_{n=-\infty}^{\infty} A_{mn}(\xi, \eta) F_D\left(\xi - \frac{m}{4}, \eta - \frac{n}{4}\right),$$

$$A_{mn} = \frac{(-1)^n}{16}\left\{1 + 2\left[\frac{3}{4}\cos(2\pi\xi) + \frac{1}{2}\cos(4\pi\xi) + \frac{1}{4}\cos(6\pi\xi)\right]\right\}$$
$$\times \left\{1 + 2\left[\frac{3}{4}\cos(2\pi\eta) + \frac{1}{2}\cos(4\pi\eta) + \frac{1}{4}\cos(6\pi\eta)\right]\right\}.$$

Color difference interpolation for the CFA pattern of Figure 1.19e provides a minor twist over the other patterns. This pattern can be viewed as consisting of diagonal pairs of like-colored pixels. Assuming the salient high spatial frequency information of the image is contained in the panchromatic channel, the option exists to treat each diagonal pair of color pixels as a single larger pixel for the purposes of noise cleaning.[4] Therefore, a strategy that largely averages adjacent diagonal pixel pairs is used for color difference interpolation. Beginning with the green channel, it is treated as the sum of two diamond-shaped neighborhoods:

$$s_D = \frac{1}{8}\text{comb}\left(\frac{x-y}{4}, \frac{x+y}{4}\right) + \frac{1}{8}\text{comb}\left(\frac{x-y}{4}, \frac{x+y-2}{4}\right).$$

As a consequence of having two diamond-shaped neighborhoods, the interpolating function must be scaled by $1/2$ as follows:

$$b_G = \frac{1}{2}\text{tri}\left(\frac{x-y}{4}, \frac{x+y}{4}\right)\text{comb}(x,y). \tag{1.38}$$

---

[4]In all the other CFA patterns of Figure 1.19, the pixels of a given color are separated by at least one panchromatic pixel. Averaging these more widely spaced pixels would introduce greater amounts of color aliasing into the demosaicked image.

This results in the following convolution kernel:

$$\mathbf{b}_G = \frac{1}{32} \begin{pmatrix} 0\,0\,0\,\ 1\ 0\,0\,0 \\ 0\,0\,3\,\ 4\ 3\,0\,0 \\ 0\,3\,8\,\ 9\ 8\,3\,0 \\ 1\,4\,9\,16\,9\,4\,1 \\ 0\,3\,8\,\ 9\ 8\,3\,0 \\ 0\,0\,3\,\ 4\ 3\,0\,0 \\ 0\,0\,0\,\ 1\ 0\,0\,0 \end{pmatrix}.$$

The resulting frequency response is very similar to that described for CFA pattern Figure 1.19c and can be written as:

$$F_D' = \sum_{m=-\infty}^{\infty} \sum_{n=-\infty}^{\infty} A_{mn}(\xi,\eta) F_D\left(\xi - \frac{m+n}{4}, \eta - \frac{-m+n}{4}\right),$$

$$\begin{aligned} A_{mn} = &\ \frac{1+(-1)^n}{16} + \frac{1+(-1)^n}{8}\left[\frac{9}{16}\cos(2\pi\xi) + \frac{1}{4}\cos(4\pi\xi) + \frac{1}{16}\cos(6\pi\xi)\right] \\ &+ \frac{1+(-1)^n}{8}\left[\frac{9}{16}\cos(2\pi\eta) + \frac{1}{4}\cos(4\pi\eta) + \frac{1}{16}\cos(6\pi\eta)\right] \\ &+ \frac{1+(-1)^n}{4}\left[\frac{1}{2}\cos(2\pi\xi, 2\pi\eta) + \frac{3}{16}\cos(2\pi\xi, 4\pi\eta) + \frac{3}{16}\cos(4\pi\xi, 2\pi\eta)\right]. \end{aligned}$$

The same approach is used for red and blue color difference interpolation. The sampling function is the sum of two rectilinear grids and the interpolating function is scaled by one-half. The red channel is considered first:

$$s_D = \frac{1}{16}\text{comb}\left(\frac{x}{4}, \frac{y-2}{4}\right) + \frac{1}{16}\text{comb}\left(\frac{x-1}{4}, \frac{y-3}{4}\right).$$

As a consequence of having two rectilinear neighborhoods, the interpolating function must be scaled by $1/2$, resulting in the following:

$$b_{RB} = \frac{1}{2}\text{tri}\left(\frac{x}{4}, \frac{y}{4}\right)\text{comb}(x,y).$$

The convolution kernel is the same for the red and blue channels and is defined as follows:

$$\mathbf{b}_{RB} = \frac{1}{32} \begin{pmatrix} 1\,2\,3\ \ 4\ \ 3\,2\,1 \\ 2\,4\,6\ \ 8\ \ 6\,4\,2 \\ 3\,6\,9\,12\,9\,6\,3 \\ 4\,8\,12\,16\,12\,8\,4 \\ 3\,6\,9\,12\,9\,6\,3 \\ 2\,4\,6\ \ 8\ \ 6\,4\,2 \\ 1\,2\,3\ \ 4\ \ 3\,2\,1 \end{pmatrix}.$$

The resulting frequency response is given by

$$F_D' = \sum_{m=-\infty}^{\infty} \sum_{n=-\infty}^{\infty} A_{mn}(\xi,\eta) F_D\left(\xi - \frac{m}{4}, \eta - \frac{n}{4}\right),$$

$$A_{mn} = \frac{(-1)^n + e^{-i\frac{\pi}{2}(m+3n)}}{32} \left[1 + \frac{3}{2}\cos(2\pi\xi) + \cos(4\pi\xi) + \frac{1}{2}\cos(6\pi\xi)\right]$$
$$\times \left[1 + \frac{3}{2}\cos(2\pi\eta) + \cos(4\pi\eta) + \frac{1}{2}\cos(6\pi\eta)\right].$$

The blue channel frequency response requires only a change to the phase term in the transfer functions.

$$A_{mn} = \frac{(-1)^m + e^{-i\frac{\pi}{2}(3m+n)}}{32} \left[1 + \frac{3}{2}\cos(2\pi\xi) + \cos(4\pi\xi) + \frac{1}{2}\cos(6\pi\xi)\right]$$
$$\times \left[1 + \frac{3}{2}\cos(2\pi\eta) + \cos(4\pi\eta) + \frac{1}{2}\cos(6\pi\eta)\right].$$

The aliasing characteristics of Figure 1.19a can be observed in Figure 1.24c and the aliasing patterns for Figure 1.19e are shown in Figure 1.24d. The predominant aliasing patterns occur half-way out from the center with Figure 1.19a having four such patterns, whereas Figure 1.19e has only two.

1.3.6 Comments

From the foregoing analysis a number of conclusions can be drawn. Even the simplest adaptive demosaicking of the luminance (i.e., green or panchromatic) channel produces greater image reconstruction fidelity than nonadaptive demosaicking, as illustrated in Figure 1.17. The best forms of adaptive demosaicking are either linear interpolation of luminance combined with appropriately weighted color Laplacians or cubic interpolation of luminance values alone, for example, Figure 1.21. In a four-channel system, color aliasing in the demosaicked image is determined by the number and arrangement of color pixels within the CFA pattern. The fewer the number of color pixels present and the more widely they are separated, the greater the resulting aliasing. Compare Figure 1.21c, which has a high number of closely spaced color pixels to Figure 1.24c, which has a low number of widely spaced color pixels. Of the four-channel CFA patterns discussed (see Figure 1.19), the pattern of Figure 1.19d demosaicked with a combination of linear and cubic interpolation strategy produces the highest overall reconstruction fidelity with the least low-frequency color aliasing. It should be noted, however, that there are other possible considerations when selecting a CFA pattern, most notably signal-to-noise performance (see Section 1.4). With the opportunity to average diagonally adjacent color pixels, the CFA pattern in Figure 1.19e can be a better choice for certain applications, for instance, low light imaging. As with all such trade-offs, the relative importance of aliasing versus signal-to-noise needs to be assessed on a case-by-case basis.

1.4 Noise and Noise Reduction

An imaging sensor captures an image through the photo-electric conversion mechanism of a silicon semiconductor. Incoming photons produce free electrons within the semiconductor in proportion to the amount of incoming photons and those electrons are gathered within the imaging chip. Image capture is therefore essentially a photon-counting process. As such, image capture is governed by the Poisson distribution, which is defined with a photon arrival rate variance equal to the mean photon arrival rate. The arrival rate variance is a source of *image noise* because if a uniformly illuminated, uniform color patch is captured with a perfect optical system and sensor, the resulting image will not be uniform but rather have a dispersion about a mean value. The dispersion is called image noise because it reduces the quality of an image when a human is observing it [84].

Image noise can also be structured, as is the case with dead pixels or optical pixel crosstalk [85]. This book chapter does not discuss structured noise, but rather focuses on the Poisson-distributed noise (also called *shot noise*) with the addition of electronic amplifier read noise, which is modeled with a Gaussian distribution [85]. A pixel value Q may be modeled as $Q = k_Q (q + g)$, where k_Q is the amplifier gain, q is a Poisson variable with mean m_q and variance σ_q^2, and g is a Gaussian variable with mean m_g and variance σ_g^2. Note that $\sigma_q^2 = m_q$ since q is a Poisson variable, and it is entirely defined by the spectral power distribution impinging upon the sensor and the channel spectral responsivities. The mean signal level of pixel Q is derived from the pixel model and is written as $m_Q = k_Q (m_q + m_g)$.

An objective measure of image noise is the signal-to-noise ratio (SNR). To increase the perceived quality of an image it is desirable to increase the SNR [86]. The SNR is defined as the signal mean level divided by the signal standard deviation and in this case the SNR of a pixel is

$$SNR_Q = \frac{k_Q (m_q + m_g)}{\left[k_Q^2 \left(\sigma_q^2 + \sigma_g^2 \right) \right]^{\frac{1}{2}}} = \frac{(m_q + m_g)}{\left(\sigma_q^2 + \sigma_g^2 \right)^{\frac{1}{2}}}. \tag{1.39}$$

If it is assumed that the read noise is negligible, this expression reduces to

$$SNR_Q = \frac{m_q}{\sigma_q} = \sqrt{m_q}. \tag{1.40}$$

That is, as the signal goes up, the SNR goes up and therefore also the perceived quality of the image. One way to increase the signal is to use good-quality optics with wide apertures. Another way is to control the illumination upon the objects to be photographed, as with a flash. A third way is to increase the intrinsic efficiency of the photon-counting process carried out within the semiconductor. If it is assumed that a system design has already taken advantage of the aforementioned ways to increase the signal or that, for whatever reason, one or more of these ways are not used (e.g., flash in a museum or good optics in a mobile phone camera), a way to increase the SNR is to use panchromatic filters in the CFA. Panchromatic filters have a passband that is wider than colored filters and typically have a higher photometric sensitivity that results in a higher responsivity, as shown in Figure 1.1.

Suppose an image of a uniformly illuminated, uniform gray patch is captured with two sensors: one that is an all-green pixel sensor and the other is an all-panchromatic pixel sensor that has a higher photometric sensitivity than the green pixel sensor. Figure 1.25 shows a simulation of the green and panchromatic images, and it can be seen that the panchromatic image has less noise than the green image. A simple image fusion scheme, as described in Section 1.3, may be employed to exploit the increased SNR of a panchromatic image while producing a color image. Even though the SNR of a panchromatic channel may be higher

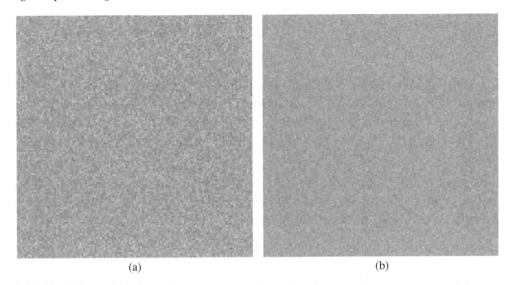

(a)   (b)

FIGURE 1.25

Simulated images of two identical sensors that have different filters: (a) green filters only, (b) panchromatic filters only.

than that of a color channel, the panchromatic channel may still exhibit visible or even objectionable noise under low exposure conditions. Additionally, the chroma information is obtained solely from the noisier color channels. It is therefore important to include noise reduction techniques for both the panchromatic and color channels in the image processing chain. Image fusion techniques may be included in the color noise reduction to again exploit the increased SNR of the panchromatic image. Such noise reduction techniques will be discussed later in this section.

1.4.1   Image Noise Propagation

Suppose an image of a uniformly illuminated, uniform gray patch is captured with a sensor that has a four-channel CFA such as those shown in Figure 1.19. Suppose that the camera exposure and electronics are such that: i) the panchromatic channels are properly exposed, ii) no signal clipping occurs, and iii) the color channels are gained up (if the panchromatic channel has a higher photometric sensitivity) or down (if the panchromatic channel has a lower photometric sensitivity) to the level of the panchromatic signal. This last supposition may be expressed as $m_P - m_g = k_C (m_C - m_g)$, where k_C is the gain. Note that if the read noise is negligible, this expression implies that

$$k_C = \sigma_P^2 / \sigma_C^2. \tag{1.41}$$

With these suppositions, the noise variance may be propagated through the demosaicking process that is described in Section 1.3. For simplicity, suppose that the color Laplacian image fusion component is not used to demosaick the panchromatic channel. Also for simplicity, only the noise variance of one color channel C is derived here. It is of course understood that the noise variance for any number of color channels may be propagated the same way.

After panchromatic demosaicking, the variance of the resulting panchromatic pixels is

$$\sigma_P'^2 = \sigma_P^2 \text{Tr}\left(\mathbf{b}\mathbf{b}^T\right),$$

where $\text{Tr}\left(\mathbf{a}\right)$ is the trace of matrix $\mathbf{a}$ and the superscript T denotes a transposed kernel. The variance of the color difference, σ_D^2, is obtained by adding the noise contribution from a color channel C, gained by k_C, to the noise contribution from the demosaicked panchromatic. Therefore,

$$\sigma_D^2 = k_C^2 \sigma_C^2 + \sigma_P^2 \text{Tr}\left(\mathbf{b}\mathbf{b}^T\right).$$

Now suppose that a kernel $\mathbf{d}$ is used to demosaick the color differences. By defining a kernel $\mathbf{e} = -\mathbf{b} * \mathbf{d}$, where $*$ is the convolution operator, the variance of the demosaicked color differences may be written as follows:

$$\sigma_D'^2 = k_C^2 \sigma_C^2 \text{Tr}\left(\mathbf{d}\mathbf{d}^T\right) + \sigma_P^2 \text{Tr}\left(\mathbf{e}\mathbf{e}^T\right).$$

To finally derive the variance for the demosaicked color, the contribution from the panchromatic pixel at the same location as the color pixel that is being demosaicked must be added to the contribution from the demosaicked color difference. This panchromatic contribution depends on whether that particular panchromatic pixel is an original sample or the result of demosaicking. Let a kernel $\mathbf{f}$ be equal to $\mathbf{b}$ (zero-padded to make it the same size as $\mathbf{e}$) if the color difference demosaicking is centered on a color pixel and equal to the discrete delta function if the color difference demosaicking is centered on a panchromatic pixel. The variance of the demosaicked color is therefore written as

$$\sigma_C'^2 = k_C^2 \sigma_C^2 \text{Tr}\left(\mathbf{d}\mathbf{d}^T\right) + \sigma_P^2 \text{Tr}\left[\left(\mathbf{e}+\mathbf{f}\right)\left(\mathbf{e}+\mathbf{f}\right)^T\right]. \tag{1.42}$$

In order for the SNR of the demosaicked color pixels to be equal to or greater than the gained original color pixels, the following constraint is introduced:

$$k_C^2 \sigma_C^2 \geq \sigma_C'^2. \tag{1.43}$$

The implication upon k_C that results from this expression depends on the negligibility of the read noise. The case where the read noise is negligible is discussed next, followed by the case where the read noise is considered.

For the case that the read noise is negligible, the gain given by Equation 1.41 is defined with Poisson variances. Equation 1.41 with Poisson variances and Equation 1.43 together imply that

$$k_C \geq \frac{\text{Tr}\left[\left(\mathbf{e}+\mathbf{f}\right)\left(\mathbf{e}+\mathbf{f}\right)^T\right]}{1 - \text{Tr}\left(\mathbf{d}\mathbf{d}^T\right)}. \tag{1.44}$$

Note that this result is independent of signal levels. As a simple example, suppose that neighbor averaging is used in only the horizontal dimension to demosaick the neighborhood shown in Figure 1.20a. Then, $\mathbf{b} = \mathbf{d} = [0.5 \ \ 0.0 \ \ 0.5]$ and $\mathbf{f} = [0 \ \ 0 \ \ 1 \ \ 0 \ \ 0]$. In this case k_C must be greater or equal than $3/4$ for the SNR of the demosaicked color pixels to be equal to or greater than the gained original color pixels. Of course, the higher k_C is made, the better the SNR of the demosaicked color pixels as compared to the SNR of the gained original color pixels.

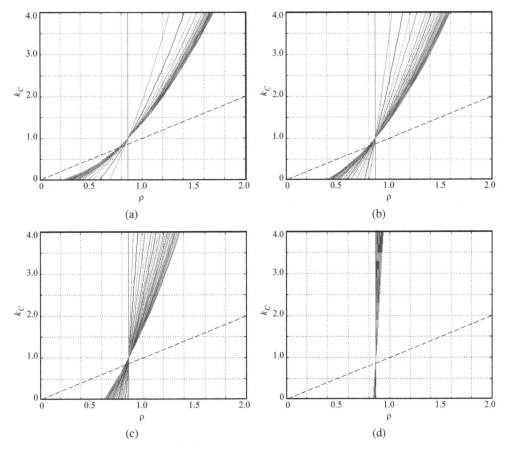

FIGURE 1.26

A graph representation of Equation 1.46 where k_C is plotted against ρ for (a) $\sigma_g = 4$, (b) $\sigma_g = 8$, (c) $\sigma_g = 16$, and (d) $\sigma_g = 64$. Each subfigure depicts the graphs of $m_C - m_g$ values between 0 and 250 in steps of 10. The demosaicking kernels are $\mathbf{b} = \mathbf{d} = [0.5\ \ 0.0\ \ 0.5]$ and $\mathbf{f} = [0\ \ 0\ \ 1\ \ 0\ \ 0]$.

When the read noise is significant, the color and panchromatic variances include both the Poisson noise and the Gaussian noise contributions. Therefore, in this case $\sigma_C^2 = \sigma_{CP}^2 + \sigma_g^2$ and $\sigma_P^2 = \sigma_{PP}^2 + \sigma_g^2$, where σ_{CP}^2 and σ_{PP}^2 are the Poisson variances for the color and panchromatic pixels, respectively. By definition of the Poisson distribution, $\sigma_{CP}^2 = m_C - m_g$ and $\sigma_{PP}^2 = m_P - m_g$, where m_C and m_P are the mean signal responses of the color and panchromatic pixels, respectively. The color gain is defined in this case as follows:

$$k_C = \frac{m_P - m_g}{m_C - m_g} = \frac{\sigma_{PP}^2}{\sigma_{CP}^2}. \tag{1.45}$$

Using Equations 1.45 and 1.43, in order to have a demosaicked color pixel SNR equal to or greater than the gained original color pixels the following inequality must hold:

$$k_C \geq \frac{\sigma_P}{\sigma_C} \sqrt{\frac{\mathrm{Tr}\left[(\mathbf{e}+\mathbf{f})(\mathbf{e}+\mathbf{f})^T\right]}{1 - \mathrm{Tr}(\mathbf{d}\mathbf{d}^T)}}. \tag{1.46}$$

This result is not independent of signal levels and the right-hand side of the inequality, which from now on is denoted as ρ, cannot be computed without first defining k_C. To illustrate the behavior of the inequality, the same demosaicking kernels as in the previous simple example are used along with $m_C - m_g$ values between 0 and 250 in steps of 10. Figure 1.26a shows plots of k_C against ρ for the case where σ_g is 4, where the graph for an $m_C - m_g$ of 0 is vertical and the graph for an $m_C - m_g$ of 250 is the one with the least slope for a given value of ρ. The dashed line indicates the values where k_C is equal to ρ, and therefore any value of k_C that intersects a graph above the dashed line will yield the SNR of the demosaicked color pixels that is equal to or greater than the gained original color pixel for the $m_C - m_g$ associated with the intersected graph. Figures 1.26b to 1.26d show the same plots for the σ_g values of 8, 16, and 64, respectively. It is evident from the graphs that at very low $m_C - m_g$, the read noise dominates over the shot noise in Equation 1.46 and in the limit that $m_C - m_g$ goes to zero, which is the same as σ_{CP} and σ_{PP} going to zero, ρ becomes independent of both k_C and σ_g because

$$\lim_{\sigma_{CP} \to 0} \frac{\sigma_P}{\sigma_C} = \frac{\sqrt{0 + \sigma_g^2}}{\sqrt{0 + \sigma_g^2}} = 1,$$

and therefore

$$\lim_{\sigma_{CP} \to 0} \rho = \sqrt{\frac{\text{Tr}\left[(\mathbf{e}+\mathbf{f})(\mathbf{e}+\mathbf{f})^T\right]}{1 - \text{Tr}(\mathbf{d}\mathbf{d}^T)}}. \tag{1.47}$$

If it is required that the SNR of the demosaicked color pixels be equal to or greater than the gained original color pixels for any color pixel value, then the gain must be chosen to be this limiting value in Equation 1.47, which is equal to $\sqrt{3/4}$ for the demosaicking kernels in the previous example calculation.

1.4.2   Image Noise Reduction

Under low exposure conditions noise reduction techniques are used to increase the SNR and therefore the perceived quality of an image. Figure 1.27a shows a simulated result of demosaicking a Bayer CFA image corrupted with Poisson noise and Gaussian noise. The captured image is a uniform gray patch, uniformly illuminated with light having a flat power spectrum. The demosaicking process is similar to that described in Section 1.3.5.3 with the green channel substituting for the panchromatic channel. Figure 1.27e shows a simulated result of demosaicking the four-channel CFA shown in Figure 1.19d corrupted with the Poisson noise and Gaussian noise that correspond to the simulated light intensity and simulated read noise of the Bayer image in Figure 1.27a. In both these experiments, the simulated red, green, and blue relative responsivities are those shown in Figure 1.1. Therefore, Figures 1.27a and 1.27e show simulations of the same image captured under the same conditions and with the same sensor but with a different CFA. Figures 1.28a and 1.28b shows the red channel noise power spectra (NPS) of the two demosaicked images and the difference between the two spectra (four-channel spectrum minus Bayer spectrum). The NPS for the blue channels are very similar to those for the red channels, and the green NPS are slightly different in shape but the trends are the same as those shown. The demosaicking

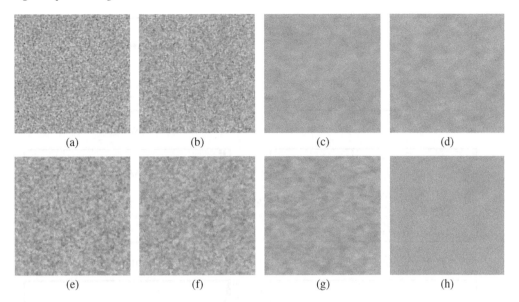

FIGURE 1.27 (See color insert.)

Simulated images demosaicked from a Bayer CFA (top row) and the four-channel CFA shown in Figure 1.19d (bottom row). The images shown are for the cases of: (a,e) no noise reduction, (b,f) median filtering only, (c,g) median and boxcar filtering only, and (d,h) median, boxcar, and low-frequency filtering.

technique with panchromatic image fusion is shown to yield a lower NPS everywhere but below about 0.035 cycles/sample. Any four-channel CFA will yield a similar situation because it necessarily has the color pixels further away from each other to make room for the panchromatic pixels. The exact frequency where the four-channel NPS becomes larger than the Bayer NPS is dependent upon the specific four-channel CFA and the demosaicking technique. Any noise reduction strategy for a four-channel CFA image must address this low-frequency noise as well as the high-frequency noise.

Given the demosaicking strategies discussed in Section 1.3, it is clear that to take the most advantage of the panchromatic channel it is best to apply noise reduction to the panchromatic pixels before any demosaicking is done. Any single-channel noise reduction strategy may be used to process the panchromatic pixels, therefore only color-pixel noise reduction techniques are discussed in this section. It is assumed for the rest of this section that the panchromatic pixels are median-filtered and boxcar-filtered. Since the green channel is used as a substitute for the panchromatic channel in the Bayer CFA comparison, this section also assumes that the Bayer green pixels are median filtered and boxcar filtered. The assumed median filter is an adaptive filter as described in Reference [87] with a window size that includes 9 of the nearest panchromatic (or green for the Bayer CFA) pixels. The assumed boxcar filter kernel is adjusted for each type of CFA image such that it includes 15 of the nearest pixels.

1.4.2.1 High-Frequency Noise Reduction

Noise with significant high-frequency content makes an image appear "grainy," and some of this appearance is due to impulse noise. Median filters may be used to remove impulse

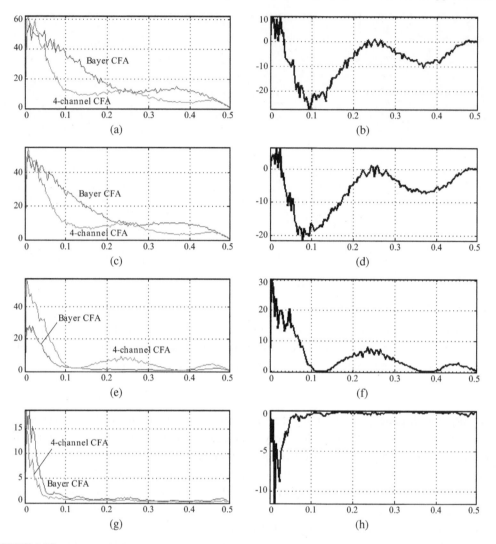

FIGURE 1.28

(left) Noise power spectra (NPS) of the demosaicked red channel from a Bayer CFA and a four-channel CFA shown in Figure 1.27 and (right) the associated difference of the four-channel NPS minus Bayer NPS. The NPS and difference plots are shown for the cases of: (a,b) no noise reduction, (c,d) median filtering only, (e,f) median and boxcar filtering only, and (g,h) median, boxcar, and low-frequency filtering. The horizontal axis is in units of cycles per sample.

noise, which has spectral power from low to very high frequencies. Figures 1.27c and 1.27f show the respective effects of median filtering the colors before demosaicking the Bayer and the four-channel CFA images with an adaptive filter as described in Reference [87] with window sizes that include nine like pixels in the filtering. The NPS and NPS-difference plots in Figures 1.28c and 1.28d show that the overall effect is of slightly lowering the NPS nearly uniformly across all frequencies. However, the appearance of the images is changed the most at the highest frequencies. It is also clear that both images could be improved with more noise reduction. A boxcar filter is a low pass filter that can be effectively used to reduce high frequencies in images. Figures 1.27c and 1.27g show the demosaicked results of boxcar filtering the Bayer and four-channel CFA images after median filtering. The NPS plots in Figures 1.28e and 1.28f show that for the image demosaicked from the Bayer CFA the noise has been reduced for all but the lowest frequencies, but for the image demosaicked from the four-channel CFA the noise still has some mid-frequency power. Again, one of the effects of having panchromatic pixels is also shown; the low-frequency portion of the NPS is much higher than that of the Bayer image because the color pixels are farther away from each other in the four-channel CFA than in the Bayer CFA. Both images again could be improved with more noise reduction.

1.4.2.2 Mid-Frequency and Low-Frequency Noise Reduction

Mid- and low-frequency noise artifacts appear as "blob" in an image and can span many pixels. Figure 1.27e shows some noise features that span about eight pixels. The average size (frequency) and standard deviation of this type of noise depend on the specific four-channel CFA and the demosaicking method. Given the large area covered by low-frequency noise, a traditional noise reduction method where each channel is processed independently will tend to remove image content. To avoid removing image content, an image fusion approach may be used. Mid- and low-frequency noise reduction can be achieved if a weighted average gradient of a color channel is set equal to a weighted average gradient of the panchromatic channel [88]. This requirement may be written in general as follows:

$$\sum_{m=-M}^{M} \sum_{n=-N}^{N} \mathbf{g}_C(m,n) \left[f_\kappa(x,y) s_C(x,y) - f_C(x-m,y-n) s_C(x-m,y-n) \right] =$$

$$\sum_{m=-M}^{M} \sum_{n=-N}^{N} \mathbf{g}_C(m,n) \left[f'_P(x,y) s_C(x,y) - f'_P(x-m,y-n) s_C(x-m,y-n) \right],$$

where $\mathbf{g}_C$ is a normalized low-pass kernel with a low cutoff frequency kernel that reduces the desired frequencies, f_κ is the noise-reduced color channel to be determined, s_C is a color sampling function, and x and y denote the location of the pixel being noise-reduced. The equality may be rewritten as

$$f_\kappa(x,y) s_C(x,y) - \sum_{m=-M}^{M} \sum_{n=-N}^{N} \mathbf{g}_C(m,n) f_C(x-m,y-n) s_C(x-m,y-n) =$$

$$f'_P(x,y) s_C(x,y) - \sum_{m=-M}^{M} \sum_{n=-N}^{N} \mathbf{g}_C(m,n) f'_P(x-m,y-n) s_C(x-m,y-n).$$

Rearranging terms and switching away from the coordinate notation, as follows:

$$\left(f_\kappa - f'_P \right) s_C = \mathbf{g}_C * \left[\left(f_C - f'_P \right) \right].$$

The quantities within parentheses are color differences so this last equality may be written as

$$f_\Delta s_C = \mathbf{g}_C * f_D s_C \tag{1.48}$$

where f_D is the noisy color-difference image, f_Δ is a noise-reduced version of f_D.

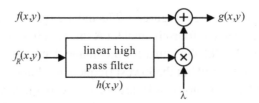

FIGURE 1.29

Linear sharpening for image enhancement.

The heuristic explanation for why this method tends to preserve image content is that f_Δ is a low-frequency signal and the final color image is derived by adding the panchromatic channel, which contains the high-frequency content. Figure 1.27h shows the result of applying Equation 1.48 to the four-channel CFA image before demosaicking and after median and boxcar filtering. For comparison, Figure 1.27c shows the equivalent noise reduction for a Bayer image, where the green channel is substituted for the panchromatic channel in Equation 1.48. For the red channel of the four-channel example, $\mathbf{g}_C$ is defined as a 33×33 kernel that when centered on a red pixel it is $1/153$ at the red pixel locations and zero elsewhere. This means that $\mathbf{g}_C$ averages together 153 red color differences (for this case, red minus panchromatic). For the red channel of the Bayer example, $\mathbf{g}_C$ is defined as a 33×17 kernel that when centered on a red pixel it is $1/153$ at the red pixel locations and zero elsewhere. This means that this version of $\mathbf{g}_C$ also averages together 153 red color differences (for this case, red minus green). Note from the examples that even though the same number of color differences are averaged together, because the panchromatic channel has a higher SNR than the Bayer green channel, the final image simulated with the four-channel CFA has much lower noise. The NPS plots in Figures 1.28g and 1.28h support the visual appearance of the images by showing that the NPS of the Bayer image is now larger than the NPS of the image from the four-channel CFA. To show that this method of noise cleaning indeed preserves image content, Section 1.5 contains a discussion and results of the full image processing chain. The noise reduction results obtained in Section 1.5 are typical of the noise reduction techniques discussed in this section.

1.4.3 Image Sharpening

Due to the low-pass nature of optical system and low-pass filtering operations involved at various stages of the image processing chain such as noise cleaning and demosaicking, the demosaicked color image appears to be blurred. Therefore, image-sharpening techniques are often employed to enhance the high frequency content, which significantly improves the visual appearance of the image [89], [90], [91], [92], [93].

The classic linear sharpening is a pixel-level image fusion approach (Figure 1.29) in which a scaled, highpass filtered record extracted from a reference image $f_R(x,y)$ is added to the blurred image $f(x,y)$ to produce sharpened image $g(x,y)$ as follows:

$$g(x,y) = f(x,y) + \lambda [f_R(x,y) * h(x,y)], \tag{1.49}$$

where $h(x,y)$, λ, and $*$ denote the high-pass filter point spread function, scale factor, and convolution operator, respectively. Note that linear unsharp masking [89] is a special case

FIGURE 1.30

Four-channel demosaicked images: (a) color image, and (b) panchromatic image.

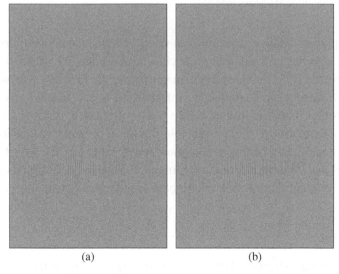

FIGURE 1.31

High-frequency information extracted from: (a) the green channel, and (b) the panchromatic image.

of linear sharpening in which the blurred image is also used as the reference image, that is, $f(x,y) = f_R(x,y)$. This method works well in many applications, however, it is extremely sensitive to noise. This leads to undesirable distortions, especially in flat regions. Typically, nonlinear coring functions are used to mitigate this type of distortion [93].

An example of linear sharpening is presented next. Figure 1.30 shows a four-channel demosaicked color and a corresponding panchromatic image. The flat, gray patches shown in the upper left quadrant of these images were inserted for measuring the noise presented in these images. It is evident that the color image shown in Figure 1.30a is noisy and soft. Therefore, a linear sharpening algorithm was applied to enhance its visual appearance.

(a)   (b)

FIGURE 1.32

Linear sharpening results: (a) sharpened using green channel high-frequency information, and (b) sharpened using panchromatic high-frequency information.

The high-frequency details extracted from the green channel of the blurred color image of Figure 1.30a and the corresponding panchromatic image are shown in Figures 1.31a and 1.31b, respectively. The sharpened image shown in Figure 1.32a was obtained by adding the green channel high-frequency information to the blurred color image. Similarly, the sharpened image of Figure 1.32b was estimated by adding the panchromatic high-frequency information to the blurred color image. The standard deviations of the gray patches before and after sharpening are summarized in Table 1.3. Clearly, the panchromatic image was less noisy than the blurred color image. Furthermore, it turned out to be a better choice for high-frequency extraction as compared to the green channel for sharpening.

1.5 Example Single-Sensor Image Fusion Capture System

The results presented in this chapter can be used to constitute a full image processing chain for achieving an improved image using a four-channel CFA pattern and image fusion techniques. Figure 1.4 shows an image processing chain that can be used in this regard.

TABLE 1.3

Standard deviations of gray patches before and after sharpening.

Operation	Red	Green	Blue	Panchromatic
Original	10.06	10.04	10.00	5.07
High-pass	N/A	7.67	N/A	3.73
Sharpened with Green	12.66	17.33	12.56	N/A
Sharpened with Panchromatic	10.73	10.65	10.62	N/A

(a)   (b)   (c)

FIGURE 1.33

Image processing chain example, Part 1: (a) original CFA image, (b) noise-cleaned image, and (c) demosaicked image.

The first step is to select the CFA pattern to be used with the sensor. In this case, the CFA pattern of Figure 1.19e will be used. Using this pattern, a simulation of a capture is generated and shown in Figure 1.33a. For the purposes of making noise measurements, a flat, gray patch is placed in the upper left quadrant of the image. The first thing that can be seen is that the image has enough noise, to the point that the underlying CFA pattern is somewhat hard to discern. Table 1.4 shows the standard deviations of the gray patch for each of the four channels present in the image. It can be seen that the panchromatic channel standard deviation is roughly half of the color channel standard deviations. The lower noise of the panchromatic channel is a consequence of its broader spectral response and associated greater light sensitivity.

Once the original CFA image is in hand, it is noise-cleaned (denoised). Using the methods described in Section 1.4, the CFA image is cleaned to produce the image shown in Figure 1.33b. The CFA pattern is now evident as a regular pattern, especially in the region of the life preserver in the lower left-hand corner of the image. The standard deviations (Table 1.4) have been reduced significantly with the panchromatic channel still having a lower amount of noise.

TABLE 1.4

Standard deviations of gray patch in image processing chain example.

Operation	Red	Green	Blue	Panchromatic
Original CFA Image	10.48	9.71	9.87	5.02
Noise Cleaned	2.39	2.37	2.35	1.49
Demosaicked	1.44	1.45	1.46	N/A
Color & Tone Corrected	1.90	1.92	1.90	N/A
Sharpened	4.99	5.04	4.98	N/A

(a)   (b)

FIGURE 1.34

Image processing chain example, Part 2: (a) color and tone corrected image, and (b) sharpened image.

After noise cleaning, the image is demosaicked as discussed in Section 1.3. Figure 1.33c shows the result; color has been restored, although it is desaturated and flat. As there are now only three color channels in the image, Table 1.4 no longer has a panchromatic entry. Of more interest is how the demosaicked color channels now have the same noise standard deviation as the noise-cleaned panchromatic channel. This is a consequence of the image fusion techniques used, specifically as accomplished through the use of color differences.

The color along with the tone scale of the image is next corrected using the techniques described in Section 1.2 with the results shown in Figure 1.34a. Since color correction and tone scaling are generally signal amplifying steps, the standard deviations of the gray patch increase. In this particular case, these corrections are relatively mild, so the noise amplification is correspondingly low.

As the image is still lacking a bit in sharpness, the final step is to sharpen the image as shown in Figure 1.34b. As indicated in Table 1.4, the standard deviations have been significantly amplified, although they are still about half of what the original CFA image had. A nonadaptive sharpening algorithm was used here. An adaptive algorithm capable of recognizing flat regions and reducing the sharpening accordingly would reduce the resulting noise amplification for the gray patch region. Still, the final image is certainly acceptable.

TABLE 1.5

Standard deviations of gray patch in Bayer image processing chain example.

Operation	Red	Green	Blue
Original CFA Image	10.09	9.70	9.76
Noise Cleaned	2.67	2.73	2.65
Demosaicked	2.65	2.66	2.63
Color & Tone Corrected	3.44	3.42	3.32
Sharpened	9.12	9.05	9.10

FIGURE 1.35

Bayer image processing chain example, Part 1: (a) original CFA Image, (b) noise-cleaned image, and (c) demosaicked image.

FIGURE 1.36

Bayer image processing chain example, Part 2: (a) color and tone corrected image, and (b) sharpened image.

As a comparison, the simulation is repeated using the Bayer CFA pattern and processing as shown in Figure 1.35 and Figure 1.36. The standard deviations of the gray patch are given in Table 1.5. It can be seen that the original CFA image starts off with the same amount of noise as in the four-channel case. Noise cleaning produces results comparable to before. The demosaicking step produces the first notable differences as the color differences cannot benefit from a significantly less noisy luminance channel. The resulting noise is amplified by the color and tone scale correction step as before. Finally, the sharpening operation is performed and the resulting noise level has almost returned to that of the original CFA image. Comparison with the four-channel system shows a double increase in the gray patch standard deviations over the four-channel example.

1.6 Conclusion

Image fusion provides a way of creating enhanced and even impossible-to-capture images through the appropriate combination of image components. These components are traditionally full-image captures acquired from either a system consisting of several specialized sensors (e.g., each with different spectral characteristics) or as part of a multicapture sequence (e.g., burst or video). This chapter describes a new approach that uses a single capture from a single sensor to produce the necessary image components for subsequent image fusion operations. This capability is achieved by the inclusion of panchromatic pixels in the color filter array pattern. Inherently, panchromatic pixels will be more light sensitive, which results in improved signal-to-noise characteristics. Additionally, being spectrally nonselective, edge and texture detail extracted from the panchromatic channel will be more complete and robust across the visible spectrum. Image fusion techniques can then be used to impart these benefits onto the color channels while still preserving color fidelity. These image fusion techniques are generally implemented as parts of the noise cleaning, demosaicking, and sharpening operations in the image processing chain. In addition to the benefits afforded requiring only one capture for enabling image fusion, the noise cleaning and demosaicking operations described in this chapter work on sparsely sampled CFA data. This reduction in the amount of data to be processed provides additional efficiency in the application of image fusion techniques.

Acknowledgment

The authors dedicate this chapter to the memories of our dear colleague Michele O'Brien and Stacy L. Moor, Efraín's well-beloved wife.

Appendix

This appendix provides a derivation of the relationship first appearing in Equation 1.15 and restated below:

$$MN \sum_{p=-\infty}^{\infty} \sum_{q=-\infty}^{\infty} \operatorname{sinc}^2 \left[M(\xi - p), N(\eta - q) \right]$$
$$= \left[1 + 2 \sum_{j=1}^{M-1} \operatorname{tri} \left(\frac{j}{M} \right) \cos(2\pi j \xi) \right] \left[1 + 2 \sum_{k=1}^{N-1} \operatorname{tri} \left(\frac{k}{N} \right) \cos(2\pi k \eta) \right].$$

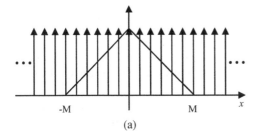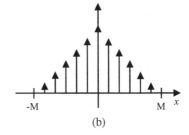

FIGURE 1.37

Discrete tri function: (a) tri and comb functions, and (b) delta functions.

Since this relationship is separable, only one dimension needs to be derived, as follows:

$$M \sum_{p=-\infty}^{\infty} \mathrm{sinc}^2 \left[M \left(\xi - p \right) \right] = 1 + 2 \sum_{j=1}^{M-1} \mathrm{tri} \left(\frac{j}{M} \right) \cos \left(2\pi j \xi \right).$$

Figure 1.37a shows the functions $\mathrm{tri}\,(x/M)$ and $\mathrm{comb}\,(x)$ superimposed on coordinate axes. The result of multiplying these two functions together is shown in Figure 1.37b. Only a finite number of delta functions remain and these are scaled by the tri function. Therefore, this discrete form of the tri function can be written as follows:

$$\mathrm{tri} \left(\frac{x}{M} \right) \mathrm{comb}\,(x) = \delta\,(x) + \sum_{j=1}^{M-1} \mathrm{tri} \left(\frac{j}{M} \right) \frac{1}{j} \delta\delta \left(\frac{x}{j} \right). \tag{1.50}$$

Taking the Fourier transform of each side produces the required relationship.

$$M\mathrm{sinc}^2 \left(M\xi \right) * \mathrm{comb}\,(\xi) = 1 + \sum_{j=1}^{M-1} \mathrm{tri} \left(\frac{j}{M} \right) 2\cos \left(2\pi j \xi \right),$$

$$M \sum_{p=-\infty}^{\infty} \mathrm{sinc}^2 \left[M \left(\xi - p \right) \right] = 1 + 2 \sum_{j=1}^{M-1} \mathrm{tri} \left(\frac{j}{M} \right) \cos \left(2\pi j \xi \right).$$

The case of Equation 1.16 is handled in a similar manner, as follows:

$$\mathrm{tri} \left(\frac{x}{2M} - \frac{y}{2N}, \frac{x}{2M} + \frac{y}{2N} \right) \mathrm{comb}\,(x)$$

$$= \delta\,(x) + \sum_{j=1}^{2M-1} \mathrm{tri}^2 \left(\frac{j}{2M} \right) \frac{1}{j} \delta\delta \left(\frac{x}{j} \right) + \sum_{k=1}^{2N-1} \mathrm{tri}^2 \left(\frac{k}{2N} \right) \frac{1}{k} \delta\delta \left(\frac{y}{k} \right)$$

$$+ \sum_{j=1}^{2M-1} \sum_{k=1}^{2N-1} \mathrm{tri} \left(\frac{j}{2M} - \frac{k}{2N}, \frac{j}{2M} + \frac{k}{2N} \right) \frac{1}{jk} \delta\delta \left(\frac{x}{2M}, \frac{y}{2N} \right).$$

Taking the Fourier transform completes the derivation.

References

[1] J.T. Compton and J.F. Hamilton Jr., "Image sensor with improved light sensitivity," U.S. Patent Application 11/191 729, February 2007.

[2] M. Kokar and K. Kim, "Review of multisensor data fusion architectures," in *Proceedings of the IEEE International Symposium on Intelligent Control*, Chicago, IL, USA, August 1993, pp. 261–266.

[3] D.L. Hall and J. Llinas, "An introduction to multisensor data fusion," *Proceedings of the IEEE*, vol. 85, no. 1, pp. 6–23, January 1997.

[4] R. Mahler, "A unified foundation for data fusion," in *Proceedings of the Data Fusion System Conference*, Laurel, MD, USA, June 1987.

[5] M.A. Abidi and R.C. Gonzalez, *Data Fusion in Robotics and Machine Intelligence*, San Diego, CA: Academic Press, October 1992.

[6] L. Waltz, J. Llinas, and E. Waltz, *Multisensor Data Fusion*, Boston, MA, USA: Artech House Publishers, August 1990.

[7] M. Kumar, "Optimal image fusion using the rayleigh quotient," in *Proceedings of the IEEE Sensors Applications Symposium*, New Orleans, LA, USA, February 2009, pp. 269–274.

[8] E. Waltz, "The principles and practice of image and spatial data fusion," in *Proceedings of the Eighth National Data Fusion Conference*, Dallas, TX, USA, March 1995, pp. 257–278.

[9] R.S. Blum, Z. Xue, and Z. Zhang, *Multi-Sensor Image Fusion and Its Applications*, ch. An overview of image fusion, R.S. Blum and Z. Liu (eds.), Boca Raton, FL: CRC Press, July 2005, pp. 1–36.

[10] Z. Zhang and R.S. Blum, "A categorization and study of multiscale-decomposition-based image fusion schemes," *Proceedings of the IEEE*, vol. 87, no. 8, pp. 1315–1326, August 1999.

[11] R.S. Blum, "Robust image fusion using a statistical signal processing approach," *Information Fusion*, vol. 6, no. 2, pp. 119–128, June 2005.

[12] R. Raskar, A. Agrawal, and J. Tumblin, "Coded exposure photography: Motion deblurring using fluttered shutter," *ACM Transactions on Graphics*, vol. 25, no. 3, pp. 795–804, July 2006.

[13] Y. Chibani, "Multisource image fusion by using the redundant wavelet decomposition," in *Proceedings of the IEEE Geoscience and Remote Sensing Symposium*, Toulouse, France, July 2003, pp. 1383–1385.

[14] M. Hurn, K. Mardia, T. Hainsworth, J. Kirkbride, and E. Berry, "Bayesian fused classification of medical images," *IEEE Transactions on Medical Imaging*, vol. 15, no. 6, pp. 850–858, December 1996.

[15] D.L Hall and S.A.H. McMullen, *Mathematical Techniques in Multisensor Data Fusion*. Boston, MA, USA: Artech House, 2nd edition, February 2004.

[16] J. Llinas and D. Hall, "A challenge for the data fusion community I: Research imperatives for improved processing," in *Proceedings of the Seventh National Symposium On Sensor Fusion*, Albuquerque, New Mexico, USA, March 1994.

[17] C. Pohl and J.L. van Genderen, "Multisensor image fusion in remote sensing: concepts, methods and applications," *International Journal of Remote Sensing*, vol. 19, no. 5, pp. 823–854, March 1998.

[18] M. Daniel and A. Willsky, "A multiresolution methodology for signal-level fusion and data assimilation with applications to remote sensing," *Proceedings of the IEEE*, vol. 85, no. 1, pp. 164–180, January 1997.

[19] E. Lallier and M. Farooq, "A real time pixel-level based image fusion via adaptive weight averaging," in *Proceedings of the Third International Conference on Information Fusion*, Paris, France, July 2000, pp. WEC3/3–WEC3/13.

[20] G. Pajares and J.M. de la Cruz, "A wavelet-based image fusion tutorial," *Pattern Recognition*, vol. 37, no. 9, pp. 1855–1872, September 2004.

[21] J. Richards, "Thematic mapping from multitemporal image data using the principal components transformation," *Remote Sensing of Environment*, vol. 16, pp. 36–46, August 1984.

[22] I. Bloch, "Information combination operators for data fusion: A comparative review with classification," *IEEE Transactions on Systems, Man, and Cybernetics. C*, vol. 26, no. 1, pp. 52–67, January 1996.

[23] Y. Gao and M. Maggs, "Feature-level fusion in personal identification," in *Proceedings of the IEEE International Conference on Computer Vision and Pattern Recognition*, San Diego, CA, USA, June 2005, pp. 468–473.

[24] V. Sharma and J.W. Davis, "Feature-level fusion for object segmentation using mutual information," in *Proceedings of Computer Vision and Pattern Recognition Workshop*, New York, USA, June 2006, pp. 139–146.

[25] A. Kumar and D. Zhang, "Personal recognition using hand shape and texture," *IEEE Transactions on Image Processing*, vol. 15, no. 8, pp. 2454–2461, August 2006.

[26] K. Fukanaga, *Introduction to Statistical Pattern Recognition*. New York, USA: Academic Press, October 1990.

[27] Y. Liao, L.W. Nolte, and L.M. Collins, "Decision fusion of ground-penetrating radar and metal detector algorithms – A robust approach," *IEEE Transactions on Geoscience and Remote Sensing*, vol. 45, no. 2, pp. 398–409, February 2007.

[28] L.O. Jimenez, A. Morales-Morell, and A. Creus, "Classification of hyperdimensional data based on feature and decision fusion approaches using projection pursuit, majority voting, and neural networks," *IEEE Transactions on Geoscience and Remote Sensing*, vol. 37, no. 3, pp. 1360–1366, May 1999.

[29] M. Fauvel, J. Chanussot, and J.A. Benediktsson, "Decision fusion for the classification of urban remote sensing images," *IEEE Transactions on Geoscience and Remote Sensing*, vol. 44, no. 10, pp. 2828–2838, October 2006.

[30] S. Foucher, M. Germain, J.M. Boucher, and G.B. Benie, "Multisource classification using ICM and Dempster-Shafer theory," *IEEE Transactions on Instrumentation and Measurement*, vol. 51, no. 2, pp. 277–281, April 2002.

[31] F. Nebeker, "Golden accomplishments in biomedical engineering," *IEEE Engineering in Medicine and Biology Magazine*, vol. 21, no. 3, pp. 17–47, May/June 2002.

[32] D. Barnes, G. Egan, G. OKeefe, and D. Abbott, "Characterization of dynamic 3-D pet imaging for functional brain mapping," *IEEE Transactions on Medical Imaging*, vol. 16, no. 3, pp. 261–269, June 1997.

[33] S. Wong, R. Knowlton, R. Hawkins, and K. Laxer, "Multimodal image fusion for noninvasive epilepsy surgery planning," *International Journal of Remote Sensing*, vol. 16, no. 1, pp. 30–38, January 1996.

[34] R. Raskar, A. Agrawal, and J. Tumblin, "Coded exposure photography: Motion deblurring using fluttered shutter," *ACM Transactions on Graphics*, vol. 25, no. 3, pp. 795–804, July 2006.

[35] A. Agrawal and R. Raskar, "Resolving objects at higher resolution from a single motion-blurred image," in *Proceedings of the IEEE International Conference on Computer Vision and Pattern Recognition*, Minneapolis, MN, USA, June 2007, pp. 1–8.

[36] M. Kumar and P. Ramuhalli, "Dynamic programming based multichannel image restoration," in *Proceedings of the IEEE International Conference on Acoustics, Speech, and Signal Processing*, Philadelphia, PA, USA, March 2005, pp. 609–612.

[37] L. Yuan, J. Sun, L. Quan, and H.Y. Shum, "Image deblurring with blurred/noisy image pairs," *ACM Transactions on Graphics*, vol. 26, no. 3, July 2007.

[38] G. Petschnigg, R. Szeliski, M. Agrawala, M. Cohen, H. Hoppe, and K. Toyama, "Digital photography with flash and no-flash image pairs," *ACM Transactions on Graphics*, vol. 23, no. 3, pp. 664–672, August 2004.

[39] P.J. Burt and R.J. Kolczynski, "Enhanced image capture through fusion," in *Proceedings of the Fourth International Conference on Computer Vision*, Berlin, Germany, May 1993, pp. 173–182.

[40] H. Li, B.S. Manjunath, and S.K. Mitra, "Multisensor image fusion using the wavelet transform," *Graphical Models and Image Processing*, vol. 57, no. 3, pp. 235–245, May 1995.

[41] B.E. Bayer, "Color imaging array," U.S. Patent 3 971 065, July 1976.

[42] Eastman Kodak Company, *Kodak KAI-11002 Image Sensor Device Performance Specification*, 2006.

[43] J.E. Adams Jr. and J.F. Hamilton Jr., "Adaptive color plan interpolation in single color electronic camera," U.S. Patent 5 506 619, April 1996.

[44] C.W. Kim and M.G. Kan, "Noise insensitive high resolution demosaicing algorithm considering cross-channel correlation," in *Proceedings of the International Conference on Image Processing*, Genoa, Italy, September 2005, pp. 1100–1103.

[45] R. Kimmel, "Demosaicing: Image reconstruction from color CCD samples," *IEEE Transactions on Image Processing*, vol. 8, no. 9, pp. 1221–1228, September 1999.

[46] R.W.G. Hunt, *The Reproduction of Colour*, 5th Edition, Kingston-upon-Thames, UK: Fountain Press, 1995.

[47] E.J. Giorgianni and T.E. Madden, *Digital Color Management Encoding Solutions*, 2nd Edition, Chichester, UK: John Wiley and Sons, Ltd., 2008.

[48] J.C. Maxwell, *On the Theory of Three Primary Colours*, Cambridge, England: Cambridge University Press, 1890.

[49] R.M. Evans, "Some notes on Maxwell's colour photograph," *Journal of Photographic Science*, vol. 9, no. 4, p. 243, July-August 1961.

[50] J.S. Friedman, *History of Color Photography*, Boston, Massachusets: The American Photographic Publishing Company, 1944.

[51] M.L. Collette, "Digital image recording device," U.S. Patent 5 570 146, May 1994.

[52] T.E. Lynch and F. Huettig, "High resolution RGB color line scan camera," in *Proceedings of SPIE, Digital Solid State Cameras: Designs and Applications*, San Jose, CA, USA, January 1998, pp. 21–28.

[53] G. Sharma and H.J. Trussell, "Digital color imaging," *IEEE Transactions on Image Processing*, vol. 6, no. 7, pp. 901–932, July 1997.

[54] R.F. Lyon, "Prism-based color separation for professional digital photography," in *Proceedings of the Image Processing, Image Quality, Image Capture, Systems Conference*, Portland, OR, USA, March 2000, pp. 50–54.

[55] R.F. Lyon and P.M. Hubel, "Eying the camera: Into the next century," in *Proceedings of the IS&T SID Tenth Color Imaging Conference*, Scottsdale, AZ, USA, November 2002, pp. 349–355.

[56] R.M. Turner and R.J. Guttosch, "Development challenges of a new image capture technology: Foveon X3 image sensors," in *Proceedings of the International Congress of Imaging Science*, Rochester, NY, USA, May 2006, pp. 175–181.

[57] K.E. Spaulding, E.J. Giorgianni, and G. Woolfe, "Optimized extended gamut color encodings for scene-referred and output-referred image states," *Journal of Imaging Science and Technology*, vol. 45, no. 5, September/October 2001, pp. 418–426.

[58] M. Anderson, R. Motta, S. Chandrasekar, and M. Stokes, "Proposal for a standard default color space for the internet: sRGB," in *Proceedinga of the Fourth IS&T/SID Color Imaging Conference*, Scottsdale, AZ, USA, November 1995, pp. 238–245.

[59] K.E. Spaulding and J. Holm, "Color encodings: sRGB and beyond," in *Proceedings of the IS&T's Conference on Digital Image Capture and Associated System, Reproduction and Image Quality Technologies*, Portland, OR, USA, April 2002, pp. 167–171.

[60] K.E. Spaulding, E.J. Giorgianni, and G. Woolfe, "Reference input/output medium metric RGB color encoding (RIMM/ROMM RGB)," in *Proceedings of the Image Processing, Image Quality, Image Capture, Systems Conference*, Portland, OR, USA, March 2000, pp. 155–163.

[61] "Adobe rgb (1998) color image encoding," Tech. Rep. http://www.adobe.com/adobergb, Adobe Systems, Inc., 1998.

[62] S. Süsstrunk, "Standard RGB color spaces," in *Proceedings of the Seventh IS&T/SID Color Imaging Conference*, Scottsdale, AZ, USA, November 1999, pp. 127–134.

[63] H.E.J. Neugebauer, "Quality factor for filters whose spectral transmittances are different from color mixture curves, and its application to color photography," *Journal of the Optical Society of America*, vol. 46, no. 10, pp. 821–824, October 1956.

[64] P. Vora and H.J. Trussel, "Measure of goodness of a set of color-scanning filters," *Journal of the Optical Society of America*, vol. 10, no. 7, pp. 1499–1508, July 1993.

[65] R.L. Baer, W.D. Holland, J. Holm, and P. Vora, "A comparison of primary and complementary color filters for ccd-based digital photography," *Proceedings of the SPIE*, pp. 16–25, January 1999.

[66] H. Kuniba and R.S. Berns, "Spectral sensitivity optimization of color image sensors considering photon shot noise," *Journal of Electronic Imaging*, vol. 18, no. 2, pp. 023002/1–14, April 2009.

[67] "Photography - electronic still picture imaging - noise measurements," Tech. Rep. ISO 15739:2003, ISO TC42/WG 18, 2003.

[68] R. Vogel, "Digital imaging device optimized for color performance," U.S. Patent 5 668 596, September 1997.

[69] M. Parmar and S.J. Reeves, "Optimization of color filter sensitivity functions for color filter array based image acquisition," in *Proceedings of the Fourteenth Color Imaging Conference*, Scottsdale, AZ, USA, November 2006, pp. 96–101.

[70] G.J.C. van der Horst, C.M.M. de Weert, and M.A. Bouman, "Transfer of spatial chromaticity-contrast at threshold in the human eye," *Journal of the Optical Society of America*, vol. 57, no. 10, pp. 1260–1266, October 1967.

[71] K.T. Mullen, "The contrast sensitivity of human colour vision to red-green and blue-yellow chromatic gratings," *Journal of Physiology*, vol. 359, pp. 381–400, February 1985.

[72] J.E. Adams Jr., M. Kumar, B.H. Pillman, and J.A. Hamilton, "Four-channel color filter array pattern," U.S. Patent Application 12/472 563, May 2009.

[73] J.E. Adams Jr., M. Kumar, B.H. Pillman, and J.A. Hamilton, "Four-channel color filter array interpolation," U.S. Patent Application 12/473 305, May 2009.

[74] J.E. Adams Jr., M. Kumar, B.H. Pillman, and J.A. Hamilton, "Color filter array pattern having four channels," U.S. Patent Application 12/478 810, June 2009.

[75] J.E. Adams Jr., M. Kumar, B.H. Pillman, and J.A. Hamilton, "Interpolation for four-channel color filter array," U.S. Patent Application 12/480 820, June 2009.

[76] J.D. Gaskill, *Linear Systems, Fourier Transforms, and Optics*, New York: John Wiley & Sons, 1978.

[77] J.D. Gaskill, *Linear Systems, Fourier Transforms, and Optics*, ch. Characteristics and Applications of Linear Filters, New York: John Wiley & Sons, 1978, pp. 279–281.

[78] R. Palum, *Single-Sensor Imaging: Methods and Applications for Digital Cameras*, ch. Optical antialiasing filters, R. Lukac (ed.), Boca Raton, FL: CRC Press / Taylor & Francis, September 2008, pp. 105–135.

[79] D. Cok, "Signal processing method and apparatus for producing interpolated chrominance values in a sampled color image signal," U.S. Patent 4 642 678, February 1987.

[80] J.F. Hamilton Jr. and J.E. Adams Jr., "Adaptive color plan interpolation in single color electronic camera," U.S. Patent 5 629 734, May 1997.

[81] P.S. Tsai, T. Acharya, and A. Ray, "Adaptive fuzzy color interpolation," *Journal of Electronic Imaging*, vol. 11, no. 3, pp. 293–305, July 2002.

[82] K. Hirakawa and T. Parks, "Adaptive homogeneity-directed demosaicking algorithm," in *Proceedings of the International Conference on Image Processing*, Barcelona, Spain, September 2003, pp. 669–672.

[83] J.E. Adams Jr., "Design of practical color filter array interpolation algorithms for digital cameras," in *Proceedings of the SPIE Conference on Real-Time Imaging*, San Jose, CA, USA, February 1997, pp. 117–125.

[84] B.W. Keelan, *Handbook of Image Quality: Characterization and Prediction*. New York, NY, USA: Marcel Dekker, March 2002.

[85] G.C. Holst and T.S. Lomheim, *CMOS/CCD sensors and camera systems*. Bellingham, WA: The International Society for Optical Engineering, October 2007.

[86] H.R.S.Z. Wang, A.C. Bovik and E.P. Simoncelli, "Image quality assessment: From error visibility to structural similarity," *IEEE Transactions on Image Processing*, vol. 13, no. 4, pp. 600–612, April 2004.

[87] J.E. Adams Jr., J.F. Hamilton Jr., and E.B. Gindele, "Noise-reducing a color filter array image," U.S. Patent Application 10/869 678, December 2005.

[88] E.O. Morales and J.F. Hamilton Jr., "Noise reduced color image using panchromatic image," U.S. Patent Application 11/752 484, November 2007.

[89] A. Polesel, G. Ramponi, and V.J. Mathews, "Image enhancement via adaptive unsharp masking," *IEEE Transactions on Image Processing*, vol. 9, no. 3, pp. 505–510, March 2000.

[90] F.P. de Vries, "Automatic, adaptive, brightness independent contrast enhancement," *Signal Processing*, vol. 21, no. 2, pp. 169–182, October 1990.

[91] G. Ramponi, N. Strobel, S.K. Mitra, and T. Yu, "Nonlinear unsharp masking methods for image contrast enhancement," *Journal of Electronic Imaging*, vol. 5, no. 3, pp. 353–366, July 1996.

[92] G. Ramponi, "A cubic unsharp masking technique for contrast enhancement," *Signal Processing*, vol. 67, no. 2, pp. 211–222, June 1998.

[93] J.E. Adams Jr. and J.F. Hamilton Jr., *Single-Sensor Imaging: Methods and Applications for Digital Cameras*, ch. Digital camera image processing chain design, R. Lukac (ed.), Boca Raton, FL: CRC Press / Taylor & Francis, September 2008, pp. 67–103.

2

Single Capture Image Fusion with Motion Consideration

James E. Adams, Jr., Aaron Deever, John F. Hamilton, Jr., Mrityunjay Kumar, Russell Palum, and Bruce H. Pillman

2.1 Introduction

In Chapter 1, an image capture system was introduced that uses a four-channel color filter array to obtain images with high color fidelity and improved signal-to-noise performance relative to traditional three-channel systems. A panchromatic (spectrally nonselective) channel was added to the digital camera sensor to decouple sensing luminance (spatial) information from chrominance (color) information. In this chapter, that basic foundation is enhanced to provide a capture system that can additionally address the issue of motion occurring during a capture to produce images with reduced motion blur.

Motion blur is a common problem in digital imaging that occurs when there is relative motion between the camera and the scene being captured. The degree of motion blur present in an image is a function of both the characteristics of the motion as well as the integration time of the sensor. Motion blur may be caused by camera motion or it may be caused by object motion within the scene. It is particularly problematic in low-light imaging, which typically requires long integration times to acquire images with acceptable signal-to-noise levels. Motion blur is also often a problem for captures taken with significant optical magnification. Not only does the magnification amplify the motion that occurs, it also decreases the amount of light reaching the sensor, causing a need for longer integration times. A familiar trade-off often exists in these situations. The integration time can be kept short to avoid motion blur, but at a cost of poor signal-to-noise performance. Conversely, the integration time can be lengthened to allow sufficient light to reach the sensor, but at a cost of increased motion blur in the image. Due to this trade-off between motion blur and noise, images are typically captured with sufficiently long exposure time to ensure satisfactory signal-to-noise levels, and signal processing techniques are used to reduce the motion blur [1], [2], [3], [4]. Reducing motion blur, especially motion blur corresponding to objects moving within the scene, is a challenging task and often requires multiple captures of the same scene [5], [6], [7], [8]. These approaches are computationally complex and memory-intensive. In contrast, the proposed four-channel image sensor architecture allows the design of computationally efficient image fusion algorithms for motion deblurring of a color image from a single capture.

By varying the length of time that different image sensor pixels integrate light, it is possible to capture a low-light image with reduced motion blur while still achieving acceptable signal-to-noise performance. In particular, the panchromatic channel of the image sensor is integrated for a shorter period of time than the color channels of the image sensor. This approach can be motivated from the perspective of spectral sensitivity. Panchromatic pixels are more sensitive to light than color (red, green and blue) pixels, and panchromatic pixel integration time can be kept shorter to reduce motion blur while still acquiring enough photons for acceptable signal-to-noise performance. This approach can also be motivated from a human visual system perspective. The human visual system has greater spatial sensitivity to high-frequency luminance information than high-frequency chrominance information [9]. It is thus desirable to keep the panchromatic pixel integration time as short as possible to retain high-frequency luminance information in the form of sharp edges and textures. High frequencies are less important in the chrominance channels, thus greater motion blur can be tolerated in these channels in exchange for longer integration and improved signal-to-noise performance. The sharp panchromatic channel information is combined with the low-noise chrominance channel information to produce an output image with reduced motion blur compared to a standard capture in which all image sensor pixels have equal integration.

An image captured with varying integration times for panchromatic and color channels has properties that require novel image processing. Motion occurring during capture can cause edges in the scene to appear out of alignment between the panchromatic and color channels. Motion estimation and compensation steps can be used to align the panchromatic and color data. After alignment, the data may still exhibit uneven motion blur between the panchromatic and color channels. A subsequent image fusion step can account for this while combining the data to produce an output image.

This chapter looks at the issues involved with using a four-channel image sensor and allowing different integration times for the panchromatic and color pixels to produce an image with reduced motion blur. Section 2.2 provides a brief introduction to the topics of motion estimation and compensation and focuses on how these techniques can be applied to align the panchromatic and color data in the proposed capture system. Section 2.3 discusses the use of image fusion techniques to combine the complementary information provided by the shorter integration panchromatic channel and longer integration color channels to produce an output image with reduced motion blur. Section 2.4 presents an example single-sensor, single-capture, image-fusion capture system. The processing path contains the steps of motion estimation and compensation, and deblurring through image fusion, as well as techniques introduced in Chapter 1. Finally, conclusions are offered in Section 2.5.

2.2 Single-Capture Motion Estimation and Compensation

This section addresses the issue of motion occurring during an image capture in which the panchromatic channel is integrated for a shorter duration than the color channels. Motion estimation and compensation are reviewed in general and are discussed with particular regard to the proposed capture system.

Imaging systems capture two-dimensional projections of a time-varying three-dimensional scene. Motion in a two-dimensional image likewise refers to the projection of the three-dimensional motion of objects in a scene onto the imaging plane. This true, projected two-dimensional motion is not always observable, however. What is observed and measured in a digital imaging system is the apparent motion. Apparent motion is measured as a change in image intensity over time. In most situations, apparent and projected motion agree, but there are situations in which they do not match. For example, a circle of uniform intensity rotating about its center has true motion but no change in intensity, and hence no observable, apparent motion. Conversely, in a scene with a change in external illumination, there is measurable apparent motion even if there is no true motion in the scene. The motion techniques in this section are all based on an analysis of observable, apparent motion, and it is assumed that this motion is equivalent to the true, projected motion.

Motion is commonly divided into two categories. A first category is global or camera motion. Global motion affects all pixels of an image and can often be modeled succinctly with just a few parameters. A six-parameter affine motion model is given below:

$$d(x,y) = \begin{bmatrix} b_1 \\ b_2 \end{bmatrix} + \begin{bmatrix} b_3 & b_4 \\ b_5 & b_6 \end{bmatrix} \begin{bmatrix} x \\ y \end{bmatrix}, \tag{2.1}$$

where (x,y) is the pixel location, $d(x,y)$ is the motion at (x,y), and $b_1, b_2, ..., b_6$ are the six parameters that define the affine model. This model can be used to represent translation, rotation, dilation, shear, and stretching. Global motion typically occurs as a result of camera unsteadiness during an exposure or between successive frames of an image sequence.

A second category of motion is local or object motion. Local motion occurs when an object within a scene moves relative to the camera. Accurate estimation of object motion requires the ability to segment an image into multiple regions or objects with different motion characteristics. Often a compromise is made sacrificing segmentation accuracy for speed by dividing an image into a regular array of rectangular tiles and computing a local motion value for each tile, as discussed below. Additional discussion on the topic of motion can be found in References [10] and [11].

Motion plays an important role in the proposed capture system, in which panchromatic pixels have a shorter integration than color pixels. Given this difference in integration, the color and panchromatic data may initially be misaligned. Should this be a source of artifacts such as halos beyond the apparent boundary of a moving object, it is desirable to

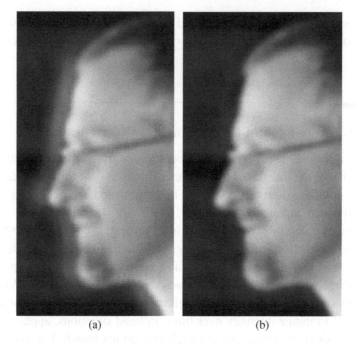

(a)   (b)

FIGURE 2.1

Captured image with short panchromatic integration and long color integration processed (a) without alignment and (b) with alignment.

align the panchromatic and color data, particularly at edge locations where misalignment is most visible. Figure 2.1 shows an example capture in which the color pixels were integrated for three times as long as the panchromatic pixels, and there was movement of the subject's head throughout the capture. The image on the left shows the result of subsequent processing without alignment, while the image on the right shows the result of equivalent processing occurring after an alignment step. The misalignment of color and panchromatic data is visible along the edge of the subject's face. This halo artifact is reduced by a preprocessing motion compensation step to align the panchromatic and color data. Note that even after alignment, the color data and panchromatic data may retain differing degrees of motion blur as a result of their differing integration times. The process of fusing the data into a single output image while accounting for this varying motion blur is discussed later in this chapter.

2.2.1 Motion Estimation Basics

Motion estimation comprises three major elements: a motion model, estimation criteria, and search strategies. The motion model represents the apparent motion of the scene. In general this model can be very complex to accurately represent the three-dimensional motion of the camera and objects in a scene as well as their projection onto a two-dimensional image plane. In practice, two simple models have been used extensively: the affine model given in Equation 2.1, and a translational model which is a special case of the affine model when $b_i = 0$, for $i = 3, 4, ..., 6$.

At one extreme, a single motion model is applied to the entire image. In this case, global motion is being modeled. Global motion estimation describes the motion of all image points with just a few parameters and is computationally efficient, but it fails to capture multiple, local motions in a scene [12], [13]. At the other extreme, a separate motion model (for instance, translational) is applied to every individual pixel. This generates a dense motion representation having at least two parameters (in the case of a translational model) for every pixel. Dense motion representations have the potential to accurately represent local motion in a scene but are computationally complex [14], [15], [16].

In between the extremes lie block-based motion models, which typically partition an image into uniform, non-overlapping rectangular blocks and estimate motion for each individual block [17]. Block-based motion models have moderate ability to represent local motion, constrained by the block boundaries imparted by the partition. Block-based translational motion models have been used extensively in digital video compression standards, for example, MPEG-1 and MPEG-2 [18], [19].

A second element of a motion estimation algorithm is the criteria that are used to determine the quality of a given motion estimate. One common strategy is to evaluate a motion vector based on a prediction error between the reference pixel(s) and the corresponding pixel(s) that are mapped to by the motion estimate. The prediction error can be written as:

$$e(x,y) = \alpha(I_k(x,y) - \hat{I}_k(x,y)),\qquad(2.2)$$

where I is the reference image, (x,y) is the pixel being predicted, $\hat{I}$ is the prediction, k represents the kth image, and α is a function that assesses a penalty for non-matching data. The prediction $\hat{I}_k$ incorporates the motion vector as $\hat{I}_k(x,y) = I_{k-1}((x,y) - d(x,y))$. In this case, the previous image, I_{k-1}, is used to form the prediction, and $d(x,y)$ is the motion vector for the given pixel (x,y).

A quadratic penalty function $\alpha(e) = e^2$ is commonly used. One of the drawbacks of the quadratic penalty function is that individual outliers with large errors can significantly affect the overall error for a block of pixels. Alternatively, an absolute value penalty function $\alpha(e) = |e|$ can be used. The absolute value penalty function is more robust to outliers and has the further advantage that it can be computed without multiplications. Another robust criterion commonly used is a cross-correlation function

$$C(d) = \sum_{(x,y)} (I_k(x,y)\hat{I}_k(x,y)).\qquad(2.3)$$

In this case, the function is maximized rather than minimized to determine the optimal motion vector.

The error function given by Equation 2.2 can be used to compare pixel intensity values. Other choices exist, including comparing pixel intensity gradients. This corresponds to comparing edge maps of images rather than comparing pixel values themselves. The use of gradients can be advantageous when the images being compared have similar image structure but potentially different mean value. This can occur, for example, when there is an illumination change, and one image is darker than another. It can also occur when the images correspond to different spectral responses.

A third element of a motion estimation algorithm is a search strategy used to locate the solution. Search strategies can vary based on the motion model and estimation criteria in

P	G	P	R
G	P	R	P
P	B	P	G
B	P	G	P

FIGURE 2.2

A color filter array pattern containing red, green, blue, and panchromatic pixels.

place. Many algorithms use iterative approaches to converge on a solution [14], [15], [16]. For motion models with only a small number of parameters to estimate and a small state space for each of the parameters, an exhaustive matching search is often used to minimize a prediction error such as given by Equation 2.2. A popular example of this approach is used with block-based motion estimation in which each block is modeled having a translational motion (only two parameters to estimate for each block of pixels), and the range of possible motion vectors is limited to integer pixel or possibly half-pixel values within a fixed-size window. Each possible motion offset is considered, and the offset resulting in the best match based on the estimation criteria is chosen as the motion estimate. Many algorithms have been proposed to reduce the complexity of exhaustive matching searches [20], [21]. These techniques focus on intelligently reducing the number of offsets searched as well as truncating the computation of error terms once it is known that the current offset is not the best match.

2.2.2 Motion Estimation for the Proposed Capture System

In the proposed capture system, color pixels are integrated for longer than panchromatic pixels, resulting in some misalignment of the data when there is motion during the overall capture. The motion estimation techniques discussed above can be used to detect this motion and compensate for it. For illustrative purposes, the specific four-channel CFA pattern shown in Figure 2.2 is assumed in the remainder of this section. Motion estimation and compensation issues are discussed with respect to this particular pattern of panchromatic, red, green, and blue pixels.

In a color digital camera, motion is usually computed between two similar images, for example, two grayscale images or two color images. In these cases, luminance information is considered sufficient for motion estimation [22], although studies have shown that additional accuracy can be obtained by including chrominance information in the motion estimation process [23], [24]. For the proposed capture system, motion must be computed between panchromatic data and color data. To facilitate this comparison, a synthetic panchromatic channel, P^{Syn}, can be computed corresponding to the color data. For a given pixel location, a synthetic panchromatic value can be computed as a linear combination of neighboring red, green, and blue pixel values:

$$P^{Syn} = \alpha R + \beta G + \gamma B. \qquad (2.4)$$

The linear weights α, β, and γ are chosen to generate an overall spectral response for P^{Syn} as similar as possible to a natural panchromatic pixel spectral response. Details are provided in the Appendix.

Depending on when motion estimation occurs in the overall image processing path, synthetic panchromatic pixel values may be computed at some or all pixel locations. The panchromatic and color pixel data are both initially available as sparse checkerboard arrays, as shown in Figure 2.3. In one processing path, the data is interpolated to generate fully populated panchromatic and color channels prior to motion estimation. In this case, synthetic panchromatic pixel values can be computed at each pixel using the available red, green, and blue values at that pixel, and motion can be estimated using the full images.

Alternatively, motion estimation can be carried out at lower resolution, with the objective of retaining a CFA image after motion compensation. Both the panchromatic and color image data are reduced to lower resolution. The color data are interpolated to form a full-color low-resolution image from which synthetic panchromatic pixel values are computed. Motion is estimated using the low-resolution images, with the results of the motion estimation applied to the original checkerboard pixel data [25]. In order to retain the original checkerboard CFA pattern, the motion estimation can be constrained to appropriate integer translational offsets.

Figure 2.3 also illustrates a scenario in which motion is estimated by directly comparing the panchromatic and color data, bypassing the need to compute a synthetic panchromatic channel. In this case, the green channel is used as an approximate match to the panchromatic channel. Each channel is fully populated by an interpolation step. Edge maps are formed and used during motion estimation to minimize the effects of spectral differences between the panchromatic and green channels [26].

The panchromatic and synthetic panchromatic channels are likely to differ in the amount of noise present, as well as the amount of motion blur present. These differences make estimation difficult for algorithms that derive individual motion vectors for every pixel. Block-based motion estimation provides some robustness to the varying noise and blur within the images while also providing some ability to detect local object motion in the scene.

With appropriate hardware capability, the panchromatic channel integration interval can align in various ways with the integration interval for the color channels, as shown in Figure 2.4 [27]. Different alignments produce different relative motion offsets under most conditions, especially with constant or nearly constant motion (both velocity and direction), as used in the following examples.

If the integration interval of the panchromatic channel is concentrically located within the integration interval of the color channels, as in Figure 2.4a, the motion offset between the channels will be very small or zero. This minimizes any need to align the color channels with the panchromatic channel during fusion of the channels, but also minimizes the information that motion offset estimation can provide regarding motion blur in the captured image. If the two integration intervals are aligned at the end of integration as in Figure 2.4b, the motion offset between the panchromatic channel and the color channels will be greater than with concentric integration intervals. This may require alignment of the color channels with the panchromatic channel to minimize artifacts during fusion of the images. In this case, motion offset estimation can also provide more information about the blur in the

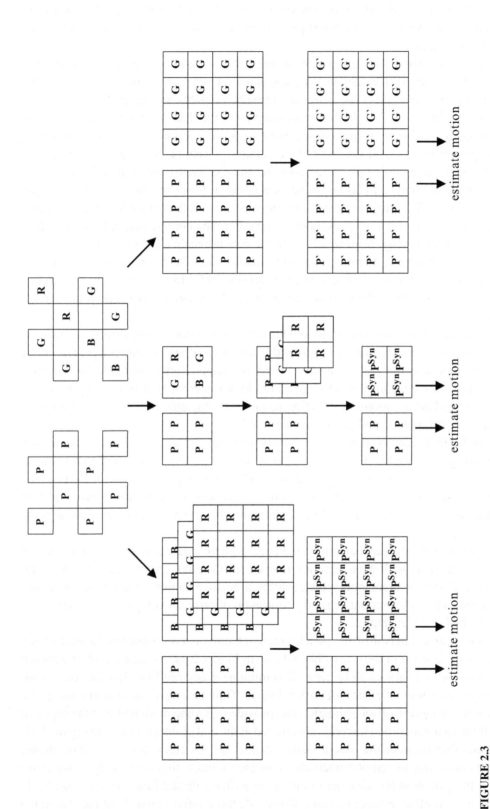

FIGURE 2.3

Three options for motion estimation in the proposed capture system. The term P^{Syn} represents the synthetic panchromatic channel whereas P' and G' are gradients of P and G, respectively.

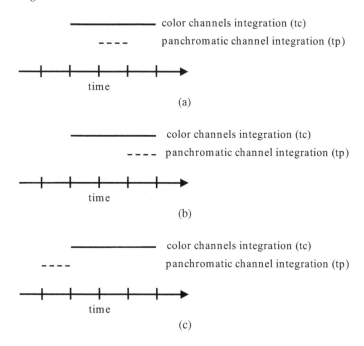

FIGURE 2.4

Integration timing options: (a) concentric integration, (b) simultaneous readout integration, and (c) non-overlapping integration.

captured image, to aid fusion or deblurring operations. If the integration intervals do not overlap, as shown in Figure 2.4c, the motion offset between the panchromatic and color channels will be still larger.

One advantage of overlapping integration intervals is to limit any motion offset between the channels and increase the correlation between the motion during the panchromatic interval and the motion during the color integration interval. The ratio of the color integration time to the panchromatic integration time, t_C/t_P, also affects the amount of motion offset. As this ratio decreases toward one, the capture converges to a standard single capture, and the relative motion offset converges to zero.

The alignment of the integration intervals has hardware implications in addition to the image processing implications just mentioned. In particular, use of end-aligned integration intervals tends to reduce the complexity of readout circuitry and buffer needs, since all pixels are read out at the same time. Concentric alignment of the integration intervals tends to maximize the complexity of readout, since the panchromatic integration interval both begins and ends at a time different from the color integration interval.

2.2.3   Motion Compensation for the Proposed Captured System

Motion compensation refers to the process of shifting the pixel data in an image according to motion information derived during an estimation step. Given a reference image and a comparison image, the motion information is used to form a shifted version of the comparison image that matches the reference image better than the original comparison image.

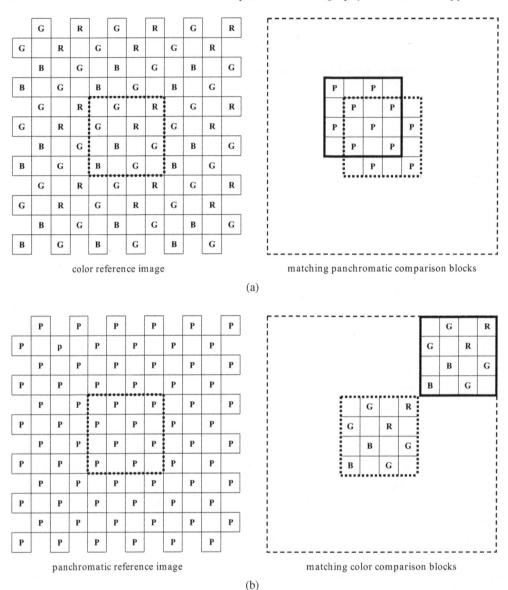

FIGURE 2.5

Motion compensation with the reference image constituted by (a) the color data and (b) the panchromatic data.

For the proposed capture system, it is possible to consider either the panchromatic data or the color data as the reference image. The choice of reference image affects which data is left untouched, and which data is shifted, and possibly interpolated. The advantage of selecting the panchromatic data as the reference image lies in keeping the sharp, panchromatic backbone of the image untouched, preserving as much strong edge and texture information as possible. The advantage of selecting the chrominance data as the reference image becomes apparent in the case that the motion compensation is performed on sparse CFA data with the intention of providing a CFA image as the output of the motion compensation step, as illustrated in Figure 2.5. In this case, shifting a block of panchromatic

FIGURE 2.6

Single capture: (a) CFA image, (b) demosaicked panchromatic image, and (c) demosaicked color image.

data to exactly fit the CFA pattern requires integer motion vectors for which the horizontal and vertical components have the same even/odd parity. Arbitrary shifts of a block of panchromatic data require an interpolation step for which the four nearest neighbors are no more than two pixels away. Shifting the chrominance data is more difficult, however, due to the sparseness of the colors. Shifting a block of chrominance data to exactly fit the CFA pattern requires motion vectors that are multiples of four, both horizontally and vertically. Arbitrary shifts of a block of chrominance data require an interpolation step for which neighboring color information may be four pixels away.

2.3 Four-Channel Single Capture Motion Deblurring

This section discusses a computationally efficient image fusion algorithm for motion deblurring of a color image from a single capture. The relative motion between the camera and the scene introduces motion blur that causes significant degradation of images [28], [29], [30], [31]. It is a well-known fact that motion blur can be minimized by reducing the pixel integration time appropriately. However, this approach leads to noisy images as imaging pixels do not get sufficient photons to faithfully represent the scene content. Due to this trade-off between motion blur and noise, images are typically captured with a sufficiently long integration time.

Signal processing techniques are also often used to reduce the motion blur [1], [2], [3], [4]. As explained in Section 2.2, in a four-channel system, it is possible to capture an image with acceptable signal-to-noise by using a relatively shorter integration time for the panchromatic pixels as compared to the color (red, green, and blue) pixels. This is a highly desirable feature for color image motion deblurring. Due to short integration time and high photometric sensitivity, panchromatic pixels do not suffer much from motion blur and at the same time produce a luminance image of the scene with high signal-to-noise, whereas a long integration time for color pixels leads to a motion blurred color image with reliable color information. An example of a four-channel CFA image is shown in Figure 2.6a. The integration time ratio of panchromatic to color pixels was set to 1:5. The corresponding demosaicked panchromatic and color images are shown in Figures 2.6b and 2.6c, respectively. The basketball is more clearly defined in the panchromatic image but appears blurred in the color image. This example illustrates that by using different integration times for panchromatic and color pixels in the four-channel imaging sensor, it is possible to generate complementary information at the sensor level, which subsequently can be exploited to generate a color image with reduced motion blur. A pixel-level fusion algorithm designed to fuse demosaicked panchromatic and color images [32] is explained below.

2.3.1 Fusion for Motion Deblurring

The pixel-level fusion algorithm presented in this section is based on the fact that the human visual system (HVS) is more sensitive to the high-frequency luminance than the corresponding chroma components. Let R, G, B, and P be the demosaicked red, green, blue, and panchromatic images, respectively, captured using a four-channel imaging system. In this fusion approach, a synthetic panchromatic image (P^{Syn}), which is comparable to the observed panchromatic image (P), is computed using Equation 2.4. The red and blue chroma images, represented by C_R and C_B, respectively, are computed as follows:

$$C_R = R - P^{Syn}, \tag{2.5}$$

$$C_B = B - P^{Syn}. \tag{2.6}$$

To restore the high-frequency luminance information of the deblurred color image, the synthetic panchromatic image P^{Syn} can be replaced with the observed panchromatic image P. However, this operation only ensures reconstruction of the luminance information. In order to restore color information, chroma images corresponding to P must be reconstructed. Note that P is a luminance image and does not contain color information. Therefore, its chroma images must be estimated from the observed RGB color image. In order to do this, a system model is determined to relate P^{Syn} and the corresponding chroma images (C_R and C_B) which in turn is used to predict chroma images for P. For the sake of simplicity and computational efficiency, the model is linear:

$$C_R = m_R P^{Syn}, \tag{2.7}$$

$$C_B = m_B P^{Syn}, \tag{2.8}$$

where m_R and m_B are model parameters.

FIGURE 2.7

Motion deblurring from a single capture: (a) panchromatic image, (b) color image, and (c) deblurred image.

Let C_R^P and C_B^P be the red and the blue chroma images, respectively, corresponding to P. Then, from Equations 2.7 and 2.8 it is apparent that

$$C_R^P = m_R P = \frac{C_R}{P^{Syn}} P, \tag{2.9}$$

$$C_B^P = m_B P = \frac{C_B}{P^{Syn}} P. \tag{2.10}$$

The new motion deblurred color image (R^N, G^N, and B^N) can be estimated as follows:

$$R^N = C_R^P + P, \tag{2.11}$$

$$B^N = C_B^P + P, \tag{2.12}$$

$$G^N = \frac{P - \alpha R^N - \gamma B^N}{\beta}. \tag{2.13}$$

ALGORITHM 2.1 Pixel-level image fusion for motion deblurring.

1. Compute synthetic panchromatic and chroma images using Equations 2.4, 2.5, and 2.6.

2. Compute chroma images corresponding to observed panchromatic image P using Equations 2.9 and 2.10.

3. Compute deblurred color image using Equations 2.11, 2.12, and 2.13.

A summary of the fusion algorithm is presented in Algorithm 2.1, whereas its feasibility is demonstrated in Figure 2.7. The integration time ratio for panchromatic image shown in Figure 2.7a and color image shown in Figure 2.7b was set to 1:5 and these two images were read out of the sensor simultaneously as illustrated in Figure 2.4b. The restored image is shown in Figure 2.7c.

2.4   Example Single-Sensor Image Fusion Capture System

Figure 2.8 shows the example image processing chain from Chapter 1 with motion estimation, motion compensation, and motion deblurring blocks added. In this example motion processing begins on the CFA image data rather than demosaicked data. The first motion operation is motion estimation which in this example is performed by comparing edge maps generated from the panchromatic and green channels from the CFA image.

Figure 2.9a shows the CFA image used as the input for edge detection. The corresponding edge maps are shown in Figure 2.9b where the edges from the CFA panchromatic channel and the CFA green channel are depicted using mid and high intensities, respectively. It can be seen that the most systematic differences between the two sets of edge maps occur along the fronts and backs of the two ball players. From these edge maps motion vectors are generated using block matching and a cross correlation penalty function (Equation 2.3). The resulting motion compensation is applied to the CFA panchromatic data using bilinear interpolation. The CFA panchromatic channels before and after motion compensation are shown in Figure 2.9a and Figure 2.10a. For illustrative purposes, the edge maps are recomputed for the motion-compensated CFA image and shown in Figure 2.10b. As a reference, the block boundaries used for computing the motion vectors are depicted in the figure.

Returning to the flowchart shown in Figure 2.8, the image is now demosaicked using the motion-compensated CFA panchromatic data along with the original color data. Afterwards, motion deblurring is applied using the image fusion techniques described in Section 2.3.1. Figure 2.11 shows the demosaicked image before motion deblurring (Fig-

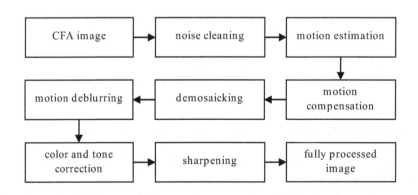

FIGURE 2.8

Example image processing chain incorporating motion estimation and compensation.

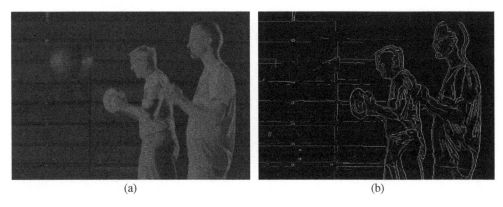

FIGURE 2.9

Edge detection using the original CFA panchromatic channel: (a) CFA panchromatic channel, and (b) edge map with mid and high intensities corresponding to the panchromatic and color pixels, respectively.

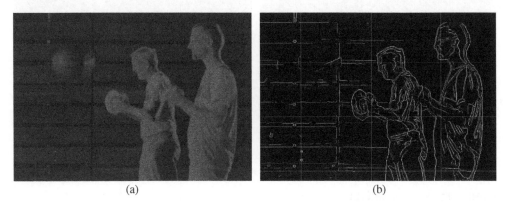

FIGURE 2.10

Edge detection using the motion compensated CFA panchromatic channel: (a) CFA panchromatic channel, and (b) edge map with block boundaries and mid and high intensities corresponding to the panchromatic and color pixels, respectively.

ure 2.11a) and after deblurring (Figure 2.11b). The differences between these two images is dramatic, especially with respect to the two balls being tossed. Finally, color and tone correction and sharpening are applied and the final result shown in Figure 2.12a. For reference, the same image without motion processing is shown in Figure 2.12b.

2.5   Conclusion

The estimation of motion and the compensation of associated artifacts classically requires the capture of a sequence of images. Image fusion techniques are then used to reduce this set of multiple images into a single image with reduced motion blur.

(a)   (b)

FIGURE 2.11 (See color insert.)

Motion deblurring of demosaicked image: (a) before motion deblurring, and (b) after motion deblurring.

(a)   (b)

FIGURE 2.12 (See color insert.)

Fully processed images: (a) with motion compensation, and (b) without motion compensation.

This chapter described a new approach to motion deblurring that uses a single capture from a single sensor, which produces the required image components for subsequent image fusion. The result is a vastly simplified hardware system that is motion-aware and provides the necessary information for performing motion deblurring. This is accomplished by the use of a four-channel color filter array consisting of three color channels and a panchromatic channel. Due to superior light sensitivity, the panchromatic pixels are exposed for a shorter duration than the color pixels during image capture. As a result, the panchromatic channel produces a luminance record of the image with significantly reduced motion blur while maintaining acceptable signal-to-noise performance. To this low motion blur luminance record, the chrominance information from the color channels is fused to produce a final motion-deblurred color image. There are a number of ways to achieve this motion-reduced result including the alignment of edges in motion between the panchromatic and color image components and the exchange of panchromatic and color-derived luminance image components. Many of these techniques can be applied to CFA data directly, thereby reducing the computational overhead of the system. Finally, apart from the motion elements, the image processing chain is as described in Chapter 1, thus allowing the system to realize most, if not all, of the advantages of the four-channel system as described there.

Appendix: Estimation of Panchromatic Coefficients

Assuming the panchromatic (*P*) and the color (*RGB*) pixels to be sensitive in the wavelength range $[\lambda_{min}, \lambda_{max}]$, the pixel value of these channels acquired at a spatial location $(x, y) \in \Omega$ can be modeled as follows:

$$P(x, y) = \int_{\lambda_{min}}^{\lambda_{max}} I(\lambda) \, Q_P(\lambda) \, S(x, y, \lambda) \, d\lambda,$$

$$R(x, y) = \int_{\lambda_{min}}^{\lambda_{max}} I(\lambda) \, Q_R(\lambda) \, S(x, y, \lambda) \, d\lambda,$$

$$G(x, y) = \int_{\lambda_{min}}^{\lambda_{max}} I(\lambda) \, Q_G(\lambda) \, S(x, y, \lambda) \, d\lambda,$$

$$B(x, y) = \int_{\lambda_{min}}^{\lambda_{max}} I(\lambda) \, Q_B(\lambda) \, S(x, y, \lambda) \, d\lambda,$$

where $I(\lambda)$ is the spectrum of the illumination as a function of wavelength, $S(x, y, \lambda)$ is the surface spectral reflectance function. The spectral quantum efficiency of *P*, *R*, *G*, and *B* sensors are represented by $Q_P(\lambda)$, $Q_R(\lambda)$, $Q_G(\lambda)$, and $Q_B(\lambda)$, respectively. The panchromatic coefficients α, β, and γ are computed by minimizing the cost function $g(\alpha, \beta, \gamma)$, as given below:

$$\min_{\alpha, \beta, \gamma} g(\alpha, \beta, \gamma) = \min_{\alpha, \beta, \gamma} \left\| \sum_{x, y \in \Omega} P(x, y) - \alpha \sum_{x, y \in \Omega} R(x, y) - \beta \sum_{x, y \in \Omega} G(x, y) - \gamma \sum_{x, y \in \Omega} B(x, y) \right\|_2^2,$$

where $\| \bullet \|_2$ denotes L_2-norm. Note that the integration times for *P*, *R*, *G*, and *B* pixels are assumed to be the same in this analysis.

References

[1] E. Shechtman, Y. Caspi, and M. Irani, "Space-time super-resolution," *IEEE Transactions on Pattern Analysis and Machine Intelligence*, vol. 27, no. 4, pp. 531–545, April 2005.

[2] Q. Shan, J. Jia, and A. Agarwala, "High-quality motion deblurring from a single image," *ACM Transactions on Graphics*, vol. 27, no. 3, pp. 1–10, August 2008.

[3] J. Jiya, "Single image motion deblurring using transparency," in *Proceedings of the IEEE Conference on Computer Vision and Pattern Recognition*, Minneapolis, MN, USA, June 2007, pp. 1–8.

[4] A. Levin, "Blind motion deblurring using image statistics," in *Proceedings of the Twentieth Annual Conference on Advances in Neural Information Processing Systems*, Vancouver, BC, Canada, December 2006, pp. 841–848.

[5] M. Kumar and P. Ramuhalli, "Dynamic programming based multichannel image restoration," in *Proceedings of the IEEE International Conference on Acoustics, Speech, and Signal Processing*, Philadelphia, PA, USA, March 2005, pp. 609–612.

[6] L. Yuan, J. Sun, L. Quan, and H.Y. Shum, "Image deblurring with blurred/noisy image pairs," *ACM Transactions on Graphics*, vol. 26, no. 3, July 2007.

[7] A. Rav-Acha and S. Peleg, "Two motion-blurred images are better than one," *Pattern Recognition Letters*, vol. 26, no. 3, pp. 311–317, February 2005.

[8] X. Liu and A.E. Gamal, "Simultaneous image formation and motion blur restoration via multiple capture," in *Proceedings of the IEEE International Conference on Acoustics, Speech, and Signal Processing*, Salt Lake City, UT, USA, May 2001, pp. 1841–1844.

[9] K.T. Mullen, "The contrast sensitivity of human colour vision to red-green and blue-yellow chromatic gratings," *Journal of Physiology*, vol. 359, pp. 381–400, February 1985.

[10] A.M. Tekalp, *Digital Video Processing*, Upper Saddle River, NJ: Prentice Hall, August 1995.

[11] A. Bovik, *Handbook of Image and Video Processing*, 2nd Edition, New York: Academic Press, June 2005.

[12] F. Dufaux and J. Konrad, "Efficient, robust, and fast global motion estimation for video coding," *IEEE Transactions on Image Processing*, vol. 9, no. 3, pp. 497–501, March 2000.

[13] J.M. Odobez and P. Bouthemy, "Robust multiresolution estimation of parametric motion models," *Journal of Visual Communication and Image Representation*, vol. 6, no. 4, pp. 348–365, December 1995.

[14] M. Black, "The robust estimation of multiple motions: Parametric and piecewise-smooth flow fields," *Computer Vision and Image Understanding*, vol. 63, no. 1, pp. 75–104, January 1996.

[15] J. Konrad and E. Dubois, "Bayesian estimation of motion vector fields," *IEEE Transactions on Pattern Analysis and Machine Intelligence*, vol. 14, no. 9, pp. 910–927, September 1992.

[16] B.K.P. Horn and B.G. Schunck, "Determining optical flow," *Artificial Intelligence*, vol. 17, pp. 185–203, August 1981.

[17] F. Dufaux and F. Moscheni, "Motion estimation techniques for digital TV: A review and a new contribution," *Proceedings of the IEEE*, vol. 83, no. 6, pp. 858–876, June 1995.

[18] "Information technology-coding of moving pictures and associated audio for digital storage media up to about 1.5 mbit/s." ISO/IEC JTC1 IS 11172-2 (MPEG-1), 1993.

[19] "Information technology–generic coding of moving pictures and associated audio." ISO/IEC JTC1 IS 13818-2 (MPEG-2), 1994.

[20] R. Li, B. Zeng, and M.L. Liou, "A new three-step search algorithm for block motion estimation," *IEEE Transactions on Circuits and Systems for Video Technology*, vol. 4, no. 4, pp. 438–442, August 1994.

[21] S. Zhu and K.K. Ma., "A new diamond search algorithm for fast block-matching motion estimation," *IEEE Transactions on Image Processing*, vol. 9, no. 2, pp. 287–290, February 2000.

[22] K.A. Prabhu and A.N. Netravali, "Motion compensated component color coding," *IEEE Transactions on Communications*, vol. 30, no. 12, pp. 2519–2527, December 1982.

[23] N.R. Shah and A. Zakhor, "Resolution enhancement of color video sequences," *IEEE Transactions on Image Processing*, vol. 8, no. 6, pp. 879–885, June 1999.

[24] B.C. Tom and A. Katsaggelos, "Resolution enhancement of monochrome and color video using motion compensation," *IEEE Transactions on Image Processing*, vol. 10, no. 2, pp. 278–287, February 2001.

[25] A.T. Deever, J.E. Adams Jr., and J.F. Hamilton Jr., "Improving defective color and panchromatic CFA image," U.S. Patent Application 12/258 389, 2009.

[26] J.E. Adams Jr., A.T. Deever, and R.J. Palum, "Modifying color and panchromatic channel CFA Image," U.S. Patent Application 12/266 824, 2009.

[27] J.A. Hamilton, J.T. Compton, and B.H. Pillman, "Concentric exposures sequence for image sensor," U.S. Patent Application 12/111 219, April 2008.

[28] M. Ben-Ezra and S.K. Nayar, "Motion-based motion deblurring," *IEEE Transactions on Pattern Analysis and Machine Intelligence*, vol. 26, no. 6, pp. 689–698, June 2004.

[29] S. Bottini, "On the visual motion blur restoration," in *Proceedings of the Second International Conference on Visual Psychophysics and Medical Imaging*, Brussels, Belgium, July 1981, p. 143.

[30] W.G. Chen, N. Nandhakumar, and W.N. Martin, "Image motion estimation from motion smear – a new computational model," *IEEE Transactions on Pattern Analysis and Machine Intelligence*, vol. 18, no. 4, pp. 412–425, April 1996.

[31] S.H. Lee, N.S. Moon, and C.W. Lee, "Recovery of blurred video signals using iterative image restoration combined with motion estimation," in *Proceedings of the International Conference on Image Processing*, Santa Barbara, CA, USA, October 1997, pp. 755–758.

[32] M. Kumar and J.E. Adams Jr., "Producing full-color image using CFA image," U.S. Patent Application Number 12/412 429, 2009.

3

Lossless Compression of Bayer Color Filter Array Images

King-Hong Chung and Yuk-Hee Chan

3.1 Introduction

Most digital cameras reduce their cost, size and complexity by using a single-sensor image acquisition system to acquire a scene in digital format [1]. In such a system, an image sensor is overlaid with a color filter array (CFA), such as the Bayer pattern [2] shown in Figure 3.1a, to record one of the three primary color components at each pixel location. Consequently, a gray-scale mosaic-like image, commonly referred to as a CFA image, is produced as the sensor output.

An imaging pipeline is required to turn a CFA image into a full-color image. Images are commonly compressed to reduce the storage requirement and store as many images as possible in a given storage medium. Figure 3.2a shows the simplified pipeline which first converts the CFA image to a full-color image using color demosaicking [3], [4], [5] and then compresses the demosaicked full-color image for storage.

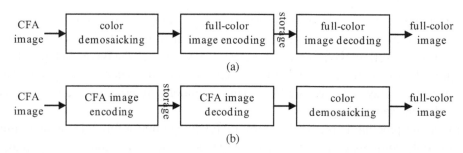

G	R	G	R	G	R
B	G	B	G	B	G
G	R	G	R	G	R
B	G	B	G	B	G
G	R	G	R	G	R
B	G	B	G	B	G

(a)   (b)   (c)

FIGURE 3.1

A Bayer color filter array with green-red-green-red phase in the first row: (a) complete pattern, (b) luminance plane, and (c) chrominance plane.

This approach reduces the burden of compression, since one can use any existing image coding scheme to compress the demosaicked full-color data in either a lossless or lossy manner. However, it may be considered suboptimal from the compression point of view because the demosaicking process always introduces some redundancy which should eventually be removed in the subsequent compression step [6], [7], [8].

To address this issue, an alternative approach, as shown in Figure 3.2b, aims at compressing the CFA image prior to demosaicking [9], [10], [11], [12]. As a result, more sophisticated demosaicking algorithms can be applied offline on a personal computer to produce a more visually pleasing full-color output. Besides, as the data size of a CFA image to be handled by the compression step is only one-third that of the corresponding demosaicked image, this alternative approach can effectively increase the pipeline throughput without degrading the output image quality.

Since this alternative pipeline has been proven to outperform the conventional one when the quality requirement of the output color image is high [7], [8], it has been adopted in many prosumer and professional grade digital cameras to serve as an optional imaging path to deliver a precise, high quality output. This motivates the demand for CFA image compression schemes accordingly.

CFA image → color demosaicking → full-color image encoding → storage → full-color image decoding → full-color image

(a)

CFA image → CFA image encoding → storage → CFA image decoding → color demosaicking → full-color image

(b)

FIGURE 3.2

Simplified imaging pipelines for single-sensor digital cameras: (a) conventional approach, (b) alternative approach.

Both lossy or lossless compression schemes can be applied to CFA image data. Lossy compression schemes, such as those presented in References [13], [14], [15], [16], [17] and [18], compress the CFA image by discarding its visually redundant information. Thus, only an approximation of the original image can eventually be reconstructed. Since loss of information is allowed, these schemes usually yield a higher compression ratio than the lossless schemes. However, lossless compression schemes preserve all the information of the original and hence allow perfect reconstruction of the original image. Therefore, they are crucial in coding the CFA images which can be seen as digital negatives and used as an ideal original archive format for producing high quality color images especially in high-end photography applications such as commercial poster production.

Certainly, many standard lossless image compression schemes such as JPEG-LS [19] and JPEG2000 (lossless mode) [20] can be used to compress a CFA image directly. However, they only achieve a fair compression performance as the spatial correlation among adjacent pixels is generally weakened in a CFA image due to its mosaic-like structure. To perform the compression more efficiently, later methods [21], [22] aim at increasing the image spatial correlation by de-interleaving the CFA image according to the red, green, and blue color channels and then compress the three subsampled color planes individually with the lossless image compression standards. Nevertheless, redundancy among the color channels still remains. Recently, some advanced lossless CFA image compression algorithms [23], [24], [25] have been reported to efficiently remove the pixel redundancy in both spatial and spectral domains. These algorithms do not rely on any single individual coding technique but rather combine various techniques to remove the data redundancy using different means. This chapter surveys relevant lossless coding techniques and presents two new lossless compression algorithms to show how to mix different techniques to achieve an effective compression. Performance comparisons, in terms of compression ratio and computational complexity, are included.

This chapter is structured as follows. Section 3.2 discusses some major concerns in the design of a lossless compression algorithm for CFA images. Section 3.3 focuses on some common coding techniques used in lossless CFA image coding. Sections 3.4 and 3.5 present two lossless compression algorithms which serve as examples to show how the various coding techniques discussed in Section 3.3 can work together to remove the redundancy in different forms. The simulation results in Section 3.6 show that remarkable compression performance can be achieved with these two algorithms. Finally, conclusions are drawn in Section 3.7.

3.2 Concerns in CFA Image Compression

Though a compression algorithm can be evaluated in a number of ways, rate-distortion performance is one of the most important factors to be considered. For lossless CFA image compression, no distortion is expected and hence rate-distortion performance can be reflected by the compression ratio or the output bit rate directly once the input is fixed.

The second consideration is the complexity of the algorithm. For in-camera compression, real-time processing is always expected. As compression is required for each image, the processing time of the compression algorithm determines the frame rate of the camera. Parallel processing support can help reduce an algorithm's processing time, but it still may not be a solution as it does not reduce the overall complexity.

The complexity of an algorithm can be measured in terms of the number of operations required to compress an image. This measure may not be able to reflect the real performance of the algorithm as number of operations is not the sole factor that determines the required processing time. The number of branch decisions, the number of data transfers involved, the hardware used to realize the algorithm, and a lot of other factors also play their roles. The impact of these factors to the processing time is hardware oriented and can fluctuate from case to case. The power consumption induced by an algorithm is also a hardware issue and is highly reliant on the hardware design. Without a matched hardware platform as the test bed of an algorithm, it is impossible to judge its real performance. Therefore, in this chapter, the processing time required to execute an algorithm in a specified general purpose hardware platform will be measured to indicate the complexity of the algorithm.

Since minimizing the complexity and maximizing the compression rate are generally mutually exclusive, a compromise is always required in practice. Note that the energy consumption increases with the complexity of any algorithm, thus affecting both the battery size and the operation time of a camera directly. There are some other factors, such as the memory requirements, that one has to consider when designing a compression algorithm, but their analysis is beyond the scope of this chapter.

3.3   Common Compression Techniques

In lossless compression, bits are reduced by removing the redundant information carried by an image. Various techniques can be used to extract and remove the redundant information by exploring i) the spatial correlation among image pixels, ii) the correlation among different color channels of the image, and iii) the statistical characteristic of selected data entities extracted from the image. The performance of an algorithm depends on how much redundant information can be removed effectively. A compression algorithm usually exploits more than one technique to achieve the goal. Entropy coding and predictive coding are two commonly used techniques to encode a CFA image nowadays.

Entropy coding removes redundancy by making use of the statistical distribution of the input data. The input data is considered as a sequence of symbols. A shorter codeword is assigned to a symbol which is more likely to occur such that the average number of bits required to encode a symbol can be reduced. This technique is widely applicable for various kinds of input of any nature and hence is generally exploited in many lossless compression algorithms as their final processing step. Entropy coding can be realized using different schemes, such as Huffman coding and arithmetic coding. When the input data follows a geometric distribution, a simpler scheme such as Rice coding can be used to reduce the complexity of the codeword assignment process.

Predictive coding removes redundancy by making use of the correlation among the input data. For each data entry, a prediction is performed to estimate its value based on the correlation and the prediction error is encoded. The spatial correlation among pixels is commonly used in predictive coding. Since color channels in a CFA image are interlaced, the spatial correlation in CFA images is generally lower than in color images. Therefore, many spatial predictive coding algorithms designed for color images cannot provide a good compression performance when they are used to encode a CFA image directly.

However, there are solutions which preprocess the CFA image to provide an output with improved correlation characteristics which is more suitable for predictive coding than the original input. This can be achieved by deinterleaving the CFA image into several separate images each of which contains the pixels from the same color channel [21], [22]. This can also be achieved by converting the data from RGB space to YC_rC_b space [10], [12]. Although a number of preprocessing procedures can be designed, not all of them are reversible and only reversible ones can be used in lossless compression of CFA images.

In transform coding, the discrete cosine transform and the wavelet transform are usually used to decorrelate the image data. Since typical images generally contain redundant edges and details, insignificant high-frequency contents can thus be discarded to save coding bits. When distortion is allowed, transform coding helps to achieve good rate-distortion performance and hence it is widely used in lossy image compression. In particular, the integer Mallat wavelet packet transform is highly suitable to decorrelate mosaic CFA data [23], [24]. This encourages the use of transform coding in lossless compression of CFA images.

Other lossless coding techniques, such as run-length coding [26], Burrows-Wheeler transform [27], and adaptive dictionary coding (e.g., LZW [28]) are either designed for a specific type of input other than CFA images (for example, run-length coding is suitable for coding binary images) or designed for universal input. Since they do not take the properties of a CFA image into account, it is expected that the redundancy in a CFA image cannot be effectively removed if one just treats the CFA image as a typical gray-level image or even a raster-scanned sequence of symbols when using these coding techniques. A preprocessing step would be necessary to turn a CFA image into a better form to improve the compression performance when these techniques are exploited.

At the moment, most, if not all, lossless compression algorithms designed for coding CFA images mainly rely on predictive, entropy, and transform coding. In the following, two dedicated lossless compression algorithms for CFA image coding are presented. These algorithms serve as examples of combining the three aforementioned coding techniques to achieve remarkable compression performance.

3.4 Compression Using Context Matching-Based Prediction

The algorithm presented in this section uses both predictive and entropy coding to compress CFA data. First, the CFA image is separated into the luminance subimage (Figure 3.1b) containing all green samples and the chrominance subimage (Figure 3.1c) containing all red and blue samples. These two subimages are encoded sequentially. Samples

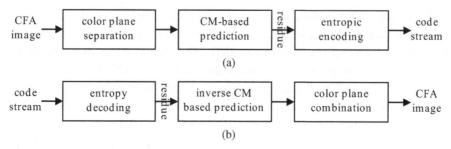

FIGURE 3.3

Structure of the context-matching-based lossless CFA image compression method: (a) encoder and (b) decoder.

in the same subimage are raster-scanned and each one of them undergoes a prediction process based on context matching and an entropy coding process as shown in Figure 3.3a. Due to the higher number of green samples in the CFA image compared to red or blue samples, the luminance subimage is encoded before encoding the chrominance subimage. When handling the chrominance subimage, the luminance subimage is used as a reference to remove the interchannel correlation.

Decoding is just the reverse process of encoding as shown in Figure 3.3b. The luminance subimage is decoded first to be used as a reference when decoding the chrominance subimage. The original CFA image is reconstructed by combining the two subimages.

3.4.1 Context Matching-Based Prediction

In the prediction process exploited here, the value of a pixel is predicted with its four closest processed neighbors in the same subimage. The four closest neighbors from the same color channel as the pixel of interest should have the highest correlation to the pixel to be predicted in different directions and hence the best prediction result can be expected. These four neighbors are ranked according to how close their contexts are to the context of the pixel to be predicted and their values are weighted according to their ranking order. Pixels with closer contexts to that of the pixel of interest contribute more to its predicted value. The details of its realization in handling the two subimages are given below.

3.4.1.1 Luminance Subimage-Based Prediction

Since pixels in the luminance subimage are processed in a raster scan order, the four nearest and already processed neighboring green samples of the pixel of interest $g(i,j)$ are $g(i,j-2)$, $g(i-1,j-1)$, $g(i-2,j)$, and $g(i-1,j+1)$ as shown in Figure 3.4a. These neighbors are ranked by comparing their contexts with the context of $g(i,j)$. In the luminance subimage, the term $S_{g(p,q)}$, which denotes the context of a sample at position (p,q), is defined as shown in Figure 3.4b. In formulation, $S_{g(p,q)} = \{g(p,q-2),g(p-1,q-1),g(p-2,q),g(p-1,q+1)\}$. The matching extent of the contexts of $g(i,j)$ and $g(m,n)$ for $g(m,n) \in \Phi_{g(i,j)} = \{g(i,j-2),g(i-1,j-1),g(i-2,j),g(i-1,j+1)\}$ is determined as follows:

$$D_1\left(S_{g(i,j)},S_{g(m,n)}\right) = |g(i,j-2) - g(m,n-2)| + |g(i-1,j-1) - g(m-1,n-1)|$$
$$+ |g(i-2,j) - g(m-2,n)| + |g(i-1,j+1) - g(m-1,n+1)|. \quad (3.1)$$

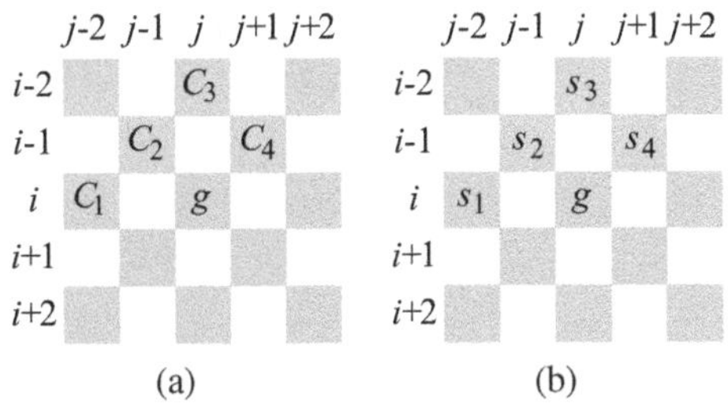

	j-2	j-1	j	j+1	j+2
i-2			C_3		
i-1		C_2		C_4	
i	C_1		g		
i+1					
i+2					

	j-2	j-1	j	j+1	j+2
i-2			s_3		
i-1		s_2		s_4	
i	s_1		g		
i+1					
i+2					

(a)   (b)

FIGURE 3.4

Prediction in the luminance subimage: (a) four closest neighbors used to predict the intensity value of sample $g(i,j)$, and (b) pixels used to construct the context of sample $g(i,j)$.

Theoretically, other metrics such as the well-known Euclidean distance can be accommodated in the above equation to enhance matching performance. However, the achieved improvement is usually not significant enough to compensate for the increased implementation complexity.

Let $g(m_k, n_k) \in \Phi_{g(i,j)}$, for $k = 1, 2, 3, 4$, represent four ranked neighbors of sample $g(i,j)$ such that $D_1\left(S_{g(i,j)}, S_{g(m_u,n_u)}\right) \leq D_1\left(S_{g(i,j)}, S_{g(m_v,n_v)}\right)$ for $1 \leq u < v \leq 4$. The value of $g(i,j)$ can then be predicted with a prediction filter as follows:

$$\hat{g}(i,j) = \text{round}\left(\sum_{k=1}^{4} w_k d(m_k, n_k)\right), \tag{3.2}$$

where w_k, for $k = 1, 2, 3, 4$, are normalized weights constrained as $w_1 + w_2 + w_3 + w_4 = 1$.

The weights are determined by quantizing the training result derived using linear regression with a set of training images. The weights are quantized to reduce the implementation complexity the predictor filter. With the training image set containing the first two rows of images shown in Figure 3.9a, the quantized training result is obtained as $\{w_1, w_2, w_3, w_4\} = \{5/8, 2/8, 1/8, 0\}$. The predicted green sample can then be determined as follows:

$$\hat{d}(i,j) = \text{round}\left(\frac{4g(m_1,n_1) + 2g(m_2,n_2) + g(m_1,n_1) + g(m_3,n_3)}{8}\right) \tag{3.3}$$

which requires only shift-add operations to implement.

3.4.1.2 Chrominance Subimage-Based Prediction

When the sample being processed is a chrominance (i.e., red or blue) sample, the prediction is carried out in the color difference domain instead of the intensity domain. This arrangement helps to remove the interchannel redundancy. Required color difference information is obtained in advance from the CFA data using the procedure from Section 3.4.2. Here, it is assumed that the estimation process has been completed and each chrominance sample $c(p,q)$ has its color difference value $d(p,q) = g(p,q) - c(p,q)$ available.

For any color difference value $d(i,j)$, its four closest neighbors in the estimated color difference plane are $d(i, j-2), d(i-2, j-2), d(i-2, j)$, and $d(i-2, j+2)$, and its context is defined as $S_{c(i,j)} = \{g(i, j-1), g(i-1, j), g(i, j+1), g(i+1, j)\}$ using its four closest

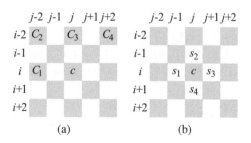

FIGURE 3.5

Prediction in the chrominance subimage: (a) four closest neighbors used to predict the color difference value of sample $c(i,j) = r(i,j)$ or $c(i,j) = b(i,j)$, and (b) pixels used to construct the context of sample $c(i,j)$.

available green samples, as shown in Figures 3.5a and 3.5b, respectively. This arrangement is based on the fact that the green channel has a double sampling rate compared to the red and blue channels in the CFA image and green samples are encoded first. As a consequence, it provides a more reliable noncausal context for matching.

Color difference values $d(i,j-2)$, $d(i-2,j-2)$, $d(i-2,j)$, and $d(i-2,j+2)$ are ranked according to the absolute difference between their context and the context of $d(i,j)$. The predicted value of $d(i,j)$ is determinable as follows:

$$\hat{d}(i,j) = \text{round}\left(\sum_{k=1}^{4} w_k d(m_k, n_k)\right), \tag{3.4}$$

where w_k is the weight associated with the the kth ranked neighbor $d(m_k, n_k)$ such that $D_2\left(S_{c(i,j)}, S_{c(m_u,n_u)}\right) \le D_2\left(S_{c(i,j)}, S_{c(m_v,n_v)}\right)$ for $1 \le u < v \le 4$, where

$$\begin{aligned} D_2\left(S_{c(i,j)}, S_{c(m,n)}\right) &= |g(i,j-1) - g(m,n-1)| + |g(i,j+1) - g(m,n+1)| \\ &\quad + |g(i-1,j) - g(m-1,n)| + |g(i+1,j) - g(m+1,n)| \end{aligned} \tag{3.5}$$

measures the difference between two contexts.

Weights w_k, for $k = 1, 2, 3, 4$, are trained similarly to the weights used in luminance signal prediction. Under this training condition, color difference prediction is obtained as follows:

$$\hat{d}(i,j) = \text{round}\left(\frac{4d(m_1,n_1) + 2d(m_2,n_2) + d(m_3,n_3) + d(m_4,n_4)}{8}\right), \tag{3.6}$$

which also involves shift-add operations only.

3.4.2   Adaptive Color Difference Estimation

When compressing the chrominance plane, prediction is carried out in the color difference domain to remove the interchannel redundancy. This implies that the color difference value for each chrominance sample has to be estimated in advance.

Let $c(m,n)$ be the intensity value of the available chrominance (either red or blue) sample at a chrominance sampling position (m,n). The corresponding color difference $d(m,n)$ is determined as

$$d(m,n) = \hat{g}(m,n) - c(m,n), \tag{3.7}$$

where $\hat{g}(m,n)$ represents an estimate of the missing luminance component at position (m,n). In particular, $\hat{g}(m,n)$ is adaptively determined according to the horizontal intensity gradient δH and the vertical intensity gradient δV at position (m,n) as follows:

$$\hat{g}(m,n) = \text{round}\left(\frac{\delta H \times G_V + \delta V \times G_H}{\delta H + \delta V}\right), \tag{3.8}$$

where G_H and G_V denote, respectively, the preliminary luminance estimates computed as $(g(i,j-1)+g(i,j+1))/2$ and $(g(i-1,j)+g(i+1,j))/2$. Gradients δH and δV are, respectively, obtained by averaging the absolute differences between all pairs of successive luminance samples along the horizontal and the vertical directions within a 5×5 supporting window centered at (m,n). Using Equation 3.8, the missing luminance value is determined in a way that a preliminary estimate contributes less if the gradient in the corresponding direction is larger. The weighting mechanism will automatically direct the estimation process along an existing edge, thus preserving important structural content of an image.

3.4.3 Subimage Coding

Entropy coding is exploited in the final stage to remove the redundancy. The prediction error $e(i,j)$ associated with pixel (i,j) is given by

$$e(i,j) = \begin{cases} \hat{g}(i,j) - g(i,j) & \text{if } (i,j) \text{ refers to luminance subimage,} \\ \hat{d}(i,j) - d(i,j) & \text{if } (i,j) \text{ refers to chrominance subimage,} \end{cases} \tag{3.9}$$

where $g(i,j)$ is the real value of the luminance sample, $d(i,j)$ is the color difference value estimated with the method described in Section 3.4.2, $\hat{g}(i,j)$ is the predicted value of $g(i,j)$, and $\hat{d}(i,j)$ is the predicted value of $d(i,j)$. The error residue, $e(i,j)$, is then mapped to a nonnegative integer via

$$E(i,j) = \begin{cases} -2e(i,j) & \text{if } e(i,j) \leq 0, \\ 2e(i,j) - 1 & \text{otherwise,} \end{cases} \tag{3.10}$$

to reshape its value distribution from a Laplacian type to a geometric one for Rice coding.

When Rice coding is used, each mapped residue $E(i,j)$ is split into a quotient $Q = \text{floor}(E(i,j)/2^\lambda)$ and a remainder $R = E(i,j)\text{mod}(2^\lambda)$, where the parameter λ is a nonnegative integer. The quotient and the remainder are then saved for storage or transmission. The length of the codeword used for representing $E(i,j)$ depends on λ and is determinable as follows:

$$L(E(i,j)|\lambda) = \text{floor}\left(\frac{E(i,j)}{2^\lambda}\right) + 1 + \lambda. \tag{3.11}$$

Parameter λ is critical for the compression performance as it determines the code length of $E(i,j)$. For a geometric source $\mathbf{S}$ with distribution parameter $\rho \in (0,1)$ (i.e. $Prob(\mathbf{S} = s) = (1-\rho)\rho^s$ for $s = 0,1,2,\ldots$), the optimal coding parameter λ is given as

$$\lambda = \max\left\{0, \text{ceil}\left(\log_2\left(\frac{\log\phi}{\log\rho^{-1}}\right)\right)\right\}, \tag{3.12}$$

where $\phi = (\sqrt{5}+1)/2$ is the golden ratio [29]. Since the expectation value of the source is $\mu = \rho/(1-\rho)$, it follows that

$$\rho = \frac{\mu}{1+\mu}. \tag{3.13}$$

As long as μ is known, parameter ρ and hence the optimal coding parameter λ for the whole source can be determined easily.

To enhance coding efficiency, μ can be estimated adaptively in the course of encoding the mapped residues $E(i,j)$ as follows:

$$\tilde{\mu} = \text{round}\left(\frac{\alpha\tilde{\mu}_p + M_{i,j}}{1+\alpha}\right), \tag{3.14}$$

where $\tilde{\mu}$ is the current estimate of μ for selecting λ to determine the codeword format of current $E(i,j)$. The weighting factor α specifies the significance of $\tilde{\mu}_p$ and $M_{i,j}$ when updating $\tilde{\mu}$. The term $\tilde{\mu}_p$, initially set as zero for all residue subplanes, is the previous estimate of $\tilde{\mu}$ which is updated for each $E(i,j)$. The term

$$M_{i,j} = \frac{1}{4}\sum_{(a,b)\in\xi_{i,j}} E(a,b) \tag{3.15}$$

denotes the local mean of $E(i,j)$ in a region defined as the set $\xi_{i,j}$ of four processed pixel locations which are closest to (i,j) and, at the same time, possess samples from the same color channel as that of (i,j). When coding the residues from the luminance subimage, $\xi_{i,j} = \{(i,j-2),(i-1,j-1),(i-2,j),(i-1,j+1)\}$. When coding the residues from the chrominance subimage, $\xi_{i,j} = \{(i,j-2),(i-2,j-2),(i-2,j),(i-2,j+2)\}$.

The value of the weighting factor α is determined empirically. Figure 3.6 shows the effect of α on the final compression ratio of the above compression algorithm. Curves "G" and "R and B" respectively show the cases when coding the residues from the luminance and the chrominance subimages. The curve marked as "all" depicts the overall performance when all residue subplanes are compressed with a common α value. This figure indicates that $\alpha = 1$ can provide good compression performance.

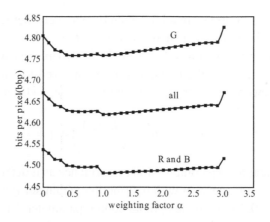

FIGURE 3.6

Average output bit rates (in bpp) versus different α values.

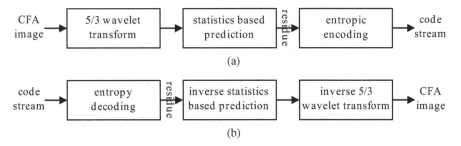

FIGURE 3.7

Workflow of the statistic-based lossless CFA image compression method: (a) encoder, and (b) decoder.

3.5 Compression Based on Statistical Learning

The algorithm presented in this section uses transform, predictive, and entropy coding. The use of transform coding follows an observation that a simple one-level two-dimensional integer Mallat wavelet packet transform can effectively decorrelate the CFA data [23], [24]. Accordingly, the algorithm converts the input CFA image to four subbands with a simple one-level reversible 5/3 wavelet transform and encodes each subband separately by scanning subband coefficients in a raster scan order and predicting each coefficient with its four causal neighboring coefficients by using a statistic-based prediction scheme. The prediction residue is encoded with adaptive Rice code. The decoding process is just the reverse process of encoding. Figure 3.7 shows the structure of this compression method.

3.5.1 Statistic-Based Prediction

As already mentioned, coefficients in each subband are handled in a raster scan order. Figure 3.8a shows the arrangement of four coefficients which are closest, in terms of their distance, to c_0, which is the coefficient under consideration. These four neighbors are used to predict the value of c_0 since they are highly correlated with c_0 and available when decoding c_0. The predicted value of c_0 is computed as

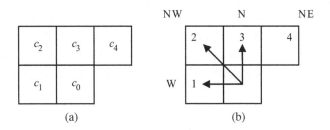

FIGURE 3.8

Subband coefficient c_0 and its causal adjacent neighbors used in statistic-based prediction: (a) causal template, and (b) four possible optimal prediction directions for c_0.

$$\hat{c}_0 = \sum_{k=1}^{4} w_k c_k, \tag{3.16}$$

where w_k is the weight associated with the coefficient c_k. These weights are constrained as $w_1 + w_2 + w_3 + w_4 = 1$ and indicate the likelihood of the neighboring coefficients to have a value closest to that of c_0 under a condition derived based on the local value distribution of the coefficients. The likelihood is estimated with the frequency of the corresponding event occurred so far while processing the coefficients in the same subband.

Using the template shown in Figure 3.8a, the optimal neighbor of c_0 is one which minimizes the difference to c_0, as follows:

$$\arg\min_{c_k} |c_k - c_0|, \text{ for } k = 1, 2, 3, 4. \tag{3.17}$$

When there are more than one optimal neighbors, one of them is randomly selected. The optimal neighbor of c_0 can be located in any of four directions shown in Figure 3.8b. The direction from c_0 to its optimal neighbor is referred to as the optimal prediction direction of c_0 and its corresponding index value is denoted as d_{c_0} hereafter.

The weights needed for predicting c_0 can be determined as follows:

$$w_k = Prob(d_{c_0} = k | d_{c_1}, d_{c_2}, d_{c_3}, d_{c_4}), \text{ for } k = 1, 2, 3, 4, \tag{3.18}$$

where d_{c_j} is the index value of the optimal prediction direction of the coefficient c_j. The term $Prob(d_{c_0} = k | d_{c_1}, d_{c_2}, d_{c_3}, d_{c_4})$ denotes the probability that the optimal prediction direction index of c_0 is k under the condition that d_{c_1}, d_{c_2}, d_{c_3}, and d_{c_4} are known.

Since d_{c_0} and hence $Prob(d_{c_0} = k | d_{c_1}, d_{c_2}, d_{c_3}, d_{c_4})$ are not available during decoding, to predict the current coefficient c_0 in this method the probability is estimated as follows:

$$Prob(d_{c_0} = k | d_{c_1}, d_{c_2}, d_{c_3}, d_{c_4}) = \frac{C(k | d_{c_1}, d_{c_2}, d_{c_3}, d_{c_4})}{\sum_{j=1}^{4} C(j | d_{c_1}, d_{c_2}, d_{c_3}, d_{c_4})}, \text{ for } k = 1, 2, 3, 4, \tag{3.19}$$

where $C(\cdot | \cdot)$ is the current value of a conditional counter used to keep track of the occurrence frequency of k being the optimal prediction direction index of a processed coefficient whose western, northwestern, northern and northeastern neighbors' optimal prediction direction indices are d_{c_1}, d_{c_2}, d_{c_3}, and d_{c_4}, respectively.

Since $d_{c_j} \in \{1, 2, 3, 4\}$ for $j = 1, 2, 3, 4$, there are totally 256 possible combinations of d_{c_1}, d_{c_2}, d_{c_3}, and d_{c_4}. Accordingly, 256×4 counters are required and a table of 256×4 entries needs to be constructed to maintain these counters. This table is initialized with all entries set to one before the compression starts and is updated in the course of the compression. As soon as the coefficient c_0 is encoded, counter $C(d_{c_0} | d_{c_1}, d_{c_2}, d_{c_3}, d_{c_4})$ is increased by one to update the table. With the table which keeps track of the occurrence frequencies of particular optimal prediction direction index values when processing the subband, the predictor can learn from experiences to improve its prediction performance adaptively.

3.5.2 Subband Coding

Entropy coding is used to remove the redundancy carried in the prediction residue of each subband. The prediction error of a subband coefficient is given by

$$e(i, j) = c(i, j) - \hat{c}(i, j), \tag{3.20}$$

where $c(i,j)$ is the real coefficient value and $\hat{c}(i,j)$ is the predicted value of $c(i,j)$. Similar to the approach presented in Section 3.4.3, the error residue $e(i,j)$ is mapped to a nonnegative integer $E(i,j)$ with Equation 3.10 and encoded with adaptive Rice code.

When estimating the expectation value of $E(i,j)$ for obtaining the optimal coding parameter λ for Rice code with Equations 3.12, 3.13, and 3.14, the local mean of $E(i,j)$ is adaptively estimated with the causal adjacent mapped errors as follows:

$$M_{i,j} = \frac{1}{4} \sum_{(m,n) \in \xi_{i,j}} E(m,n), \qquad (3.21)$$

where $\xi_{i,j} = \{E(i-1,j), E(i-1,j-1), E(i-1,j+1), E(i,j-1)\}$. Note that the members of $\xi_{i,j}$ are derived based on the coefficients in the same subband and they all have the same nature. This estimation thus differs from the approach presented in Section 3.4.3 where the members of $\xi_{i,j}$ are obtained using samples from different color channels and hence should not be used to derive the local mean of $E(i,j)$.

With $M_{i,j}$ available, the expectation value of $E(i,j)$ is estimated adaptively using Equation 3.14. Through a training process similar to that discussed in Section 3.4.3, the weighting factor α is set to one. The term $\tilde{\mu}$ is updated for each $E(i,j)$.

3.6 Simulation Results

Simulations were carried out to evaluate the performance of various lossless coding algorithms for CFA images. In the first experiment, twenty-four full-color images (Figure 3.9a) of size 768×512 pixels with eight bits per color component representation from the Kodak test database [30] were sampled using the Bayer pattern to generate the test CFA images. In the second experiment, sixteen real raw Bayer sensor images (Figure 3.9b) captured with Nikon D200 camera were used as the test images. This camera has a 10M-pixel CCD image sensor which can output a raw Bayer sensor image with resolution of 3898×2614 pixels and twelve bits per color component representation. DCRAW software [31] was used to extract raw image data from native Nikon Electronic Format (NEF).

Coding approaches considered here can be classified into three categories:

- Standard or *de facto* standard approaches for compressing a gray-level or full-color image: These approaches exploit various kinds of coding techniques such as LZW coding and wavelet transform coding to reduce the bit rate. However, none of these algorithms is dedicated to handling CFA images.

- Preprocessing-driven approaches which transform a CFA image into a form that can be handled using the first category of approaches: Quincunx separation was used here as the preprocessing step. The approaches based on JPEG-LS and JPEG2000 after quincunx separation are respectively referred to as (S+JLS) [21] and (S+J2K) [32].

- Dedicated compression algorithms for CFA images: This category includes algorithms presented in References [23] and [24], Section 3.4, and Section 3.5 which are referred here to as LCCMI, CP and SL, respectively.

(a)

(b)

FIGURE 3.9

Test images: (a) Kodak image set – referred to as Images 1 to 24 in raster scan order, and (b) real raw Bayer image set captured with a Nikon D200 camera – referred to as Images 25 to 40 in raster scan order.

3.6.1 Bit Rate Comparison

Table 3.1 shows the output bit rates achieved by various coding approaches for the CFA images extracted from the Kodak image set. The coding performance is measured in terms of bits per pixel (bpp). It can be seen that the three dedicated compression algorithms outperform all other evaluated methods remarkably due to their CFA-centric design characteristics.

It can further be observed that the preprocessing step can significantly improve the performance of JPEG-LS [19]. This may be due to the fact that JPEG-LS relies heavily on prediction under the assumption that the spatial correlation of the input image is very high. The preprocessing step makes the assumption valid and hence helps. However, the same preprocessing step does not work when JPEG2000 is used. JPEG2000 is based on wavelet transform coding which is not as sensitive to the validity of the assumption as predictive coding.

TABLE 3.1

Bit rates (in bpp) achieved by various lossless coding approaches for the Kodak image set. The original input is an 8-bit CFA image of size 768×512 pixels.

Image	Lossless coding approaches								
	LZW	PNG	JPEG-LS	JPEG2000	S+JLS	S+J2K	LCCMI	CP	SL
1	8.773	6.472	6.403	5.825	5.944	6.185	5.824	5.497	5.613
2	7.796	6.425	6.787	5.142	4.632	4.847	4.629	4.329	4.513
3	6.843	6.003	5.881	4.225	4.117	4.311	3.965	3.744	3.882
4	8.209	6.621	6.682	4.941	4.827	4.966	4.606	4.367	4.514
5	8.849	6.567	6.470	5.956	6.187	6.516	5.859	5.427	5.586
6	7.598	6.015	5.870	5.219	5.220	5.412	5.139	4.894	4.966
7	7.222	6.085	5.974	4.509	4.426	4.716	4.299	3.989	4.193
8	8.969	6.507	6.295	5.908	6.044	6.408	5.966	5.635	5.716
9	6.590	5.556	5.074	4.400	4.440	4.597	4.319	4.192	4.285
10	6.998	5.768	5.395	4.565	4.558	4.810	4.415	4.226	4.377
11	7.447	5.951	5.370	4.995	5.070	5.319	4.952	4.693	4.790
12	6.942	6.076	5.628	4.494	4.404	4.617	4.308	4.097	4.228
13	9.325	6.762	6.747	6.381	6.568	6.795	6.503	6.130	6.188
14	8.324	6.428	6.288	5.565	5.740	6.011	5.487	5.169	5.261
15	7.622	6.406	6.317	4.666	4.335	4.591	4.396	4.098	4.312
16	6.632	5.448	5.289	4.561	4.724	4.913	4.521	4.387	4.407
17	6.857	5.277	4.965	4.556	4.801	5.018	4.499	4.286	4.384
18	8.672	6.474	6.184	5.579	5.766	5.961	5.538	5.274	5.328
19	7.727	5.905	5.470	4.918	5.084	5.219	4.898	4.747	4.795
20	5.338	4.202	4.317	4.035	3.402	3.685	4.054	3.542	3.902
21	7.292	5.944	5.467	5.048	5.073	5.260	4.983	4.804	4.843
22	8.311	6.410	6.188	5.227	5.239	5.417	5.060	4.842	4.938
23	7.902	7.014	6.827	4.536	4.097	4.205	3.960	3.839	3.936
24	7.857	5.971	5.719	5.232	5.401	5.759	5.257	4.895	5.010
Average	7.671	6.095	5.900	5.020	5.004	5.231	4.893	4.629	4.749

Table 3.2 lists bit rates for various coding approaches evaluated on real raw Bayer sensor images. Similar to the previous experiment, CP provides the lowest output bit rate among all the evaluated coding approaches. Note that the original bit rate is twelve bits per pixel for this test set. The compression ratio is hence around two to one. As compared with the original storage format (NEF), CP can save up to 1.558M bytes per image on average.

For this test database, quincunx separation helps both JPEG-LS and JPEG2000 to reduce the output bit rates; achieving significant improvements against JPEG-LS and even outperforming LCCMI. This can be explained based on the observation that the spatial correlation increases with image resolution. The larger the image size, the higher the spatial correlation of the output of quincunx separation is and the more suitable it is for JPEG-LS compression. Based on this finding, it may be worth exploring if there is a better preprocessing step to allow better coding performance using JPEG-LS.

3.6.2 Complexity Comparison

Table 3.3 allows comparison of various coding approaches in terms of their execution time on a personal computer equipped with a 3.0 GHz Pentium IV processor and 2 GB RAM. For each evaluated coding approach, all CFA images in the same testing set were

TABLE 3.2

Bit rates (in bpp) achieved by various lossless coding approaches for the real raw Bayer sensor image set. The original input is a 12-bit CFA image of size 3898×2614 pixels.

Image	Lossless coding approaches									
	LZW	PNG	JPEG-LS	JPEG2000	S+JLS	S+J2K	LCCMI	CP	SL	NEF
25	11.613	10.828	7.551	6.584	6.535	6.726	6.523	6.172	6.304	7.615
26	10.596	10.344	8.637	5.948	5.470	5.521	5.522	5.382	5.403	6.331
27	11.136	9.839	8.438	6.551	6.108	6.277	6.321	6.085	6.132	7.364
28	11.511	10.447	8.058	6.381	6.024	6.186	6.247	6.014	6.035	7.163
29	10.454	8.935	6.981	5.489	5.137	5.211	5.387	5.295	5.244	6.867
30	12.202	11.147	8.796	6.993	6.742	6.958	6.804	6.483	6.582	7.633
31	11.658	10.242	7.013	6.528	6.364	6.379	6.623	6.408	6.368	6.805
32	11.962	10.944	7.994	6.373	6.198	6.332	6.220	5.970	6.015	7.375
33	11.546	10.725	7.290	6.295	6.145	6.360	6.124	5.840	5.952	7.551
34	12.040	10.727	7.920	6.812	6.596	6.679	6.769	6.537	6.523	7.466
35	12.669	10.882	8.667	6.933	6.621	6.670	6.779	6.505	6.595	8.051
36	11.668	10.809	8.708	6.784	6.526	6.717	6.660	6.395	6.435	7.413
37	11.446	10.530	8.585	6.948	6.633	6.804	6.852	6.607	6.609	7.448
38	10.106	9.925	6.866	5.660	5.460	5.589	5.476	5.288	5.347	6.991
39	11.864	10.955	7.705	6.364	6.192	6.390	6.201	5.925	6.001	7.339
40	11.713	10.552	7.752	6.566	6.534	6.735	6.461	6.127	6.252	8.158
Average	11.512	10.489	7.935	6.450	6.206	6.346	6.311	6.065	6.112	7.348

individually compressed ten times and the average processing time per image was recorded. As can be seen, JPEG-LS is the the most efficient approach among considered approaches. On average, it is almost twice as fast as other approaches among which SL shows the highest efficiency when handling large real raw CFA images. Finally, it should be noted that comparing the processing time for JPEG-LS and S+JLS indicates low complexity of quincunx separation.

3.7 Conclusion

Compressing the CFA image can improve the efficiency of in-camera processing as one can skip the demosaicking process to eliminate the overhead. Without the demosaicking step, no extra redundant information is added to the image to increase the loading of the subsequent compression process. Since digital camera images are commonly stored in the so-called "raw" format to allow their high quality processing on a personal computer, lossless compression of CFA images becomes necessary to avoid information loss. Though a number of lossy compression algorithms have been proposed for coding CFA images [6], [7], [10], [11], [12], [13], [14], [15], [16], [17], [18], only a few lossless compression algorithms have been reported in literature [23], [24], [25], [33].

This chapter revisited some common lossless image coding techniques and presented two new CFA image compression algorithms. These algorithms rely on predictive and entropy coding to remove the redundancy using the spatial correlation in the CFA image and the

TABLE 3.3

Average execution time (in seconds per frame) for compressing images from two test sets.

	Lossless coding approaches						
Image Set	JPEG-LS	JPEG2000	S+JLS	S+J2K	LCCMI	CP	SL
Kodak Set	0.0294s	0.0689s	0.0357s	0.0751s	0.0533s	0.0687s	0.0583s
Real Raw Set	0.7571s	1.7343s	0.9626s	1.9711s	1.8215s	1.8409s	1.6954s

TABLE 3.4

Approaches used in algorithms CP and SL to remove the image redundancy.

	Approach	
Redundancy	CP	SL
Interchannel redundancy	Operates in color difference domain	Uses integer Mallat wavelet packet transform
Spatial redundancy	Linear prediction with 4 neighbors, the neighbor whose context is closer to that of the pixel of interest is weighted more	Linear prediction with 4 neighbors, the neighbor whose value is more likely to be the closest to that of the pixel of interest is weighted more
Statistical redundancy	Adaptive Rice code	Adaptive Rice code

statistical distribution of the prediction residue. Table 3.4 highlights the approaches used in the proposed CP and SL algorithms to remove various kinds of data redundancy.

Extensive experimentation showed that CP provides the best average output bit rates for various test image databases. An interesting finding is that quincunx separation greatly enhances the performance of JPEG-LS. When the input CFA image is large enough, quincunx separation produces gray-level images with strong spatial correlation characteristics and hence JPEG-LS can compress them easily. Considering its low computational complexity, S+JLS could be a potential rival for dedicated CFA image compression algorithms. On the other hand, since S+JLS does not remove the interchannel redundancy during compression, well-designed dedicated CFA coding algorithms which take the interchannel redundancy into account should be able to achieve better compression ratios than S+JLS.

Compressing raw mosaic-like single-sensor images constitutes a rapidly developing research field. Despite recent progress, a number of challenges remain in the design of low-complexity high-performance lossless coding algorithms. It is therefore expected that there will be new CFA image coding algorithms proposed in the near future.

Acknowledgment

We would like to thank the Research Grants Council of the Hong Kong Special Administrative Region, China (PolyU 5123/08E) and Center for Multimedia Signal Processing,

The Hong Kong Polytechnic University, Hong Kong (G-U413) for supporting our research project on the topics presented in this chapter.

References

[1] R. Lukac (ed.), *Single-Sensor Imaging: Methods and Applications for Digital Cameras*. Boca Raton, FL: CRC Press / Taylor & Francis, September 2008.

[2] B.E. Bayer, "Color imaging array," U.S. Patent 3 971 065, July 1976.

[3] K. Hirakawa and T.W. Parks, "Adaptive homogeneity-directed demosaicing algorithm," *IEEE Transactions on Image Processing*, vol. 14, no. 3, pp. 360–369, March 2005.

[4] B.K. Gunturk, J. Glotzbach, Y. Altunbasak, R.W. Schafer, and R.M. Mersereau, "Demosaicking: Color filter array interpolation," *IEEE Signal Processing Magazine*, vol. 22, no. 1, pp. 44–54, January 2005.

[5] K.H. Chung and Y.H. Chan, "Color demosaicing using variance of color differences," *IEEE Transactions on Image Processing*, vol. 15, no. 10, pp. 2944–2955, October 2006.

[6] C.C. Koh, J. Mukherjee, and S.K. Mitra, "New efficient methods of image compression in digital cameras with color filter array," *IEEE Transactions on Consumer Electronics*, vol. 49, no. 4, pp. 1448–1456, November 2003.

[7] N.X. Lian, L. Chang, V. Zagorodnov, and Y.P. Tan, "Reversing demosaicking and compression in color filter array image processing: Performance analysis and modeling," *IEEE Transactions on Consumer Electronics*, vol. 15, no. 11, pp. 3261–3278, November 2006.

[8] D. Menon, S. Andriani, G. Calvagno, and T. Eirseghe, "On the dependency between compression and demosaicing in digital cinema," in *Proceedings of the IEE European Conference on Visual Media Production*, London, UK, December 2005, pp. 104–111.

[9] N.X. Lian, V. Zagorodnov, and Y.P. Tan, *Single-Sensor Imaging: Methods and Applications for Digital Cameras*, ch. Modelling of image processing pipelines in single-sensor digital cameras, R. Lukac (ed.), Boca Raton, FL: CRC Press / Taylor & Francis, September 2008, pp. 381–404.

[10] S.Y. Lee and A. Ortega, "A novel approach of image compression in digital cameras with a Bayer color filter array," in *Proceedings of the IEEE International Conference on Image Processing*, Thessaloniki, Greece, October 2001, pp. 482–485.

[11] T. Toi and M. Ohta, "A subband coding technique for image compression in single ccd cameras with bayer color filter arrays," *IEEE Transactions on Consumer Electronics*, vol. 45, no. 1, pp. 176–180, February 1999.

[12] G.L.X. Xie, Z. Wang, D.L.C. Zhang, and X. Li, "A novel method of lossy image compression for digital image sensors with bayer color filter arrays," in *Proceedings of the IEEE International Symposium on Circuits and Systems*, Kobe, Japan, May 2005, pp. 4995–4998.

[13] Y.T. Tsai, "Color image compression for single-chip cameras," *IEEE Transactions on Electron Devices*, vol. 38, no. 5, pp. 1226–1232, May 1991.

[14] A. Bruna and F. Vella, A. Buemi, and S. Curti, "Predictive differential modulation for CFA compression," in *Proceedings of the 6th Nordic Signal Processing Symposium*, Espoo, Finland, June 2004, pp. 101–104.

[15] S. Battiato, A. Bruna, A. Buemi, and F. Naccari, "Coding techniques for CFA data images," in *Proceedings of IEEE International Conference on Image Analysis and Processing*, Mantova, Italy, September 2003, pp. 418–423.

[16] A. Bazhyna, A. Gotchev, and K. Egiazarian, "Near-lossless compression algorithm for Bayer pattern color filter arrays," *Proceedings of SPIE*, vol. 5678, pp. 198–209, January 2005.

[17] B. Parrein, M. Tarin, and P. Horain, "Demosaicking and JPEG2000 compression of microscopy images," in *Proceedings of the IEEE International Conference on Image Processing*, Singapore, October 2004, pp. 521–524.

[18] R. Lukac and K.N. Plataniotis, "Single-sensor camera image compression," *IEEE Transactions on Consumer Electronics*, vol. 52, no. 2, pp. 299–307, May 2006.

[19] "Information technology - lossless and near-lossless compression of continuous-tone still images (JPEG-LS)," ISO/IEC Standard 14495-1, 1999.

[20] "Information technology - JPEG 2000 image coding system - Part 1: Core coding system," INCITS/ISO/IEC Standard 15444-1, 2000.

[21] A. Bazhyna, A.P. Gotchev, and K.O. Egiazarian, "Lossless compression of Bayer pattern color filter arrays," in *Proceedings of SPIE*, vol. 5672, pp. 378–387, January 2005.

[22] X. Xie, G. Li, and Z. Wang, "A low-complexity and high-quality image compression method for digital cameras," *ETRI Journal*, vol. 28, no. 2, pp. 260–263, April 2006.

[23] N. Zhang and X.L. Wu, "Lossless compression of color mosaic images," *IEEE Transactions on Image Processing*, vol. 15, no. 6, pp. 1379–1388, June 2006.

[24] N. Zhang, X. Wu, and L. Zhang, *Single-Sensor Imaging: Methods and Applications for Digital Cameras*, ch. Lossless compression of color mosaic images and video, R. Lukac (ed.), Boca Raton, FL: CRC Press / Taylor & Francis, September 2008, pp. 405–428.

[25] K.H. Chung and Y.H. Chan, "A lossless compression scheme for Bayer color filter array images," *IEEE Transactions on Image Processing*, vol. 17, no. 2, pp. 134–144, February 2008.

[26] S. Golomb, "Run-length encodings," *IEEE Transactions on Information Theory*, vol. 12, no. 3, pp. 399–401, July 1966.

[27] M. Burrows and D.J. Wheeler, "A block sorting lossless data compression algorithm," Technical Report 124, Digital Equipment Corporation, 1994.

[28] T.A. Welch, "A technique for high-performance data compression," *Computer*, vol. 17, no. 6, pp. 8–19, June 1984.

[29] A. Said, "On the determination of optimal parameterized prefix codes for adaptive entropy coding," Technical Report HPL-2006-74, HP Laboratories Palo Alto, 2006.

[30] R. Franzen, "Kodak lossless true color image suite." Available online: http://r0k.us/graphics/kodak.

[31] D. Coffin, "Decoding raw digital photos in Linux." Available online: http://www.cybercom.net/ dcoffin/dcraw.

[32] A. Bazhyna and K. Egiazarian, "Lossless and near lossless compression of real color filter array data," *IEEE Transactions on Consumer Electronics*, vol. 54, no. 4, pp. 1492–1500, November 2008.

[33] S. Andriani, G. Calvagno, and D. Menon, "Lossless compression of Bayer mask images using an optimal vector prediction technique," in *Proceedings of the 14th European Signal Processing Conference*, Florence, Italy, September 2006.

4

Color Restoration and Enhancement in the Compressed Domain

Jayanta Mukherjee and Sanjit K. Mitra

4.1   Introduction

The quality of an image captured by a camera is influenced primarily by three main factors; namely, the three-dimensional (3D) scene consisting of the objects present in it, the illuminant(s) or radiations received by the scene from various sources, and the camera characteristics (of its optical lenses and sensors). In a typical scenario, the 3D scene and the camera may be considered as invariants, whereas the illumination varies depending on the nature of the illuminant. For example, the same scene may be captured at different times of the day. The pixel values of these images then would be quite different from each other and the colors may also be rendered differently in the scene.

Interestingly, a human observer is able to perceive the true colors of the objects even in complex scenes with varying illumination. Restoration of colors from varying illumination is also known as solving for color constancy of a scene. The objective of the computation of color constancy is to derive an illumination-independent representation of an image, so that it could be suitably rendered with a desired illuminant. The problem has two components, one in estimating the spectral component of the illuminant(s) and the other one in performing the color correction for rendering the image with a target illumination. The latter task is usually carried out by following the Von Kries equation of diagonal correction [1].

Another factor involved in the visualization of a color image is the color reproduction capability of a display device depending upon its displayable range of color gamut and the brightness range it can handle. The captured image may also have a poor dynamic range of brightness values due to the presence of strong background illumination. In such situations, for good color rendition in a display device, one may have to enhance an image. This enhancement is mostly done independent of solving for color constancy; as usually, the illuminants are assumed to be invariant conventional sources. However, one may require to apply both color correction and enhancement for the display of color images, when the scene suffers from both widely varying spectral components and brightness of illuminants.

Several methods have been reported in the literature for solving these problems, mostly in the spatial representation of images. However, as more and more imaging devices are producing end results in the compressed domain, namely in the block discrete cosine transform (DCT) space of Joint Photographic Experts Group (JPEG) compression, it is of interest to study these methods in that domain. The primary objective for processing these images directly in the compressed domain is to reduce the computational and storage requirements. Due to the processing of the compressed images in their own domain, the computational overhead on inverse and forward transforms of the spatial domain data into a compressed domain gets eliminated by this process. In particular, processing in the DCT domain has drawn significant attention of the researchers due to its use in the JPEG and Moving Picture Experts Group (MPEG) compression standards. There are also other advantages of using compressed domain representation. One may exploit the spectral separation of the DCT coefficients in designing these algorithms.

This chapter discusses the two above aspects of color restoration. Unlike previous work which dealt with color correction and color enhancement of images represented in the block DCT space of JPEG compression independently, this chapter presents the color restoration task as a combination of these two computational stages. Here, restoration of colors is not considered from a noisy environment, the attention is rather focused on the limitation of sensors and display devices due to varying illumination of a scene.

The following section presents the fundamentals related to the block DCT space. These are required to understand and to design algorithms in this space. Next, the color constancy problem is introduced and different methods for solving this problem with the DCT coefficients [2] are discussed. This is followed by the discussion on color enhancement in the compressed domain. A simple approach based on scaling of DCT coefficients is also elaborated here. Finally, some examples of color restoration using both color correction and enhancement are shown and discussed.

4.2 DCT: Fundamentals

The DCT of an $N \times N$ block of a two-dimensional (2D) image $\{x(m,n), 0 \leq m \leq N - 1, 0 \leq n \leq N - 1\}$ is given by

$$C(k,l) = \frac{2}{N} \alpha(k)\alpha(l) \sum_{m=0}^{N-1} \sum_{n=0}^{N-1} \left(x(m,n) \cdot \cos\left(\frac{(2m+1)\pi k}{2N}\right) \cos\left(\frac{(2n+1)\pi l}{2N}\right) \right), \quad (4.1)$$

where $0 \leq k, l \leq N - 1$ and $\alpha(p)$ is defined as

$$\alpha(p) = \begin{cases} \sqrt{\frac{1}{2}} & \text{for } p = 0, \\ 1 & \text{otherwise.} \end{cases} \quad (4.2)$$

The $C(0,0)$ coefficient is the DC coefficient and the rest are called AC coefficients for a block. The normalized transform coefficients $\hat{c}(k,l)$ are defined as

$$\hat{c}(k,l) = \frac{C(k,l)}{N}. \quad (4.3)$$

Let μ and σ denote the mean and standard deviation of an $N \times N$ image. Then μ and σ are related to the normalized DCT coefficients as given below:

$$\mu = \hat{c}(0,0), \quad (4.4)$$

$$\sigma = \sqrt{\sum_{k=0}^{N-1} \sum_{l=0}^{N-1} \hat{c}(k,l)^2 - \mu^2}. \quad (4.5)$$

In fact, from Equation 4.5, it is obvious that the sum of the square of the normalized AC coefficients provides the variance of the image. Hence, any change in the DC component does not have any bearing on its standard deviation (σ). These two statistical measures computable directly in the compressed domain, are quite useful for designing algorithms of color constancy and enhancement. Moreover, there exist two interesting relationships between the block DCT coefficients, namely the relationship between the coefficients of adjacent blocks [3] and between the higher order coefficients and the lower ones (or subband relationship) [4]. Using these relationships, one may efficiently compose or decompose DCT blocks, or perform interpolation or decimation operations. For details, refer to the discussion in Reference [5].

4.3 Computation of Color Constancy

The major challenge in the computation of the color constancy is to estimate the spectral components of the illuminant, mainly three components in the Red (R), Green (G) and Blue (B) spectral zones. Many techniques have been reported to address this problem;

comprehensive surveys are available in References [6], [7], and [8]. All these techniques solve the color constancy problem in the spatial representation of the images in the *RGB* color space. In this chapter, the solution of this problem is considered in the block DCT space. Moreover, since the color space in JPEG compression is *YCbCr*, the Von Kries model will be adapted in the YCbCr space and demonstrated how this model could be further simplified for obtaining reasonably good results with less computation.

In a simplified model [6], assuming all the reflecting bodies as ideal 2D flat Lambertian surfaces, the brightness $I(x)$ at an image coordinate x is related to the illuminant property of the surface and camera sensor as follows:

$$I(x) = \int_\omega E(\lambda) R^X(\lambda) S(\lambda) d\lambda, \qquad (4.6)$$

where $E(\lambda)$ is the spectral power distribution (*SPD*) of the incident illuminant, X is the surface point projected on x, $R^X(\lambda)$ represents the surface reflectance spectrum at that point, and $S(\lambda)$ is the relative spectral response of the sensor. The responses are accumulated over the range of wavelength ω on which the sensors are active. In this chapter, it is assumed that there is a single illuminant for a scene.

In an optical color camera with three sensors, each sensor operates on different zones of the optical wavelengths namely with small wavelengths (Blue zone), mid wavelength range (Green zone) and large wavelengths (Red zone). Computation for color constancy involves estimating $E(\lambda)$ from these *three* responses. Typically, the SPD of the illuminant for the three different zones of the optical range of wavelengths needs to be estimated. It is explained in Reference [6] that the problem is underconstrained (the number of unknowns are more than the number of observations). That is why many researchers have taken additional assumption to reduce the number of unknowns.

Namely, in the *gray world assumption* [9], [10], average reflectance of all surfaces is taken as gray or achromatic. Hence, the average of color components provides the colors of the incident illuminant. This approach has been also extended in the *gradient space* of images [11], where it is assumed that the average edge difference in a scene is achromatic. This hypothesis is termed the *gray edge hypothesis*. Some researchers [12] assume the existence of a white object in the scene; this assumption is referred to as the *white world assumption*. In this case, the maximum values of individual color components provide the colors of the incident illuminant. However, the method is very much sensitive over the dynamic ranges of the sensors; although, given a scene whose dynamic range of brightness distribution is in accordance with the linear response of the sensor, this assumption works well in many cases. More recent trend on solving color constancy problem is to use statistical estimation techniques with prior knowledge on the distribution of pixels in a color space given known camera sensors and source of illumination. In these techniques, an illuminant (or a set of illuminants) is chosen from a select set of canonical illuminants based on certain criteria. There are *color gamut mapping* approaches both in the 3D [13] and the 2D [14] color spaces, where one tries to maximize the evidence of color maps with known color maps of canonical illuminants. Reference [6] reports a *color by correlation* technique which attempts to maximize a likelihood of an illuminant given the distribution of pixels in the 2D chromatic space. The same work shows that many other algorithms, like the *gamut mapping* method in the 2D chromatic space, could be implemented under the same framework. Note that all these techniques require significant amount of storage space for storing the statistics of each canonical illuminant. As one of the motivations in this chapter is to reduce the storage requirement for efficient implementation in the block DCT space, a simple nearest neighbor (NN) classification approach for determining the canonical illuminant has also been explored. Interestingly, it is found that the *nearest neighbor* classification in the 2D chromatic space performs equally well as other existing techniques such as color by correlation [6] or gamut mapping [14].

Once the SPDs of the illuminant in three spectral zones are estimated, the next step is to convert the pixel values to those under a target illuminant (which may be fixed to a standard illumination). This computation is performed by the diagonal color correction following the Von Kries model [1]. Let R_s, G_s and B_s be the spectral components for the source illuminant (for Red, Green and Blue zones). Let the corresponding spectral components for the target illuminant be R_d, G_d and B_d. Then, given a pixel in RGB color space with R, G, and B as its corresponding color components, the updated color components, R_u, G_u, and B_u, are expressed as follows:

$$k_r = \frac{R_d}{R_s}, \quad k_g = \frac{G_d}{G_s}, \quad k_b = \frac{B_d}{B_s},$$
$$f = \frac{R+G+B}{k_r R + k_g G + k_b B}, \tag{4.7}$$
$$R_u = f k_r R, \quad G_u = f k_g G, \quad B_u = f k_b B.$$

The next section discusses the usage of these techniques in the block DCT domain.

4.4 Color Constancy with DCT coefficients

All the spatial domain techniques for solving the color constancy could be extended in the block DCT space by treating the array of DC coefficients only, which is a low resolution or subsampled version of the original image. However, there are two things one needs to take care of during this treatment. First, the color space used in the JPEG standard is $YCbCr$. Next, one should take care of visible blocking artifacts, if any, due to the independent treatment of the DC coefficients in each block. However, as the estimation of the spectral components of the illuminant takes into account of all the DC coefficents together, blocking artifacts do not make their appearances in most cases. These can be established by using a measure of the quality of JPEG images in the experimentation.

4.4.1 Color Constancy in the YCbCr Color Space and Proposed Variations

The $YCbCr$ color space is related to the RGB space as given below:

$$Y = 0.502G + 0.098R + 0.256B,$$
$$Cb = -0.290G + 0.438R - 0.148B + 128, \tag{4.8}$$
$$Cr = -0.366G - 0.071R + 0.438B + 128,$$

assuming eight bits for each color component.

For implementing the gray world algorithm in the $YCbCr$ space, one can directly obtain the *mean* values by computing the means of the DC coefficients in individual Y, Cb and Cr components. However, finding the maximum of a color component is not a linear operation. Hence, for the white world algorithm, one needs to convert all the DC coefficients in the RGB space and then compute their maximum values. To this end, a simple heuristic can be used; it is assumed here that the color of the maximum luminance value is the color of the illuminant. This implies that only the maximum in the Y component is computed and the corresponding Cb and Cr values at that point provide the color of the illuminant. This significantly reduces the computation, as it does not require conversion of DC values from the $YCbCr$ space to RGB space. Further, the maximum finding operation is restricted to one component only. This assumption is referred to as white world in YCbCr.

With regard to the statistical techniques, the color by correlation technique [6] and the gamut mapping approach in 2D chromatic space [14] were adapted here for use in the YCbCr space. Naturally, CbCr was chosen as the chrominance space instead of rg space as used in Reference [6], where $r = R/(R+G+B)$ and $g = G/(R+G+B)$. This space was discretized into 32×32 cells to accumulate the distribution of pixels. A new statistical approach based on the nearest neighbor classification was also explored; this approach will be described in the following subsection. Note that there are other techniques, such as neural network-based classification [15], probabilistic approaches [16], [17], [18], and gamut mapping in the 3D color space [13], which are not considered in this study.

4.4.2   Color by Nearest Neighbor Classification

In this method, the computation is only performed in the 2D chromatic space. Let $C \in Cb \times Cr$ denote the SPD of an illuminant in the $CbCr$ space which follows a 2D Gaussian distribution as described below:

$$p(C) = \frac{1}{2\pi|\Sigma|^{\frac{1}{2}}} e^{-\frac{1}{2}(C-\mu)\Sigma^{-1}(C-\mu)^t}, \tag{4.9}$$

where $\mu (= [\mu_{Cb}\ \mu_{Cr}])$ is the mean of the distribution and Σ is the covariance matrix defined as

$$\Sigma (= \begin{bmatrix} \sigma_{Cb}^2 & \sigma_{CbCr} \\ \sigma_{CbCr} & \sigma_{Cr}^2 \end{bmatrix}).$$

Following the Bayesian classification rule and assuming that all the illuminants are equally probable, a minimum distance classifier can be designed. Let the mean chromatic components of an image be C_m. Then, for an illuminant L with the mean μ_L and the covariance matrix Σ_L, the distance function for the nearest neighbor classifier is nothing but the *Mahalanobis distance function* [19], as defined below:

$$d(C_m, \mu_L) = (C_m - \mu_L)\Sigma_L^{-1}(C_m - \mu_L)^t. \tag{4.10}$$

4.4.3   Color Correction in the YCbCr space

In the JPEG standard, the DCT coefficients of the color components are in the YCbCr color space. As a result, it would be necessary to convert every DC coefficient to the

RGB color space for applying the diagonal correction of Equation 4.7. Additionally it also would be necessary to transform back the updated color values to the YCbCr space. The color space transformations can be avoided by performing the diagonal correction directly in the YCbCr space as outlined in the following theorem.

Theorem 4.1

Let k_r, k_g, and k_b be the parameters for diagonal correction as defined in Equation 4.7. Given a pixel with color values in the YCbCr color space, the updated color values Y_u, C_{bu}, and C_{ru} are expressed by the following equations:

$$C_b' = C_b - 128,$$
$$C_r' = C_r - 128,$$
$$f = \frac{3.51Y + 1.63C_b' + 0.78C_r'}{1.17(k_r + k_g + k_b)Y + (2.02k_b - 0.39k_g)C_b' + (1.6k_r - 0.82k_g)C_r'},$$
$$Y_u = f((0.58k_g + 0.12k_b + 0.30k_r)Y + 0.2(k_b - k_g)C_b' + 0.41(k_r - k_g)C_r'),$$
$$C_{bu} = f((0.52k_b - 0.34k_g - 0.18k_r)Y + (0.11k_g + 0.89k_b)C_b' + 0.24(k_g - k_r)C_r') + 128,$$
$$C_{ru} = f((0.52k_r - 0.43k_g - 0.09k_b)Y + 0.14(k_g - k_b)C_b' + (0.3k_g + 0.7k_r)C_r') + 128.$$

The proof of the above theorem is straightforward. However, one should note that the number of multiplications and additions in the above equations does not get reduced compared to the diagonal correction method applied in the RGB color space. □

4.4.4 Color Correction by Chromatic Shift

Because the chromatic components are more decorrelated in the YCbCr space, one may apply the heuristics of color correction using the shift in the chromatic components of a target illuminant with respect to the source illuminant. In fact, Reference [20] reports that with this simple measure good quality of color rendition (or transfer) is possible. Let Y_d, C_{bd} and C_{rd} be the color components of a target illuminant in the YCbCr space and the corresponding components in the source illuminant are Y_s, C_{bs} and C_{rs}. Then color correction by the *chromatic shift* (CS) is expressed by the following equation.

$$\begin{aligned} Y_u &= Y, \\ C_{bu} &= C_b + C_{bd} - C_{bs}, \\ C_{ru} &= C_r + C_{rd} - C_{rs}. \end{aligned} \tag{4.11}$$

4.5 Benchmark Dataset and Metrics of Comparison

The image dataset [21] captured using different illuminants was used to evaluate the performances of different algorithms. This dataset is available at http://www.cs.sfu.ca/~colour/data. For every scene, it also provides an estimate of the spectral components of the illuminant in the RGB color space. These were used here for collecting the statistics (means and covariance matrices of the SPD of the illuminants) related to the nearest neighbor classification-based technique. Further, from the chromatic components of the images

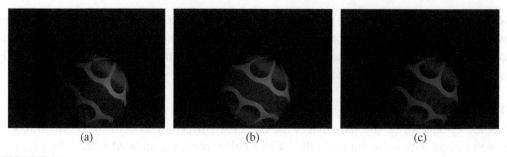

(a)   (b)   (c)

FIGURE 4.1

Images of the same scene (*ball*) captured under three different illuminants: (a) ph-ulm, (b) syl-50mr16q, and (c) syl-50mr16q+3202.

captured under different illuminants statistics related to the color by correlation and gamut mapping techniques are formed. It should be mentioned that all these techniques provide the estimate of an illuminant as its mean SPD in the RGB or YCbCr color space.

Experiments were performed using the images with objects having minimum specularity. Though the scenes are captured at different instances, it was observed that the images are more or less registered. The images are captured by three different fluorescent lights, four different incandescent lights and also each of them in conjunction with a blue filter (Roscolux 3202). In these experiments, Sylvania 50MR16Q is taken as the target illuminant, as it is quite similar to a regular incandescent lamp. The list of different illuminants is given in Table 4.1. Figure 4.1 shows a typical set of images for the same scene under some of these illuminants.

Different metrics were used to compare the performances of all techniques in the block DCT domain as well as their performances with respect to different spatial domain algorithms. Four metrics described below, reported earlier in References [7] and [8], were also used for studying the performances of different algorithms in estimating the spectral components of illuminants. Let the target illuminant T be expressed by the spectral component triplet in the RGB colors-pace as (R_T, G_T, B_T) and let the corresponding estimated illuminant be represented by $E = (R_E, G_E, B_E)$. The respective illuminants in the (r, g) chromatic

TABLE 4.1
List of illuminants.

Name	Nature of Source	Short Name
Philips Ultralume	Fluorescent	ph-ulm
Sylvania cool white	Fluorescent	syl-cwf
Sylvania warm white	Fluorescent	syl-wwf
Sylvania 50MR16Q	Incandescent	syl-50mr16q
Sylvania 50MR16Q with blue filter		syl-50mr16q+3202
Lamp at 3500K temperature	Incandescent	solux-3500
Lamp at 3500K with blue filter		solux-3500+3202
Lamp at 4100K temperature	Incandescent	solux-4100
Lamp at 4100K with blue filter		solux-4100+3202
Lamp at 4700K temperature	Incandescent	solux-4700
Lamp at 4700K with blue filter		solux-4700+3202

space can be expressed as $(r_T, g_T) = (R_T/S_T, G_T/S_T)$ and $(r_E, g_E) = (R_E/S_E, G_E/S_E)$, where $S_T = R_T + G_T + B_T$ and $S_E = R_E + G_E + B_E$. Then, different performance metrics can be defined as follows:

$$
\begin{aligned}
\Delta\theta &= cos^{-1}\left(\frac{T \circ E}{|T||E|}\right), \\
\Delta_{rg} &= |(r_T - g_T, r_E - g_E)|, \\
\Delta_{RGB} &= |T - E|, \\
\Delta_L &= |S_T - S_E|,
\end{aligned}
\tag{4.12}
$$

where Δ_θ, Δ_{rg}, Δ_{RGB}, and Δ_L denote the angular, rg, RGB, and luminance error, respectively. In the above definitions, $\circ$ denotes the dot product between two vectors and $|.|$ denotes the magnitude of the vector.

Next, the performances on image rendering were studied for different algorithms after applying the color correction with the estimated illuminants. It was observed that the images in the dataset are roughly registered. In the experiment, the image captured at the target illuminant (syl-50mr16q) is considered to be the reference image. The images obtained by applying different color constancy algorithms are compared with respect to this image. Two different measures were used for this purpose; the usual PSNR measure and the so-called WBQM similarity measure proposed in Reference [22]. The latter measure was used because the reference images are not strongly registered. For two distributions x and y, the WBQM between these two distributions is defined as follows:

$$
WBQM(x, y) = \frac{4\sigma_{xy}^2 \bar{x}\bar{y}}{(\sigma_x^2 + \sigma_y^2)(\bar{x}^2 + \bar{y}^2)},
\tag{4.13}
$$

where σ_x and σ_y are the standard deviations of x and y, respectively, $\bar{x}$ with $\bar{y}$ denoting their respective means, and σ_{xy}^2 is the covariance between x and y. It may be noted that this measure takes into account the correlation between the two distributions and also their proximity in terms of brightness and contrast. The WBQM values should lie in the interval $[-1, 1]$. Processed images with WBQM values closer to one are more similar in quality according to human visual perception. Applying WBQM independently to each component in the YCbCr space provides Y-WBQM, Cb-WBQM, and Cr-WBQM, respectively.

Reference [23] suggests another *no reference* metric, called here as *JPEG quality metric* (JPQM), for judging the image quality reconstructed from the block DCT space to take into account of visible blocking and blurring artifacts. To measure the quality of the images obtained by DCT domain algorithms, the source code available at http://anchovy.ece.utexas.edu/~zwang/research/nr_jpeg_quality/index.html was used to compute JPQM values. It should be noted that for an image with a good visual quality, the JPQM value should be close to ten.

4.6 Experimental Results on Computation of Color Constancy

For syl-50mr16q as the target illuminant, different color constancy algorithms were used to render the images captured under other illuminants. Table 4.2 lists the algorithms implemented either in spatial domain or block DCT domain. In subsequent discussion, these

TABLE 4.2
List of algorithms for estimating the color components of an illuminant. © 2009 IEEE

Algorithmic Approach	Domain	Short Name
Gray world	Spatial	GRW
Gray world	Block DCT	GRW-DCT
White world	Spatial	MXW
White world in RGB	Block DCT	MXW-DCT
White world in YCbCr	Block DCT	MXW-DCT-Y
Color by correlation	Spatial	COR
Color by correlation	Block DCT	COR-DCT
Gamut mapping	Spatial	GMAP
Gamut mapping	Block DCT	GMAP-DCT
Nearest neighbor	Spatial	NN
Nearest neighbor	Block DCT	NN-DCT

algorithms are referred to by their short names as given in the table. These are followed by any one of the two color correction techniques, which is either the diagonal correction (DGN) method or the chromatic shift (CS) method, as discussed previously.

4.6.1 Estimation of Illuminants

Tables 4.3 to 4.6 compare the performance of different algorithms in estimating the illuminants using errors defined earlier in Equation 4.12. The average errors are presented for each category of illuminant separately, making the performances of different techniques for estimating illuminants of different nature easier to compare. Note that the compressed domain techniques are grouped in the upper portion of the tables whereas the results related to the spatial domain techniques are presented in the bottom of the tables. For each group the best performances are highlighted by bold numerals.

TABLE 4.3
Average Δ_θ for different techniques and various illuminants (IL).[†]

Method	IL_1	IL_2	IL_3	IL_4	IL_5	IL_6	IL_7	IL_8	IL_9	IL_{10}	IL_{11}
GRW-DCT	13.28	11.57	14.90	11.22	22.09	13.19	28.07	15.44	31.70	10.39	18.87
MXW-DCT	11.28	9.98	12.16	7.32	21.52	12.83	25.47	17.47	28.30	7.62	19.55
MXW-DCT-Y	15.26	12.94	17.80	8.30	22.52	13.55	27.55	16.34	29.33	6.04	19.18
COR-DCT	**4.30**	8.84	**5.91**	**5.74**	19.38	9.02	21.96	**10.08**	20.31	**0.43**	12.69
GMAP-DCT	4.65	9.99	6.57	6.88	16.40	**8.77**	16.71	11.42	16.17	**0.43**	**10.09**
NN-DCT	7.69	**6.94**	9.32	10.78	**14.65**	12.86	**12.91**	12.56	**14.33**	**0.43**	13.15
GRW	10.79	11.77	15.13	10.33	21.31	12.45	27.69	15.31	29.38	9.64	18.56
MXW	29.83	26.65	30.66	26.71	29.90	26.07	32.46	27.30	33.73	27.73	28.30
COR	4.64	**10.03**	7.28	7.04	21.32	10.06	23.39	**9.97**	22.05	**0.43**	15.89
GMAP	**4.36**	10.93	**6.93**	**5.98**	20.74	11.45	21.00	12.42	20.84	**0.43**	15.09
NN	8.90	13.17	7.50	7.77	**11.71**	**9.60**	**12.22**	13.35	**12.87**	**0.43**	**10.01**

[†] IL_1: ph-ulm, IL_2: syl-cwf, IL_3: syl-wwf, IL_4: syl-50mr16q, IL_5: syl-50mr16q+3202, IL_6: solux-3500, IL_7: solux-3500+3202, IL_8: solux-4100, IL_9: solux-4100+3202, IL_{10}: solux-4700, IL_{11}: solux-4700+3202.

TABLE 4.4
Average Δ_{rg} for different techniques and various illuminants (IL).[†]

Method	IL_1	IL_2	IL_3	IL_4	IL_5	IL_6	IL_7	IL_8	IL_9	IL_{10}	IL_{11}
GRW-DCT	0.093	0.081	0.119	0.087	0.161	0.100	0.205	0.114	0.239	0.075	0.136
MXW-DCT	0.081	0.082	0.089	0.058	0.197	0.112	0.244	0.160	0.278	0.056	0.188
MXW-DCT-Y	0.100	0.080	0.146	0.058	0.157	0.093	0.197	0.113	0.213	0.039	0.131
COR-DCT	**0.036**	0.060	**0.056**	**0.044**	0.129	0.062	0.148	**0.071**	0.137	**0.003**	0.086
GMAP-DCT	0.038	0.068	0.061	0.051	0.108	**0.060**	0.109	0.079	0.107	**0.003**	**0.066**
NN-DCT	0.058	**0.054**	0.085	0.082	**0.102**	0.093	**0.090**	0.089	**0.099**	**0.003**	0.092
GRW	0.078	0.081	0.135	0.078	0.146	0.087	0.191	0.107	0.206	0.077	0.127
MXW	0.199	0.186	0.211	0.176	0.271	0.195	0.317	0.217	0.339	0.177	0.240
COR	0.038	**0.064**	0.070	0.049	0.143	**0.065**	0.158	**0.066**	0.151	**0.003**	0.103
GMAP	**0.037**	0.073	**0.064**	**0.045**	0.139	0.078	0.142	0.086	0.142	**0.003**	0.100
NN	0.073	0.096	0.074	0.064	**0.082**	0.072	**0.085**	0.095	**0.090**	**0.003**	**0.073**

TABLE 4.5
Average Δ_{RGB} for different techniques and various illuminants (IL).[†]

Method	IL_1	IL_2	IL_3	IL_4	IL_5	IL_6	IL_7	IL_8	IL_9	IL_10	IL_11
GRW-DCT	247.5	243.6	243.8	217.1	246.2	228.1	259.7	250.2	259.4	229.9	251.3
MXW-DCT	114.6	121.7	108.7	93.4	155.1	122.9	173.9	151.9	176.6	96.6	155.1
MXW-DCT-Y	139.7	145.0	143.0	105.2	179.8	140.4	205.2	171.3	209.8	109.0	181.0
COR-DCT	**60.0**	71.2	**70.1**	63.9	117.6	**74.1**	128.3	**75.0**	122.8	**52.9**	92.5
GMAP-DCT	60.6	71.0	75.2	65.1	102.8	75.8	103.2	83.4	102.1	**52.9**	**83.4**
NN-DCT	81.4	**65.0**	74.3	78.6	**99.6**	89.6	**87.4**	87.4	**89.8**	**52.9**	91.7
GRW	241.5	237.1	239.7	209.0	241.0	220.4	254.7	244.1	254.1	224.5	246.2
MXW	171.4	164.4	162.9	147.9	172.1	148.4	190.0	172.8	194.0	150.5	171.2
COR	60.7	76.5	**71.6**	67.2	122.2	78.9	135.1	**73.4**	131.7	**52.9**	98.9
GMAP	**60.3**	**75.7**	77.6	**64.5**	130.4	**77.9**	126.2	84.4	126.0	**52.9**	103.8
NN	79.9	88.1	72.6	71.5	**84.8**	78.3	**86.3**	90.9	**85.2**	**52.9**	81.8

TABLE 4.6
Average Δ_L for different techniques and various illuminants (IL).[†]

Method	IL_1	IL_2	IL_3	IL_4	IL_5	IL_6	IL_7	IL_8	IL_9	IL_10	IL_11
GRW-DCT	426.8	420.4	420.8	374.0	420.4	391.8	443.5	429.5	443.7	397.1	432.8
MXW-DCT	179.6	191.8	169.3	147.0	224.6	187.9	259.4	231.0	259.8	156.9	244.4
MXW-DCT-Y	218.0	232.4	224.9	167.4	279.9	224.3	328.3	274.4	335.1	182.6	301.6
COR-DCT	95.0	87.0	105.9	95.6	135.0	95.4	146.5	**95.1**	146.3	**91.3**	119.1
GMAP-DCT	**93.5**	**81.0**	114.1	**89.9**	**116.4**	94.6	114.8	105.4	120.2	**91.3**	**108.3**
NN-DCT	125.4	89.2	**94.0**	95.1	121.5	107.6	**105.9**	107.8	**96.5**	**91.3**	113.1
GRW	416.7	409.2	413.9	359.9	410.8	378.3	434.1	418.7	433.7	388.0	423.8
MXW	189.5	194.8	159.3	129.3	200.3	159.7	237.4	204.3	255.7	127.4	223.0
COR	**93.3**	90.8	**100.5**	**91.2**	133.7	96.6	153.3	**90.1**	155.5	**91.3**	116.3
GMAP	94.8	**87.9**	117.2	93.3	158.5	**84.4**	150.0	102.0	150.3	**91.3**	126.7
NN	114.2	102.9	101.2	100.5	**101.9**	102.7	**111.9**	110.5	**103.8**	**91.3**	**109.7**

† IL_1: ph-ulm, IL_2: syl-cwf, IL_3: syl-wwf, IL_4: syl-50mr16q, IL_5: syl-50mr16q+3202, IL_6: solux-3500, IL_7: solux-3500+3202, IL_8: solux-4100, IL_9: solux-4100+3202, IL_{10}: solux-4700, IL_{11}: solux-4700+3202.

TABLE 4.7
Overall average performances of different techniques on estimating illuminants. © 2009 IEEE

Technique	Error Measure			
	Δ_θ	Δ_{rg}	Δ_{RGB}	Δ_L
GRW-DCT	17.344	0.128	243.390	418.313
MXW-DCT	15.777	0.141	133.729	204.739
MXW-DCT-Y	17.169	0.121	157.243	251.764
COR-DCT	10.790	0.076	84.446	110.245
GMAP-DCT	**9.829**	**0.068**	**79.654**	**102.726**
NN-DCT	10.515	0.077	81.642	104.362
GRW	16.582	0.119	264.797	407.961
MXW	29.036	0.230	167.849	197.400
COR	12.014	0.083	88.150	110.300
GMAP	11.839	0.083	89.105	114.263
NN	**9.780**	**0.073**	**79.338**	**104.661**

Comparing different error measures in estimating the illuminants reveals that statistical techniques perform better than the others in both the compressed domain and spatial domain. It is also noted that recovering illuminants in conjunction with the blue filter is more difficult than those without it. Moreover, with the blue filter, in most cases proposed nearest neighbor classification-based techniques (NN and NN-DCT) perform better than the others. The proposed technique is found to be equally good with respect to the other statistical techniques such as COR and GMAP. Finally, as shown in Table 4.7 which indicates the overall performances of all considered techniques in estimating the illuminants, the GMAP technique is found to have the best performance in the block DCT domain values whereas the NN algorithm tops the list in all respects in the spatial domain.

4.6.2 Performances After Color Correction

The following presents a few typical results for transferring images to the target illuminant. In order to demonstrate the quality of rendering, examples for rendering scenes of three typical illuminants are considered; namely, ph-ulm which is a fluorescent light source, solux-4100 which is an incandescent light source, and syl-50mr16q+3202 which is

    (a)   (b)   (c)

FIGURE 4.2 (See color insert.)

Target reference images captured under syl-50mr16q: (a) *ball*, (b) *books*, and (c) *Macbeth*.

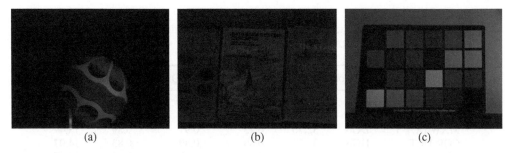

FIGURE 4.3 (See color insert.)

Source images captured under different illuminations: (a) *ball* under solux-4100, (b) *books* under syl-50mr16q+3202, and (c) *Macbeth* under ph-ulm.

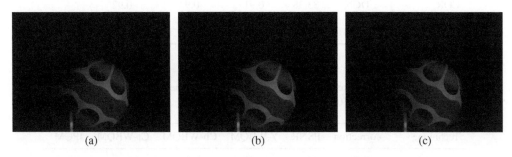

FIGURE 4.4 (See color insert.)

Color corrected images for the illuminant solux-4100: (a) MXW-DCT-Y, (b) COR, and (c) COR-DCT.

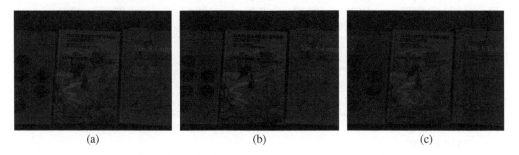

FIGURE 4.5

Color corrected images for the illuminant syl-50mr16q+3202: (a) MXW-DCT-Y, (b) COR, and (c) COR-DCT.

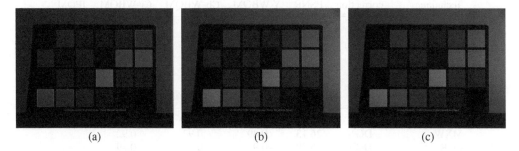

FIGURE 4.6 (See color insert.)

Color corrected images for the illuminant ph-ulm: (a) MXW-DCT-Y, (b) COR, and (c) COR-DCT.

TABLE 4.8
Performances of different techniques for color correction of the *ball* from the illuminant solux-4100 to syl-50mr16q.

Technique	correction	PSNR	Y-WBQM	Cb-WBQM	Cr-WBQM	JPQM
GRW-DCT	DGN	32.85	0.93	0.91	0.90	13.32
MXW-DCT	DGN	32.23	0.93	0.85	0.88	12.60
MXW-DCT-Y	DGN	32.76	0.93	0.91	0.89	13.47
COR-DCT	DGN	31.29	0.88	0.89	0.83	14.07
GMAP-DCT	DGN	31.29	0.88	0.89	0.83	14.07
NN-DCT	DGN	32.79	0.93	0.90	0.90	12.54
GRW	DGN	33.49	0.93	0.92	0.92	-
MXW	DGN	33.47	0.93	0.92	0.92	-
COR	DGN	33.38	0.91	0.92	0.92	-
GMAP	DGN	33.43	0.93	0.92	0.92	-
NN	DGN	33.43	0.93	0.92	0.92	-

TABLE 4.9
Performances of different techniques for color correction of the *books* from the illuminant syl-50mr16q+3202 to syl-50mr16q.

Technique	correction	PSNR	Y-WBQM	Cb-WBQM	Cr-WBQM	JPQM
GRW-DCT	DGN	27.50	0.89	0.66	0.83	10.15
MXW-DCT	DGN	27.02	0.88	0.34	0.96	11.23
MXW-DCT-Y	DGN	29.51	0.73	0.82	0.94	11.93
COR-DCT	DGN	27.07	0.46	0.63	0.97	12.78
GMAP-DCT	DGN	27.07	0.46	0.63	0.97	12.78
NN-DCT	DGN	29.43	0.71	0.79	0.96	12.13
GRW	DGN	28.39	0.94	0.73	0.84	-
MXW	DGN	29.65	0.67	0.83	0.97	-
COR	DGN	29.30	0.58	0.84	0.98	-
GMAP	DGN	29.30	0.58	0.84	0.98	-
NN	DGN	29.79	0.73	0.81	0.97	-

TABLE 4.10
Performances of different techniques for color correction of the *Macbeth* from the illuminant ph-ulm to syl-50mr16q.

Technique	correction	PSNR	Y-WBQM	Cb-WBQM	Cr-WBQM	JPQM
GRW-DCT	DGN	25.97	0.91	0.88	0.92	12.59
MXW-DCT	DGN	25.65	0.88	0.89	0.92	12.48
MXW-DCT-Y	DGN	16.94	0.38	0.41	0.61	9.54
COR-DCT	DGN	26.21	0.92	0.92	0.89	12.95
GMAP-DCT	DGN	26.21	0.92	0.92	0.89	12.95
NN-DCT	DGN	26.21	0.92	0.92	0.89	12.95
GRW	DGN	26.32	0.91	0.91	0.92	-
MXW	DGN	26.13	0.89	0.92	0.92	-
COR	DGN	26.39	0.92	0.92	0.89	-
GMAP	DGN	26.39	0.92	0.92	0.89	-
NN	DGN	26.39	0.92	0.92	0.89	-

the target illuminant itself in conjunction with the blue filter. The reference images captured under syl-50mr16q are shown in Figure 4.2 for three different objects — *ball*, *books*, and *Macbeth*. The corresponding source images are shown in Figure 4.3. The results are first presented for the diagonal color correction method. This is followed by the results obtained using the chromatic shift color correction method.

4.6.2.1 Diagonal Correction

Figures 4.4 to 4.6 show a few typical color corrected images after applying some of the representative compressed domain and spatial domain techniques. Tables 4.8 to 4.10 summarize the corresponding performance measures for all techniques.

One could observe that for most cases the PSNR values obtained in the compressed domain are quite close to those of equivalent spatial domain techniques. However, in some cases, there are significant degradations. For example, the techniques such as MXW-DCT, COR-DCT, and GMAP-DCT suffer in rendering the *ball* and *books* images. The fact is also corroborated with a larger drop in PSNR values from the corresponding values in the spatial domain. Interestingly, the proposed MXW-DCT-Y technique performs quite well in these two cases while failing in the third scene (*Macbeth* under ph-ulm illuminant). Though there exists a distinctive white patch in that image (see Figure 4.2c), the proposed white world assumption in the YCbCr space does not hold well in this case. In all such cases the JPQM values for the compressed domain techniques are more than 10. This indicates that the algorithms in the compressed domain do not seriously suffer from visible blocking artifacts or blurring in the reconstruction of images from the block DCT domain.

Similar experiments were performed with the complete dataset, that is, 223 images of 21 objects. The averages of the performance metrics are shown in Tables 4.11 to 4.15. It can be seen that the results corresponding to the block DCT domain are quite close to those obtained in the spatial domain. Moreover, though the errors of estimation of illuminants using the gray world techniques (GRW and GRW-DCT) are relatively higher than the statistical techniques, their performances as reflected by different measures, are not so poor. In fact,

TABLE 4.11

Average PSNR for different techniques (under diagonal correction) and various illuminant (IL).[†]

Method	IL_1	IL_2	IL_3	IL_4	IL_5	IL_6	IL_7	IL_8	IL_9	IL_{10}
GRW-DCT	23.99	23.46	24.20	26.75	**22.76**	27.26	**21.06**	24.35	20.28	22.71
MXW-DCT	24.57	23.60	24.82	27.39	22.36	26.72	20.65	24.35	20.71	22.63
MXW-DCT-Y	22.66	23.20	23.25	26.77	22.65	27.27	21.06	**24.45**	**20.96**	22.94
COR-DCT	**24.93**	**23.81**	**25.24**	27.46	22.50	26.45	20.93	24.04	20.73	22.62
GMAP-DCT	24.82	23.46	25.21	**27.96**	22.47	26.83	20.53	24.02	20.51	22.42
NN-DCT	24.49	23.37	25.06	26.44	22.17	**27.38**	20.14	24.31	20.17	**22.94**
GRW	24.24	23.65	24.50	27.42	**22.91**	28.22	21.21	24.55	**20.99**	22.82
MXW	24.82	23.69	25.36	28.04	22.62	27.28	21.03	24.29	20.93	23.08
COR	24.95	**24.05**	25.44	28.53	22.80	27.08	21.09	24.21	20.87	23.16
GMAP	**25.02**	23.58	25.16	**29.22**	22.85	27.55	20.80	24.19	20.74	23.09
NN	24.01	23.58	**25.48**	27.09	21.90	27.09	20.20	**24.62**	20.35	22.72

† IL_1: ph-ulm, IL_2: syl-cwf, IL_3: syl-wwf, IL_4: solux-3500, IL_5: solux-3500+3202, IL_6: solux-4100, IL_7: solux-4100+3202, IL_8: solux-4700, IL_9: solux-4700+3202, IL_{10}: syl-50mr16q+3202

TABLE 4.12
Average Y-WBQM for different techniques (under diagonal correction) and various illuminant (IL).[†]

Method	IL_1	IL_2	IL_3	IL_4	IL_5	IL_6	IL_7	IL_8	IL_9	IL_{10}
GRW-DCT	0.789	0.750	0.814	0.861	0.683	0.850	0.604	0.764	0.540	0.693
MXW-DCT	0.812	**0.788**	0.852	**0.937**	**0.733**	**0.885**	**0.655**	**0.801**	**0.580**	**0.751**
MXW-DCT-Y	0.725	0.718	0.720	0.916	0.647	0.841	0.548	0.734	0.464	0.657
COR-DCT	0.838	0.768	**0.879**	0.936	0.608	0.834	0.469	0.690	0.343	0.585
GMAP-DCT	0.836	0.767	0.876	0.929	0.574	0.844	0.395	0.699	0.296	0.549
NN-DCT	**0.842**	0.759	0.863	0.875	0.580	0.849	0.360	0.734	0.262	0.627
GRW	0.792	0.739	0.808	0.867	**0.655**	0.836	**0.566**	**0.744**	**0.481**	**0.663**
MXW	0.817	0.763	0.848	0.944	0.615	0.853	0.498	0.728	0.412	0.633
COR	0.836	**0.795**	**0.874**	0.950	0.653	**0.859**	0.502	0.685	0.383	0.648
GMAP	**0.837**	0.783	0.873	**0.952**	**0.655**	0.861	0.463	0.710	0.360	0.635
NN	0.820	0.767	0.867	0.907	0.544	0.841	0.358	0.740	0.276	0.569

TABLE 4.13
Average Cb-WBQM for different techniques (under diagonal correction) and various illuminant (IL).[†]

Method	IL_1	IL_2	IL_3	IL_4	IL_5	IL_6	IL_7	IL_8	IL_9	IL_{10}
GRW-DCT	0.796	0.785	0.795	0.917	0.717	0.876	0.586	0.795	0.462	0.698
MXW-DCT	0.781	0.739	0.714	0.936	0.541	0.829	0.334	0.670	0.241	0.489
MXW-DCT-Y	0.713	0.750	0.718	0.925	0.720	0.862	0.610	0.784	0.534	0.708
COR-DCT	0.818	**0.800**	0.769	**0.937**	**0.778**	0.899	**0.650**	0.817	**0.542**	0.714
GMAP-DCT	0.815	0.792	0.790	0.933	0.775	**0.902**	0.634	**0.823**	0.528	**0.727**
NN-DCT	**0.819**	0.780	**0.823**	0.921	0.734	0.900	0.569	0.815	0.480	0.710
GRW	**0.824**	**0.827**	**0.848**	0.922	**0.788**	0.914	**0.691**	**0.859**	**0.602**	**0.781**
MXW	0.813	0.809	0.802	**0.953**	0.744	0.904	0.647	0.819	0.567	0.733
COR	0.819	0.817	0.816	0.939	0.785	**0.923**	0.657	0.837	0.550	0.770
GMAP	0.821	0.806	0.813	0.951	0.777	0.919	0.640	0.828	0.537	0.754
NN	0.820	0.812	0.841	0.940	0.726	0.917	0.574	0.842	0.478	0.720

TABLE 4.14
Average Cr-WBQM for different techniques (under diagonal correction) and various illuminant (IL).[†]

Method	IL_1	IL_2	IL_3	IL_4	IL_5	IL_6	IL_7	IL_8	IL_9	IL_{10}
GRW-DCT	0.789	0.839	0.811	0.906	0.866	**0.934**	0.790	0.893	0.707	0.851
MXW-DCT	**0.819**	0.846	**0.856**	0.901	0.889	0.931	0.861	**0.933**	0.853	0.904
MXW-DCT-Y	0.733	0.826	0.778	0.875	0.853	0.911	0.798	0.905	0.781	0.858
COR-DCT	0.796	0.848	0.828	0.910	0.870	0.929	0.854	0.917	0.865	0.894
GMAP-DCT	0.795	0.835	0.817	0.919	0.886	0.932	0.888	0.913	0.889	**0.911**
NN-DCT	0.726	**0.851**	0.790	**0.933**	**0.903**	0.930	**0.910**	0.909	**0.903**	0.905
GRW	0.804	**0.853**	0.827	**0.936**	0.874	0.946	0.797	0.902	0.761	0.855
MXW	**0.816**	**0.853**	**0.850**	0.921	0.899	0.947	0.840	0.931	0.818	0.901
COR	0.802	0.832	0.826	0.933	0.856	0.936	0.830	**0.939**	0.834	0.877
GMAP	0.803	0.831	0.809	0.930	0.839	0.939	0.831	0.916	0.842	0.867
NN	0.738	0.839	0.824	0.935	**0.921**	**0.949**	**0.918**	0.915	**0.915**	**0.925**

† IL_1: ph-ulm, IL_2: syl-cwf, IL_3: syl-wwf, IL_4: solux-3500, IL_5: solux-3500+3202, IL_6: solux-4100, IL_7: solux-4100+3202, IL_8: solux-4700, IL_9: solux-4700+3202, IL_{10}: syl-50mr16q+3202

TABLE 4.15

Average JPQM for different techniques in the compressed domain (under diagonal correction) and various illuminant (IL).[†]

Method	IL_1	IL_2	IL_3	IL_4	IL_5	IL_6	IL_7	IL_8	IL_9	IL_{10}
GRW-DCT	12.06	12.25	12.05	12.12	11.24	12.02	10.79	12.00	10.04	11.84
MXW-DCT	12.31	12.26	12.33	12.35	11.06	12.00	10.72	11.75	10.70	11.61
MXW-DCT-Y	11.65	12.16	11.69	12.25	11.41	12.06	11.15	11.93	11.16	11.93
COR-DCT	12.51	12.39	12.53	12.38	11.73	12.35	11.73	12.38	12.03	12.37
GMAP-DCT	12.50	12.28	12.51	12.30	12.00	12.37	12.17	12.28	12.38	12.62
NN-DCT	12.30	12.50	12.36	12.11	12.07	12.03	12.50	12.19	12.64	12.35

† IL_1: ph-ulm, IL_2: syl-cwf, IL_3: syl-wwf, IL_4: solux-3500, IL_5: solux-3500+3202, IL_6: solux-4100, IL_7: solux-4100+3202, IL_8: solux-4700, IL_9: solux-4700+3202, IL_{10}: syl-50mr16q+3202

TABLE 4.16

Overall average performances of different techniques on rendering color corrected images across different illumination.

Technique	PSNR	Y-WBQM	Cb-WBQM	Cr-WBQM	JPQM
GRW-DCT	23.737	0.735	0.743	0.839	11.645
MXW-DCT	23.784	**0.779**	0.627	**0.879**	11.713
MXW-DCT-Y	23.525	0.697	0.732	0.832	11.744
COR-DCT	**23.876**	0.675	**0.772**	0.871	12.245
GMAP-DCT	23.827	0.677	**0.772**	**0.879**	**12.346**
NN-DCT	23.652	0.675	0.755	0.876	12.311
GRW	24.057	0.715	**0.806**	0.856	-
MXW	24.116	0.711	0.779	0.878	-
COR	**24.244**	**0.719**	0.791	0.867	-
GMAP	**24.244**	0.713	0.785	0.861	-
NN	23.708	0.669	0.767	**0.888**	-

TABLE 4.17

Average PSNR for different techniques (under CS) and various illuminant (IL).[†]

Method	IL_1	IL_2	IL_3	IL_4	IL_5	IL_6	IL_7	IL_8	IL_9	IL_{10}
GRW-DCT	23.05	**22.47**	**23.21**	**24.10**	**21.19**	**25.11**	**19.78**	**22.86**	**19.89**	**21.74**
MXW-DCT	19.25	20.68	19.24	23.07	17.73	21.59	16.61	19.37	16.06	18.04
MXW-DCT-Y	18.17	20.91	19.16	23.63	19.40	23.09	18.26	21.27	18.63	20.43
COR-DCT	**23.25**	20.48	22.18	22.94	18.69	22.26	17.56	21.09	17.88	19.47
GMAP-DCT	23.04	20.22	21.93	22.99	19.68	22.44	18.45	21.04	18.53	20.09
NN-DCT	22.13	20.64	20.80	21.30	19.19	20.82	18.60	20.22	18.09	19.09
GRW	**23.55**	**22.85**	**23.70**	25.11	**21.32**	**25.84**	19.82	**23.18**	**19.88**	**21.91**
MXW	20.80	22.15	21.23	24.72	19.65	23.62	18.19	21.27	18.37	20.19
COR	23.14	20.90	21.87	23.51	18.76	22.53	17.47	21.33	17.91	19.24
GMAP	23.27	20.91	21.34	24.27	19.06	22.06	18.17	20.72	18.14	19.30
NN	20.87	19.26	21.41	22.45	19.83	22.35	19.14	20.09	18.59	19.70

† IL_1: ph-ulm, IL_2: syl-cwf, IL_3: syl-wwf, IL_4: solux-3500, IL_5: solux-3500+3202, IL_6: solux-4100, IL_7: solux-4100+3202, IL_8: solux-4700, IL_9: solux-4700+3202, IL_{10}: syl-50mr16q+3202

FIGURE 4.7 (See color insert.)

Chromaticity shift corrected images by GRW-DCT: (a) *ball*, (b) *books*, and (c) *Macbeth*.

in many categories of illuminants, their performance measures have topped the list. This is also observed for the MXW-DCT-Y technique which has achieved the highest average PSNR among other techniques for the illuminants solux-4700 and solux-4700+3202. For judging the recovery of colors, one may look into the values corresponding to Cb-WBQM and Cr-WBQM. It is generally observed that with the blue filter, the reconstruction of Cb component is relatively poorer. This may be the reason lower performance measures are obtained in such cases. In the compressed domain, the statistical techniques performed better than the others in reconstructing the color components.

Table 4.16 summarizes the overall performances of all the techniques in transferring the images to the target illuminant. One may observe that in the block DCT domain the overall performances of the COR-DCT are usually better than the others. In the spatial domain, the GMAP and the COR techniques have better performance indices in most cases. It is interesting to note that though the errors of estimation of illuminants in the block DCT domain are usually less than those in spatial domain (see Table 4.7), the end results after color correction provide higher PSNR values in the spatial domain. It is felt that the color correction to all the pixels in spatial domain, as compared to those made with only DC coefficients in the block DCT domain, makes the rendering more successful.

4.6.2.2 Chromatic Shift Correction

Focusing on the performances of the color constancy algorithms followed by the chromatic shift (CS) for color correction reveals that the quality of the rendered images is significantly degraded. However, in most cases the GRW and GRW-DCT techniques perform better than the others (see Table 4.17). In this case, the degradation of the PSNR values (with respect to the diagonal correction) remains around 0.5 dB to 3.5 dB.

Figure 4.7 shows typical results on image rendering of same source images using the GRW-DCT technique. It is trivial to note that the chromatic shift correction technique is much faster than the diagonal correction technique.

4.7 Complexity of Computation of Color Constancy

This section discusses the computational complexity of different techniques. The computational complexity can be expressed by the number of multiplications, additions, and comparisons. In the employed notation, $\alpha M + \beta A + \gamma C$ represents α number of multiplications, β number of additions, and γ number of comparisons. It is also considered that the computational cost for multiplication and division are equivalent.

It can be easily shown that for a total number n of pixels, both the GRW and the MXW techniques require $3M + 3(n-1)A$ and $3(n-1)C$ computations, respectively. In the implementation approach for using statistical techniques, there is an overhead of converting pixels from the *RGB* color space to the *YCbCr* color space at a cost of $9nM + 6nA$ for n pixels. The techniques also depend on the number of illuminants, n_l, and the fractional coverage, f, over the discretized chromaticity space. Let the size of the discretized space be n_c (here, it is 32×32). For both COR and GMAP techniques the number of computations for n pixels is $f n_c n_l A + (n_l - 1)C$. However, one may note that the GMAP algorithm performs only integer additions. While computing with the *NN* technique, for each illuminant it is necessary to determine the Mahalanobis distance with its chromatic mean (of dimension 2×1) and covariance matrices (which are symmetric and of dimension 2×2). This computation can be performed by $8M + 4A$ computations. Hence, the total number of computations become $8n_l M + 4n_l A$ for n pixels.

For storing the statistics, every statistical technique needs to store the mean of the illuminant (of dimension 3×1). In addition, the NN technique would be required to store 3 elements of the covariance matrix per illuminant. It makes its minimum storage requirement of $6n_l$ number of elements. But both the COR and GMAP require $n_c n_l$ number of elements per illuminant for storing the chromatic mapping. In addition, they also require to store the mean illuminant (3 per illuminant).

In the block DCT space, color constancy techniques handle only the DC coefficients. As a result of this, the input data size itself gets reduced by the size of the block (in JPEG compression, it is $8 \times 8 = 64$ times). This speeds up all the nonstatistical techniques. On the other hand, the computation using the statistical techniques is relatively independent of the input data size. However in such cases, the conversion from the RGB color space to the YCbCr color space is not required. This makes computation of statistics in the chromatic space faster in the compressed domain. It should be noted that in the compressed domain, the additional overhead of reverse and forward transforms to and from the spatial domain is also avoided in the block DCT-based approaches. As can be shown, the MXW-DCT-Y technique is the fastest among all such techniques, as it performs comparisons only with the Y components. On the other hand, the MXW-DCT algorithm is relatively slower as it requires the conversion of each DC coefficient to the RGB color space from the YCbCr color space. The computational and storage complexities of all techniques are summarized in Table 4.18.

4.8 Color Enhancement in the Compressed Domain

For better visualization of images, particularly when the scene suffers from both widely varying spectral components and brightness of illuminants, one has to also consider the right combination of brightness, contrast, and colors of pixels in an image. One may con-

TABLE 4.18

Algorithm complexities given n_l number of illuminants, n_c as the size of the 2D chromaticity space, and n number of image pixels. © 2009 IEEE

Algorithms	Computational Complexity	Storage Complexity
GRW	$3M + 3(n-1)A$	-
GRW-DCT	$3M + 3(\frac{n}{64} - 1)A$	-
MXW	$3(n-1)C$	-
MXW-DCT	$\frac{9}{64}nM + \frac{6}{64}nA + 3(\frac{n}{64} - 1)C$	-
MXW-DCT-Y	$(\frac{n}{64} - 1)C$	-
COR	$9nM + (6n + fn_cn_l)A + (n_l - 1)C$	$n_cn_l + 3n_l$
COR-DCT	$fn_cn_lA + (n_l - 1)C$	$n_cn_l + 3n_l$
GMAP	$9nM + (6n + fn_cn_l)A + (n_l - 1)C$	$n_cn_l + 3n_l$
GMAP-DCT	$fn_cn_lA + (n_l - 1)C$	$n_cn_l + 3n_l$
NN	$(9n + 8n_l)M + (6n + 4n_l)A$	$6n_l$
NN-DCT	$8n_lM + 4n_lA$	$6n_l$

jecture that for natural images the inherent combination of brightness, contrast, and colors should be preserved in the process of enhancement. This leads to the development of a *contrast* and *color preserving* enhancement algorithm in the compressed domain [24], which is elaborated below.

4.8.1 Preservation of Contrast and Color

The contrast measure has been defined in various ways in the literature. One of the intuitive definitions comes from the model based on the Weber's law. This measure can be used to define contrast at a pixel. Let μ and σ denote the mean and standard deviation of an $N \times N$ image, respectively. The contrast ζ of an image as modeled by the Weber's law is given by $\zeta = \Delta L/L$, where ΔL is the difference in luminance between a stimulus and its surround, and the L is the luminance of the surround [25]. As μ provides a measure for surrounding luminance and σ is strongly correlated with ΔL, the contrast ζ of an image can be redefined as follows:

$$\zeta = \frac{\sigma}{\mu}. \tag{4.14}$$

The above definition can be restricted to a $N \times N$ block of an image in order to simplify the processing in the block DCT domain.

The following theorem states how the contrast of an image is related to the scaling of its DCT coefficients. The proof of this theorem is given in Reference [24].

Theorem 4.2

Let κ_{dc} be the scale factor for the normalized DC coefficient and κ_{ac} be the scale factor for the normalized AC coefficients of an image Y of size $N \times N$, such that the DCT coefficients in the processed image $\widetilde{Y}$ are given by:

$$\widetilde{Y}(i,j) = \begin{cases} \kappa_{dc}Y(i,j) & \text{for } i = j = 0, \\ \kappa_{ac}Y(i,j) & \text{otherwise.} \end{cases} \tag{4.15}$$

The contrast of the processed image then becomes κ_{ac}/κ_{dc} times the contrast of the original image. $\square$

One may note here that in the block DCT space, to preserve the local contrast of the image, scale factor should be kept as $\kappa_{dc} = \kappa_{ac} = \kappa$ in a block. However, though the above operations with the Y component of an image preserve the contrast, they do not preserve the colors or color vectors of the pixels. Hence, additional operations with the chromatic components, that is, Cr and Cb components of the image in the compressed domain, need to be carried out. Theorem 4.3 states how colors could be preserved under the uniform scaling operation. The proof of this theorem is given in Reference [24].

Theorem 4.3

Let $U = \{U(k,l)|0 \le k,l \le (N-1)\}$ and $V = \{V(k,l)|0 \le k,l \le (N-1)\}$ be the DCT coefficients of the Cb and Cr components, respectively. If the luminance (Y) component of an image is uniformly scaled by a factor κ, the colors of the processed image with $\widetilde{Y}$, $\widetilde{U}$ and $\widetilde{V}$ are preserved by the following operations:

$$\widetilde{U}(i,j) = \begin{cases} N(\kappa(\frac{U(i,j)}{N} - 128) + 128) & \text{for } i = j = 0, \\ \kappa U(i,j) & \text{otherwise,} \end{cases} \tag{4.16}$$

$$\widetilde{V}(i,j) = \begin{cases} N(\kappa(\frac{V(i,j)}{N} - 128) + 128) & \text{for } i = j = 0, \\ \kappa V(i,j) & \text{otherwise.} \end{cases} \tag{4.17}$$

$\square$

4.8.2 Color Enhancement by Scaling DCT Coefficients

It is straightforward to design an algorithm for enhancing a color image in the block DCT domain by scaling its coefficients by making use of the above theorems. In the process of scaling different types of coefficients (of different components), one would carry out the following operations:

1. *Adjustment of local background illumination.* This adjustment is made by scaling the DC coefficients of each block using a global monotonic function. A few typical examples of these functions [26], [27], [28] are given below:

$$\tau(x) = x(2 - x), \quad 0 \le x \le 1, \tag{4.18}$$

$$\eta(x) = \frac{(x^{\frac{1}{\gamma}} + (1 - (1 - x)^{\frac{1}{\gamma}}))}{2}, \quad 0 \le x \le 1, \tag{4.19}$$

$$\psi(x) = \begin{cases} n(1 - (1 - \frac{x}{m})^{p_1}) & \text{for } 0 \le x \le m, \\ n + (1 - n)(\frac{x-m}{1-m})^{p_2} & \text{for } m \le x \le 1, 0 \le m \le n \le 1, p_1, p_2 > 0. \end{cases} \tag{4.20}$$

Note that the same scale factor is used for scaling other coefficients subsequently as discussed below.

2. *Preservation of local contrast.* Once the scale factor is obtained from the mapping of the DC coefficient using any of the functions mentioned above, the same factor is used for scaling the AC coefficients of the block (according to Theorem 4.2). However, while performing this operation, there is a risk of crossing the maximum

displayable brightness value at a pixel in that block. To restrict this overflow, the scale factor is clipped by a value as stated in Theorem 4.4. The proof of this theorem is given in Reference [24].

3. *Preservation of colors.* Finally, the colors are preserved by performing the scaling of the DCT coefficients of the Cb and Cr components as described in Theorem 4.3.

Theorem 4.4

If the values in a block are assumed to lie within $\mu \pm \lambda \sigma$, the scaled values will not exceed the maximum displayable brightness value B_{max} if $1 \leq \kappa \leq B_{max}/(\mu + \lambda \sigma)$.   □

Due to the independent processing of blocks, blocking artifacts near edges or near sharp changes of brightness and colors of pixels can occur. To suppress this effect, in such cases the same computation is carried out in blocks of smaller size. These smaller blocks are obtained by decomposing the given block using the block decomposition operation [3]. Similarly, the resulting scaled coefficients of smaller blocks are recomposed into the larger one using the block composition operation [3]. It may be noted that both these operations can be performed efficiently in the block DCT domain. Hence, the algorithm for color enhancement remains totally confined within the block DCT domain. For detecting a block requiring these additional operations, the *standard deviation* σ of that block was used here as a measure. If σ is greater than a threshold (say, σ_{th}), the 8×8 blocks are decomposed into four subblocks to perform the scaling operations. The typical value of σ_{th} in these experiments was empirically chosen as 5.

4.8.3 Results

This section presents the results of the enhancement algorithms and compares these results with that of other compressed domain techniques reported in the literature. Table 4.19 lists considered techniques including the algorithms discussed in this section, referred under the category of *color enhancement by scaling* (CES). The details of these techniques and the description of its different parameters can be found in the literature cited in the table. For the sake of completeness, parameter values are also presented in the table.

TABLE 4.19
List of techniques considered for comparative study.

Techniques	Short Names	Parameter Set
Alpha rooting [29]	AR	$\alpha = 0.98$
Multi-Contrast Enhancement [30]	MCE	$\lambda = 1.95$
with Dynamic Range Compression [27]	MCEDRC	$\gamma = 1.95, \tau_1 = 0.1, \tau_2 = 1.95$
CES using $\tau(x)$	TW-CES-BLK	$B_{max} = 255, k = 1.0,$
with blocking artifact removal [24]		$\sigma_{th} = 15$
CES using $\eta(x)$	DRC-CES-BLK	$\lambda = 1.95, B_{max} = 255,$
with blocking artifact removal [24]		$k = 1.0, \sigma_{th} = 5$
CES using $\psi(x)$		$B_{max} = 255, k = 1.0,$
with blocking artifact removal [24]	SF-CES-BLK	$\sigma_{th} = 5, m = n = 0.5, p_1 = 1.8, p_2 = 0.8$

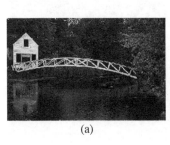

(a)

(b)

(c)

FIGURE 4.8

Original images used in the implementation of the color enhancement algorithms: (a) *Bridge*, (b) *Mountain*, and (c) *Under-water*.

To evaluate these techniques, two measures for judging the quality of reconstruction in the compressed domain were used. One, based on JPEG-quality-metric (JPQM) [23], is related to the visibility of blocking artifacts, whereas the other one, a no-reference metric of Reference [25], is related to the enhancement of colors in images. The definition for this latter metric in the RGB color space is given below.

Let the red, green and blue components of an image I be denoted by R, G, and B, respectively. Let $\alpha = R - G$ and $\beta = (\frac{R+G}{2}) - B$. Then the colorfulness of the image is defined as follows:

$$CM(I) = \sqrt{\sigma_\alpha^2 + \sigma_\beta^2} + 0.3\sqrt{\mu_\alpha^2 + \mu_\beta^2}, \qquad (4.21)$$

where σ_α and σ_β are standard deviations of α and β, respectively. Similarly, μ_α and μ_β are their means. In this comparison, however, the ratio of CMs between the enhanced image and its original for observing the color enhancement factor was used and it is referred here as *color enhancement factor* (*CEF*).

Typical examples of the enhanced images are presented here for the set of images shown in Figure 4.8. The average JPQM and CEF values obtained on these set of images are presented in the Table 4.20. From this table, one can observe that the color enhancement performance indicated by the measure CEF is quite improved using the color enhancement by scaling (CES) algorithms. In particular, the TW-CES-BLK algorithm, which uses a very simple mapping function (Equation 4.18), is found to provide excellent color rendition in the enhanced images. However, the JPQM measure indicates that its performance in suppressing blocking artifacts is marginally poorer than the other schemes.

TABLE 4.20

Average performance measures obtained by different color enhancement techniques.

	Techniques					
Measures	AR	MCE	MCEDRC	TW-CES-BLK	DRC-CES-BLK	SF-CES-BLK
JPQM	9.80	9.38	9.67	9.06	9.50	9.41
CEF	1.00	0.91	1.00	1.63	1.32	1.37

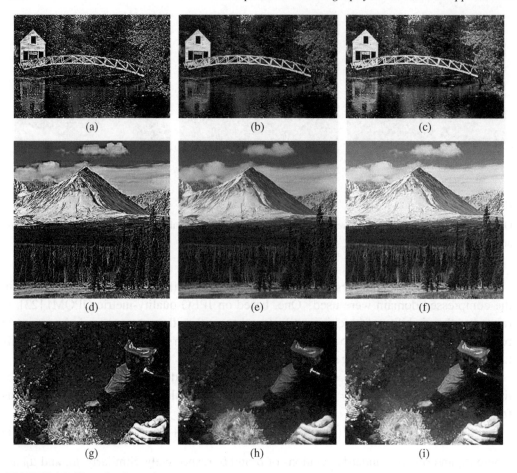

FIGURE 4.9 (See color insert.)

Color enhancement of the images: (a,d,g) MCE, (b,e,h) MCEDRC, and (c,f,i) TW-CES-BLK.

FIGURE 4.10

Color enhancement of the image *Bridge* with increasing number of iterations: (a) 2, (b) 3, and (c) 4.

Figure 4.9 shows the enhancement results obtained by three typical techniques. Namely, these techniques are MCE, MCEDRC, and TW-CES-BLK. One may observe that in all these cases the TW-CES-BLK provides better visualization of these images than other two techniques.

One may further enhance images by iteratively subjecting the resulting image to the next stage of enhancement using the same algorithm. Figure 4.10 shows a typical example

TABLE 4.21
Performance measures after iterative application of the TW-CES-BLK algorithm on the image *Bridge*.

Measures	number of iterations			
	2	3	4	5
CEF	2.76	3.48	3.79	3.74
JPQM	7.10	6.48	6.06	5.81

of iterative enhancement of the image *Bridge*. It was observed that for a few iterations the CEF measure shows improvements with respect to the original image. However, the process suffers from the risk of making blocking artifacts more and more visible. This is also evidenced by an increased degradation of the JPQM values with an increase in the number of iterations, as shown in Table 4.21.

4.9 Color Enhancement with Color Correction

There are situations when color restoration demands both color correction and color enhancement for better visualization of an image. In particular, if an image is captured under artificial illumination, color enhancement will not provide the desired effect, unless its colors are corrected or the image is recomputed for a target illumination. A typical example is shown in the Figure 4.11. Namely, Figure 4.11a shows the original image of a *manufacturing unit* scene. It is evident that this image is captured under the illumination constrained by its environment. If the image itself is subjected to color enhancement, the resulting image does not have proper color rendition, as shown in Figure 4.11b. However, the enhanced result shows significant improvement, if the color correction is performed through the com-

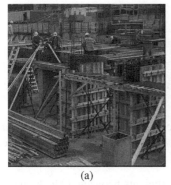  

| (a) | (b) | (c) |

FIGURE 4.11 (See color insert.)

Color restoration through color correction followed by enhancement: (a) original image, (b) enhanced image without color correction, and (c) enhanced image with color correction.

putation of color constancy, as demonstrated in Figure 4.11c. In this case, the illumination is transferred to the Syl-50mr16q (see Section 4.3) using the COR-DCT method [6] followed by the TW-CES-BLK enhancement algorithm applied to the color corrected image.

4.10   Conclusion

This chapter discussed the restoration of colors under varying illuminants and illumination in the block DCT domain. The basic computational task involved in this process is to obtain an illuminant independent representation by solving the problem of color constancy. Once the spectral components of the (global) illuminant of a scene are computed, one is required to transfer the image under a canonical illuminant. This computational task is marked as color correction. However, due to wide variation of illumination in a scene and the limited dynamic range of displaying colors, one would be required to further modify the colors of pixels by the process of enhancement. It may be noted that in an ordinary situation color enhancement may not necessarily follow the color correction stage.

This chapter reviewed several color constancy algorithms and discussed the extension of these algorithms in the block DCT domain. It was observed that many algorithms are quite suitable for computation considering only the DC coefficients of the blocks. This chapter further discussed the enhancement algorithm in the block DCT domain. As the computations are restricted only in the compressed domain, overhead of compression and decompression is avoided. This makes the algorithms fast and less memory intensive.

Acknowledgment

Tables 4.2, 4.7, and 4.18 are reprinted from Reference [2], with the permission of IEEE.

References

[1] V. Kries, *Handbuch der Physiologic des Menschen*, vol. 3. Braunschweig, Germany: Viewieg und Sohn, 1905.

[2] J. Mukhopadhyay and S.K. Mitra, "Color constancy in the compressed domain," in *Proceedings of the IEEE International Conference on Image Processing*, Cairo, Egypt, November 2009.

[3] J. Jiang and G. Feng, "The spatial relationships of DCT coefficients between a block and its sub-blocks," *IEEE Transactions on Signal Processing*, vol. 50, no. 5, pp. 1160–1169, May 2002.

[4] S.H. Jung, S.K. Mitra, and D. Mukherjee, "Subband DCT: Definition, analysis and appli-

cations," *IEEE Transactions on Circuits and Systems for Video Technology*, vol. 6, no. 3, pp. 273–286, June 1996.

[5] J. Mukherjee and S.K. Mitra, *Color Image Processing: Methods and Applications*, ch. "Resizing of color images in the compressed domain," R. Lukac and K.N. Plataniotis (eds.), Boca Raton, FL: CRC Press / Taylor & Francis, October 2006, pp. 129–156.

[6] G. Finlayson, S. Hordley, and P. Hubel, "Color by correlation: A simple, unifying framework for color constancy," *IEEE Transactions on Pattern Analysis and Machine Intelligence*, vol. 23, no. 11, pp. 1209–1221, November 2001.

[7] K. Barnard, V. Cardei, and B. Funt, "A comparison of computational color constancy algorithms – Part I: Methodology and experiments with synthesized data," *IEEE Transactions on Image Processing*, vol. 11, no. 9, pp. 972–984, September 2002.

[8] K. Barnard, V. Cardei, and B. Funt, "A comparison of computational color constancy algorithms – Part II: Experiments with image data," *IEEE Transactions on Image Processing*, vol. 11, no. 9, pp. 985–996, September 2002.

[9] G. Buchsbaum, "A spatial processor model for object colour perception," *Journal of Franklin Inst.*, vol. 310, pp. 1–26, 1980.

[10] R. Gershon, A. Jepson, and J. Tsotsos, "From [r,g,b] to surface reflectance: Computing color constant descriptors in images," *Perception*, pp. 755–758, 1988.

[11] J. Van de Weijer, T. Gevers, and A. Gijsenij, "Edge-based color constancy," *IEEE Transactions on Image Processing*, vol. 16, no. 9, pp. 2207–2214, September 2007.

[12] E. Land, "The retinex theory of color vision," *Scientific American*, vol. 3, pp. 108–129, 1977.

[13] D. Forsyth, "A novel algorithm for color constancy," *Int. Journal of Computer Vision*, vol. 5, no. 1, pp. 5–36, August 1990.

[14] G. Finlayson, "Color in perspective," *IEEE Transactions on Pattern Analysis and Machine Intelligence*, vol. 18, no. 10, pp. 1034–1038, October 1996.

[15] B. Funt, V. Cardei, and K. Barnard, "Learning color constancy," *Proceedings of the 4th IS&T/SID Color Imaging Conference*, Scottsdale, AZ, USA, November 1996, pp. 58–60.

[16] G. Sapiro, "Color and illuminant voting," *IEEE Transactions on Pattern Analysis and Machine Intelligence*, vol. 10, no. 11, pp. 1210–1215, November 1999.

[17] G. Sapiro, "Bilinear voting," in *Proceedings of the Sixth International Conference on Computer Vision*, Bombay, India, January 1998, pp. 178–183.

[18] D. Brainard and W. Freeman, "Bayesian color constancy," *Journal of Optical Society of America, A,*, vol. 14, no. 7, pp. 1393–1411, July 1997.

[19] P. Mahalanobis, "On the generalised distance in statistics," *Proceedings of the National Institute of Science of India*, vol. 12, pp. 49–55, 1936.

[20] M. Ebner, G. Tischler, and J. Albert, "Integrating color constancy into JPEG2000," *IEEE Transactions on Image Processing*, vol. 16, no. 11, pp. 2697–2706, November 2007.

[21] K. Barnard, L. Martin, B. Funt, and A. Coath, "A data set for color research," *Color Research and Applications*, vol. 27, no. 3, pp. 148–152, June 2000.

[22] Z. Wang and A. Bovik, "A universal image quality index," *IEEE Signal Processing Letters*, vol. 9, no. 3, pp. 81–84, March 2002.

[23] Z. Wang, H. Sheikh, and A. Bovik, "No-reference perceptual quality assessment of JPEG compressed images," in *Proceeding of the IEEE International Conference on Image Processing*, vol. I, September 2002, pp. 477–480.

[24] J. Mukherjee and S.K. Mitra, "Enhancement of color images by scaling the DCT coefficients," *IEEE Transactions on Image Processing*, vol. 17, no. 10, pp. 1783–1794, October 2008.

[25] S. Susstrunk and S. Winkler, "Color image quality on the Internet," *Proceedings of SPIE*, vol. 5304, pp. 118–131, January 2004.

[26] S.K. Mitra and T.H. Yu, "Transform amplitude sharpening: A new method of image enhancement," *Computer Vision Graphics and Image Processing*, vol. 40, no. 2, pp. 205–218, November 1987.

[27] S. Lee, "An efficient content-based image enhancement in the compressed domain using retinex theory," *IEEE Transactions on Circuits and Systems for Video Technology*, vol. 17, no. 2, pp. 199–213, February 2007.

[28] T. De, "A simple programmable S-function for digital image processing," in *Proceedings of the Fourth IEEE Region 10 International Conference*, Bombay, India, November 1989, pp. 573–576.

[29] S. Aghagolzadeh and O. Ersoy, "Transform image enhancement," *Optical Engineering*, vol. 31, no. 3, pp. 614–626, March 1992.

[30] J. Tang, E. Peli, and S. Acton, "Image enhancement using a contrast measure in the compressed domain," *IEEE Signal Processing Letters*, vol. 10, no. 10, pp. 289–292, October 2003.

5

Principal Component Analysis-Based Denoising of Color Filter Array Images

Rastislav Lukac and Lei Zhang

5.1   Introduction

In the last decade, advances in hardware and software technology have allowed for massive replacement of conventional film cameras with their digital successors. This reflects the fact that capturing and developing photos using chemical and mechanical processes cannot provide users with the conveniences of digital cameras which record, store and manipulate photographs electronically using image sensors and built-in computers. The ability to display an image immediately after it is recorded, to store thousands of images on a small memory device and delete them from this device in order to allow its further re-use, to edit captured visual data, and to record images and video with sound makes digital cameras very attractive consumer electronic products.

To create an image of a scene, digital cameras use a series of lenses that focus light onto a sensor which samples the light and records electronic information which is subsequently converted into digital data. The sensor is an array of light-sensitive cells which record the

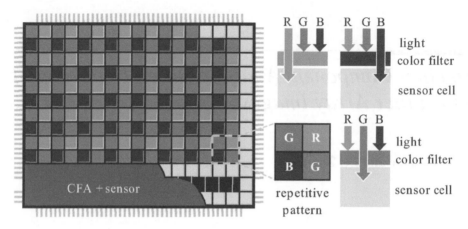

FIGURE 5.1

The concept of acquiring the visual information using the image sensor with a color filter array.

total intensity of the light that strikes their surfaces. Various image analysis and processing steps are then typically needed to transform digital sensor image data into a full-color fully processed image, commonly referred to as a digital photograph.

This chapter focuses on denoising of image data captured using a digital camera equipped with a color filter array and a monochrome image sensor (Figure 5.1). The chapter presents a principal component analysis-driven approach which takes advantage of local similarities that exist among blocks of image data in order to improve the estimation accuracy of the principal component analysis transformation matrix. This adaptive calculation of a covariance matrix and the utilization of both spatial and spectral correlation characteristics of a CFA image allow effective signal energy clustering and efficient noise removal with simultaneous preservation of local image structures such as edges and fine details.

The chapter begins with Section 5.2, which briefly discusses digital color camera imaging fundamentals and relevant denoising frameworks. Section 5.3 presents principal component analysis basics and notations used throughout this chapter, and outlines the concept of a spatially adaptive denoising method using principal component analysis of color filter array mosaic data. Section 5.4 is devoted to the design of the denoising method. Included examples and experimental results indicate that principal component analysis-driven denoising of color filter array mosaic images constitutes an attractive tool for a digital camera image processing pipeline, since it yields good performance and produces images of reasonable visual quality. The chapter concludes with Section 5.5, which summarizes the main camera image denoising ideas.

5.2 Digital Color Camera Imaging Fundamentals

Digital cameras acquire a scene by first focusing and then actuating the shutter to allow light through the optical system. Once the light reaches the sensor surface, it is sampled by the sensor and transformed from an analog signal to a digital output in order to obtain the

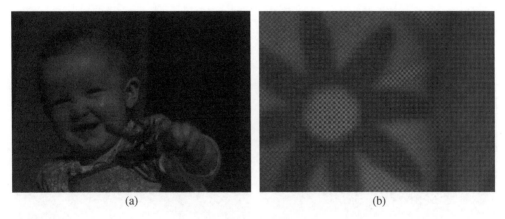

(a)   (b)

FIGURE 5.2

Color filter array imaging: (a) acquired image and (b) cropped region showing the mosaic layout.

corresponding digital representation of the sensor values. Since common image sensors, such as *charge-coupled devices* (CCD) [1], [2] and *complementary metal oxide semiconductor* (CMOS) sensors [3], [4], are monochromatic devices, digital camera manufacturers place a color filter on top of each sensor cell to capture color information. Figure 5.1 shows a typical solution, termed as a *color filter array* (CFA), which is a mosaic of color filters with different spectral responses. Both the choice of a color system and the arrangement of color filters in the array have significant impacts on the design, implementation and performance characteristics of a digital camera. Detailed discussion on this topic can be found in References [5], [6], [7], and [8].

The acquired CFA sensor readings constitute a single plane of data. Figure 5.2 shows an example. The image has a mosaic-like structure dictated by the CFA. It can either be stored as a so-called *raw camera file* [9] together with accompanying metadata containing information about the camera settings to allow its processing on a personal computer (PC) or directly undergo in-camera image processing realized by an application-specific integrated circuit (ASIC) and a microprocessor. In the former case, which is typical for a digital *single lens reflex* (SLR) camera, the *Tagged Image File Format for Electronic Photography* (TIFF-EP) [10] is used to compress image data in a lossless manner. This application scenario allows for developing high-quality digital photographs on a PC using sophisticated solutions, under different settings, and reprocessing the image until certain quality criteria are met. In the latter case, the captured image is completely processed in a camera under *real-time* constraints to produce the final image which is typically stored using lossy *Joint Photographic Experts Group* (JPEG) compression [11] in the *Exchangeable Image File* (EXIF) format [12] together with the metadata. Image compression methods suitable for these tasks can be found in References [13], [14], [15], and [16].

In either case, extensive processing is needed to faithfully restore full-color information required by common output media such as displays, image storage systems, and printers. Various image processing and analysis techniques usually operate based on the assumption of noise-free CFA data. Unfortunately, this assumption does not hold well in practice. Noise is an inherent property of image sensors and cannot be eliminated in the digital camera design.

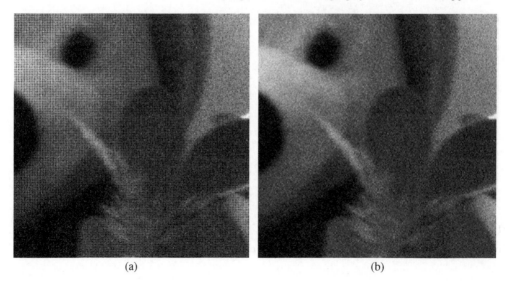

(a)   (b)

FIGURE 5.3

Example digital camera images corrupted by sensor noise: (a) CFA image, and (b) demosaicked image.

5.2.1   Digital Camera Image Noise

Noise in digital camera images usually appears as random speckles in otherwise smooth regions (Figure 5.3). Typically, noise is caused by random sources associated with quantum signal detection, signal independent fluctuations, and inhomogeneity of the sensor elements' responses. The appearance of noise in images varies amongst different digital camera models. Noise increases with the *sensitivity* (ISO) setting in the camera, length of *exposure*, and *temperature*. It can vary within an individual image; darker regions usually suffer more from noise than brighter regions. The level of noise also depends on characteristics of the camera electronics and the physical size of photosites in the sensor [17]. Larger photosites usually have better light-gathering abilities, thus producing a stronger signal and higher signal-to-noise (SNR) ratio.

As shown in Figure 5.3, noisy pixels deviate from their neighbors. In full-color RGB images, such as images output by a digital camera or images produced on a personal computer using supporting software able to handle raw camera image formats, noise can be seen as fluctuations in intensity and color [5]. Noise significantly degrades the value of the captured visual information, altering the desired image characteristics and decreasing the perceptual quality and image fidelity. It also complicates other image processing and analysis tasks. To overcome the problem, image denoising, which refers to the process of estimating the original image information from noisy data, is an essential part of the digital camera image processing pipeline [5], [18], [19].

5.2.2   Camera Image Processing Pipeline

Early steps in the camera image processing pipeline aim at compensating for sensor non-linearities and nonidealities, such as a nonlinear pixel response, thermal gradients, defective pixels, and dark current noise [19], [20], [21]. These preprocessing routines are followed by various image restoration and color manipulation steps such as *demosaicking* [22], [23],

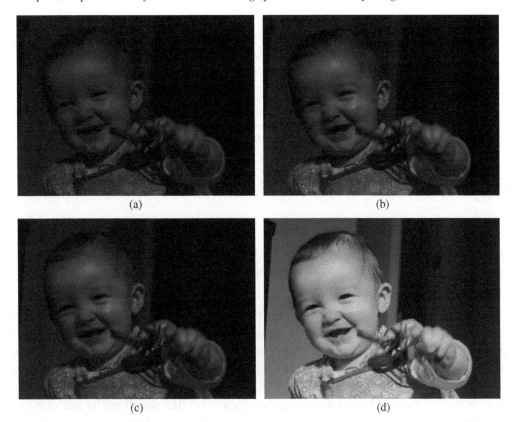

FIGURE 5.4 (See color insert.)

Different stages of the camera image processing pipeline: (a) demosaicked image, (b) white-balanced image, (c) color-corrected image, and (d) tone / scale-rendered image. The results correspond to Figure 5.2a.

[24], [25] to restore full-color information from CFA mosaic data, *white balancing* [26], [27] to compensate for the scene illuminant, *color correction* [19] to achieve visually pleasing scene reproduction, and *tone / scale rendering* to transform the color data from an unrendered to a rendered space and make the tonality of a captured image match the nonlinear characteristics of the human visual system [18]. Figure 5.4 illustrates the effect of these steps on the image as it progresses through the pipeline. Visual quality of the captured image also is highly dependent on *denoising* [28] to suppress noise and various outliers, *image sharpening* [29] to enhance structural content such as edges and color transitions, and *exposure correction* [30] to compensate for inaccurate exposure settings.

In addition to the above processing operations, a *resizing* step [31] can be employed to produce images of dimensions different from those of the sensor. Red-eye removal [32] is used to detect and correct defects caused by the reflection of the blood vessels in the retina due to flash exposure. *Face detection* [33], [34] can help to improve auto-focus, optimize exposure and flash settings, and allow more accurate color manipulation to produce photographs with enhanced color and tonal quality. *Image stabilization* [35] compensates for undesired camera movements, whereas *image deblurring* [36] removes the blurring effect caused by the camera optical system, lack of focus, or camera motion during exposure. Advanced imaging systems can enhance *resolution* [37], [38] and *dynamic-range* [39].

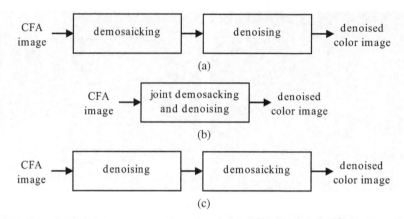

FIGURE 5.5

Pipelining the demosaicking and denoising steps: (a) demosaicked image denoising, (b) joint demosaicking and denoising, and (c) color filter array image denoising.

The overall performance of the imaging pipeline can vary significantly depending on the choice and order of the processing steps. The way an image processing pipeline is constructed usually differs among camera manufacturers due to different design characteristics, implementation constraints, and preferences regarding the visual appearance of digital photographs. Note that there is no ideal way of cascading individual processing steps; therefore, the problem of designing the pipeline is often simplified by analyzing just a very few steps at the time. Detailed discussion on pipelining the image processing and analysis steps can be found in References [5], [18], and [19].

5.2.3   Denoising Strategies for Single-Sensor Digital Color Cameras

Since demosaicking is an inseparable component of most consumer digital color cameras, the position of any processing step employed in the single-sensor imaging pipeline can be related to the position of demosaicking. Practically any processing step can be used *before* or *after* demosaicking. Performing steps before demosaicking can allow significant computational savings due to the grayscale nature of CFA image data, as opposed to performing the same operation on demosaicked color data which basically increases the number of calculations three-fold. Some processing operations can also be implemented in a *joint* manner with demosaicking, thus potentially reducing the cost of implementation, enhancing performance of processing tasks, and producing higher visual quality. Implementing various steps in a joint process is usually possible if they employ similar digital signal processing concepts. Additional information about these three processing frameworks can be found in Reference [5].

Figure 5.5 shows three simplified pipelines which use a CFA image as the input to produce a denoised demosaicked image as the output. Note that a CFA image is basically a mosaic-like grayscale image as opposed to a full-color demosaicked image. This fundamental difference in representation of these two images and the different order of demosaicking and denoising operations in each of the three simplified pipelines suggest the following characteristics:

- The framework shown in Figure 5.5a performs *denoising after demosaicking*. Algorithms directly adopted from grayscale imaging, such as various median [40], [41], averaging [42], [43], multiresolution [44], and wavelet [45], [46], [47] filters, process each color channel of the demosaicked image separately, whereas modern filters for digital color imaging process color pixels as vectors to preserve the essential spectral correlation and avoid new color artifacts in the output image [41], [48], [49]. Unfortunately, the CFA sensor readings corresponding to different color channels have different noise statistics and the demosaicking process blends the noise contributions across channels, thus producing compound noise that is difficult to characterize and remove by traditional filtering approaches.

- Figure 5.5b shows the framework which produces the output image by performing *demosaicking and image denoising simultaneously*. Approaches designed within this framework include additive white noise assumption-driven demosaicking using minimum mean square error estimation [50], bilateral filter-based demosaicking [51], and joint demosaicking and denoising using total least square estimation [52], wavelets [53], [54], color difference signals [55], and local polynomial approximation-based nonlinear spatially adaptive filtering [56]. At the expense of increased complexity of the design, performing the two estimation processes jointly avoids the problem associated with the other two processing frameworks which tend to amplify artifacts created in the first processing step.

- Finally, the framework depicted in Figure 5.5c addresses *denoising before demosaicking*. This approach aims at restoring the desired signal for subsequent color interpolation, thus enhancing the performance of the demosaicking process which can fail in edge regions in the presence of noise. Existing denoising methods for grayscale images cannot be directly used on the CFA image due to its underlying mosaic structure. However, these methods are applicable to subimages [13], [15] extracted from the CFA image; the denoised CFA image is obtained by combining the denoised subimages. This approach often results in various color shifts and artifacts due to the omission of the essential spectral characteristics during processing. Therefore, recent methods for denoising the CFA mosaic data exploit both spatial and spectral correlations to produce color artifact-free estimates without the need to extract subimages [57].

This chapter focuses on denoising the CFA image, that is, the framework depicted in Figure 5.5c, as this is the most natural way of handling the denoising problem in the digital cameras under consideration. The framework can effectively suppress noise while preserving color edges and details. Since it performs denoising before color restoration and manipulation steps, it gives the camera image processing pipeline that uses this strategy an advantage of less noise-caused color artifacts. Moreover, since CFA images consist of three times less data compared to demosaicked images, this framework has an obvious potential to achieve high processing rates.

5.3 Principal Component Analysis

Principal component analysis (PCA) [58], [59] is a popular decorrelation technique used for dimensionality reduction with direct applications in pattern recognition, data compression and noise reduction. This technique, as formally described below, will be adapted to solve the problem at hand.

5.3.1 Notation and Basics

Let $\mathbf{x} = [x_1 \; x_2 \; \cdots \; x_m]^T$ be an m-component vector variable and

$$\mathbf{X} = \begin{bmatrix} X_1 \\ X_2 \\ \vdots \\ X_m \end{bmatrix} = \begin{bmatrix} x_1^1 & x_1^2 & \cdots & x_1^n \\ x_2^1 & x_2^2 & \cdots & x_2^n \\ \vdots & \vdots & \vdots & \vdots \\ x_m^1 & x_m^2 & \cdots & x_m^n \end{bmatrix}$$

be the sample matrix of $\mathbf{x}$, with x_i^j denoting the discrete samples of variable x_i and $X_i = [x_i^1 \; x_i^2 \; \cdots \; x_i^n]$ denoting the sample vector of x_i, for $i = 1, 2, , m$ and $j = 1, 2, , n$. The centralized version $\bar{\mathbf{X}}$ of the sample matrix $\mathbf{X}$ can be written as

$$\bar{\mathbf{X}} = \begin{bmatrix} \bar{X}_1 \\ \bar{X}_2 \\ \vdots \\ \bar{X}_m \end{bmatrix} = \begin{bmatrix} \bar{x}_1^1 & \bar{x}_1^2 & \cdots & \bar{x}_1^n \\ \bar{x}_2^1 & \bar{x}_2^2 & \cdots & \bar{x}_2^n \\ \vdots & \vdots & \vdots & \vdots \\ \bar{x}_m^1 & \bar{x}_m^2 & \cdots & \bar{x}_m^n \end{bmatrix},$$

where $\bar{x}_i^j = x_i^j - \mu_i$ is obtained using the mean value of x_i estimated as follows:

$$\mu_i = \mathrm{E}[x_i] \approx \frac{1}{n} \sum_{j=1}^{n} X_i(j).$$

A set of all such mean values gives $\boldsymbol{\mu} = \mathrm{E}[\mathbf{x}] = [\mu_1 \; \mu_2 \; \cdots \; \mu_m]^T$ which is the mean value vector of $\mathbf{x}$. This mean vector is used to express the centralized vector as $\bar{\mathbf{x}} = \mathbf{x} - \boldsymbol{\mu}$, with the elements of $\bar{\mathbf{x}}$ defined as $\bar{x}_i = x_i - \mu_i$ and the corresponding sample vectors as $\bar{X}_i = \bar{X}_i - \mu_i = [\bar{x}_i^1 \; \bar{x}_i^2 \; \cdots \; \bar{x}_i^n]$, where $\bar{x}_i^j = x_i^j - \mu_i$. Accordingly, the covariance matrix of $\bar{\mathbf{x}}$ is calculated as $\boldsymbol{\Omega} = \mathrm{E}[\bar{\mathbf{x}}\bar{\mathbf{x}}^T] \approx \frac{1}{n}\bar{\mathbf{X}}\bar{\mathbf{X}}^T$.

The goal of principal component analysis is to find an orthonormal transformation matrix $\mathbf{P}$ to decorrelate $\bar{\mathbf{x}}$. This transformation can be written as $\bar{\mathbf{y}} = \mathbf{P}\bar{\mathbf{x}}$, with the covariance matrix of $\bar{\mathbf{y}}$ being diagonal. Since $\boldsymbol{\Omega}$ is symmetrical, its singular value decomposition (SVD) can be expressed as follows:

$$\boldsymbol{\Omega} = \boldsymbol{\Phi}\boldsymbol{\Lambda}\boldsymbol{\Phi}^T,$$

where $\boldsymbol{\Phi} = [\phi_1 \; \phi_2 \; \cdots \; \phi_m]$ denotes the $m \times m$ orthonormal eigenvector matrix and $\boldsymbol{\Lambda} = \mathrm{diag}\{\lambda_1 \; \lambda_2 \; \cdots \; \lambda_m\}$ is the diagonal eigenvalue matrix with $\lambda_1 \geq \lambda_2 \geq \cdots \geq \lambda_m$. By setting $\mathbf{P} = \boldsymbol{\Phi}^T$, the vector $\bar{\mathbf{x}}$ can be decorrelated, resulting in $\bar{\mathbf{Y}} = \mathbf{P}\bar{\mathbf{X}}$ and $\boldsymbol{\Lambda} = \mathrm{E}[\bar{\mathbf{y}}\bar{\mathbf{y}}^T] \approx \frac{1}{n}\bar{\mathbf{Y}}\bar{\mathbf{Y}}^T$.

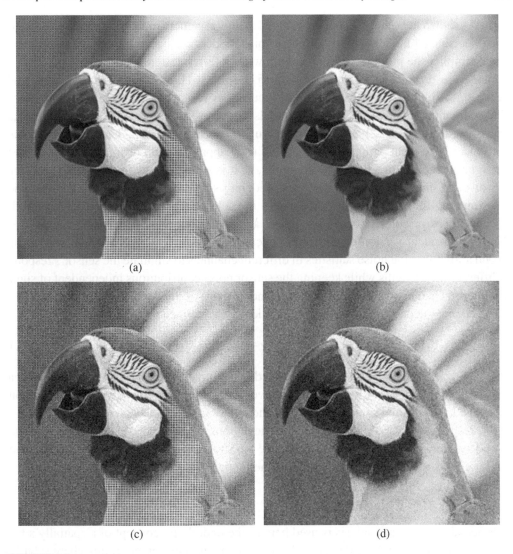

FIGURE 5.6

Digital camera noise modeling: (a) original, noise-free CFA image and (b) its demosaicked version; (c) noised CFA image and (d) its demosaicked version.

Principal component analysis not only decorrelates the data, but it is also an optimal way to represent the original signal using a subset of principal components. This property, known as optimal dimensionality reduction [59], refers to the situations when the k most important eigenvectors are used to form the transformation matrix $\mathbf{P}^T = [\phi_1 \ \phi_2 \ \cdots \ \phi_k]$, for $k < m$. In this case, the transformed dataset $\bar{\mathbf{Y}} = \mathbf{P}\bar{\mathbf{X}}$ will be of dimensions $k \times n$, as opposed to the original dataset $\bar{\mathbf{X}}$ of dimensions $m \times n$, while preserving most of the energy of $\bar{\mathbf{X}}$.

5.3.2 Noise Modeling

To design a filter capable of removing noise and simultaneously preserving image edges and details, the effect of noise on the desired signal is usually studied in simulated conditions. As shown in Figure 5.6, noise observed in digital camera images can be approximated using specialized models. Many popular noise models are based on certain assumptions which simplify the problem at hand and allow for a faster design.

Image sensor noise is signal-dependent [52], [60], [61], as the noise variance depends on the signal magnitude. Reference [60] argues that various techniques, such as Poisson, film-grain, multiplicative, and speckle models can be used to approximate such noise characteristics. Reference [52] proposes simulating the noise effect using Gaussian white noise and sensor dependent parameters. A widely used approximation of image sensor noise is the signal-independent additive noise model, as it is simple to use in the design and analysis of denoising algorithms and allows modeling signal-dependent noise characteristics by estimating the noise variance adaptively in each local area [50]. By considering the different types of color filters in the image acquisition process, it is reasonable to use a channel-dependent version of the signal-independent additive noise model [55]. This approach allows varying noise statistics in different channels to simulate the sensor's response in different wavelengths while keeping the sensor noise contributions independent of signal within each channel to simplify the design and analysis of the denoising algorithm [57].

The channel-dependent additive noise model can be defined as follows [56], [57]:

$$\tilde{r} = r + \upsilon_r, \quad \tilde{g} = g + \upsilon_g, \quad \tilde{b} = b + \upsilon_b, \tag{5.1}$$

where υ_r, υ_g and υ_b are mutually uncorrelated noise signals in the red, green and blue locations of the CFA image. Following the additive nature of this model, the noise contributions are added to the desired sample values r, g and b to obtain the noisy (acquired) signals $\tilde{r}$, $\tilde{g}$, and $\tilde{b}$, respectively. Note that the standard deviations σ_r, σ_g, and σ_b corresponding to υ_r, υ_g and υ_b, may have different values. Figure 5.6 shows that this noise model can produce similar effects as can be seen in real-life camera images in Figure 5.3.

5.3.3 Block-Based Statistics

Principal component analysis is used here as the underlying concept of a spatially adaptive denoising method for CFA mosaic data. The method presented in this chapter builds on the approach introduced in Reference [62] which uses the optimal dimensionality reduction property of principal component analysis in the design of a denoising algorithm for monochromatic images. To fully exploit the correlation among the samples acquired using different types of color filters, the so-called *variable block* is defined to include at least one pixel for each type of color filter used in the acquisition process [57]. The pixels inside this block are used as the variables in PCA training.

Each variable vector is associated with the so-called *training block* which contains additional samples for training [57]. The training block should be much bigger than the variable block in order to ensure that the statistics of the variables can be reasonably calculated. If any part of the training block matches the variable block, the pixels of that part are considered as the samples of the variable vector. Since such parts occupy different spatial locations in the image, the samples can be assumed to be independent draws of the variable.

For the sake of simplicity in the following discussion, Figure 5.7 shows the minimum size variable block and the corresponding training block for the mosaic data captured using the Bayer CFA [63]. This variable block can be written as a four-element vector $\mathbf{x} =$

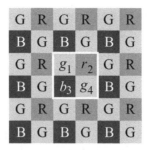

FIGURE 5.7

Illustration of the 6×6 variable block and 2×2 training block (g_1, r_2, b_3, and g_4) in the spatially adaptive PCA-based CFA image denoising method.

$[g_1 \ r_2 \ b_3 \ g_4]^T$. Practical implementations, however, can use larger size variable blocks. The whole dataset of $\mathbf{x}$ can be written as $\mathbf{X} = [G_1^T \ R_2^T \ B_3^T \ G_4^T]^T$, where G_1, R_2, B_3, and G_4 denote the row vectors containing all the samples associated with g_1, r_2, b_3, and g_4, respectively.

The mean values μ_{g_1}, μ_{r_2}, μ_{b_3}, and μ_{g_4} of variables g_1, r_2, b_3, and g_4 can be estimated as the average of all the samples in G_1, R_2, B_3, and G_4, respectively. These mean values constitute the mean vector $\boldsymbol{\mu} = [\mu_{g_1} \ \mu_{r_2} \ \mu_{b_3} \ \mu_{g_4}]^T$ of the variable vector $\mathbf{x}$. Using the mean vector $\boldsymbol{\mu}$, the centralized version of $\mathbf{x}$ and $\mathbf{X}$ can be expressed as $\bar{\mathbf{x}} = \mathbf{x} - \boldsymbol{\mu}$ and $\bar{\mathbf{X}} = [G_1^T - \mu_{g_1} \ R_2^T - \mu_{r_1} \ B_3^T - \mu_{b_3} \ G_4^T - \mu_{g_4}]^T$, respectively.

Using the additive noise model, the noisy observation of $\mathbf{x}$ can be expressed as $\tilde{\mathbf{x}} = \mathbf{x} + \mathbf{v}$, where $\mathbf{v} = [\upsilon_{g_1} \ \upsilon_{r_2} \ \upsilon_{b_3} \ \upsilon_{g_4}]^T$ is the noise variable vector. Assuming additive noise with zero mean, the mean vectors of $\tilde{\mathbf{x}}$ and $\mathbf{x}$ are identical, that is, $\mathrm{E}[\tilde{\mathbf{x}}] = \mathrm{E}[\mathbf{x}] = \boldsymbol{\mu}$. Since $\mathbf{x}$ is unavailable in practice, $\boldsymbol{\mu}$ is calculated from the samples of $\tilde{\mathbf{x}}$, resulting in $\bar{\tilde{\mathbf{x}}} = \tilde{\mathbf{x}} - \boldsymbol{\mu} = \bar{\mathbf{x}} + \mathbf{v}$ as the centralized vector of $\tilde{\mathbf{x}}$.

The whole dataset of additive channel-dependent noise $\mathbf{v}$ can be written as $\mathbf{V} = [V_{g_1}^T \ V_{r_2}^T \ V_{b_3}^T \ V_{g_4}^T]^T$, where V_{r_2} comes from the red channel noise υ_r, V_{g_1} and V_{g_4} come from the green channel noise υ_g, and V_{b_3} comes from the blue channel noise υ_b. The available measurements of the noise-free dataset $\mathbf{X}$ can thus be expressed as $\tilde{\mathbf{X}} = \mathbf{X} + \mathbf{V}$. Subtracting the mean vector $\boldsymbol{\mu}$ from $\tilde{\mathbf{X}}$ provides the centralized dataset $\bar{\tilde{\mathbf{X}}} = \bar{\mathbf{X}} + \mathbf{V}$ of the vector $\bar{\tilde{\mathbf{x}}}$.

The problem can now be seen as estimating $\bar{\mathbf{X}}$ from the noisy measurement $\bar{\tilde{\mathbf{X}}}$; the use of PCA to complete this task is discussed in the next section. Assuming that $\hat{\bar{\mathbf{X}}}$, which is the estimated dataset of $\bar{\mathbf{X}}$, is available, then the samples in the training block are denoised. Since pixels located far away from the location under consideration have usually very little or even no influence on the denoising estimate, the central part of the training block can be used as the denoising block [57]. The CFA image is denoised by moving the denoising block across the pixel array to affect all the pixels in the image.

5.4 Denoising in the PCA Domain

Since the energy of a signal usually concentrates in a small subset of the PCA transformed dataset, as opposed to the energy of noise that spreads evenly over the whole dataset, the optimal dimensionality reduction property of principal component analysis can be used in noise removal. Namely, keeping the most important subset of the transformed dataset to conduct the inverse PCA transform can significantly reduce noise while still being able to restore the desired signal.

5.4.1 Denoising Procedure

The covariance matrix $\Omega_{\bar{x}}$ of $\bar{x}$ can be used to obtain the corresponding optimal PCA transformation matrix $P_{\bar{x}}$ using the approach discussed in Section 5.3.1. Since the available dataset $\bar{\bar{X}}$ is corrupted by noise, $\Omega_{\bar{x}}$ cannot be directly computed; instead it can be estimated using the linear noise model $\bar{\bar{x}} = \bar{x} + v$. Assuming that n training samples are available for each element of $\bar{\bar{x}}$, this can be realized via the maximum likelihood estimation (MLE), a method that fits a statistical model to the data to obtain model's parameters, as follows [57]:

$$\Omega_{\bar{\bar{x}}} = E\left[(\bar{\bar{x}} - E[\bar{\bar{x}}])(\bar{\bar{x}} - E[\bar{\bar{x}}])^T\right], \tag{5.2}$$

$$\approx \frac{1}{n}\bar{\bar{X}}\bar{\bar{X}}^T = \frac{1}{n}(\bar{X}\bar{X}^T + \bar{X}V^T + V\bar{X}^T + VV^T). \tag{5.3}$$

Given the fact that the signal term $\bar{X}$ and the noise term V are uncorrelated, $\bar{X}V^T$ and $V\bar{X}^T$ are negligible. The above equation can thus be reduced to the following:

$$\Omega_{\bar{\bar{x}}} = \Omega_{\bar{x}} + \Omega_V \approx \frac{1}{n}(\bar{X}\bar{X}^T + VV^T), \tag{5.4}$$

where $\Omega_{\bar{x}} \approx \frac{1}{n}\bar{X}\bar{X}^T$ and $\Omega_V \approx \frac{1}{n}VV^T$ are the covariance matrices of $\bar{x}$ and v, respectively.

Since the elements of the noise vector $v = [v_{g_1}\ v_{r_2}\ v_{b_3}\ v_{g_4}]^T$ are uncorrelated with each other, Ω_V can be expressed as follows:

$$\Omega_V = E\left[vv^T\right] = diag\left\{\sigma_g^2, \sigma_r^2, \sigma_b^2, \sigma_g^2\right\}, \tag{5.5}$$

where σ_g, σ_r, and σ_b are standard deviations of channel-dependent noise in v_g, v_r and v_b in Equation 5.1. The covariance of $\bar{x}$ can thus be calculated as $\Omega_{\bar{x}} = \Omega_{\bar{\bar{x}}} - \Omega_V$, with possible negative values in the diagonal positions replaced with zeroes.

By decomposing $\Omega_{\bar{x}}$ as $\Omega_{\bar{x}} = \Phi_{\bar{x}}\Lambda_{\bar{x}}\Phi_{\bar{x}}^T$ where $\Phi_{\bar{x}} = [\phi_1\ \phi_2\ \phi_3\ \phi_4]$ is the 4×4 orthonormal eigenvector matrix and $\Lambda_{\bar{x}} = diag\{\lambda_1, \lambda_2, \lambda_3, \lambda_4\}$ is the diagonal eigenvalue matrix with $\lambda_1 \geq \lambda_2 \geq \lambda_3 \geq \lambda_4$, the orthonormal PCA transformation matrix for $\bar{X}$ can be expressed as $P_{\bar{x}} = \Phi_{\bar{x}}^T$. For the identical noise levels of v_r, v_g, and v_b in Equation 5.1, that is, $\sigma_r = \sigma_g = \sigma_b$, the covariance matrix Ω_V will reduce to an identity matrix with a scaling factor σ_g^2. In this case, the singular value decomposition of $\Omega_{\bar{\bar{x}}}$ and $\Omega_{\bar{x}}$ will give the same eigenvector matrix and hence the same PCA transformation matrix $P_{\bar{x}}$. In all other cases, the singular value decomposition of $\Omega_{\bar{\bar{x}}}$ and $\Omega_{\bar{x}}$ will yield different eigenvector matrices.

Applying $P_{\bar{x}}$ to the noisy dataset $\bar{\bar{X}}$ gives $\bar{\bar{Y}} = P_{\bar{x}}\bar{\bar{X}} = P_{\bar{x}}(\bar{X} + V)$. This can be equivalently written as $\bar{\bar{Y}} = \bar{Y} + V_Y$, where $\bar{Y} = P_{\bar{x}}\bar{X}$ and $V_Y = P_{\bar{x}}V$ denote the decorrelated dataset for signal and the transformed dataset for noise, respectively. Since $\bar{Y}$ and V_Y are uncorrelated, the covariance matrix of $\bar{\bar{Y}}$ can be expressed as follows [57]:

$$\Omega_{\bar{\bar{y}}} = \Omega_{\bar{y}} + \Omega_{v_y} \approx \frac{1}{n}\bar{\bar{Y}}\bar{\bar{Y}}^T, \tag{5.6}$$

where $\Omega_{\bar{y}} = \Lambda_{\bar{x}} \approx \frac{1}{n}\bar{\mathbf{Y}}\bar{\mathbf{Y}}^T$ and $\Omega_{v_y} = P_{\bar{x}}\Omega_v P_{\bar{x}}^T \approx \frac{1}{n}\mathbf{V}_{\mathbf{Y}}\mathbf{V}_{\mathbf{Y}}^T$ are the covariance matrices of $\bar{\mathbf{Y}}$ and $\mathbf{V}_{\mathbf{Y}}$, respectively.

Given the fact that most of the energy of $\bar{\mathbf{Y}}$ concentrates in the first several rows of $\tilde{\bar{\mathbf{Y}}}$ whereas the energy of $\mathbf{V}_{\mathbf{Y}}$ is distributed in $\tilde{\bar{\mathbf{Y}}}$ much more evenly, setting the last several rows of $\tilde{\bar{\mathbf{Y}}}$ to zero preserves the signal $\bar{\mathbf{Y}}$ while removing the noise $\mathbf{V}_{\mathbf{Y}}$. Unaltered rows of $\tilde{\bar{\mathbf{Y}}}$ constitute the so-called dimension reduced dataset $\tilde{\bar{\mathbf{Y}}}'$. It holds that $\tilde{\bar{\mathbf{Y}}}' = \bar{\mathbf{Y}}' + \mathbf{V}_{\mathbf{Y}'}$ where $\bar{\mathbf{Y}}'$ and $\mathbf{V}_{\mathbf{Y}'}$ represent the dimension reduced datasets of $\bar{\mathbf{Y}}$ and $\mathbf{V}_{\mathbf{Y}}$, respectively. The corresponding covariance matrices relate as $\Omega_{\tilde{\bar{y}}'} = \Omega_{\bar{y}'} + \Omega_{v_{y'}}$.

Further denoising of $\tilde{\bar{\mathbf{Y}}}'$ can be achieved via linear minimum mean square error estimation (LMMSE) applied to individual rows, as follows [57]:

$$\hat{\bar{\mathbf{Y}}}'_i = c_i \cdot \tilde{\bar{\mathbf{Y}}}'_i, \ \text{ for } c_i = \Omega_{\bar{y}'}(i,i)/(\Omega_{\bar{y}'}(i,i)+\Omega_{v_{y'}}(i,i)), \tag{5.7}$$

where i denotes the row index. Repeating the estimation procedure for each nonzero row of $\bar{\mathbf{Y}}'$ yields the denoised dataset $\hat{\bar{\mathbf{Y}}}'$. The denoised version of the original dataset $\bar{\mathbf{X}}$, which represents the estimate of an unknown noiseless dataset $\bar{\mathbf{X}}$, can be obtained as $\hat{\bar{\mathbf{X}}} = P_{\bar{x}}^{-1}\hat{\bar{\mathbf{Y}}}'$ by performing the transform from the PCA domain to the time domain. The denoised CFA block is produced by reformatting $\hat{\bar{\mathbf{X}}}$.

5.4.2 Performace Improvements

The proposed denoising method effectively removes noise in CFA images [57]. Unfortunately, the method can produce noise residual-like effects in smooth areas with low local signal to noise ratio or low contrast. This phenomenon is caused by the lack of structural content in the image, resulting in less significant principal components and hence less effective discrimination between noise and signals. It should also be noted that mean value estimation can be biased due to the availability of only one color component in each spatial location of the CFA image, resulting in an estimation bias of the covariance matrix and hence the PCA transformation matrix. Another problem is the occurrence of phantom artifacts along edge boundaries with smooth backgrounds. This can be attributed to the difference between the sample structure and the local training block statistics. Depending on the color and structural characteristics of the captured image, the presence of the above-described effects can significantly reduce visual quality of the final image.

5.4.2.1 CFA Image Decomposition

Noise residual-like effects can be removed by tuning the denoising parameters in the proposed method. Another possible solution is to decompose [57] the noisy CFA image I_v into a low-pass smooth image $I_v^l = I_v * G$ and a high-pass image $I_v^h = I_v - I_v^l$, where G denotes the convolution kernel of a two dimensional Gaussian low-pass filter defined as

$$G(x,y) = \frac{1}{\sqrt{2\pi}s} \exp\left(-\frac{x^2+y^2}{2s^2}\right). \tag{5.8}$$

Choosing a suitable value of the scaling parameter s allows for I_v^l with blurred structural content and almost completely removed noise. This implies that a complementary image

I_v^h contains almost complete high-frequency content, including the essential edge information to be preserved and undesired structures which are attributed to noise. Since noise is dominant in the flat areas of I_v^h, it can be effectively suppressed by LMMSE filtering in the PCA domain using the method described in the previous section. The denoising procedure outputs the image $\hat{I}_v^h$ which can be used to produce the denoised CFA image $\hat{I} = I_v^l + \hat{I}_v^h$.

5.4.2.2 Training Sample Selection

Phantom artifacts around edge boundaries with smooth backgrounds are caused by inappropriate training samples in the training block. In PCA training, this problem can be overcome by using only blocks which are similar to the underlying variable block, resulting in better estimates of the covariance matrix of the variable block and hence outputting a more accurate PCA transformation matrix [57].

By following the description in Section 5.3.3, the variable block $\tilde{\mathbf{x}}$ in I_v^h is associated with the training dataset $\tilde{\mathbf{X}}$ generated from the training block centered on $\tilde{\mathbf{x}}$. Each column of $\tilde{\mathbf{X}}$ can thus be seen as a sample vector of $\tilde{\mathbf{x}}$. Selecting the best samples from $\tilde{\mathbf{X}}$ for PCA transformation requires to evaluate the difference between the sample vector $\vec{\tilde{x}}_0$ containing the samples of the variable block $\tilde{\mathbf{x}}$ and the vector $\vec{\tilde{x}}_k$ which denotes the k-th column of $\tilde{\mathbf{X}}$, for $k = 1, 2, ..., n$. This can be done using the Euclidean distance as follows [57]:

$$d_k = \frac{1}{m} \sum_{i=1}^{m} \left(\vec{\tilde{x}}_k(i) - \vec{\tilde{x}}_0(i) \right)^2 \tag{5.9}$$

$$\approx \frac{1}{m} \sum_{i=1}^{m} \left(\vec{x}_k(i) - \vec{x}_0(i) \right)^2 + \sigma_a^2, \tag{5.10}$$

where m denotes the vector length and $\sigma_a = (\sigma_r^2 + 2\sigma_g^2 + \sigma_b^2)^{1/2}/2$. Vectors $\vec{x}_k$ and $\vec{x}_0$ are the noiseless counterparts of $\vec{\tilde{x}}_k$ and $\vec{\tilde{x}}_0$, respectively. Obviously, the smaller the distance d_k is, the more similar $\vec{x}_k$ is to $\vec{x}_0$.

The training sample selection criterion can be defined as follows [57]:

$$d_k \leq T^2 + \sigma_a^2, \tag{5.11}$$

where T is a predetermined parameter. If the above condition is met, then $\vec{\tilde{x}}_k$ is selected as one training sample of $\tilde{\mathbf{x}}$. Note that a high number of sample vectors may be required in practice to guarantee a reasonable estimation of the covariance matrix of $\tilde{\mathbf{x}}$. Assuming that $\tilde{\mathbf{X}}_b$ denotes the dataset composed of the sample vectors that give the smallest distance to $\vec{\tilde{x}}_0$, the algorithms described in Section 5.4.1 should be applied to $\tilde{\mathbf{X}}_b$, instead of the original dataset $\tilde{\mathbf{X}}$.

5.4.3 Experimental Results

The size of the variable and training blocks is an important design parameter in the proposed method to achieve good performance. Note that the size of the denoising block should not exceed the size of the variable block. Different block size settings lead to different results and have an impact on the computational efficiency. In general, denoising low-resolution images requires a small variable block since the spatial correlation in such images is also low. It was found empirically that 4×4, 6×6, or 8×8 variable block size

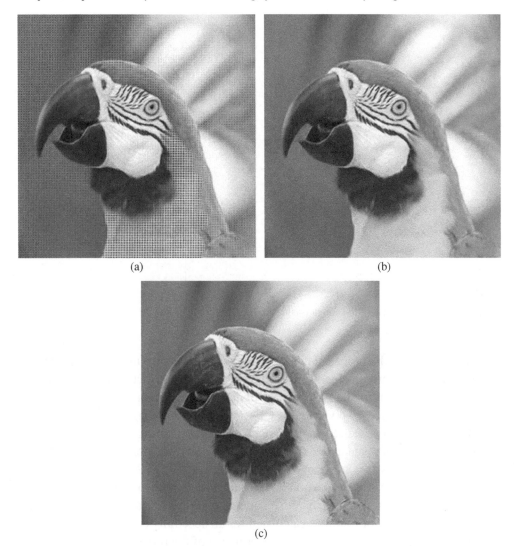

(a)   (b)

(c)

FIGURE 5.8

PCA-driven denoising of an artificially noised image shown in Figure 5.6c: (a) denoised CFA image, (b) its demosaicked version, and (c) original, noise-free image.

settings can produce good results in most situations, while the size of training block should be at least 16 times larger, that is, 24×24 or 30×30 for a 6×6 variable block.

Figure 5.8a shows the result when the proposed method is applied to the CFA image with simulated noise. Comparing this image with its noisy version shown in Figure 5.6c reveals that noise is effectively suppressed in both smooth and edge regions while there is no obvious loss of the structural contents. Visual inspection of the corresponding images demosaicked using the method of Reference [64] confirms what was expected. Namely, as shown in Figure 5.6d, demosaicking the noisy CFA image with no denoising produces poor results; in some situations the noise level actually increases due to blending the noise contributions across channels. This is not the case of the demosaicked denoised image in Figure 5.8b which is qualitatively similar to the original test image shown in Figure 5.8c.

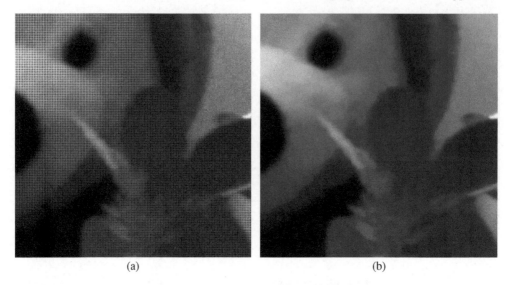

(a)   (b)

FIGURE 5.9

PCA-driven denoising of a real-life camera image shown in Figure 5.3a: (a) denoised CFA image and (b) its demosaicked version.

The performance of the proposed method will now be evaluated using digital camera images with real, non-approximated noise. Denoising raw sensor images using the proposed method requires calculating the noise energy of each channel from the acquired CFA data. This can be accomplished by dividing the CFA image into subimages and then processing each subimage using the one-stage orthogonal wavelet transform [65]. Assuming that $\mathbf{w}$ denotes the diagonal subband at the decomposed first stage, the noise level in each subimage can be estimated as $\sigma = \text{median}(\mathbf{w})/0.6475$ [66] or $\sigma = ((MN)^{-1} \sum_i^M \sum_j^N \mathbf{w}^2(i,j))^{0.5}$, where (i,j) denotes the spatial location and M and N denote the subband dimensions. In the situations when there is more than one subimage per color channel, the noise level is estimated as the average of σ values calculated for all spectrally equivalent subimages.

Figure 5.9 and 5.10 demonstrate good performance of the proposed method in environments with the presence of real image sensor noise. Comparing the denoised CFA images with the acquired ones clearly shows that the proposed method efficiently uses spatial and spectral image characteristics to suppress noise and simultaneously preserve edges and image details. The same conclusion can be made when visually inspecting the corresponding demosaicked images.

Full-color results presented in this chapter are available at http://www4.comp.polyu.edu. hk/~cslzhang/paper/cpPCA.pdf. Additional results and detailed performance analysis can be found in Reference [57].

5.5   Conclusion

This chapter presented image denoising solutions for digital cameras equipped with a color filter array placed on top of a monochrome image sensor. Namely, taking into con-

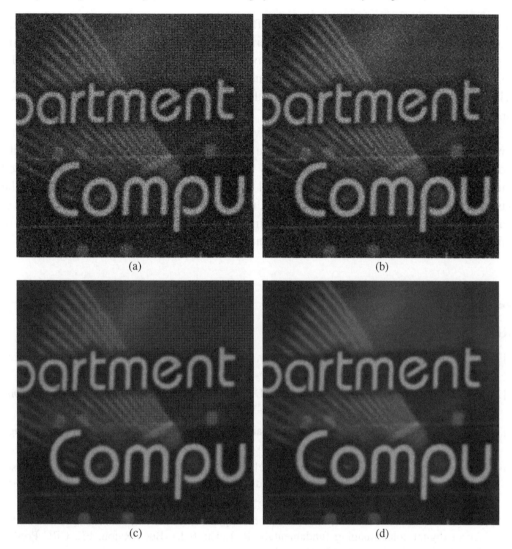

FIGURE 5.10

PCA-driven denoising of a real-life digital camera image: (a) acquired CFA sensor image and (b) its demosaicked version; (c) denoised CFA image and (d) its demosaicked version.

sideration the fundamentals of single-sensor color imaging and digital camera image processing, the chapter identified three pipelining frameworks that can be used to produce a denoised image. These frameworks differ in the position of the denoising step with respect to the demosaicking step in the camera image processing pipeline, thus having their own design, performance, and implementation challenges.

The framework that performs denoising before demosaicking was the main focus of this chapter. Denoising the color filter array mosaic data is the most natural way of handling the image noise problem in the digital cameras under consideration. The framework can effectively suppress noise and preserve color edges and details, while having the potential to achieve high processing rates. This is particularly true for the proposed principal component analysis-driven approach that adaptively calculates covariance matrices to allow effective signal energy clustering and efficient noise removal. The approach utilizes both spatial and spectral correlation characteristics of the captured image and takes advantage of local similarities that exist among blocks of color filter array mosaic data in order to improve the estimation accuracy of the principal component analysis transformation matrix. This constitutes a basis for achieving the desired visual quality using the proposed approach.

Obviously, image denoising solutions have an extremely valuable position in digital imaging. The trade-off between performance and efficiency makes many denoising methods indispensable tools for digital cameras and their applications. Since the proposed method is reasonably robust in order to deal with the infinite number of variations in the visual scene and varying image sensor noise, it can play a key role in modern imaging systems and consumer electronic devices with image-capturing capabilities which attempt to mimic human perception of the visual environment.

References

[1] P.L.P. Dillon, D.M. Lewis, and F.G. Kaspar, "Color imaging system using a single CCD area array," *IEEE Journal of Solid-State Circuits*, vol. 13, no. 1, pp. 28–33, February 1978.

[2] B.T. Turko and G.J. Yates, "Low smear CCD camera for high frame rates," *IEEE Transactions on Nuclear Science*, vol. 36, no. 1, pp. 165–169, February 1989.

[3] A.J. Blanksby and M.J. Loinaz, "Performance analysis of a color CMOS photogate image sensor," *IEEE Transactions on Electron Devices*, vol. 47, no. 1, pp. 55–64, January 2000.

[4] D. Doswald, J. Haflinger, P. Blessing, N. Felber, P. Niederer, and W. Fichtner, "A 30-frames/s megapixel real-time CMOS image processor," *IEEE Journal of Solid-State Circuits*, vol. 35, no. 11, pp. 1732–1743, November 2000.

[5] R. Lukac, *Single-Sensor Imaging: Methods and Applications for Digital Cameras*, ch. Single-sensor digital color imaging fundamentals, R. Lukac (ed.), Boca Raton, FL: CRC Press / Taylor & Francis, September 2008, pp. 1–29.

[6] R. Lukac and K.N. Plataniotis, "Color filter arrays: Design and performance analysis," *IEEE Transactions on Consumer Electronics*, vol. 51, no. 4, pp. 1260–1267, November 2005.

[7] FillFactory, "Technology - image sensor: The color filter array faq." Available online: http://www.fillfactory.com/htm/technology/htm/rgbfaq.htm.

[8] K. Hirakawa and P.J. Wolfe, *Single-Sensor Imaging: Methods and Applications for Digital Cameras*, ch. Spatio-spectral sampling and color filter array design, R. Lukac (ed.), Boca Raton, FL: CRC Press / Taylor & Francis, September 2008, pp. 137–151.

[9] K.A. Parulski and R. Reisch, *Single-Sensor Imaging: Methods and Applications for Digital Cameras*, ch. Digital camera image storage formats, R. Lukac (ed.), Boca Raton, FL: CRC Press / Taylor & Francis, September 2008, pp. 351–379.

[10] Technical Committee ISO/TC 42, Photography, "Electronic still picture imaging - removable memory, part 2: Image data format - TIFF/EP," ISO 12234-2, January 2001.

[11] "Information technology - digital compression and coding of continuous-tone still images: Requirements and guidelines." ISO/IEC International Standard 10918-1, ITU-T Recommendation T.81, 1994.

[12] Japan Electronics and Information Technology Industries Association, "Exchangeable image file format for digital still cameras: Exif Version 2.2," Technical Report, JEITA CP-3451, April 2002.

[13] N. Zhang, X. Wu, and L. Zhang, *Single-Sensor Imaging: Methods and Applications for Digital Cameras*, ch. Lossless compression of color mosaic images and videos, R. Lukac (ed.), Boca Raton, FL: CRC Press / Taylor & Francis, September 2008, pp. 405–428.

[14] N.X. Lian, V. Zagorodnov, and Y.P. Tan, *Single-Sensor Imaging: Methods and Applications for Digital Cameras*, ch. Modelling of image processing pipelines in single-sensor digital cameras, R. Lukac (ed.), Boca Raton, FL: CRC Press / Taylor & Francis, September 2008, pp. 381–404.

[15] C.C. Koh, J. Mukherjee, and S.K. Mitra, "New efficient methods of image compression in digital cameras with color filter array," *IEEE Transactions on Consumer Electronics*, vol. 49, no. 4, pp. 1448–1456, November 2003.

[16] R. Lukac and K.N. Plataniotis, "Single-sensor camera image compression," *IEEE Transactions on Consumer Electronics*, vol. 52, no. 2, pp. 299–307, May 2006.

[17] S.T. McHugh, "Digital camera image noise." Available online: http://www. cambridgein-colour.com/tutorials/noise.htm.

[18] K. Parulski and K.E. Spaulding, *Digital Color Imaging Handbook*, ch. Color image processing for digital cameras, G. Sharma (ed.), Boca Raton, FL: CRC Press, 2002, pp. 728–757.

[19] J.E. Adams and J.F. Hamilton, *Single-Sensor Imaging: Methods and Applications for Digital Cameras*, ch. Digital camera image processing chain design, R. Lukac (ed.), Boca Raton, FL: CRC Press / Taylor & Francis, September 2008, pp. 67–103.

[20] R. Ramanath, W.E. Snyder, Y. Yoo, and M.S. Drew, "Color image processing pipeline," *IEEE Signal Processing Magazine*, Special Issue on Color Image Processing, vol. 22, no. 1, pp. 34–43, January 2005.

[21] R. Lukac and K.N. Plataniotis, *Color Image Processing: Methods and Applications*, ch. Single-sensor camera image processing, R. Lukac and K.N. Plataniotis (eds.), Boca Raton, FL: CRC Press / Taylor & Francis, October 2006, pp. 363–392.

[22] B.K. Gunturk, J. Glotzbach, Y. Altunbasak, R.W. Schaffer, and R.M. Murserau, "Demosaicking: Color filter array interpolation," *IEEE Signal Processing Magazine*, Special Issue on Color Image Processing, vol. 22, no. 1, pp. 44–54, January 2005.

[23] E. Dubois, *Single-Sensor Imaging: Methods and Applications for Digital Cameras*, ch. Color filter array sampling of color images: Frequency-domain analysis and associated demosaicking algorithms, R. Lukac (ed.), Boca Raton, FL: CRC Press / Taylor & Francis, September 2008, pp. 183–212.

[24] D. Alleysson, B.C. de Lavarène, S. Süsstrunk, and J. Hérault, *Single-Sensor Imaging: Methods and Applications for Digital Cameras*, ch. Linear minimum mean square error demosaicking, R. Lukac (ed.), Boca Raton, FL: CRC Press / Taylor & Francis, September 2008, pp. 213–237.

[25] L. Zhang and W. Lian, *Single-Sensor Imaging: Methods and Applications for Digital Cameras*, ch. Video-demosaicking, R. Lukac (ed.), Boca Raton, FL: CRC Press / Taylor & Francis, September 2008, pp. 485–502.

[26] E.Y. Lam and G.S.K. Fung, *Single-Sensor Imaging: Methods and Applications for Digital Cameras*, ch. Automatic white balancing in digital photography, R. Lukac (ed.), Boca Raton, FL: CRC Press / Taylor & Francis, September 2008, pp. 267–294.

[27] R. Lukac, "New framework for automatic white balancing of digital camera images," *Signal Processing*, vol. 88, no. 3, pp. 582–593, March 2008.

[28] K. Hirakawa, *Single-Sensor Imaging: Methods and Applications for Digital Cameras*, ch. Color filter array image analysis for joint demosaicking and denoising, R. Lukac (ed.), Boca Raton, FL: CRC Press / Taylor & Francis, September 2008, pp. 239–266.

[29] R. Lukac and K.N. Plataniotis, "A new image sharpening approach for single-sensor digital cameras," *International Journal of Imaging Systems and Technology*, Special Issue on Applied Color Image Processing, vol. 17, no. 3, pp. 123–131, June 2007.

[30] S. Battiato, G. Messina, and A. Castorina, *Single-Sensor Imaging: Methods and Applications for Digital Cameras*, ch. Exposure correction for imaging devices: An overview, R. Lukac (ed.), Boca Raton, FL: CRC Press / Taylor & Francis, September 2008, pp. 323–349.

[31] R. Lukac, *Single-Sensor Imaging: Methods and Applications for Digital Cameras*, ch. Image resizing solutions for single-sensor digital cameras, R. Lukac (ed.), Boca Raton, FL: CRC Press / Taylor & Francis, September 2008, pp. 459–484.

[32] F. Gasparini and R. Schettini, *Single-Sensor Imaging: Methods and Applications for Digital Cameras*, ch. Automatic red-eye removal for digital photography, R. Lukac (ed.), Boca Raton, FL: CRC Press / Taylor & Francis, September 2008, pp. 429–457.

[33] P. Viola and M.J. Jones, "Robust real-time object detection," Tech. Rep. CRL 2001/01, Compaq Cambridge Research Laboratory, Cambridge, Massachusetts, February 2001.

[34] R.L. Hsu, M. Abdel-Mottaleb, and A.K. Jain, "Face detection in color images," *IEEE Transactions on Pattern Analysis and Machine Intelligence*, vol. 24, no. 5, pp. 696–706, May 2002.

[35] W.C. Kao and S.Y. Lin, *Single-Sensor Imaging: Methods and Applications for Digital Cameras*, ch. An overview of image/video stabilization techniques, R. Lukac (ed.), Boca Raton, FL: CRC Press / Taylor & Francis, September 2008, pp. 535–561.

[36] M. Ben-Ezra and S.K. Nayar, "Motion-based motion deblurring," *IEEE Transactions on Pattern Analysis and Machine Intelligence*, vol. 26, no. 6, pp. 689–698, June 2004.

[37] S. Farsiu, D. Robinson, M. Elad, and P. Milanfar, *Single-Sensor Imaging: Methods and Applications for Digital Cameras*, ch. Simultaneous demosaicking and resolution enhancement from under-sampled image sequences, R. Lukac (ed.), Boca Raton, FL: CRC Press / Taylor & Francis, September 2008, pp. 503–533.

[38] S.G. Narasimhan and S.K. Nayar, "Enhancing resolution along multiple imaging dimensions using assorted pixels," *IEEE Transactions on Pattern Recognition and Machine Intelligence*, vol. 27, no. 4, pp. 518–530, April 2005.

[39] E. Reinhard, G. Ward, S. Pattanaik, and P. Debevec, *High Dynamic Range Imaging*. San Francisco, CA, USA: Morgan Kaufmann Publishers, November 2005.

[40] I. Pitas and A.N. Venetsanopoulos, "Order statistics in digital image processing," *Proceedings of the IEEE*, vol. 80, no. 12, pp. 1892–1919, December 1992.

[41] R. Lukac and K.N. Plataniotis, *Advances in Imaging and Electron Physics*, ch. A taxonomy of color image filtering and enhancement solutions, pp. 187–264, San Diego, CA: Elsevier / Academic Press, February/March 2006.

[42] J.S. Lee, "Digital image smoothing and the sigma filter," *Graphical Models and Image Processing*, vol. 24, no. 2, pp. 255–269, November 1983.

[43] C. Tomasi and R. Manduchi, "Bilateral filtering for gray and color images," in *Proceedings of the IEEE International Conference on Computer Vision*, Bombay, India, January 1998, pp. 839–846.

[44] M. Zhang and B.K. Gunturk, "Multiresolution bilateral filtering for image denoising," *IEEE Transactions on Image Processing*, vol. 17, no. 12, pp. 2324–2333, December 2008.

[45] S.G. Chang, B. Yu, and M. Vetterli, "Spatially adaptive wavelet thresholding with context modeling for image denoising," *IEEE Transactions on Image Processing*, vol. 9, no. 9, pp. 1522–1531, September 2000.

[46] A. Pizurica and W. Philips, "Estimating the probability of the presence of a signal of interest in multiresolution single- and multiband image denoising," *IEEE Transactions on Image Processing*, vol. 15, no. 3, pp. 654–665, March 2006.

[47] J. Portilla, V. Strela, M.J. Wainwright, and E.P. Simoncelli, "Image denoising using scale mixtures of Gaussians in the wavelet domain," *IEEE Transactions on Image Processing*, vol. 12, no. 11, pp. 1338–1351, November 2003.

[48] R. Lukac, B. Smolka, K. Martin, K. N. Plataniotis, and A. N. Venetsanopulos, "Vector filtering for color imaging," *IEEE Signal Processing Magazine*, Special Issue on Color Image Processing, vol. 22, no. 1, pp. 74–86, January 2005.

[49] K.N. Plataniotis and A.N. Venetsanopoulos, *Color Image Processing and Applications*. New York: Springer Verlag, 2000.

[50] H.J. Trussell and R.E. Hartwig, "Mathematics for demosaicking," *IEEE Transactions on Image Processing*, vol. 11, no. 4, pp. 485–492, April 2002.

[51] R. Ramanath and W.E. Snyder, "Adaptive demosaicking," *Journal of Electronic Imaging*, vol. 12, no. 4, pp. 633–642, October 2003.

[52] K. Hirakawa and T.W. Parks, "Joint demosaicking and denoising," *IEEE Transactions on Image Processing*, vol. 15, no. 8, pp. 2146–2157, August 2006.

[53] K. Hirakawa, X.L. Meng, and P.J. Wolfe, "A framework for wavelet-based analysis and processing of color filter array images with applications to denoising and demosaicing," in *Proceedings of the IEEE International Conference on Acoustics, Speech, and Signal Processing*, Honolulu, Hawai, USA, April 2007, vol. 1, pp. 597–600.

[54] K. Hirakawa and X.L. Meng, "An empirical Bayes EM-wavelet unification for simultaneous denoising, interpolation, and/or demosaicing," in *Proceedings of the IEEE International Conference on Image Processing*, Atlanta, GA, USA, October 2006, pp. 1453–1456.

[55] L. Zhang, X. Wu and D. Zhang, "Color reproduction from noisy CFA data of single sensor digital cameras," *IEEE Transactions on Image Processing*, vol. 16, no. 9, pp. 2184–2197, September 2007.

[56] D. Paliy, V. Katkovnik, R. Bilcu, S. Alenius and K. Egiazarian, "Spatially adaptive color filter array interpolation for noiseless and noisy data," *International Journal of Imaging Systems and Technology, Special Issue on Applied Color Image Processing*, vol. 17, no. 3, pp. 105–122, June 2007.

[57] L. Zhang, R. Lukac, X. Wu, and D. Zhang, "PCA-based spatially adaptive denoising of CFA images for single-sensor digital cameras," *IEEE Transactions on Image Processing*, vol. 18, no. 4, pp. 797–812, April 2009.

[58] S. Haykin, *Neural Networks: A Comprehensive Foundation*. 2nd Edition, India: Prentice Hall, July 1998.

[59] K. Fukunaga, *Introduction to Statistical Pattern Recognition*. 2nd Edition, San Diego, CA: Academic Press, October 1990.

[60] A. Foi, S. Alenius, V. Katkovnik, and K. Egiazarian, "Noise measurement for raw-data of digital imaging sensors by automatic segmentation of nonuniform targets," *IEEE Sensors Journal*, vol. 7, no. 10, pp. 1456–1461, October 2007.

[61] A. Foi, V. Katkovnik, D. Paliy, K. Egiazarian, M. Trimeche, S. Alenius, R. Bilcu, and M. Vehvilainen, "Apparatus, method, mobile station and computer program product for noise estimation, modeling and filtering of a digital image," U.S. Patent Application No. 11/426,128, 2006.

[62] D.D. Muresan and T.W. Parks, "Adaptive principal components and image denoising," in *Proceedings of the IEEE International Conference on Image Processing*, Barcelona, Spain, September 2003, vol. 1, pp. 101–104.

[63] B.E. Bayer, "Color imaging array," U.S. Patent 3 971 065, July 1976.

[64] L. Zhang and X. Wu, "Color demosaicking via directional linear minimum mean square-error estimation," *IEEE Transactions on Image Processing*, vol. 14, no. 12, pp. 2167–2178, December 2005.

[65] S. Mallat, *A Wavelet Tour of Signal Processing* 2nd Edition, New York: Academic Press, September 1999.

[66] D.L. Donoho and I.M. Johnstone, "Ideal spatial adaptation via wavelet shrinkage," *Biometrika*, vol. 81, no. 3, pp. 425-455, September 1994.

6

Regularization-Based Color Image Demosaicking

Daniele Menon and Giancarlo Calvagno

6.1   Introduction

Demosaicking is the process of reconstructing the full color representation of an image acquired by a digital camera equipped with a color filter array (CFA). Most demosaicking approaches have been designed for the Bayer pattern [1] shown in Figure 6.1a. However, a number of different patterns [2], [3], [4], such as those shown in Figures 6.1b to 6.1d, have been recently proposed to enhance color image acquisition and restoration processes.

Various demosaicking methods have been surveyed in References [5], [6], [7], and [8]. Popular demosaicking approaches rely on directional filtering [9], [10], [11], wavelet [12], [13], frequency-domain analysis [14], and reconstruction [15] methods. In particular, an effective strategy consists in considering demosaicking as an inverse problem which can be solved by making use of some prior knowledge about the natural color images. This approach, generally known as *regularization*, has been also exploited for demosaicking [16], [17], [18], [19], [20], [21], as it allows to design algorithms that are suitable for any CFA.

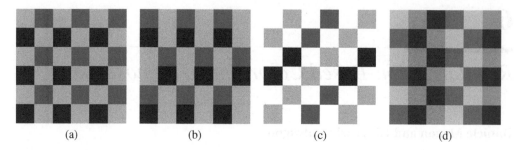

(a)   (b)   (c)   (d)

FIGURE 6.1 (See color insert.)

Examples of existing CFAs: (a) Bayer [1], (b) Lukac [2], (c) Hamilton [3], and (d) Hirakawa [4]. © 2009 IEEE

This chapter presents regularization methods for demosaicking. Namely, Section 6.2 focuses on the problem formulation and introduces the notation to be used throughout the chapter. Section 6.3 surveys existing regularization methods for sole demosaicking or jointly with super-resolution. Section 6.4 presents a new regularization technique which allows noniterative demosaicking. Section 6.5 presents performance comparisons of different methods. Finally, conclusions are drawn in Section 6.6.

6.2 Problem Statement

Given a continuous color image $\mathscr{I}_c(\boldsymbol{u}) = [R_c(\boldsymbol{u}), G_c(\boldsymbol{u}), B_c(\boldsymbol{u})]$, where $\boldsymbol{u} \in \mathbb{R}^2$ and $R_c(\boldsymbol{u})$, $G_c(\boldsymbol{u})$, and $B_c(\boldsymbol{u})$ denote the red, green, and blue color component, a digital camera aims at acquiring a discrete image $\mathscr{I}(\boldsymbol{n}) = [R(\boldsymbol{n}), G(\boldsymbol{n}), B(\boldsymbol{n})]$, with $\boldsymbol{n} \in \Gamma$, where Γ is the desired sampling lattice. It is assumed here that $\mathscr{I}_c(\boldsymbol{u})$ is bandlimited, such that sampling in the lattice Γ does not generate aliasing.

However, existing imaging systems are subject to some design requirements and technology limits which do not allow acquisition of the desired image. Therefore, the image acquisition process in a digital camera is as not straightforward as the ideal sampling (see Figure 6.2). The continuous color components $X_c(\boldsymbol{u})$, for $X = R$, G, and B, are captured at each pixel $\boldsymbol{n} \in \Gamma$ through spatial integration over the pixel sensor that can be represented as a convolution with a low-pass filter $p_X(\cdot)$ as follows:

$$X_p(\boldsymbol{n}) = p_X * X_c(\boldsymbol{n}), \text{ for } \boldsymbol{n} \in \Gamma. \tag{6.1}$$

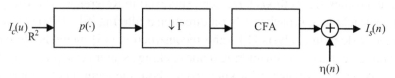

FIGURE 6.2

Image acquisition in a digital camera. © 2009 IEEE

Common practical models for the impulse response of the prefilter are the Gaussian or the rect functions. However, today's digital cameras use a CFA placed in front of the sensor to capture one color component in each pixel location [1], [2]. Moreover, some CFAs capture colors which can be expressed as a linear combination of the traditional red, green and blue components [3], [4]. The CFA-sampled image $\mathscr{I}_s(\boldsymbol{n})$ can thus be expressed as

$$\mathscr{I}_s(\boldsymbol{n}) = \sum_{X=R,G,B} c_X(\boldsymbol{n})X_p(\boldsymbol{n}) + \eta(\boldsymbol{n}), \tag{6.2}$$

where the *acquisition functions* $c_X(\boldsymbol{n})$, for $X = R$, G, and B, are periodic and for any pixel $\boldsymbol{n} \in \Gamma$ constrained as $c_R(\boldsymbol{n}) + c_G(\boldsymbol{n}) + c_B(\boldsymbol{n}) = 1$. The term $\eta(\boldsymbol{n})$ characterizes sensor noise introduced during image acquisition. It is assumed here that $\eta(\boldsymbol{n})$ is uncorrelated with respect to the acquired image $\mathscr{I}_s(\boldsymbol{n})$. More complex models for the noise can be found in Reference [22].

To this end, the image acquisition process can be represented as in Figure 6.2, and the relation between $\mathscr{I}_s(\boldsymbol{n})$, for $\boldsymbol{n} \in \Gamma$, and the continuous image $\mathscr{I}_c(\boldsymbol{u})$ is given by

$$\mathscr{I}_s(\boldsymbol{n}) = \sum_{X=R,G,B} c_X(\boldsymbol{n})p_X * X_c(\boldsymbol{n}) + \eta(\boldsymbol{n}), \text{ for } \boldsymbol{n} \in \Gamma. \tag{6.3}$$

Based on these considerations, frequency-domain analysis of the acquired image $\mathscr{I}_s(\boldsymbol{n})$ can be carried out. Since the acquisition functions $c_X(\boldsymbol{n})$ are periodic, they can be represented as a finite sum of harmonically related complex exponentials using the discrete Fourier series as follows:

$$c_X(\boldsymbol{n}) = \sum_{k \in \mathscr{P}(\Lambda)} \alpha_X(\boldsymbol{k})e^{j\boldsymbol{k}^T 2\pi V^{-1}\boldsymbol{n}}, \tag{6.4}$$

where V is the periodicity matrix and Λ is the lattice[1] generated by V. The term $\mathscr{P}(\Lambda)$ is a fundamental parallelepiped of the lattice Λ and the coefficients $\alpha(\boldsymbol{k})$ are complex, but $\alpha(\boldsymbol{k}) = \bar{\alpha}(-\boldsymbol{k})$, in order to ensure real values for $c(\boldsymbol{n})$.

Equation 6.2 can thus be rewritten as

$$\mathscr{I}_s(\boldsymbol{n}) = \sum_X \sum_k \alpha_X(\boldsymbol{k})e^{j\boldsymbol{k}^T 2\pi V^{-1}\boldsymbol{n}}X_p(\boldsymbol{n}) + \eta(\boldsymbol{n}), \tag{6.5}$$

where $X = R,G,B$ and $\boldsymbol{k} \in \mathscr{P}(\Lambda)$. The Fourier transform of the acquired image can be expressed as follows:

$$\mathscr{I}_s(\boldsymbol{\omega}) = \sum_{X=R,G,B} \sum_{k \in \mathscr{P}(\Lambda)} \alpha_X(\boldsymbol{k})X_p(\boldsymbol{\omega} - \boldsymbol{k}^T 2\pi V^{-1}) + \eta(\boldsymbol{\omega}), \tag{6.6}$$

where $X_p(\boldsymbol{\omega})$ denotes the Fourier transform of the color component $X_p(\boldsymbol{n})$. As a consequence, the spectrum of a CFA image has at most $|\mathscr{P}(V)|$ frequency peaks and each peak is a linear combination of the spectra of the three color components. In particular, the baseband component is given by a weighted positive sum of the three color channels, where each weight $\alpha_X(\boldsymbol{0})$ is equivalent to the ratio between the number of acquired samples for

---

[1]When the acquisition functions are periodic on different lattices, Λ indicates the densest lattice over which all the $c(\boldsymbol{n})$, for $X = R,G,B$, are periodic. This lattice denotes also the periodicity of the CFA.

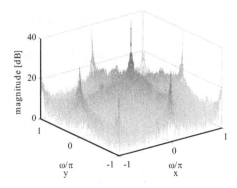

FIGURE 6.3

Spectrum of the test image *Lighthouse* sampled using the Bayer pattern. © 2009 IEEE

the component X and the number of the pixels in the image. In many demosaicking strategies it is suitable to define this weighted sum as the luminance component of the image, representing the achromatic information of the original scene.

For instance, in the case of the Bayer pattern, under the assumption of ideal impulse response and zero noise, (see also Reference [23]):

$$\mathscr{I}_s(\omega_1,\omega_2) = \frac{1}{4}\left[B(\omega_1 \pm \pi,\omega_2) - R(\omega_1 \pm \pi,\omega_2)\right] + \frac{1}{4}\left[R(\omega_1,\omega_2 \pm \pi) - B(\omega_1,\omega_2 \pm \pi)\right]$$
$$+ \frac{1}{4}\left[-R(\omega_1 \pm \pi,\omega_2 \pm \pi) + 2G(\omega_1 \pm \pi,\omega_2 \pm \pi) - B(\omega_1 \pm \pi,\omega_2 \pm \pi)\right]$$
$$+ \frac{1}{4}\left[R(\omega_1,\omega_2) + 2G(\omega_1,\omega_2) + B(\omega_1,\omega_2)\right], \tag{6.7}$$

where ω_1 and ω_2 indicate the horizontal and vertical frequencies. Therefore, it is possible to identify nine regions containing the signal energy, where the central region corresponds to the luminance and the others to the modulated chrominance components. Due to the correlation between the high-frequencies of the color components, the chrominance replicas have limited support, while the luminance overlies a large part of the frequency plane. This is clearly shown in Figure 6.3 which depicts the spectrum of the Kodak test image *Lighthouse* sampled with the Bayer pattern.

6.3 Regularization-Based Methods

The desired image $\mathscr{I}(\boldsymbol{n})$ has to be estimated from the acquired data $\mathscr{I}_s(\boldsymbol{n})$. It is evident that this is an ill-posed problem due to the loss of information introduced by the CFA-sampling process. A general principle for dealing with the instability of the problem is that of *regularization* [24], [25].

The *regularized solution* $\hat{\mathscr{I}}$ is defined as the solution to the following problem:

$$\hat{\mathscr{I}} = \arg\min_{\mathscr{I}} \left\{ \Psi(\mathscr{I},\mathscr{I}_s) + \sum_k \lambda_k J_k(\mathscr{I}) \right\}, \tag{6.8}$$

where the first term $\Psi(\mathscr{I}, \mathscr{I}_s)$, called the *data-fidelity term*, denotes a measure of the distance between the estimated image and the observed data, while the terms $J_k(\mathscr{I})$ denote regularizing constraints based on *a priori* knowledge of the original image. The regularization parameters λ_k control the tradeoff between the various terms.

In the matrix notation used in this chapter, r, g, and b denote the stacking vectors of the three full-resolution color components $R(n)$, $G(n)$, and $B(n)$, respectively. The term i denotes the vector obtained by stacking the three color component vectors, that is, $i^T = [r^T, g^T, b^T]$, whereas i_s is the stacking version of the image $\mathscr{I}_s(n)$ acquired by the sensor and η is the stacking vector of the noise. Then, the relation between the acquired samples i_s and the full-resolution image i is described as follows:

$$i_s = Hi + \eta, \tag{6.9}$$

where the matrix H is given by

$$H = [C_R P_R, C_G P_G, C_B P_B]. \tag{6.10}$$

The square matrices P_X account for the impulse response of the filters $p_X(n)$, respectively. The entries of the diagonal matrices C_X are obtained by stacking the acquisition functions $c_X(n)$.

Using the above notation, the solution to the regularization problem in Equation 6.8 can be found as follows:

$$\hat{i} = \arg\min_i \left\{ \Psi(i, i_s) + \sum_k \lambda_k J_k(i) \right\}. \tag{6.11}$$

The data-fidelity term $\Psi(i, i_s)$ is usually defined according to least-squares approaches, using the residual norm, that is

$$\Psi(i, i_s) = \|i_s - Hi\|_2^2, \tag{6.12}$$

where $\|\cdot\|_2$ indicates the ℓ_2 norm. However, in some techniques, the characteristics of the sensor noise are considered and the weighted ℓ_2 norm

$$\Psi(i, i_s) = \|i_s - Hi\|_{R_\eta^{-1}}^2, \tag{6.13}$$

where R_η denotes the autocorrelation matrix of the noise η, is preferred.

The main differences among the several regularization-based demosaicking methods relate to the regularizing terms considered in Equation 6.11. Reference [16] uses the total-variation technique [26] to impose smoothness on each color component and estimate the three color channels independently. Therefore, the prior terms are

$$\sum_{i=1}^{N_x} \sqrt{|\nabla_i^h x|^2 + |\nabla_i^v x|^2 + \beta^2}, \text{ for } x = r, g, b, \tag{6.14}$$

where $\nabla_i^h x$ and $\nabla_i^v x$ are discrete approximations to the horizontal and vertical first order difference, respectively, at the pixel i. The term N_x denotes the length of the vector x, and β is a constant that is included to make the term differentiable in zero. Then, the energy functional in Equation 6.11 is minimized with an iterative algorithm for any color component. References [18] and [19] improve this approach using the total-variation technique to impose smoothness also to the color difference signals $G - R$ and $G - B$ and the color sum signals $G + R$ and $G + B$. This is the same as requiring strong color correlation in high-frequency regions, which is a common assumption in many demosaicking approaches [6]. The estimation of the green component, which has the highest density in the Bayer pattern, is obtained as in Reference [16]. The missing red and blue values are estimated with an iterative procedure taking into account total-variation terms both of the color components and of the color differences and sums.

Reference [17] introduces a novel regularizing term to impose smoothness on the chrominances. This demosaicking algorithm uses suitable edge-directed weights to avoid excessive smoothing in edge regions. The prior term is defined as follows:

$$\sum_{n_1=1}^{N_1} \sum_{n_2=1}^{N_2} \sum_{l=-1}^{1} \sum_{m=-1}^{1} \tilde{e}_{n_1,n_2}^{l,m} \left[(X_{cb}(n_1,n_2) - X_{cb}(n_1+l,n_2+m))^2 \right.$$
$$\left. + (X_{cr}(n_1,n_2) - X_{cr}(n_1+l,n_2+m))^2 \right], \tag{6.15}$$

where X_{cb} and X_{cr} are the chrominances obtained using the following linear transform:

$$X_{cb} = -0.169R - 0.331G + 0.5B,$$
$$X_{cr} = 0.5R - 0.419G - 0.081B. \tag{6.16}$$

The terms N_1 and N_2 denote the width and the height of the image, respectively. The weights $\tilde{e}_{n_1,n_2}^{l,m}$, computed using the values of the CFA samples in the locations (n_1,n_2) and (n_1+2l,n_2+2m), are used to discourage smoothing across the edges. Using vector notation, Equation 6.15 can be rewritten as follows:

$$\sum_{l=-1}^{1} \sum_{m=-1}^{1} \left(\|x_{cb} - Z^{l,m}x_{cb}\|_{W_{is}^{l,m}}^2 + \|x_{cr} - Z^{l,m}x_{cr}\|_{W_{is}^{l,m}}^2 \right), \tag{6.17}$$

where x_{cb} and x_{cr} denote the vector representation of the chrominances, and $Z^{l,m}$ is the operator such that $Z^{l,m}x$ corresponds to shifting the image x by l pixels horizontally and m pixels vertically. The matrix $W_{is}^{l,m}$ is diagonal; its diagonal values correspond to the vector representation of the weights $\tilde{e}_{n_1,n_2}^{l,m}$. This constraint set is convex and a steepest descent optimization is used to find the solution.

Reference [20] exploits the *sparse* nature of the color images to design the prior terms. Instead of imposing smoothness on the intensity and chrominance values, it is assumed that for any natural image i there exists a sparse linear combination of a limited number of fixed-size patches that approximates it well. Given a dictionary D, which is a $3N_x \times k$ matrix which contains the k prototype patches, this implies that for any image there exists $\alpha \in \mathbb{R}^k$ such that $i \simeq D\alpha$ and $\|\alpha\|_0 \ll 3N_x$ where $\| \cdot \|_0$ denotes the ℓ_0-quasi norm which counts the number of nonzero elements. Based on these assumptions, novel regularizing terms can be proposed and an estimate $\hat{i}$ of the original image can be found using an iterative method that incorporates the K-SVD (single value decomposition) algorithm, as presented in Reference [27].

6.3.1 Joint Demosaicking and Super-Resolution

An approach to enhance the spatial resolution is to fuse together several low-resolution images of the same scene in order to produce a high-resolution image (or a sequence of them). This approach is called *super-resolution* image reconstruction and has proved to be useful in many practical situations where multiple frames of the same scene are available, including medical imaging, satellite imaging, video-surveillance and other video applications. Many techniques for super-resolution imaging have been proposed and an overview of them can be found in Reference [28]. Since the low-resolution images are often acquired by digital cameras equipped with a CFA, it may be desired to enhance image resolution by directly using the raw-data in a joint demosaicking and super-resolution reconstruction process.

The problem can be formulated by relating each observed image i_{s_k} to the original scene i as follows:

$$i_{s_k} = H_k i + \eta_k, \qquad (6.18)$$

where H_k takes into account the effects of the motion, the point-spread function, the downsampling, and the CFA-sampling. The term η_k denotes the noise affecting the kth-image. Given K observed images i_{s_k}, the goal is to estimate the original image i.

A solution to this problem is proposed in Reference [29] using a method which exploits two regularization constraints. The first constraint is obtained through high-pass operators S_d which evaluate directional smoothness of the luminance component of YIQ data in the horizontal, vertical, diagonal, and isotropic directions as follows:

$$J_1(i) = \sum_d \|\Lambda_d S_d x_Y\|_2^2, \qquad (6.19)$$

where the diagonal matrix Λ_d contains suitable weights determined by detecting the edge orientation at each pixels.

The second constraint imposes isotropic smoothness on the chrominance components with a high-pass filter S as follows:

$$J_2(i) = \|S x_I\|_2^2 + \|S x_Q\|_2^2, \qquad (6.20)$$

with x_I and x_Q denoting the chrominance components in the YIQ space.

Another joint demosaicking and super-resolution approach is proposed in Reference [30]. The first regularization constraint imposes smoothness to the luminance component with the bilateral total-variation technique, an extension of the total-variation criterion [26]. It is defined as

$$J_1(i) = \sum_{l=-P}^{P} \sum_{m=-P}^{P} \alpha^{|m|+|l|} \|x_Y - Z^{l,m} x_Y\|_1, \qquad (6.21)$$

where the matrices $Z^{l,m}$ are defined as in Equation 6.17 and the scalar weight $0 < \alpha < 1$ is applied to give a decreasing effect when l and m increase. The quadratic penalty term in Equation 6.20 is used to describe the bandlimited characteristic of the chrominances, and another term penalizes the mismatch between locations or orientation of edges across the color bands using the element by element multiplication operator $\odot$, as follows:

$$J_3(i) = \sum_{l=-1}^{1} \sum_{m=-1}^{1} \left[\|g \odot Z^{l,m}b - b \odot Z^{l,m}g\|_2^2 \right.$$

$$\left. + \|b \odot Z^{l,m}r - r \odot Z^{l,m}g\|_2^2 + \|r \odot Z^{l,m}g - g \odot Z^{l,m}r\|_2^2 \right]. \qquad (6.22)$$

The data fidelity term measuring the similarity between the resulting high-resolution image and the original low-resolution images is based on the ℓ_1 norm. A steepest descent optimization is used to minimize the cost function expressed as the sum of the three terms.

6.4 Non-Iterative Demosaicking

As described, the techniques presented in the previous section use iterative approaches to find an approximation of the solution to the inverse problem of Equation 6.8. However, this strategy could be too computationally demanding for real-time applications. Therefore, this section presents a non-iterative demosaicking strategy [21] that is more efficient than previous approaches.

6.4.1 Quadratic Estimation

Regularization constraints $J_k(i)$ can be designed according to the Tikhonov method. This method incorporates prior information about the image $\mathscr{I}$ through the inclusion of quadratic terms, $J_k(i) = \|M_k i\|_2^2$, where M_k are appropriate matrices.

Using the characteristics of natural color images, two constraints are included in the regularization method. The first constraint $J_1(i)$ is chosen in order to impose smoothness to each single color component, that is $M_1 = I_3 \otimes S_1$, where I_3 is the 3×3 identity matrix, $\otimes$ denotes the Kronecker operator, and S_1 represents a high-pass filter. As discussed in Section 6.3, S_1 is commonly chosen as discrete approximations of two-dimensional gradient (e.g., Laplacian operators), but two directional filters can also be used. In this latter case, $S_1^T = \left[S_{1h}^T, S_{1v}^T \right]$ where S_{1h} and S_{1v} represent a horizontal and a vertical high-pass filter S_h and S_v, respectively.

The second constraint forces smoothness on the differences between the three color components. Therefore, $J_2(i)$ can be expressed as follows:

$$J_2(i) = \|S_2 r - S_2 g\|_2^2 + \|S_2 g - S_2 b\|_2^2 + \|S_2 r - S_2 b\|_2^2, \qquad (6.23)$$

where S_2 represents a high-pass filter with a cut-off frequency lower than S_1 since color differences are smoother than the color channels[2]. Equation 6.23 can be reformulated as

$$J_2(i) = i^T \begin{bmatrix} 2S_2^T S_2 & -S_2^T S_2 & -S_2^T S_2 \\ -S_2^T S_2 & 2S_2^T S_2 & -S_2^T S_2 \\ -S_2^T S_2 & -S_2^T S_2 & 2S_2^T S_2 \end{bmatrix} i = i^T \left\{ \begin{bmatrix} 2 & -1 & -1 \\ -1 & 2 & -1 \\ -1 & -1 & 2 \end{bmatrix} \otimes S_2^T S_2 \right\} i. \qquad (6.24)$$

---

[2]In Reference [21], a finite impulse response (FIR) filter with coefficients $[0.2, -0.5, 0.65, -0.5, l0.2]$ for S_h and S_v is chosen. For the second constraint the filter coefficients $[-0.5, 1, -0.5]$ are used.

Then, exploiting the properties of the Kronecker product, this constraint can be expressed as $J_2(i) = \|M_2 i\|_2^2$, where

$$M_2 = \text{sqrt}\left(\begin{bmatrix} 2 & -1 & -1 \\ -1 & 2 & -1 \\ -1 & -1 & 2 \end{bmatrix} \otimes S_2^T S_2\right) = \begin{bmatrix} 1.547 & -0.577 & -0.577 \\ -0.577 & 1.547 & -0.577 \\ -0.577 & -0.577 & 1.547 \end{bmatrix} \otimes S_2. \quad (6.25)$$

Using the two regularizing constraints $J_1(i)$ and $J_2(i)$ defined as above and the data-fidelity term defined in Equation 6.13, the solution of Equation 6.11 can be obtained by solving the problem

$$\left(H^T R_\eta^{-1} H + \lambda_1 M_1^T M_1 + \lambda_2 M_2^T M_2\right)\hat{i} - H^T R_\eta^{-1} i_s = 0, \quad (6.26)$$

that is, $\hat{i} = \mathscr{G} i_s$, with

$$\mathscr{G} = \left(H^T R_\eta^{-1} H + \lambda_1 M_1^T M_1 + \lambda_2 M_2^T M_2\right)^{-1} H^T R_\eta^{-1}. \quad (6.27)$$

The coefficients of the filters that estimate the three color components from the CFA sampled image can be extracted from the matrix $\mathscr{G}$. In fact, $\mathscr{G}$ can be written as $\mathscr{G} = [\mathscr{G}_R, \mathscr{G}_G, \mathscr{G}_B]$, where the submatrices $\mathscr{G}_R$, $\mathscr{G}_G$, and $\mathscr{G}_B$ are the representation (according to the matrix notation introduced in Section 6.2) of the filters that estimate the red, green and blue components from the CFA image. Due to the data sampling structure of the CFA, the resulting filters are periodically space-varying and the number of different states depends on the periodicity of the CFA.

6.4.1.1 Estimation of the Luminance

An alternative and effective demosaicking approach focuses on finding an estimate $\hat{L}(n)$ of the luminance component of the image, from which the three color components can be easily computed by interpolating the color differences $R - \hat{L}$, $G - \hat{L}$, and $B - \hat{L}$, as proposed in Reference [14]. By introducing $A = [\alpha_R(0)I, \alpha_G(0)I, \alpha_B(0)I]$, where I is the identity matrix, the stacking vector of the luminance component can be expressed as $\ell = Ai$. Therefore, an estimate of the luminance from the CFA-sampled data is given by

$$\hat{\ell} = \mathscr{G}_\ell i_s, \quad (6.28)$$

where $\mathscr{G}_\ell = A\mathscr{G}$ represents the filter that estimates the luminance component from the CFA data.

As for the estimation of the red, green, and blue components, the resulting filter is periodically space-varying. Reference [14] describes the luminance estimation process for the Bayer pattern with the sensor PSFs assumed to be ideal impulses, that is, $p_X(n) = \delta(n)$. In this case, inspecting the rows of the matrix $\mathscr{G}_\ell$ reveals two different states for the reconstruction filter. For pixels corresponding to the red and blue locations of the Bayer pattern the resulting filter has the frequency response as shown in Figure 6.4a. For pixels corresponding to the green locations in the Bayer pattern, the frequency response of the filter is as shown in Figure 6.4b.

Recalling the frequency analysis of a Bayer-sampled image reported in Section 6.2, the estimation of the luminance corresponds to eliminating the chrominance replicas using

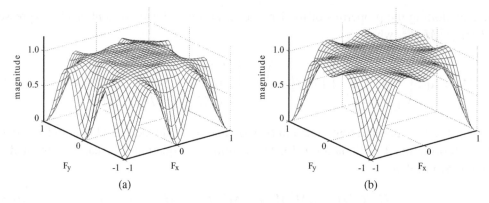

FIGURE 6.4

(a) Frequency response of the 9×9 filter used for the luminance estimation in the red/blue pixels. (b) Frequency response of the 5×5 filter used for the luminance estimation in the green pixels. © 2009 IEEE

appropriate low-pass filters. For green pixels in the quincunx layout, the color difference terms modulated at $(0, \pm\pi)$ and $(\pm\pi, 0)$ vanish, as reported in Reference [14]. Thus, only the chrominance components modulated at $(\pm\pi, \pm\pi)$ have to be eliminated. This is not the case of the red and blue locations, where all the chrominance components have to be filtered and the spectrum of the image is as shown in Figure 6.3. Therefore, the frequency response of the low-pass filters follows Figure 6.4. Since the requirements imposed on the filter design are less demanding in the green locations than in the red and blue locations, a smaller number of filter coefficients is usually sufficient. A similar analysis can be carried out for other CFA arrangements.

6.4.2 Adaptive Estimation

In the previous section, some global properties of the images are exploited and a general approach is applied to solve the demosaicking problem. In particular, the addition of quadratic penalties $J_k(i) = \|M_k i\|_2^2$ to the least-squares (and hence quadratic) data fidelity criterion permits an efficient computational method. However, natural images are often inhomogeneous and contain abrupt changes in both intensity and color due to the presence of edges. Hence, quadratic approaches produce unacceptable artifacts in high-frequency regions, as opposed to adaptive approaches that usually give better results.

Local adaptivity can be included in the regularization framework by considering some non-quadratic regularizing constraints of the following form:

$$J_k(i) = \|M_k i\|_{W_i}^2, \tag{6.29}$$

where W_i is a diagonal matrix estimated from the image in order to adapt the penalty term to the local features of the image [31]. If a regularization term of this type is considered together with the two quadratic penalties $J_1(i)$ and $J_2(i)$ proposed in the previous section, the solution to Equation 6.11 is found by solving

$$\left(H^T R_\eta^{-1} H + \lambda_1 M_1^T M_1 + \lambda_2 M_2^T M_2 + \lambda_3 M_3^T W_i M_3\right) i - H^T R_\eta^{-1} i_s = 0. \tag{6.30}$$

Since W_i depends on i, Equation 6.30 is nonlinear, and often is solved with a Landweber fixed point iteration [25], [31]. However, a large number of iterations can be required before convergence is reached, precluding fast implementations.

An alternative approach is proposed below. An initial estimate, $\tilde{i}$, of the original image is used such that the value of $M_3^T W_{\tilde{i}} M_3 \tilde{i}$ approximates $M_3^T W_i M_3 i$. Therefore, the resulting image $\hat{i}$ is obtained as follows:

$$\hat{i} = \left(H^T R_\eta^{-1} H + \lambda_1 M_1^T M_1 + \lambda_2 M_2^T M_2\right)^{-1} \left(H^T R_\eta^{-1} i_s - \lambda_3 M_3^T W_{\tilde{i}} M_3 \tilde{i}\right). \tag{6.31}$$

Since the operator M_3 is set to be equivalent to the first quadratic operator M_1, and both the matrices are designed using two directional filters, it can be written that

$$M_1 = M_3 = I_3 \otimes S_1 = I_3 \otimes \begin{bmatrix} S_{1h} \\ S_{1v} \end{bmatrix}, \tag{6.32}$$

in order to detect the discontinuities of the image along horizontal and vertical directions, respectively. The diagonal entries of $W_{\tilde{i}}$ depend on the horizontal and vertical high frequencies of the estimated image $\tilde{i}$. In fact, $W_{\tilde{i}} = \text{diag}\left(W_{r,h}, W_{r,v}, W_{g,h}, W_{g,v}, W_{b,h}, W_{b,v}\right)$, where $\text{diag}(\cdot)$ denotes the diagonal entries and $W_{x,h}$ and $W_{x,v}$, for $x = r, g, b$, are diagonal matrices with their values defined as follows:

$$\{W_{x,h}\}_j = \xi\left(\frac{\{e_{x,v}\}_j}{\{e_{x,h}\}_j + \{e_{x,v}\}_j}\right), \tag{6.33}$$

$$\{W_{x,v}\}_j = \xi\left(\frac{\{e_{x,h}\}_j}{\{e_{x,h}\}_j + \{e_{x,v}\}_j}\right), \tag{6.34}$$

where $\{e_{x,h}\}_j$ and $\{e_{x,v}\}_j$ are the energies of the j-th value of $S_{1h}x$ and $S_{1v}x$, respectively, and $\xi(\cdot)$ is a function defined as

$$\xi(y) = \begin{cases} 0 & \text{if } y < \varepsilon \\ \dfrac{y - \varepsilon}{1 - 2\varepsilon} & \text{if } \varepsilon \leq y \leq 1 - \varepsilon \\ 1 & \text{if } y > 1 - \varepsilon \end{cases} \tag{6.35}$$

with $0 \leq \varepsilon \leq 1/2$ (in Reference [21] $\varepsilon = 0.25$ is used). In this way, when $\{S_{1h}x\}_j \gg \{S_{1v}x\}_j$ the presence of a vertical edge can be assumed; therefore, $\{W_{x,h}\}_j = 0$ and the constraint of smoothness of the color components is not considered along the horizontal direction, while it is preserved for the vertical direction. The same analysis holds when horizontal edges are found. Finally, when $\{S_h x\}_j$ and $\{S_v x\}_j$ have similar energies, smoothing is imposed along both horizontal and vertical directions.

A similar approach was adopted in Reference [31], where a *visibility function* was applied to compute the diagonal values of W_i. The visibility function depends on the local variance of the image and goes to zero near the edges. However, this technique does not discriminate between horizontal and vertical edges, so the high-frequency penalty is disabled for both directions. Moreover, this approach is applied in iterative restoration methods.

It can be pointed out that there are two smoothing penalties in Equation 6.31, as the adaptive term $J_3(i)$ is included together with the quadratic constraint $J_1(i)$. In fact, $J_1(i)$

cannot be removed, as the matrix $H^T R_\eta^{-1} H + \lambda_2 M_2^T M_2$ is not invertible since $\ker(H^T H) \cap \ker(M_2^T M_2) \neq \{0\}$. Therefore, the regularization process with respect to the spatial smoothness of the color components uses two constraints, where the quadratic one allows inverting the matrix $H^T R_\eta^{-1} H + \lambda_1 M_1^T M_1 + \lambda_2 M_2^T M_2$ and the second one includes adaptivity in the solution of the problem. The same approach is applied also in the half-quadratic minimization methods in the additive form [32]. However, in these approaches, the diagonal submatrix $W_{x,h}$ for the horizontal details and $W_{x,v}$ for the vertical details does not accommodate the vertical frequencies $S_{1v} x$ and the horizontal frequencies $S_{1h} x$, respectively. Thus, the local adaptivity is not based on the comparison between $S_{1h} x$ and $S_{1v} x$ as in Equations 6.33 and 6.34, and convergence to the optimal solution is therefore reached more slowly after many iterations.

As for the initial estimate $\tilde{i}$ used in Equation 6.31, an efficient solution is to apply the quadratic approach described in Section 6.4.1, leading to $\tilde{i} = \mathscr{G} i_s$. In this way, the approximation $M_3^T W_{\tilde{i}} M_3 \tilde{i} \simeq M_3^T W_i M_3 i$ is verified, and the proposed scheme provides a reliable estimate of the color image i, as proved by the experimental results reported in Reference [21] and in Section 6.5.

6.4.2.1 Luminance Estimation Using an Adaptive Scheme

As discussed in Section 6.4.1, estimating the luminance instead of the three color components usually reduces the computational cost of the algorithm. In fact, the three color components can be computed through bilinear interpolation of the color differences using the estimated luminance as a reference component.

Considering that the luminance can be expressed as $\ell = A i$, an estimate of the luminance using the spatially adaptive scheme of Equation 6.31 is given by

$$\hat{\ell} = A \left(H^T R_\eta^{-1} H + \lambda_1 M_1^T M_1 + \lambda_2 M_2^T M_2 \right)^{-1} \left(H^T R_\eta^{-1} i_s - \lambda_3 M_3^T W_{\tilde{i}} M_3 \tilde{i} \right). \quad (6.36)$$

Since $M_3 = I_3 \otimes S_1$ and $W_{\tilde{i}} = \mathrm{diag}\left(W_{\tilde{r}}, W_{\tilde{g}}, W_{\tilde{b}} \right)$, with $W_{\tilde{x}} = \mathrm{diag}\left(W_{\tilde{x},h}, W_{\tilde{x},v} \right)$, it can be written that

$$M_3^T W_{\tilde{i}} M_3 \tilde{i} = \begin{bmatrix} S_1^T W_{\tilde{r}} S_1 \tilde{r} \\ S_1^T W_{\tilde{g}} S_1 \tilde{g} \\ S_1^T W_{\tilde{b}} S_1 \tilde{b} \end{bmatrix}, \quad (6.37)$$

and, considering that the high frequencies of the three color components are highly correlated with those of the luminance, the following approximation holds:

$$S_1^T W_{\tilde{r}} S_1 \tilde{r} \simeq S_1^T W_{\tilde{g}} S_1 \tilde{g} \simeq S_1^T W_{\tilde{b}} S_1 \tilde{b} \simeq S_1^T W_{\tilde{\ell}} S_1 \tilde{\ell}. \quad (6.38)$$

and the term in Equation 6.37 can therefore be replaced with

$$M_3^T W_{\tilde{i}} M_3 \tilde{i} = I_3 \otimes S_1^T W_{\tilde{\ell}} S_1 \tilde{\ell}. \quad (6.39)$$

By introducing the matrix

$$F_\ell = A \left(H^T R_\eta^{-1} H + \lambda_1 M_1^T M_1 + \lambda_2 M_2^T M_2 \right)^{-1} [I, I, I]^T,$$

Equation 6.36 becomes

$$\hat{\ell} = \mathscr{G}_\ell i_s - \lambda_3 F_\ell S_1^T W_{\tilde{\ell}} S_1 \tilde{\ell} \quad (6.40)$$

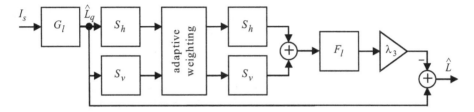

FIGURE 6.5

Adaptive luminance estimation scheme. © 2009 IEEE

where $\mathscr{G}_\ell$ is defined in the previous section. If the initial estimate of the luminance $\tilde{\ell}$ is computed with the quadratic approach described in Section 6.4.1, that is, $\tilde{\ell} = \mathscr{G}_\ell i_s$, Equation 6.40 can be written as

$$\hat{\ell} = \left(\boldsymbol{I} - \lambda_3 \boldsymbol{F}_\ell \boldsymbol{S}_1^T \boldsymbol{W}_{\tilde{\ell}} \boldsymbol{S}_1\right) \mathscr{G}_\ell i_s. \tag{6.41}$$

This equation indicates the procedure to compute the luminance from the CFA-sampled image using the proposed adaptive method. The resulting scheme is depicted in Figure 6.5, where G_ℓ is the space-varying filter designed in Section 6.4.1 and the filter F_ℓ is obtained by matrix $\boldsymbol{F}_\ell$. Filters S_h and S_v are the horizontal and vertical high-pass filters represented by matrices $\boldsymbol{S}_{1h}$ and $\boldsymbol{S}_{1v}$, respectively.[3]

6.5 Performance Comparisons

This section presents experimental results and performance comparisons of various regularization-based approaches. Twenty test images (Figure 6.6) from the Kodak dataset, each with resolution of 512×768 pixels coded using 24 bits per pixel were sampled using selected CFAs and the full color representation is reconstructed using several demosaicking approaches. The demosaicked images are evaluated by comparing them to the original image using the *color peak signal-to-noise ratio* (CPSNR):

$$\text{CPSNR} = 10\log_{10} \frac{255^2}{\dfrac{1}{3N_1N_2}\sum_X\sum_{n_1}\sum_{n_2}\left(\hat{X}(n_1,n_2) - X(n_1,n_2)\right)^2}, \tag{6.42}$$

where N_1 and N_2 denote image dimensions, $X = R, G, B$ denotes the color channel, and $n_1 = 1, 2, ..., N_1$ and $n_2 = 1, 2, ..., N_2$ denote the pixel coordinates.

Performance of regularization-based algorithms was tested with four different acquisition models to produce: i) noise-free Bayer CFA data, ii) noise-free CFA data generated using the CFA with panchromatic sensors from Reference [4], iii) the Bayer CFA data corrupted

---

[3]In Figure 6.5, it is assumed that filters S_h and S_v are even-symmetric since in this case $\boldsymbol{S}_{1h}^T = \boldsymbol{S}_{1h}$ and $\boldsymbol{S}_{1v}^T = \boldsymbol{S}_{1v}$. Instead, if S_h and S_v are odd-symmetric, that is, $\boldsymbol{S}_{1h}^T = -\boldsymbol{S}_{1h}$ and $\boldsymbol{S}_{1v}^T = -\boldsymbol{S}_{1v}$, after the adaptive weighting S_h and S_v have to be replaced with $-S_h$ and $-S_v$, respectively.

FIGURE 6.6

The test images of the Kodak dataset used in the experiments. © 2009 IEEE

by Gaussian noise, and iv) CFA data generated using the Bayer pattern and sensors with a non-ideal impulse response. In addition to these simulated scenarios, raw images acquired by a Pentax *ist DS2 digital camera are also used for performance comparisons.

6.5.1 Bayer CFA Images

Various demosaicking methods are first tested on images sampled with the popular Bayer CFA (see Figure 6.1a). This comparison includes some well-known methods [9], [10], [11], [14] and regularization-based methods [19], [20], [21] described in the previous sections. In the approach presented in Reference [21], the adaptive estimate of the luminance is obtained using the procedure depicted in Figure 6.5 and bilinear interpolation is performed to obtain full color information. The complete procedure of the proposed adaptive reconstruction method from the CFA samples is depicted in Figure 6.7. The regularization parameters λ_k and the filter dimensions are set as reported in Reference [21].

Table 6.1 presents CPSNR values of demosaicked images. As can be seen, the method proposed in References [20] achieves the highest average CPSNR value. Visual inspection of the restored images shown in Figure 6.8 reveals that regularization-based approaches preserve edges and finest details during demosaicking.

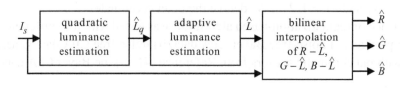

FIGURE 6.7

Complete scheme of the adaptive color reconstruction described in Section 6.4.2. © 2009 IEEE

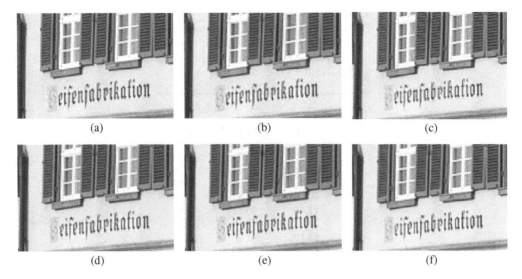

(a)   (b)   (c)

(d)   (e)   (f)

FIGURE 6.8

Portion of the image #6 of the Kodak set: (a) original image, and (b-f) demosaicked images obtained using the methods presented in (b) Reference [9], (c) Reference [10], (d) Reference [19], (e) Reference [20], and (f) Reference [21]. © 2009 IEEE

TABLE 6.1

CPSNR (dB) evaluation of demosaicking methods using images sampled by the Bayer CFA shown in Figure 6.1a.

Image	Method / Reference						
	[9]	[10]	[11]	[14]	[19]	[20]	[21]
1	38.36	38.58	36.91	37.58	39.04	**39.32**	38.04
2	40.42	40.18	40.16	40.26	37.98	**40.66**	38.16
3	38.05	38.08	37.44	38.12	36.17	**38.67**	38.17
4	40.03	**40.10**	39.24	38.11	38.56	39.98	39.66
5	42.03	42.15	41.61	42.69	39.73	**42.70**	42.39
6	35.96	**36.50**	35.42	35.37	35.17	36.39	36.01
7	42.76	**43.11**	42.34	42.72	41.26	43.03	42.38
8	41.77	42.60	42.14	**42.69**	40.44	42.64	42.46
9	40.10	40.07	39.30	39.47	38.38	**40.16**	39.78
10	42.81	**43.54**	42.95	42.77	41.93	43.38	43.07
11	34.73	35.03	33.33	33.89	**36.05**	35.27	34.62
12	39.17	39.61	39.01	39.53	38.28	**40.14**	39.18
13	43.26	**43.87**	43.05	41.30	42.01	43.47	43.28
14	41.43	41.74	40.77	41.58	40.44	**41.84**	41.59
15	36.48	36.97	36.10	36.76	35.88	**37.47**	37.01
16	40.46	40.65	39.82	40.01	39.08	**41.14**	39.97
17	37.80	**41.06**	39.87	40.54	38.22	40.86	40.63
18	38.93	39.37	37.93	38.70	39.12	**39.69**	39.16
19	38.55	38.38	37.74	38.64	36.91	**38.83**	38.47
20	34.91	35.07	34.21	34.64	33.53	**35.59**	35.30
Ave.	39.40	39.83	38.97	39.27	38.41	**40.06**	39.47

TABLE 6.2

CPSNR (dB) evaluation of demosaicking methods using images sampled by the panchromatic CFA shown in Figure 6.1d. © 2009 IEEE

Image	Method / Reference		
	[33]	[21] quad	[21] adap
1	38.97	**39.71**	39.53
2	**38.96**	37.47	37.86
3	36.76	36.33	**36.93**
4	40.13	**40.69**	40.63
5	**41.40**	40.12	40.97
6	36.87	37.19	**37.29**
7	**42.10**	41.42	41.88
8	**42.21**	41.83	42.20
9	39.82	39.81	**40.02**
10	**43.00**	42.42	42.86
11	34.58	**36.38**	35.69
12	**39.42**	38.56	38.92
13	44.00	**44.33**	44.29
14	41.11	**41.84**	41.72
15	36.77	36.69	**36.83**
16	40.63	40.57	**40.67**
17	**40.47**	40.01	40.35
18	39.77	39.76	**39.81**
19	**38.73**	38.04	38.40
20	35.91	36.09	**36.12**
Ave.	39.58	39.46	**39.65**

6.5.2 Panchromatic CFA Images

This section focuses on demosaicking performance on the data obtained using the CFA with panchromatic filters, that is, color filters which allow acquiring the light corresponding to a linear combination of the red, green and blue components. A number of demosaicking approaches proposed for the Bayer pattern cannot be extended to deal with the panchromatic data. Reference [4] presents a few panchromatic CFAs; the demosaicking procedure suitable for such CFAs is proposed in Reference [33]. Table 6.2 compares the performances of this approach and the algorithms presented in Sections 6.4.1 and 6.4.2 for the CFA data generated using the pattern shown in Figure 6.1d. It can be seen that the adaptive regularization method improves the quality of the reconstructed images by obtaining higher average CPSNR value.

6.5.3 Noisy Bayer CFA Images

In a digital camera, the image acquisition process is rarely noise-free. To demonstrate the robustness of the different demosaicking methods, various demosaicking algorithms are now tested on Bayer CFA images corrupted with Gaussian noise η with standard deviation $\sigma_\eta = 5$. Table 6.3 summarizes achieved numerical results. As expected, the CPSNR values are lower than the results obtained in the noise-free case reported in Table 6.1. The

TABLE 6.3

CPSNR (dB) evaluation of demosaicking methods using images sampled by the Bayer CFA shown in Figure 6.1 and subsequently corrupted with Gaussian noise of $\sigma_\eta = 5$. © 2009 IEEE

Image	Method / Reference						
	[9]	[10]	[11]	[14]	[19]	[20]	[21]
1	33.06	33.01	32.26	32.91	33.20	**33.40**	32.95
2	33.62	33.41	33.13	**33.67**	32.97	33.57	33.42
3	33.03	32.88	32.48	33.15	32.29	**33.23**	33.17
4	33.54	33.44	33.02	33.21	33.19	**33.62**	33.59
5	33.97	33.79	33.44	34.20	33.47	34.04	**34.33**
6	32.19	32.33	31.73	32.00	31.86	**32.42**	32.06
7	34.07	33.94	33.53	34.18	33.78	34.07	**34.30**
8	33.92	33.89	33.52	34.21	33.62	33.98	**34.33**
9	33.52	33.40	32.96	33.54	33.11	33.58	**33.63**
10	34.07	34.02	33.67	34.24	33.92	34.16	**34.40**
11	31.71	31.72	30.71	31.33	**32.20**	31.95	31.29
12	33.48	33.53	33.13	33.70	33.27	33.72	**33.74**
13	34.05	33.96	33.57	33.93	33.88	34.12	**34.29**
14	33.89	33.79	33.33	34.03	33.61	33.92	**34.16**
15	32.52	32.56	32.02	32.71	32.22	**32.82**	32.73
16	33.62	33.56	33.12	33.63	33.30	**33.74**	33.69
17	33.53	34.43	33.96	34.49	33.73	34.38	**34.68**
18	33.31	33.28	32.64	33.37	33.30	**33.49**	33.46
19	33.19	32.99	32.58	33.35	32.63	33.26	**33.41**
20	31.85	31.75	31.23	31.80	31.07	**32.09**	32.05
Ave.	33.31	33.28	32.80	33.38	33.03	**33.48**	**33.48**

regularization-based techniques described in Sections 6.3 and 6.4 produce high average CPSNR values and seem to be more robust on noisy images than edge-adaptive approaches which tend to fail due to inaccurate edge detection on noisy data. The quality of the demosaicked images can be improved by applying a denoising procedure. An alternative strategy [34] could be to use techniques that perform demosaicking jointly with denoising.

6.5.4 Blurry and Noisy Images

The quality of demosaicked images strongly depends on the sensor's characteristics. In fact, sensors with a bandlimited frequency response reduce aliasing and high-frequency content of an image, thus making it easier for demosaicking. Therefore, a deblurring and/or sharpening step is included in the imaging pipeline to sharpen the demosaicked image. The techniques presented in this chapter can overcome this problem since the design of the filters used for color reconstruction takes into account the impulse response of the sensor (see Section 6.2), thus adapting demosaicking to its characteristics and performing deblurring jointly with demosaicking.

Table 6.4 reports the performances of different approaches in scenarios where the sensor is equipped with the Bayer CFA and pixels have a Gaussian impulse response with standard deviation $\sigma_s = 0.5$. In this experiment, the demosaicking methods proposed in References [9] and [10] followed by various deblurring algorithms are directly compared with

TABLE 6.4

CPSNR (dB) evaluation of demosaicking methods using images sampled by the Bayer CFA shown in Figure 6.1 and the sensor with a Gaussian impulse response of $\sigma_s = 0.5$. © 2009 IEEE

Image	Method / Reference							
	[9]	[9]+[35]	[9]+[36]	[10]	[10]+[35]	[10]+[36]	[21] quad	[21] adap
1	33.47	38.26	**38.85**	33.44	38.70	38.71	37.03	38.07
2	38.06	**40.42**	39.78	38.03	39.41	39.37	37.86	38.24
3	33.24	38.08	37.05	33.12	36.86	36.60	36.98	**38.16**
4	34.81	**40.03**	39.83	34.85	39.43	39.51	37.58	39.81
5	39.08	42.03	40.77	39.03	39.90	40.34	41.20	**42.71**
6	30.97	35.90	36.19	31.08	36.26	36.17	33.66	**36.32**
7	39.16	**42.76**	42.04	39.19	42.11	42.07	41.02	42.59
8	38.50	41.77	41.06	38.70	41.52	41.47	41.38	**42.60**
9	35.81	**40.10**	39.74	35.80	39.18	39.25	38.56	39.78
10	39.41	42.81	42.03	39.56	42.12	42.46	41.33	**43.35**
11	30.29	34.59	**35.06**	30.32	34.75	34.75	34.28	34.25
12	36.87	39.17	38.41	36.98	38.46	38.48	38.28	**39.37**
13	38.37	43.26	42.95	38.58	43.10	43.12	40.79	**43.50**
14	37.95	41.43	40.98	38.00	40.93	40.90	41.01	**41.54**
15	33.54	36.45	36.13	33.62	35.93	36.03	36.55	**36.72**
16	35.49	**40.46**	40.23	35.46	39.15	39.61	38.14	40.33
17	35.89	37.80	37.43	37.33	39.53	40.33	39.11	**40.72**
18	35.10	38.93	39.07	35.15	38.69	38.98	38.20	**39.12**
19	36.02	**38.56**	37.77	35.90	36.98	37.09	37.69	38.53
20	32.14	34.91	34.59	32.23	33.98	34.44	34.90	**35.15**
Ave.	35.71	39.39	39.00	35.82	38.85	38.98	38.28	**39.54**

the total variation-based image deconvolution [35] and the deconvolution using a sparse prior [36] method.[4] The performance of these methods is also compared with the results obtained with the non-iterative regularization method described in Section 6.4.

Using no deblurring, the methods presented in References [9] and [10] produce images with a poor quality. Employing the restoration method after the demosaicking step considerably improves performance, providing average CPSNR improvements up to 3.6 dB. The adaptive approach is able to produce sharp demosaicked images, thus making use of enhancement procedures unnecessary. The average CPSNR value obtained by the regularization-based method is higher compared to values achieved using the demosaicking methods [9], [10] followed by computationally demanding deblurring methods of References [35] and [36]. Figure 6.9 allows visual comparisons of different methods, with the original image shown in Figure 6.8a. As can be seen, the image shown in Figure 6.9a reconstructed without sharpening is blurred. Figure 6.9b shows the image with demosaicking artifacts amplified by the deblurring algorithm. The image shown in Figure 6.9c is excessively smoothed in the homogeneous regions. The best compromise between sharpness and absence of demosaicking artifacts demonstrated by the regularization approach of Reference [21], with output image shown in Figure 6.9d.

---

[4]The source codes of the deblurring algorithms presented in References [35] and [36] are available at `http://www.lx.it.pt/~bioucas`, and `http://groups.csail.mit.edu/graphics/CodedAperture/DeconvolutionCode.html`, respectively.

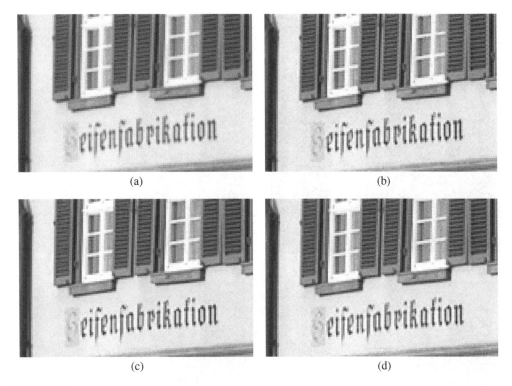

(a)   (b)

(c)   (d)

FIGURE 6.9

Portion of the image #6 of the Kodak set: (a) image reconstructed using the method of Reference [9] and without deblurring, (b) image reconstructed using the method of Reference [9] and sharpened using the method of Reference [35], (c) image reconstructed using the method of Reference [9] and sharpened with the method of Reference [36], (d) image reconstructed by the adaptive method of Reference [21]. © 2009 IEEE

6.5.5 Raw CFA Data

In addition to experiments with simulated CFA data, real raw CFA images captured using a Pentax *ist DS2 camera equipped with a 6.1 megapixel CCD sensor and the Bayer CFA were used. Figure 6.10 shows a portion of an image demosaicked using the methods presented in References [9], [14], and [21]. As can be seen, the regularization approach excels also in the edge regions, since it avoids introducing the zipper effect and produces images which are sharper than those produced using other methods.

6.6 Conclusion

This chapter presented demosaicking methods based on the concept of regularization. Demosaicking is considered as an inverse problem and suitable regularization terms are designed using the characteristics of natural images. As demonstrated in this chapter, taking advantage of assumptions based on the smoothness of the color components and the high-frequency correlation between the color channels allows the design of efficient algo-

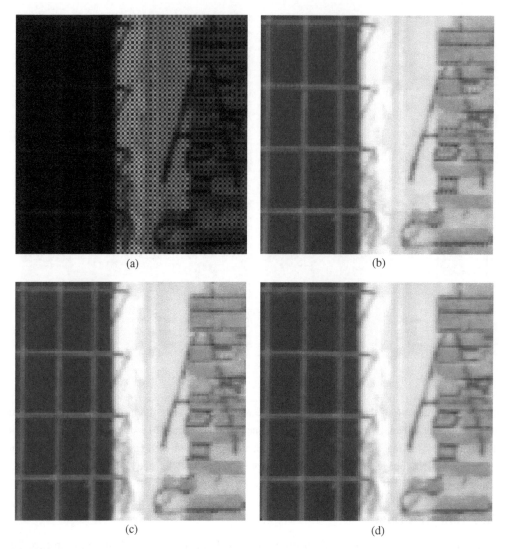

FIGURE 6.10

Portion of an image captured using a Pentax *ist DS2 camera: (a) CFA image; (b-d) images reconstructed using the method of (b) Reference [9], (c) Reference [14], and (d) Reference [21].

rithms. The regularization-based methods are easily applicable to any CFA and can demosaick efficiently images acquired with sensors having a non-ideal impulse response, since the characteristics of the PSF are taken into account in the reconstruction method. Moreover, the regularization-based strategy permits coupling demosaicking with other frequent problems in image reconstruction and restoration.

Acknowledgment

Figures 6.1 to 6.9 and Tables 6.2 to 6.4 are reprinted from Reference [21], with the permission of IEEE.

References

[1] B.E. Bayer, "Color imaging array," U.S. Patent 3 971 065, July 1976.

[2] R. Lukac and K.N. Plataniotis, "Color filter arrays: Design and performance analysis," *IEEE Transactions on Consumer Electronics*, vol. 51, no. 4, pp. 1260–1267, November 2005.

[3] J.F. Hamilton and J.T. Compton, "Processing color and panchromatic pixels," U.S. Patent Application 0024879, February 2007.

[4] K. Hirakawa and P.J. Wolfe, "Spatio-spectral color filter array design for optimal image recovery," *IEEE Transactions Image Processing*, vol. 17, no. 10, pp. 1876–1890, October 2008.

[5] R. Lukac (ed.), *Single-Sensor Imaging: Methods and Applications for Digital Cameras*. Boca Raton, FL: CRC Press / Taylor & Francis, September 2008.

[6] B.K. Gunturk, J. Glotzbach, Y. Altunbasak, R.W. Schafer, and R.M. Mersereau, "Demosaicking: Color filter array interpolation," *IEEE Signal Processing Magazine*, vol. 22, no. 1, pp. 44–54, January 2005.

[7] X. Li, B.K. Gunturk, and L. Zhang, "Image demosaicing: A systematic survey," *Proceedings of SPIE*, vol. 6822, pp. 68221J:1–15, January 2008.

[8] S. Battiato, M. Guarnera, G. Messina, and V. Tomaselli, "Recent patents on color demosaicing," *Recent Patents on Computer Science*, vol. 1, no. 3, pp. 94–207, November 2008.

[9] L. Zhang and X. Wu, "Color demosaicking via directional linear minimum mean square-error estimation," *IEEE Transactions Image Processing*, vol. 14, no. 12, pp. 2167–2177, December 2005.

[10] K.H. Chung and Y.H. Chan, "Color demosaicing using variance of color differences," *IEEE Transactions Image Processing*, vol. 15, no. 10, pp. 2944–2955, October 2006.

[11] D. Menon, S. Andriani, and G. Calvagno, "Demosaicing with directional filtering and *a posteriori* decision," *IEEE Transactions on Image Processing*, vol. 16, no. 1, pp. 132–141, January 2007.

[12] B.K. Gunturk, Y. Altunbasak, and R.M. Mersereau, "Color plane interpolation using alternating projections," *IEEE Transactions on Image Processing*, vol. 11, no. 9, pp. 997–1013, September 2002.

[13] X. Li, "Demosaicing by successive approximation," *IEEE Transactions on Image Processing*, vol. 14, no. 3, pp. 370–379, March 2005.

[14] N.X. Lian, L. Chang, Y.P. Tan, and V. Zagorodnov, "Adaptive filtering for color filter array demosaicking," *IEEE Transactions on Image Processing*, vol. 16, no. 10, pp. 2515–2525, October 2007.

[15] D. Taubman, "Generalized Wiener reconstruction of images from colour sensor data using a scale invariant prior," in *Proceedings of IEEE International Conference on Image Processing*, Vancouver, BC, Canada, September 2000, pp. 801–804.

[16] T. Saito and T. Komatsu, "Sharpening-demosaicking method with a total-variation-based super-resolution technique," *Proceedings of SPIE*, vol. 5678, pp. 1801–1812, January 2005.

[17] O.A. Omer and T. Tanaka, "Image demosaicking based on chrominance regularization with region-adaptive weights," in *Proceedings of the International Conference on Information, Communications, and Signal Processing*, Singapore, December 2007, pp. 1–5.

[18] T. Saito and T. Komatsu, "Demosaicing method using the extended color total-variation regularization," *Proceedings of SPIE*, vol. 6817, pp. 68170C:1–12, January 2008.

[19] T. Saito and T. Komatsu, "Demosaicing approach based on extended color total-variation regularization," in *Proceedings of the IEEE International Conference on Image Processing*, San Diego, CA, USA, September 2008, pp. 885–888.

[20] J. Mairal, M. Elad, and G. Sapiro, "Sparse representation for color image restoration," *IEEE Transactions on Image Processing*, vol. 17, no. 1, pp. 53–69, January 2008.

[21] D. Menon and G. Calvagno, "Regularization approaches to demosaicking," *IEEE Transactions onb Image Processing*, vol. 18, no. 10, pp. 2209–2220, October 2009.

[22] K. Hirakawa, *Single-Sensor Imaging: Methods and Applications for Digital Cameras*, ch. Color filter array image analysis for joint denoising and demosaicking, R. Lukac (ed.), Boca Raton, FL: CRC Press/ Taylor & Francis, September 2008, pp. 239–266.

[23] D. Alleysson, S. Süsstrunk, and J. Hérault, "Linear demosaicing inspired by the human visual system," *IEEE Transactions on Image Processing*, vol. 14, no. 4, pp. 439–449, April 2005.

[24] G. Demoment, "Image reconstruction and restoration: Overview of common estimation structures and problems," *IEEE Transactions on Acoustics, Speech, and Signal Processing*, vol. 37, no. 12, pp. 2024–2036, December 1989.

[25] W.C. Karl, *Handbook of Image and Video Processing*, ch. Regularization in image restoration and reconstruction, A. Bovik (ed.), San Diego, CA: Academic Press, June 2000, pp. 141–160.

[26] L.I. Rudin, S. Osher, and E. Fatemi, "Nonlinear total variation based noise removal algorithms," *Physica D*, vol. 60, pp. 259–268, November 1992.

[27] M. Aharon, M. Elad, and A. Bruckstein, "The K-SVD: An algorithm for designing overcomplete dictionaries for sparse representation," *IEEE Transactions on Signal Processing*, vol. 54, no. 11, pp. 4311–4322, November 2006.

[28] S.C. Park, M.K. Park, and M.G. Kang, "Super-resolution image reconstruction: A technical overview," *IEEE Signal Processing Magazine*, vol. 20, no. 5, pp. 21–36, May 2003.

[29] T. Gotoh and M. Okutomi, "Direct super-resolution and registration using raw CFA images," in *Proceedings of the IEEE International Conference on Computer Vision and Pattern Recognition*, Washington, DC, USA, July 2004, pp. 600–607.

[30] S. Farsiu, M. Elad, and P. Milanfar, "Multiframe demosaicing and super-resolution of color images," *IEEE Transactions on Image Processing*, vol. 15, no. 1, pp. 141–159, January 2006.

[31] A.K. Katsaggelos, J. Biemond, R.W. Schafer, and R.M. Mersereau, "A regularized iterative image restoration algorithm," *IEEE Transactions on Signal Processing*, vol. 39, no. 4, pp. 914–929, April 1991.

[32] M. Nikolova and M.K. Ng, "Analysis of half-quadratic minimization methods for signal and image recovery," *SIAM Journal on Scientific Computing*, vol. 27, no. 3, pp. 937–966, June 2005.

[33] K. Hirakawa and P.J. Wolfe, "Second-generation color filter array and emosaicking designs," *Proceedings of SPIE*, vol. 6822, pp. 68221P:1–12, January 2008.

[34] L. Zhang, X. Wu, and D. Zhang, "Color reproduction from noisy CFA data of single sensor digital cameras," *IEEE Transactions Image Processing*, vol. 16, no. 9, pp. 2184–2197, September 2007.

[35] J.M. Bioucas-Dias, M.A.T. Figueiredo, and J.P. Oliveira, "Total variation-based image deconvolution: A majorization-minimization approach," in *Proceedings of the IEEE International Conference on Acoustics, Speech and Signal Processing*, Toulouse, France, May 2006, pp. 861–864.

[36] A. Levin, R. Fergus, F. Durand, and W.T. Freeman, "Image and depth from a conventional camera with a coded aperture," *ACM Transactions on Graphics*, vol. 26, no. 3, July 2007.

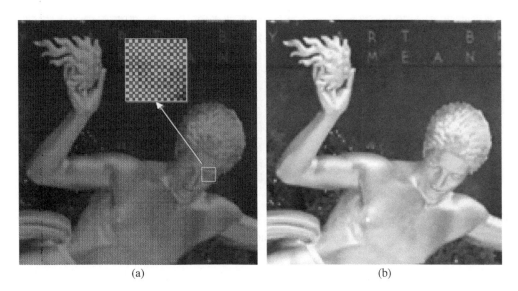

(a)   (b)

FIGURE 1.3

CFA-based digital imaging: (a) Bayer CFA image, and (b) full-color reconstructed image.

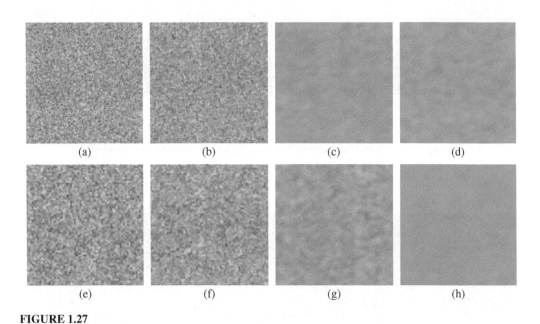

(a)   (b)   (c)   (d)

(e)   (f)   (g)   (h)

FIGURE 1.27

Simulated images demosaicked from a Bayer CFA (top row) and the four-channel CFA shown in Figure 1.19d (bottom row). The images shown are for the cases of: (a,e) no noise reduction, (b,f) median filtering only, (c,g) median and boxcar filtering only, and (d,h) median, boxcar, and low-frequency filtering.

(a)   (b)

FIGURE 2.11

Motion deblurring of demosaicked image: (a) before motion deblurring, and (b) after motion deblurring.

(a)   (b)

FIGURE 2.12

Fully processed images: (a) with motion compensation, and (b) without motion compensation.

(a)   (b)   (c)

FIGURE 4.2

Target reference images captured under syl-50mr16q: (a) *ball*, (b) *books*, and (c) *Macbeth*.

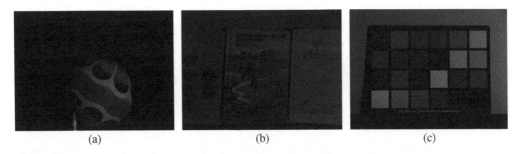

(a)   (b)   (c)

FIGURE 4.3

Source images captured under different illuminations: (a) *ball* under solux-4100, (b) *books* under syl-50mr16q+3202, and (c) *Macbeth* under ph-ulm.

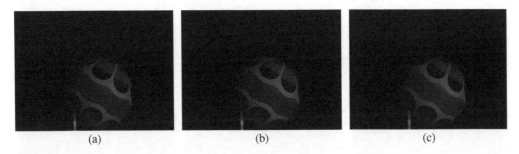

(a)   (b)   (c)

FIGURE 4.4

Color corrected images for the illuminant solux-4100: (a) MXW-DCT-Y, (b) COR, and (c) COR-DCT.

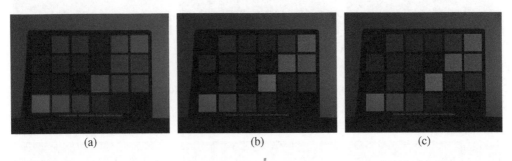

(a)   (b)   (c)

FIGURE 4.6

Color corrected images for the illuminant ph-ulm: (a) MXW-DCT-Y, (b) COR, and (c) COR-DCT.

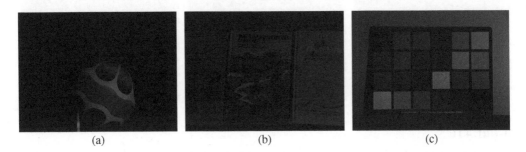

(a)   (b)   (c)

FIGURE 4.7

Chromaticity shift corrected images by GRW-DCT: (a) *ball*, (b) *books*, and (c) *Macbeth*.

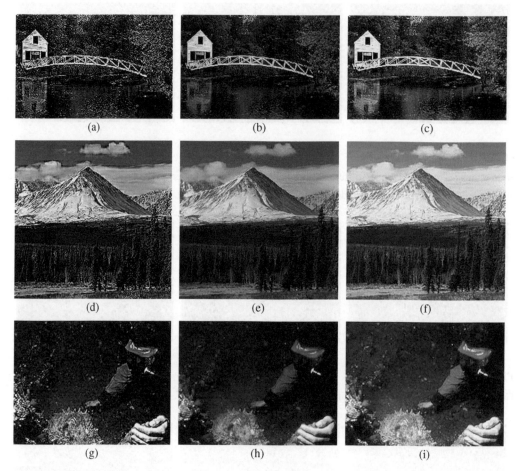

FIGURE 4.9

Color enhancement of the images: (a,d,g) MCE, (b,e,h) MCEDRC, and (c,f,i) TW-CES-BLK.

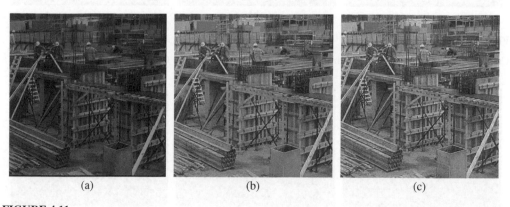

FIGURE 4.11

Color restoration through color correction followed by enhancement: (a) original image, (b) enhanced image without color correction, and (c) enhanced image with color correction.

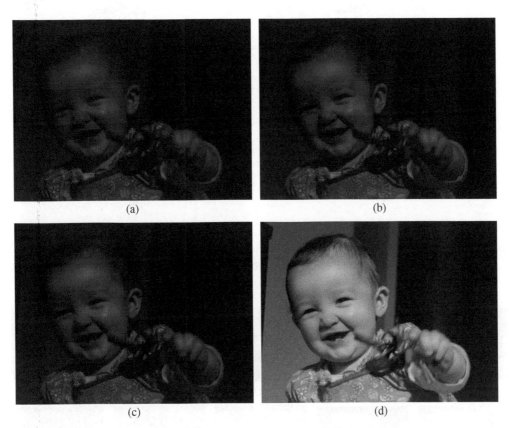

(a)   (b)

(c)   (d)

FIGURE 5.4

Different stages of the camera image processing pipeline: (a) demosaicked image, (b) white-balanced image, (c) color-corrected image, and (d) tone / scale-rendered image. The results correspond to Figure 5.2a.

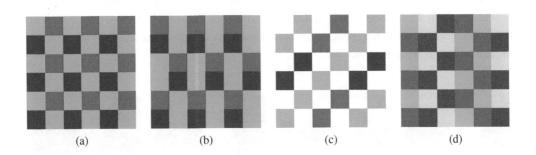

(a)   (b)   (c)   (d)

FIGURE 6.1

Examples of existing CFAs: (a) Bayer [1], (b) Lukac [2], (c) Hamilton [3], and (d) Hirakawa [4]. © 2009 IEEE

(a)   (b)

FIGURE 7.7

Least square method-based image restoration: (a) bilinearly interpolated observation from a sequence of 50 frames, and (b) restored image using the least squares SR algorithm.

(a)   (b)   (c)   (d)   (e)

(f)   (g)   (h)   (i)   (j)

FIGURE 7.11

Comparison of the affine and nonlinear photometric conversion using differently exposed images. The images are geometrically registered: (b) is photometrically mapped on (a), and (g) is photometrically mapped on (f). Images in (c) and (h) are the corresponding residuals when the affine photometric model is used. Images in (d) and (i) are the residuals when the nonlinear photometric model is used. Images in (e) and (j) are the residuals multiplied by the weighting function.

FIGURE 7.12

A subset of 22 input images showing different exposure times and camera positions.

(a)   (b)

(c)   (d)

FIGURE 7.13

Image restoration using the method in Reference [84]: (a,c) input images, and (b,d) their restored versions.

FIGURE 7.14

High-dynamic-range high-resolution image obtained using the method in Reference [3].

FIGURE 7.17

Image demosaicking: (a) bilinear interpolation, (b) edge-directed interpolation [94], (c) multi-frame demosaicking [91], and (d) super-resolution restoration [91].

FIGURE 8.5

Image deblurring using a blurred and noisy image pair: (a) blurred input image, (b) noisy input image, and (c) output image obtained using two input images.

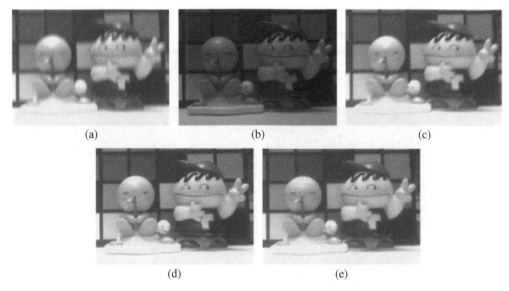

FIGURE 8.9

Comparison of two deblurring approaches: (a-c) three test images, (d) output produced using two first input images, and (e) output produced using all three input images.

FIGURE 9.1

HDR imaging: (a) single exposure from a clearly HDR scene, (b) HDR image linearly scaled to fit the normal 8-bit display interval — the need for tone mapping is evident, and (c) tone mapped HDR image showing a superior amount of information compared with the two previous figures.

(a)   (b)

FIGURE 9.8

The HDR image is composed in luminance-chrominance space and transformed: (a) directly to RGB for tone mapping, (b) to RGB utilizing the presented saturation control, with subsequent tone mapping is done in RGB.

(a)   (b)

FIGURE 9.9

An HDR image of a stained glass window from Tampere Cathedral, Finland: (a) the image is linearly scaled for display, and (b) the tone mapping was done with the method described in Section 9.5.2.1.

FIGURE 9.12

Four exposures from the real LDR sequence used to compose an HDR image. The exposure times of the frames are 0.01, 0.0667, 0.5, and 5 seconds.

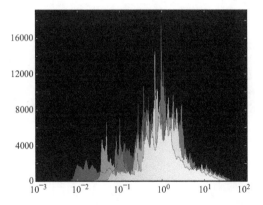

FIGURE 9.13

The histograms of the HDR image composed in luminance-chrominance space and transformed into RGB. Histogram is displayed in logarithmic scale.

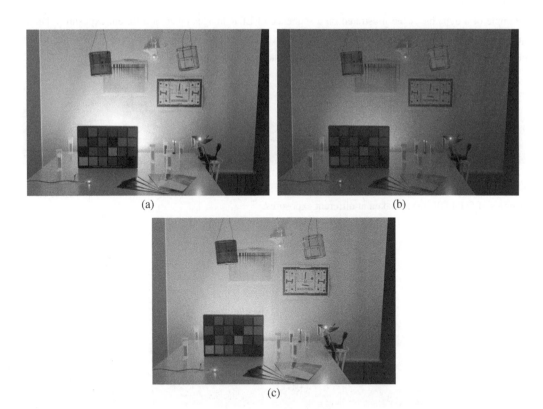

FIGURE 9.14

The HDR image composed in luminance-chrominance space from real data and tone mapped using: (a) the histogram adjustment technique presented in Section 9.5.2.2, (b) the anchoring technique presented in Section 9.5.2.1, and (c) the adaptive logarithmic technique presented in Reference [22] applied in RGB.

FIGURE 10.1

Example of a dynamic scene illustrated on a sequence of LDR images taken at different exposures. On the quay, pedestrians stroll. The boat, floating on the water, oscillates with the water movement. The water and the clouds are subjected to the wind. Therefore water wrinkles change from one image to another.

FIGURE 10.2

A series of five LDR images taken at different exposures.

FIGURE 10.4

The HDR image obtained using traditional methods, (a) from the images shown in Figure 10.2 after alignment, (b) from the image sequence shown in Figure 10.1. (Images built using HDRShop [8].)

FIGURE 10.7

Movement removal using variance and uncertainty [16]: (a) sequence of LDR images captured with different exposure times; several people walk through the viewing window, (b) variance image *VI*, (c) uncertainty image *UI*, (d) HDR image after object movement removal using the variance image, and (e) HDR image after movement removal using the uncertainty image. © 2008 IEEE

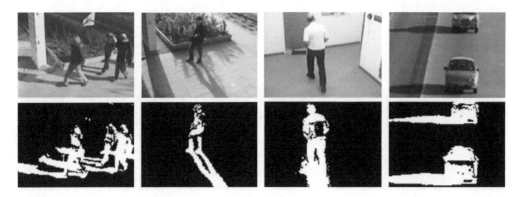

FIGURE 11.1

Results of background subtraction using the algorithm of Reference [7]. Object silhouettes are strongly corrupted, and multiple moving objects cannot be separated due to cast shadows.

FIGURE 11.2

Built-in area extraction using cast shadows: (left) input image, (middle) output of a color-based shadow filter with red areas indicating detected shadows, and (right) built-in areas identified as neighboring image regions of the shadows blobs considering the sun direction.

(a)   (b)   (c)   (d)

FIGURE 11.5

Different parts of the day on *Entrance* sequence with the corresponding segmentation results: (a) *morning am*, (b) *noon*, (c) *afternoon pm*, and (d) *wet weather*.

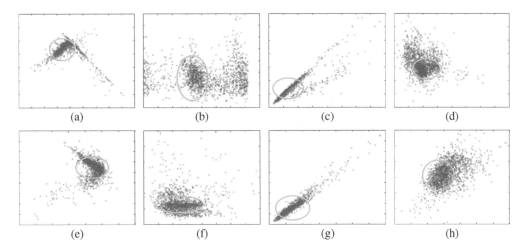

FIGURE 11.10

Two dimensional projection of foreground (red) and shadow (blue) $\overline{\psi}$ values in the *Entrance pm* test sequence: (a) $C_1 - C_2$, (b) $H - S$, (c) $R - G$, (d) $L - u$, (e) $C_2 - C_3$, (f) $S - V$, (g) $G - B$, and (h) $u - v$. The ellipse denotes the projection of the optimized shadow boundary.

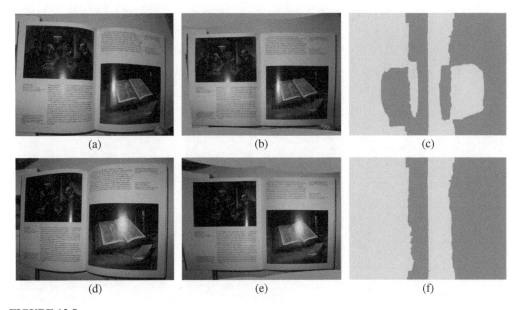

FIGURE 12.5

Document rectification from a stereo pair: (a,d) input stereo pair, (b,e) rectified images, and (c,f) stitching boundary using Equations 12.20 and 12.21. © 2009 IEEE

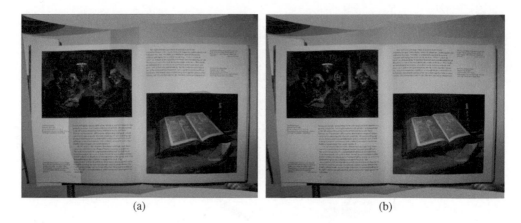

FIGURE 12.6

Composite image generation: (a) without blending, and (b) with blending (final result). © 2009 IEEE

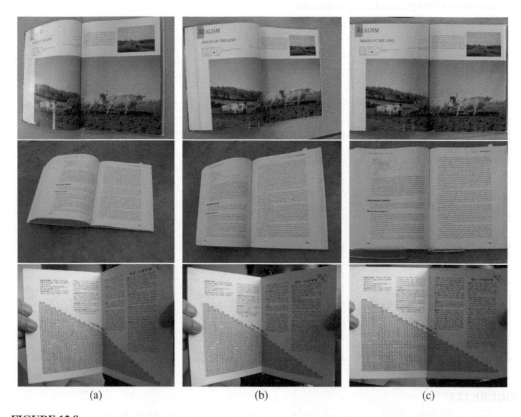

FIGURE 12.8

Additional experimental results: (a,b) input stereo pairs, and (c) final results. © 2009 IEEE

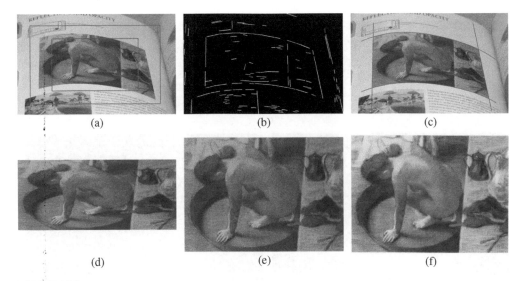

FIGURE 12.9

Single-view rectification procedure: (a) user interaction, the inner box is displayed just for the intuitive under-standing of the system and is ignored after user interaction, (b) feature extraction, one of three feature maps is shown, (c) result of the presented segmentation method, (d) result of References [13] and [27] using the segmented image, aspect ratio 2.32, (e) result of the presented rectification method using the segmented image, aspect ratio 1.43, and (f) figure scanned by a flatbed scanner for comparison, aspect ratio 1.39.

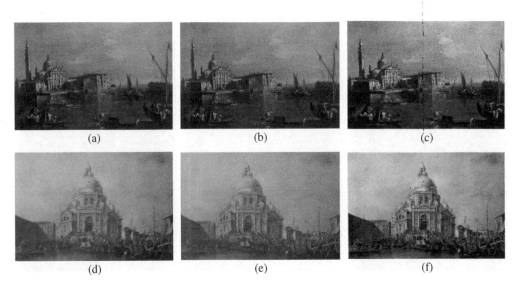

FIGURE 12.16

Performance comparison of the two approaches: (a,d) results of the method presented in Section 12.2 with aspect ratio 1.52 and 1.50, (b,e) results of the method presented in Section 12.3 with aspect ratio 1.52 and 1.53, and (c,f) scanned images with aspect ratio 1.54 and 1.51.

FIGURE 13.12

High definition range imaging using the bilateral filter: (a) a set of four input images, (b) linearly scaled HDR image, and (c) image obtained using the tone-mapping method of Reference [10].

FIGURE 13.14

Image fusion using the bilateral filter: (a) no-flash image, (b) flash image, and (c) fusion of the flash and no-flash images.

(a)   (b)   (c)

FIGURE 14.5

The effect of $\mathbf{v}(\mathbf{r}) \perp \nabla I(\mathbf{r})$ on image quality: (a) input image, (b) synthetic painting obtained by imposing the condition $\mathbf{v}(\mathbf{r}) \perp \nabla I(\mathbf{r})$ on the whole image, and (c) synthetic painting obtained by imposing the condition $\mathbf{v}(\mathbf{r}) \perp \nabla I(\mathbf{r})$ only on high contrast edge points.

(a)   (b)

FIGURE 14.8

Morphological color image processing: (a) input image and (b) output image obtained via close-opening applied separately to each RGB component.

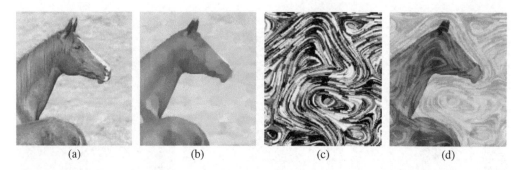

FIGURE 14.16

Artistic image generation for the example of Figure 14.15: (a) input image $I(\mathbf{r})$, (b) edge preserving smoothing output $I_{EPS}(\mathbf{r})$, (c) associated synthetic painterly texture, and (d) final output $y(\mathbf{r})$. © 2009 IEEE

FIGURE 14.17

Comparison of various artistic effects: (a) input image, (b) Glass pattern algorithm, (c) artistic vision [15], and (d) impressionistic rendering [9]. © 2009 IEEE

FIGURE 14.19

Examples of cross continuous Glass patterns: (a) input image, (b) output corresponding to $\mathbf{v}(\mathbf{r}) = [x,y]/\sqrt{x^2+y^2}$, (c) output corresponding to $\mathbf{v}(\mathbf{r}) = [-y,x]/\sqrt{x^2+y^2}$, and (d) output corresponding to $\mathbf{v}(\mathbf{r}) \perp \nabla I(\mathbf{r})$. The last image is perceptually similar to a painting.

FIGURE 15.1

Failure of standard colorization algorithms in the presence of texture: (left) manual initialization, (right) result of Reference [1] with the code available at http://www.cs.huji.ac.il/~yweiss/Colorization/. Despite the general efficiency of this simple method, based on the mean and the standard deviation of local intensity neighborhoods, the texture remains difficult to deal with. Hence texture descriptors and learning edges from color examples are required.

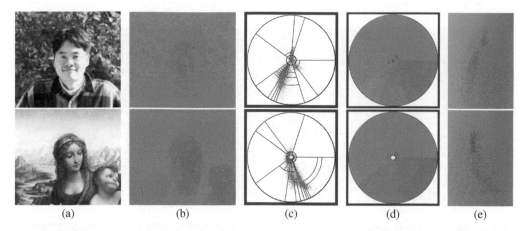

| (a) | (b) | (c) | (d) | (e) |

FIGURE 15.2

Examples of color spectra and associated discretizations: (a) color image, (b) corresponding 2D colors, (c) the location of the observed 2D colors in the *ab*-plane (a red dot for each pixel) and the computed discretization in color bins, (d) color bins filled with their average color, and (e) continuous extrapolation with influence zones of each color bin in the *ab*-plane (each bin is replaced by a Gaussian, whose center is represented by a black dot; red circles indicate the standard deviation of colors within the color bin, blue ones are three times larger).

FIGURE 15.3

Coloring a painting given another painting by the same painter: (a) training image, (b) test image, (c) image colored using Parzen windows — the border is not colored because of the window size needed for SURF descriptors, (d) color variation predicted — white stands for homogeneity and black for color edge, (e) most probable color at the local level, and (f) 2D color chosen by graph cuts.

FIGURE 15.4

Landscape example with Parzen windows: (a) training image, (b) test image, (c) output image, (d) predicted color variation, (e) most probable color locally, (f) 2D color chosen by graph cuts, (g) colors obtained after refinement step.

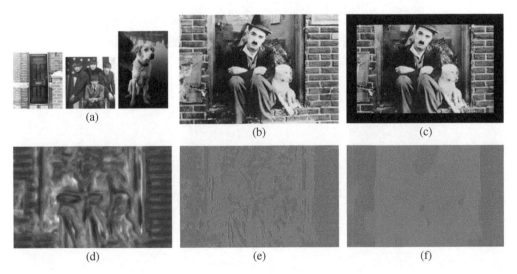

FIGURE 15.7

Image colorization using Parzen windows: (a) three training images, (b) test image, (c) colored image, (d) prediction of color variations, (e) most probable colors at the local level, and (f) final colors.

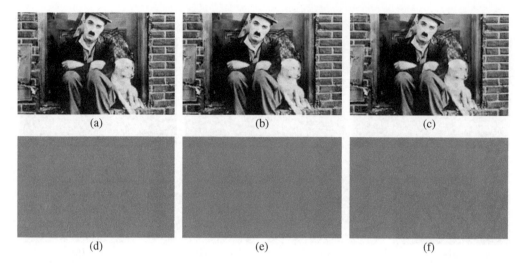

(a)   (b)   (c)

(d)   (e)   (f)

FIGURE 15.8

SVM-driven colorization of Charlie Chaplin frame using the training set of Figure 15.7: (a,d) SVM, (b,e) SVM with spatial regularization — Equation 15.7, and (c,f) SVMstruct.

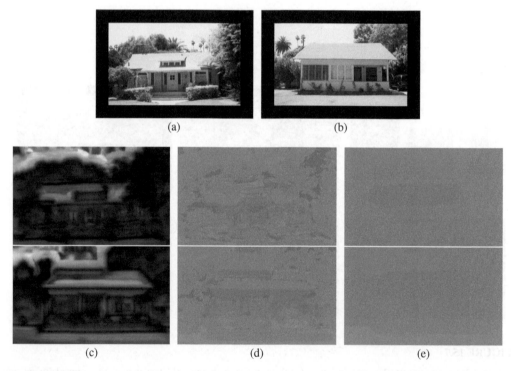

FIGURE 15.10

Colorization of the 21st and 22nd images from the Pasadena houses Caltech dataset: (a) 21st image colored, (b) 22nd image colored, (c) predicted edges, (d) most probable color at the pixel level, and (e) colors chosen.

FIGURE 15.11

SVM-driven colorization of the 21st and 22nd images from the Pasadena houses Caltech dataset; results and colors chosen are displayed: (a-c) 21st image, (d-f) 22nd image; (a,d) SVM, (b,e) SVM with spatial regularization, and (c,f) SVMstruct.

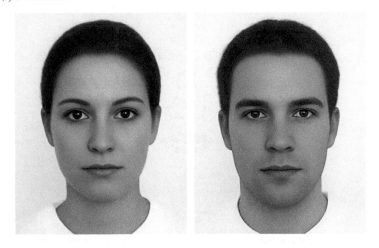

FIGURE 16.3

Female and male face-composites, each averaging sixteen faces. It has been empirically shown that average faces tend to be considerably more attractive than the constituent faces.

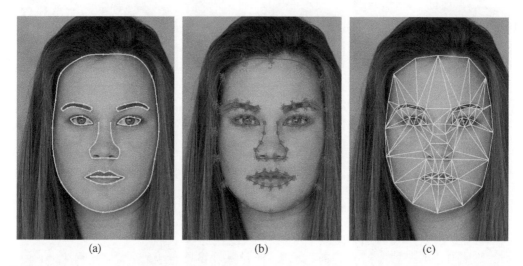

FIGURE 16.4

An example of the eight facial features (two eyebrows, two eyes, the inner and outer boundaries of the lips, the nose, and the boundary of the face), composed of a total of 84 feature points, used in the proposed algorithm: (a) output feature points from the active shape model search, (b) scatter of the aligned 84 landmark points of 92 sample training data and their average, and (c) 234 distances between these points. © 2008 ACM

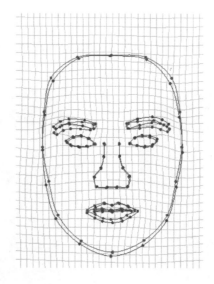

FIGURE 16.7

The warp field is defined by the correspondence between the source feature points (in blue) and the beautified geometry (in red). © 2008 ACM

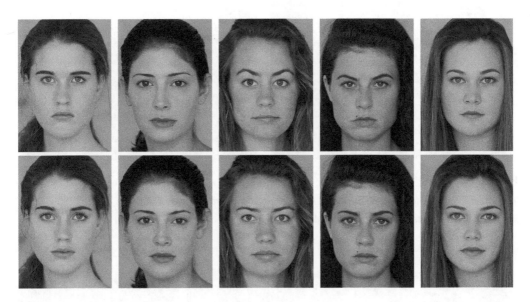

FIGURE 16.8

Beautification examples: (top) input portraits and (bottom) their beautified versions. © 2008 ACM

FIGURE 17.3

Performance improvement by multiplexing: (a) light field image captured without multiplexing, (b) demultiplexed light field image, (c) image captured with multiplexing, that is, $M_u(x)$ in Equation 17.4, and (d) enlarged portions of (a) and (b). The insets in (a-c) show the corresponding multiplexing patterns.

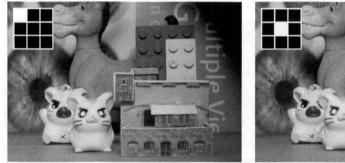

FIGURE 17.5

The effect of the photometric distortion. The images shown here are two of the nine light field images of a static scene. The insets show the corresponding aperture shape.

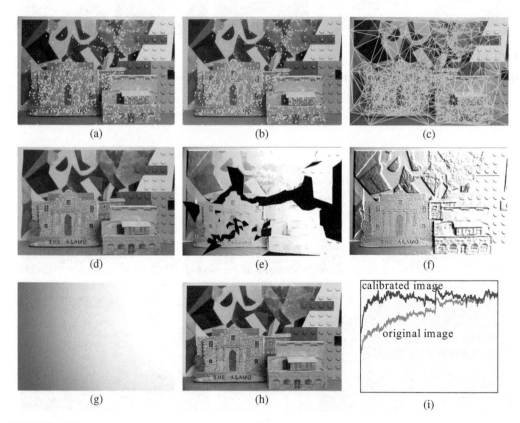

(a)

(b)

(c)

(d)

(e)

(f)

(g)

(h)

(i)

FIGURE 17.7

Photometric calibration results: (a) input image I_u^d to be corrected — note the darker left side, (b) reference image I_0, (c) triangulation of the matched features marked in the previous two images, (d) image I_u^w warped using the reference image based on the triangular mesh, (e) approximated vignetting field with suspicious areas removed, (f) vignetting field approximated without warping, (g) estimated parametric vignetting field, (h) calibrated image I_u, and (i) intensity profile of the 420th scanline before and after the calibration.

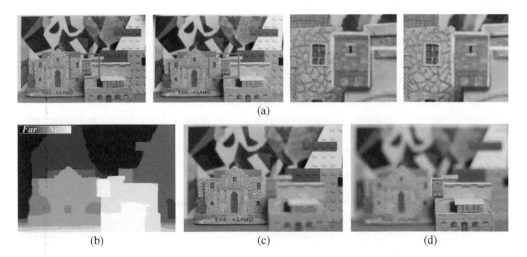

FIGURE 17.11

(a) Two demultiplexed light field images generated by the proposed system; the full 4D resolution is $4 \times 4 \times 3039 \times 2014$. (b) The estimated depth map of the left image of (a). (c,d) Postexposure refocused images generated from the light field and the depth maps.

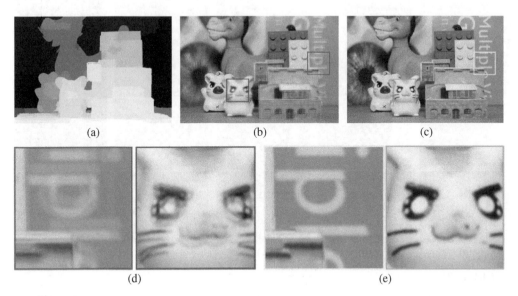

FIGURE 17.13

(a) An estimated depth map. (b) Image interpolated without depth information. (c) Image interpolated with depth information. (d-e) Close-up of (b-c). The angular resolution of the light field is 3×3.

FIGURE 17.14

(a) An estimated depth map. (b) Digital refocused image with the original angular resolution 4×4. (c) Digital refocused image with the angular resolution 25×25 boosted by view interpolation. (d-e) Close-up of (b-c).

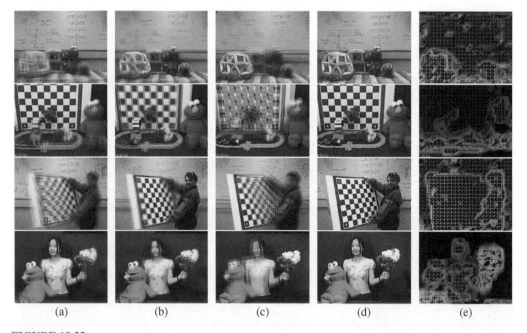

FIGURE 18.22

Scenes captured and rendered with the proposed camera array: (a) rendering with a constant depth at the background, (b) rendering with a constant depth at the middle object, (c) rendering with a constant depth at the closest object, (d) rendering with the proposed method, and (e) multi-resolution 2D mesh with depth reconstructed on-the-fly, brighter intensity means smaller depth. Captured scenes from top to bottom: *toys*, *train*, *girl and checkerboard*, and *girl and flowers*.

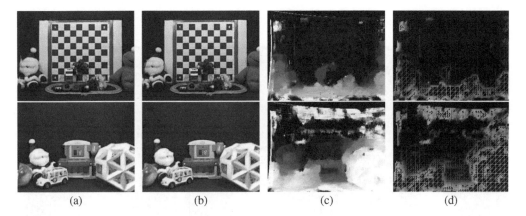

<table>
<tr><td>(a)</td><td>(b)</td><td>(c)</td><td>(d)</td></tr>
</table>

FIGURE 18.23

Real-world scenes rendered with the proposed algorithm: (top) scene *train*, (bottom) scene *toys*; (a,c) used per-pixel depth map to render, and (b,d) the proposed adaptive mesh to render.

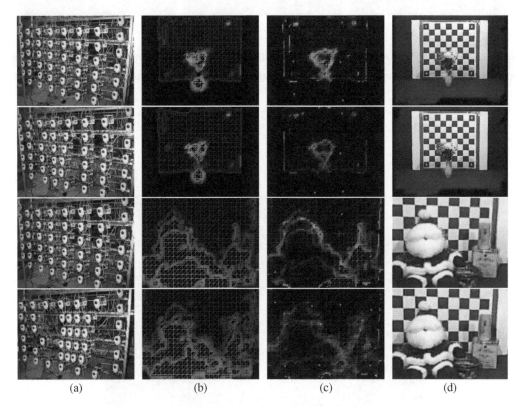

(a)　　　　　(b)　　　　　(c)　　　　　(d)

FIGURE 18.24

Scenes rendered by reconfiguring the proposed camera array: (a) the camera arrangement, (b) reconstructed depth map, brighter intensity means smaller depth, (c) CCV score of the mesh vertices and the projection of the camera positions to the virtual imaging plane denoted by dots — darker intensity means better consistency, and (d) rendered image. Captured scenes from top to bottom: *flower* with cameras evenly spaced, *flower* with cameras self-reconfigured (6 epochs), *Santa* with cameras are evenly spaced, and *Santa* with cameras self-reconfigured (20 epochs). © 2007 IEEE

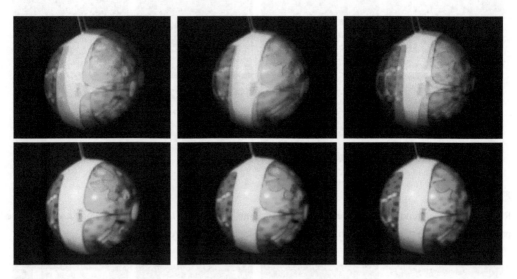

FIGURE 18.31

Synthesized results of the toy ball sequence: (top) traditional LFR with unsynchronized frames and (bottom) space-time LFR using estimated temporal offsets among input sequences. Note that the input data set does not contain global time stamps.

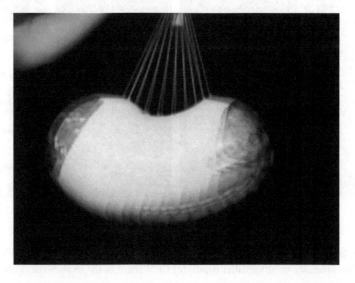

FIGURE 18.32

Visualization of the trajectory.

7

Super-Resolution Imaging

Bahadir K. Gunturk

7.1   Introduction

Super-resolution (SR) image restoration is the process of producing a high-resolution image (or a sequence of high-resolution images) from a set of low-resolution images [1], [2], [3]. The process requires an image acquisition model that relates a high-resolution image to multiple low-resolution images and involves solving the resulting inverse problem. The acquisition model includes aliasing, blurring, and noise as the main sources of information

FIGURE 7.1

A subset of 21 low-resolution input images.

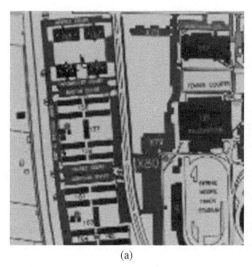 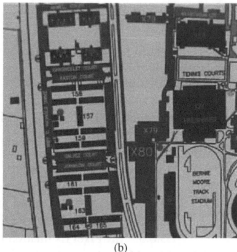

(a)   (b)

FIGURE 7.2

Interpolation vs. super-resolution: (a) one of the input images resized with bilinear interpolation, and (b) high-resolution image obtained with the SR restoration algorithm presented in Reference [3].

loss. A super-resolution algorithm increases the spatial detail in an image, and equivalently recovers the high-frequency information that is lost during the imaging process.

There is a wide variety of application areas for SR image restoration. In biomedical imaging, multiple images can be combined to improve the resolution, which may help in diagnosis. In surveillance systems, the resolution of a video sequence can be increased to obtain critical information, such as license plate or facial data. High-definition television (HDTV) sets may utilize SR image restoration to produce and display higher quality video from a standard definition input signal. High-quality prints from low-resolution images can be made possible with SR image restoration. Imaging devices such as video cameras and microscopes might be designed to create intentional sensor shifts to produce high-resolution images [1]. Similarly, super-resolution projectors can be realized by superimposing multiple shifted images on a screen [2]. Other application areas include satellite and aerial imaging, astronomy, video coding, and radar imaging. Figure 7.1 shows a number of low-resolution images. A set of these images is used to create a high-resolution image; Figure 7.2 shows that the restored image is de-aliased and more legible than the bilinearly interpolated input.

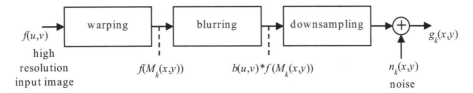

FIGURE 7.3

Imaging model.

SR image restoration has been an active research field for more than two decades. As revealed in various survey studies [4], [5], [6], [7], [8], [9], there are both well-established methods and open problems to be addressed.

This chapter aims to serve as an introductory material for SR imaging, and also provide a comprehensive review of SR methods. Section 7.2 describes a commonly used imaging model and two implementation approaches. Section 7.3 presents the main SR methods. Registration and parameter estimation issues are discussed in Section 7.4. Variations on SR imaging model are presented in Section 7.5. Finally, conclusions and research directions are given in Section 7.6.

7.2 Observation Model

There are two main factors that cause loss of detail in images. The first factor is blurring, which may be due to atmospheric blur, optical blur (out-of-focus, diffraction, and lens aberrations), sensor blur (spatial averaging on photosensitive pixel sites on sensor), and motion blur (which is pronounced when the exposure time is long considering the motion in the scene). The blurring effects are typically combined into a space-invariant point spread function (PSF) and implemented as the convolution of the ground truth image with the PSF. Most SR algorithms do not consider motion blur, and use an isotropic convolution kernel. An exception to this model is presented in Reference [10], where motion blur is included in the model, resulting in a space-variant PSF. The second factor that reduces resolution in images is aliasing, which is due to spatial sampling on the sensor. Spatial sampling is modeled as downsampling in image formation. In addition to these factors, noise also corrupts the observed data, and is included in most imaging models. There are several causes of noise, such as charges generated in the sensor independent of the incoming photons, and quantization of signals in the electronics of the imaging device. Observation noise is typically modeled as an additive term.

In SR image restoration, the goal is to improve the resolution of an image by combining multiple images. Therefore, the imaging model relates a high-resolution (unobserved) image to multiple low-resolution observations. Mathematically, the imaging process can be formulated as follows (Figure 7.3):

$$g_k(x,y) = D[b(u,v) * f(M_k(x,y))] + n_k(x,y), \qquad (7.1)$$

where g_k is the kth low-resolution image, D is the downsampling operator, b is the PSF,

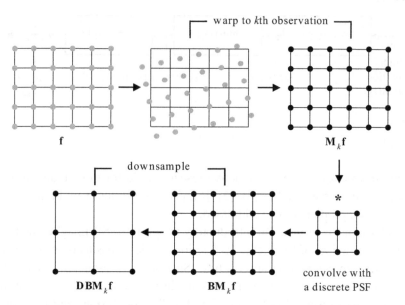

FIGURE 7.4

The high-resolution image is warped onto the kth frame, convolved with a discrete PSF, and downsampled to form the observation.

f is the high-resolution image, M_k is geometric mapping between f and the kth observation, and n_k denotes observation noise. The term $M_k(x, y)$ relates the coordinates of the kth observation and the high-resolution image; in other words, $f(M_k(x, y))$ is the warped high-resolution image and is the higher resolution version of g_k. This model makes several assumptions, including a shift-invariant PSF, additive noise, and constant illumination conditions. Later in the chapter, three enhancements of this model, through motion modeling, photometric modeling, and modeling of color filter array sampling, will be discussed.

The discretized imaging model can be written in matrix form as follows:

$$\mathbf{g}_k = \mathbf{H}_k \mathbf{f} + \mathbf{n}_k, \tag{7.2}$$

where $\mathbf{f}$ is the vectorized version of the high-resolution image, $\mathbf{g}_k$ is the kth vectorized observation, $\mathbf{n}_k$ is the kth vectorized noise, and $\mathbf{H}_k$ is the matrix that includes the linear operations, that is, geometric warping, convolution with the PSF, and downsampling.

Sometimes, all N observations are stacked to form a simplified representation of the problem:

$$\underbrace{\begin{bmatrix} \mathbf{g}_1 \\ \mathbf{g}_2 \\ \vdots \\ \mathbf{g}_N \end{bmatrix}}_{\mathbf{g}} = \underbrace{\begin{bmatrix} \mathbf{H}_1 \\ \mathbf{H}_2 \\ \vdots \\ \mathbf{H}_N \end{bmatrix}}_{\mathbf{H}} \mathbf{f} + \underbrace{\begin{bmatrix} \mathbf{n}_1 \\ \mathbf{n}_2 \\ \vdots \\ \mathbf{n}_N \end{bmatrix}}_{\mathbf{n}} \implies \mathbf{g} = \mathbf{H}\mathbf{f} + \mathbf{n} \tag{7.3}$$

There are two main approaches to implement the forward imaging process. In the first method, the high-resolution image is warped to align with the low-resolution observation, convolved with a discrete PSF, and downsampled to simulate the observation. This method

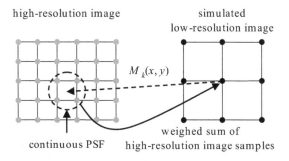

FIGURE 7.5

For a low-resolution position (x,y), a continuous PSF is placed at location (u,v) on the high-resolution image based on the geometric mapping $M_k(x,y)$. The weights of the PSF corresponding to the high-resolution pixels under the kernel support are calculated. The weighted sum of the high-resolution pixels produces the intensity at low-resolution image position (x,y). This process is repeated for all low-resolution image locations.

is illustrated in Figure 7.4 and basically corresponds to consecutive application of three matrices:

$$\mathbf{H}_k = \mathbf{DBM}_k, \tag{7.4}$$

where $\mathbf{D}$ denotes the downsampling matrix, $\mathbf{B}$ denotes the matrix for convolution with the discrete PSF, and $\mathbf{M}_k$ is the matrix for geometric mapping from f to kth observation.

In the second method, a continuous PSF is placed on the high-resolution image using the motion vector from a low-resolution sample location. Then, the weights of the PSF corresponding to the high-resolution image samples are obtained, and finally the weighted sum of the high-resolution samples is calculated to simulate the low-resolution sample. By repeating this process for all low-resolution sample locations, the low-resolution image is obtained. This method is illustrated in Figure 7.5.

7.3 Super-Resolution Methods

7.3.1 Frequency Domain Approach

One of the earliest SR methods is the frequency domain method presented in Reference [11]. This method relates the discrete Fourier transform (DFT) coefficients of the low-resolution images and the continuous Fourier transform (CFT) of the high-resolution image, and constructs a linear system in terms of the Fourier coefficients.

A continuous band-limited high-resolution image $f(u,v)$ is shifted by (δ_u, δ_v) to produce a shifted version $f_k(u,v) = f(u + \delta_u, v + \delta_v)$. The shifted image is then sampled with period (T_x, T_y) to produce the discrete low-resolution image $g_k[n_x, n_y] = f_k(T_x n_x, T_y n_y)$, where $n_x = 0, ..., N_x - 1$ and $n_y = 0, ..., N_y - 1$. The DFT of g_k and CFT of f_k are related as follows:

$$G_k[m_x, m_y] = \frac{1}{T_x T_y} \sum_{p_x} \sum_{p_y} F_k \left(\frac{m_x + N_x p_x}{N_x T_x}, \frac{m_y + N_y p_y}{N_y T_y} \right) \tag{7.5}$$

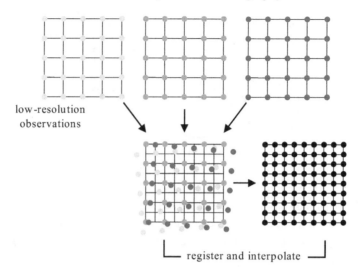

low-resolution
observations

register and interpolate

FIGURE 7.6

Low-resolution observations are registered and interpolated on a high-resolution grid. This process is followed by a deconvolution process.

Combining this equation with the shifting property of Fourier transform $F_k(w_x, w_y) = e^{j2\pi(T_x w_x + T_y w_y)} F(w_x, w_y)$ results in a set of linear equations relating the DFT of the observations with the samples of the CFT of the high-resolution image. The CFT samples are then solved to form the high-resolution image.

This frequency domain algorithm was later extended to include blur and noise in the model [12]. A total least squares version was presented in Reference [13] for regularizing against registration errors; and a DCT domain version was presented in Reference [14] to reduce the computational cost. The frequency domain approach has the advantage of low-computational complexity and an explicit dealiasing mechanism. Among the disadvantages are the limitation to global translational motion, limitation to shift-invariant blur, and limited capability of incorporating spatial domain priors for regularization.

7.3.2   Interpolation-Deconvolution Method

In the interpolation-deconvolution method, all observations are first registered and interpolated on a high-resolution grid as illustrated in Figure 7.6. This requires interpolation from nonuniformly distributed samples. A simple nonuniform interpolation algorithm is to take a weighted sum of the samples at each grid location, where the weights are inversely proportional to the distance between the grid location and the sample location [15], [16]. The weights can be chosen as $\exp(-\alpha\delta)$, where δ is the distance between the grid and sample locations, and α is a constant that controls the decay of the weights. Another nonuniform interpolation-based SR algorithm is presented in Reference [17], where the interpolation method is based on the Delaunay triangulation. The methods presented in References [18] and [19] are among those that do nonuniform interpolation. The interpolation step is followed by a deconvolution step to create a high-resolution image. While the interpolation-deconvolution method does not limit the type of motion among the images, the optimality of separating the interpolation and deconvolution steps is an open question. A general observation is that the interpolation-deconvolution method does not perform as well as the methods that do not separate interpolation from deconvolution.

7.3.3 Least Squares Estimation

A high-resolution image can be obtained by solving the imaging model given in Equation 7.2 by least squares estimation. The least squares solution $\mathbf{f}_{ls}$ minimizes the cost function $C(\mathbf{f}) = \sum_k ||\mathbf{g}_k - \mathbf{H}_k\mathbf{f}||^2$, and it can be calculated directly by setting the derivative of the cost function to zero:

$$\frac{\partial C(\mathbf{f})}{\partial \mathbf{f}} = -2\sum_k \mathbf{H}_k^T (\mathbf{g}_k - \mathbf{H}_k\mathbf{f}) = 0, \tag{7.6}$$

resulting in the following:

$$\mathbf{f}_{ls} = \left(\sum_k \mathbf{H}_k^T \mathbf{H}_k\right)^{-1} \left(\sum_k \mathbf{H}_k^T \mathbf{g}_k\right). \tag{7.7}$$

In practice, the direct solution is not computationally feasible due to the sizes of the matrices involved. If $\mathbf{H}_k$ was a block-circulant matrix, which is not the case in general, an efficient frequency domain implementation would be possible. Therefore, iterative methods, such as the steepest descent and conjugate gradient methods, are adopted. These methods start with an initial estimate and update it iteratively until a convergence criterion is reached. The convergence criterion could be, for instance, the maximum number of iterations or the rate of change between two successive iterations. An iteration of the steepest descent method is the following:

$$\begin{aligned}\mathbf{f}^{(i+1)} &= \mathbf{f}^{(i)} - \alpha \left.\frac{\partial C(\mathbf{f})}{\partial \mathbf{f}}\right|_{\mathbf{f}^{(i)}}, \\ &= \mathbf{f}^{(i)} + 2\alpha \sum_k \mathbf{H}_k^T \left(\mathbf{g}_k - \mathbf{H}_k\mathbf{f}^{(i)}\right),\end{aligned} \tag{7.8}$$

where $\mathbf{f}^{(i)}$ is the ith estimate. The step size α should be small enough to guarantee convergence; on the other hand, the convergence would be slow if it is too small. The value of α could be fixed or adaptive, changing at each iteration. A commonly used method to choose the step size is the *exact line search* method. In this method, defining $\mathbf{d} = \partial C(\mathbf{f})/\partial \mathbf{f}|_{\mathbf{f}^{(i)}}$ at the ith iteration, the step size that minimizes the next cost $C\left(\mathbf{f}^{(i+1)}\right) = C\left(\mathbf{f}^{(i)} - \alpha\mathbf{d}\right)$ is given by

$$\alpha = \frac{\mathbf{d}^T\mathbf{d}}{\mathbf{d}^T \left(\sum_k \mathbf{H}_k^T \mathbf{H}_k\right) \mathbf{d}}. \tag{7.9}$$

The algorithm suggested by Equation 7.8 can be taken literally by converting images to vectors and obtaining the matrix forms of warping, blurring, and downsampling operations. There is an alternative implementation, which involves simple image manipulations. The key is to understand the image domain operations implied by the transpose matrices $\mathbf{H}_k^T = (\mathbf{D}\mathbf{B}\mathbf{W}_k)^T = \mathbf{W}_k^T\mathbf{B}^T\mathbf{D}^T$. Through the analysis of the matrices, it can be seen that

(a)   (b)

FIGURE 7.7 (See color insert.)

Least square method-based image restoration: (a) bilinearly interpolated observation from a sequence of 50 frames, and (b) restored image using the least squares SR algorithm.

$\mathbf{D}^T$ corresponds to upsampling by zero insertion; $\mathbf{B}^T$ corresponds to convolution with the flipped PSF kernel; and finally, since $\mathbf{W}_k$ corresponds to warping from reference frame to kth frame, $\mathbf{W}_k^T$ is implemented by back warping from kth frame to reference frame. It should be noted that $\mathbf{W}_k^T$ is not exactly equal to $\mathbf{W}_k^{-1}$ unless $\mathbf{W}_k$ is a permutation matrix. This is not surprising because an image warp operation involves interpolation, which is a lossy process. Also note that the adaptive step size α in Equation 7.9 can be calculated with similar image operations as well [20].

A sample result is shown in Figure 7.7. The iterative least squares algorithm is essentially identical to the iterated backpropagation algorithm presented in Reference [21]. The main difference in the iterated backpropagation algorithm is that the backprojection PSF is not necessarily the flipped version of the forward imaging PSF. Reference [22] proposes to take the median of the backprojections coming from different images in order to eliminate outliers.

One issue with the least squares approach is the sensitivity to small perturbations in the data. SR restoration is an ill-posed inverse problem; small perturbations, such as noise and registration errors, could get amplified and dominate the solution. Regularization is therefore needed to make the least squares approach more robust. Tikhonov regularization [23], [24] is a commonly used technique; the cost function to be minimized is modified to $C(\mathbf{f}) = \sum_k ||\mathbf{g}_k - \mathbf{H}_k\mathbf{f}||^2 + \lambda ||\mathbf{L}\mathbf{f}||^2$, where λ is the regularization parameter, and $\mathbf{L}$ is an identity matrix or a high-pass filter matrix used to impose smoothness on the solution. The maximum entropy [23], total variation [25], and L_p norm [26] are some of the other regularization approaches.

7.3.4 Bayesian Approach

Bayesian estimation provides an elegant statistical perspective to the SR image restoration problem. The unknown image, noise, and in some cases other parameters, such as motion vectors, are viewed as random variables. The maximum likelihood (ML) estimator seeks the solution that maximizes the probability $P(\mathbf{g}|\mathbf{f})$, while the maximum *a posteriori* (MAP) estimator maximizes $P(\mathbf{f}|\mathbf{g})$. Using the Bayes rule, the MAP estimator can be written in terms of the conditional probability $P(\mathbf{g}|\mathbf{f})$ and the prior probability of $P(\mathbf{f})$ as follows:

$$\mathbf{f}_{map} = \arg\max_{\mathbf{f}} P(\mathbf{f}|\mathbf{g}) = \arg\max_{\mathbf{f}} \frac{P(\mathbf{g}|\mathbf{f})P(\mathbf{f})}{P(\mathbf{g})}. \tag{7.10}$$

In most SR restoration problems, image noise is modeled to be a zero-mean independent identically distributed (iid) Gaussian random variable. Thus, the probability of an observation given the high-resolution image can be expressed as follows:

$$P(\mathbf{g}_k|\mathbf{f}) = \prod_{x,y} \frac{1}{\sqrt{2\pi}\sigma} \exp\left(-\frac{1}{2\sigma^2}(g_k(x,y)-\hat{g}_k(x,y))^2\right),$$

$$= \frac{1}{(\sqrt{2\pi}\sigma)^M} \exp\left(-\frac{1}{2\sigma^2}\|\mathbf{g}_k-\mathbf{H}_k\mathbf{f}\|^2\right), \tag{7.11}$$

where σ is the noise standard deviation, $\hat{g}_k(x,y)$ is the predicted low-resolution pixel value using $\mathbf{f}$ and the forward imaging process $\mathbf{H}_k$, and M is the total number of pixels in the low-resolution image.

The probability of all N observations given the high-resolution image can then be expressed as

$$P(\mathbf{g}|\mathbf{f}) = \prod_{k=1}^{N} P(\mathbf{g}_k|\mathbf{f}) = \frac{1}{(\sqrt{2\pi}\sigma)^{NM}} \exp\left(-\frac{1}{2\sigma^2}\sum_k \|\mathbf{g}_k-\mathbf{H}_k\mathbf{f}\|^2\right),$$

$$= \frac{1}{(\sqrt{2\pi}\sigma)^{NM}} \exp\left(-\frac{1}{2\sigma^2}\|\mathbf{g}-\mathbf{H}\mathbf{f}\|^2\right). \tag{7.12}$$

$$\tag{7.13}$$

Substituting this into Equation 7.10, and by taking the logarithm and neglecting the irrelevant terms, provides the following:

$$\mathbf{f}_{map} = \arg\max_{\mathbf{f}} \{\log P(\mathbf{g}|\mathbf{f}) + \log P(\mathbf{f})\},$$

$$= \arg\max_{\mathbf{f}} \left\{-\frac{1}{2\sigma^2}\|\mathbf{g}-\mathbf{H}\mathbf{f}\|^2 + \log P(\mathbf{f})\right\}. \tag{7.14}$$

This MAP approach has been used in many SR algorithms, including those in References [27], [28], and [29]. The main difference is the prior model chosen. A commonly used prior model is the Gaussian model:

$$P(\mathbf{f}) \propto \exp\left(-(\mathbf{f}-\mu)^T \mathbf{Q}(\mathbf{f}-\mu)\right), \tag{7.15}$$

where μ is the average image and $\mathbf{Q}$ is the inverse of the covariance matrix. When both the data fidelity and the prior terms are quadratic, a straightforward analytical implementation is possible. Another popular quadratic model is $P(\mathbf{f}) \propto \exp(-\|\mathbf{L}\mathbf{f}\|^2)$, which leads to the following optimization problem:

$$\mathbf{f}_{map} = \arg\min_{\mathbf{f}} \left\{\|\mathbf{g}-\mathbf{H}\mathbf{f}\|^2 + \lambda \|\mathbf{L}\mathbf{f}\|^2\right\}, \tag{7.16}$$

where $\mathbf{L}$ is typically a discrete approximation of a derivative operator, Laplacian or identity matrix. With $\mathbf{L}$ being a high-pass filter, the $\|\mathbf{L}\mathbf{f}\|$ term penalizes the high-frequency content and leads to a smoother solution. The regularization parameter λ controls the relative contribution of the prior term.

The MAP estimate in Equation 7.16 is equivalent to the least squares estimation solution with Tikhonov regularization; and the direct solution is $\mathbf{f}_{map} = (\mathbf{H}^T\mathbf{H} + \lambda\mathbf{L}^T\mathbf{L})^{-1}\mathbf{H}^T\mathbf{g}$. As discussed earlier, this direct solution is not feasible in practice due to the sizes of the matrices involved. Instead, an iterative approach should be taken.

Another commonly used prior model is the Gibbs distribution [30]:

$$P(\mathbf{f}) = \frac{1}{Z}\exp\left(-\beta E(\mathbf{f})\right) = \frac{1}{Z}\exp\left(-\beta\sum_{c\in\mathscr{C}} V_c(\mathbf{f})\right), \tag{7.17}$$

where Z is a normalization constant, β is a positive constant that controls the peakiness of the distribution, $E(\cdot)$ is the energy function, c is a *clique*, $\mathscr{C}$ is the set of all cliques, and $V_c(\cdot)$ is the potential function of clique c. A *clique* is a single site or a set of site pairs in a neighborhood. Through the way the cliques and the clique potentials are defined, different structural properties can be modeled. The Gibbs distribution is an exponential distribution; the resulting optimization problem can be expressed as follows:

$$\mathbf{f}_{map} = \arg\min_{\mathbf{f}}\left\{\|\mathbf{g} - \mathbf{H}\mathbf{f}\|^2 + \lambda\sum_{c\in\mathscr{C}} V_c(\mathbf{f})\right\}. \tag{7.18}$$

A sample potential function at a pixel location is defined as

$$V_c(\mathbf{f}) = \sum_{n=1}^{4} d_n^2(x,y) \tag{7.19}$$

with

$$\begin{aligned}
d_1(x,y) &= f(x+1,y) - 2f(x,y) + f(x-1,y), \\
d_2(x,y) &= f(x,y+1) - 2f(x,y) + f(x,y-1), \\
d_3(x,y) &= f(x+1,y+1)/2 - f(x,y) + f(x-1,y-1)/2, \\
d_4(x,y) &= f(x+1,y-1)/2 - f(x,y) + f(x+1,y-1)/2,
\end{aligned} \tag{7.20}$$

where the clique potentials $d_n(x,y)$ measure the spatial activity in horizontal, vertical and diagonal directions using second order derivatives. This prior is an example of the Gauss Markov Random Field (GMRF). For this model, the energy function can be obtained by linear filtering; for example, the clique potentials for $d_1(x,y)$ can be obtained by convolving the image with the filter $[1,-2,1]$. Therefore, the energy can be written as

$$E(\mathbf{f}) = \sum_{c\in\mathscr{C}} V_c(\mathbf{f}) = \sum_{n=1}^{4} \|\Phi_n\mathbf{f}\|^2 = \|\Phi\mathbf{f}\|^2, \tag{7.21}$$

where Φ_n is the convolution matrix for $d_n(x,y)$, and $\Phi = \left[\Phi_1^T\ \Phi_2^T\ \Phi_3^T\ \Phi_4^T\right]^T$ is constructed by stacking Φ_n. Substituting Equation 7.21 into Equation 7.18 reveals that the implementation approaches discussed previously for quadratic cost functions can be applied here as well.

A criticism of the GMRF prior is the over-smoothing effect. In Reference [28], the Huber function is used to define the Huber Markov Random Field (HMRF), with clique potentials

$$V_c(\mathbf{f}) = \sum_{n=1}^{4} \rho_\phi(d_n(x,y)), \tag{7.22}$$

where the Huber function is

$$\rho_\phi(z) = \begin{cases} z^2 & \text{if } |z| \le \phi, \\ 2\phi|z| - \phi^2 & \text{otherwise.} \end{cases} \tag{7.23}$$

The resulting optimization problem is nonquadratic:

$$\mathbf{f}_{map} = \arg\min_{\mathbf{f}} \left\{ \|\mathbf{g} - \mathbf{Hf}\|^2 + \lambda \mathbf{1}^T \rho_\phi(\Phi\mathbf{f}) \right\}, \tag{7.24}$$

where $\mathbf{1}$ is a vector of ones. This optimization problem can be solved using a gradient descent algorithm, which requires the gradient of the cost function:

$$\frac{\partial}{\partial \mathbf{f}} \left\{ \|\mathbf{g} - \mathbf{Hf}\|^2 + \lambda \mathbf{1}^T \rho_\theta(\Phi\mathbf{f}) \right\} = -2\mathbf{H}^T(\mathbf{g} - \mathbf{Hf}) + \lambda \Phi^T \rho_\theta'(\Phi\mathbf{f}), \tag{7.25}$$

where the derivative of the Huber function is

$$\rho_\phi'(z) = \begin{cases} 2z & \text{if } |z| \le \phi, \\ 2\phi\,\text{sign}(z) & \text{otherwise.} \end{cases} \tag{7.26}$$

The GMRF and HMRF prior formulations above lead to efficient analytical implementations. In general, the use of Gibbs prior requires numerical methods, such as the simulated annealing and iterated conditional modes, for optimization.

Another commonly used prior model is based on total variation, where the L_1 norm (i.e., the sum of the absolute value of the elements) is used instead of the L_2 norm:

$$P(\mathbf{f}) \propto \exp(-\|\mathbf{Lf}\|_1). \tag{7.27}$$

The total variation regularization has been shown to preserve edges better than the Tikhonov regularization [25]. A number of SR algorithms, including those presented in References [31] and [32], have adopted this prior. Recently, Reference [33] proposed a bilateral total variation prior:

$$P(\mathbf{f}) \propto \exp\left(-\sum_{l=-R}^{R}\sum_{m=-R}^{R} \gamma^{|m|+|l|}\|\mathbf{f} - S_x^l S_y^m \mathbf{f}\|_1\right), \tag{7.28}$$

where γ is a regularization constant in the range $(0,1)$, S_x^l and S_y^m shift an image by integer amounts l and m in x and y directions, respectively, and R is the maximum shift amount. This prior term penalizes intensity differences at different scales, and γ gives a spatial decay effect. The gradient descent solution requires the derivative of the prior, which is

$$\frac{\partial \log P(\mathbf{f})}{\partial \mathbf{f}} \propto \sum_l \sum_m \gamma^{|l|+|m|} \left(I - S_y^{-m} S_x^{-l}\right) \text{sign}\left(\mathbf{f} - S_x^l S_y^m \mathbf{f}\right) \tag{7.29}$$

where $S_y^{-m} S_x^{-l}$ are the transposes of $S_y^m S_x^l$, respectively.

It should be noted that in case of the iid Gaussian model, the MAP estimation could be interpreted under the regularized least estimation context, and vice versa. However, there are non-Gaussian cases, where the MAP approach and the least squares approach lead to different solutions. For example, in tomographic imaging, the observation noise is better modeled with a Poisson distribution; and the MAP approach works better than the least squares approach. Another advantage of the MAP approach is that it provides an elegant and effective formulation when some uncertainties (for example, in registration or PSF estimation) are modeled as random variables.

7.3.5 Projection onto Convex Sets

Another SR approach is based on the projection onto convex sets (POCS) technique [34] which consists of iterative projection of an initial estimate onto predefined constraint sets. When the constraint sets are convex and not disjoint, the technique guarantees convergence to a solution that is consistent with all the constraint sets. The solution is not unique; depending on the initial estimate and the order of the projections, the iterations may lead to different solutions. One advantage of the POCS technique is the ease of incorporating space-variant PSF into the restoration [10].

Data fidelity constraint set is a commonly used constraint set to ensure consistency with observations. The set is defined at each observation pixel and constrains the difference between the predicted pixel value and the actual pixel value:

$$C_D[g_k(x,y)] = \{f : |r_k(x,y)| \le T_k(x,y)\}, \qquad (7.30)$$

where

$$r_k(x,y) = g_k(x,y) - \sum_{u,v \in S_{xy}} h_k(x,y;u,v)f(u,v) \qquad (7.31)$$

is the residual, S_{xy} is the set of pixels in the high-resolution image that contributes to the pixel (x,y) in the kth observation, $h_k(x,y;u,v)$ is the contribution of the pixel $f(u,v)$, and $T_k(x,y)$ is the threshold value that controls the data fidelity constraint. The threshold should be chosen considering the noise power [35]. If the noise power is small, the threshold could be chosen small; otherwise it should be large to allow space for disturbances caused by the noise. The projection operation onto the data fidelity constraint set is illustrated in Figure 7.8, and formulated as follows [10], [35]:

$$f^{(i+1)}(u,v) = \begin{cases} f^{(i)}(u,v) + \frac{(r_k(x,y)-T_k(x,y))h_k(x,y;u,v)}{\sum\limits_{u,v \in S_{xy}} h_k^2(x,y;u,v)} & \text{for } r_k(x,y) > T_k(x,y), \\ f^{(i)}(u,v) & \text{for } |r_k(x,y)| \le T_k(x,y), \\ f^{(i)}(u,v) + \frac{(r_k(x,y)+T_k(x,y))h_k(x,y;u,v)}{\sum\limits_{u,v \in S_{xy}} h_k^2(x,y;u,v)} & \text{for } r_k(x,y) < -T_k(x,y), \end{cases} \qquad (7.32)$$

where S_{xy} is the set of pixels under the support of the PSF, centered by the mapping $M_k(x,y)$. The projection operation is repeated for all low-resolution pixels.

In addition to the data fidelity constraint set, it is possible to define other constraint sets. For example, the amplitude constraint set, $C_A = \{f : 0 \le f(u,v) \le 255\}$, ensures that the resulting image has pixel values within a certain range. A smoothness constraint set could be defined as $C_S = \{f : |f(u,v) - \bar{f}(u,v)| \le \delta_S\}$, where $\bar{f}(u,v)$ is the average image and δ_S is a nonnegative threshold.

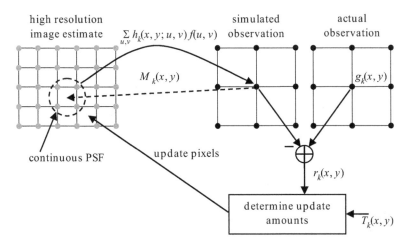

FIGURE 7.8

POCS implementation.

7.3.6 Training-Based Super Resolution

Training-based SR has recently become popular mainly because it is possible to design algorithms that do not require motion estimation, and class specific priors can be incorporated into the restoration. Motion estimation is a computationally demanding and probably the most critical part of a traditional super-resolution algorithm. Methods that do not require motion estimation are more appropriate for real-time applications. These methods are mostly single-image methods; however, extension to multiple frames is possible. Class specific priors, when usable, are more powerful than generic image priors. If the image to be restored is from a certain class, for example a face image or a text, class specific priors could be utilized. Three main training-based approaches are discussed below.

7.3.6.1 Constraining Solution to a Subspace

Suppose that the image to be restored is a face image, and there is a set of training images $\mathbf{f}_1, ..., \mathbf{f}_M$. Using the *principal component analysis* (PCA) technique, a low-dimensional representation of the face space can be obtained. Specifically, a face image $\mathbf{f}$ can be represented in terms of the average image μ and the basis vectors $\Lambda = [\mathbf{v}_1, ..., \mathbf{v}_K]$ as follows:

$$\widetilde{\mathbf{f}} = \mu + \Lambda \mathbf{e}, \tag{7.33}$$

where $\mathbf{e}$ keeps the contributions of the basis vectors and is calculated by taking the inner product with the basis vectors as follows:

$$\mathbf{e} = \Lambda^T (\mathbf{f} - \mu). \tag{7.34}$$

The size of the vector $\mathbf{e}$ is typically much smaller than the total number of pixels in $\mathbf{f}$. The difference between the image $\mathbf{f}$ and its subspace representation $\widetilde{\mathbf{f}}$ can be made as small as needed by increasing the number of basis vectors.

With the subspace representation, the SR problem can be formulated in terms of the representation vector $\mathbf{e}$. For example, the least squares approach finds the optimal representation vector [36]:

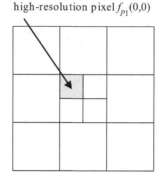

low-resolution image patch   high-resolution pixel $f_{p_1}(0,0)$

FIGURE 7.9

High-resolution pixel and the corresponding local patch in Reference [38].

$$\mathbf{e}_{ls} = \arg\min_{\mathbf{e}} \| \mathbf{g} - \mathbf{H}(\mu + \Lambda \mathbf{e}) \|^2, \tag{7.35}$$

from which the high-resolution image is obtained as follows:

$$\mathbf{f}_{ls} = \mu + \Lambda \mathbf{e}_{ls}. \tag{7.36}$$

With this approach, the solution is constrained to the subspace spanned by the average image μ and the basis vectors $[\mathbf{v}_1, ..., \mathbf{v}_K]$. Noise that is orthogonal to the subspace is eliminated automatically. This may turn out to be very helpful in some applications, for example, face recognition from low-resolution surveillance video [37].

7.3.6.2 Subspace-Based Regularization

Unlike the previous method, the subspace-based regularization approach restores the image in image space without constraining it to a subspace. However, it forces the difference between the solution and its representation in the subspace to be small. That is, the training-based prior is still used, but the solution can be outside the subspace. For the same problem defined in the previous section, the least squares solution in this case is as follows [36]:

$$\mathbf{f}_{ls} = \arg\min_{\mathbf{f}} \| \mathbf{g} - \mathbf{H}\mathbf{f} \|^2 + \lambda \| \mathbf{f} - \widetilde{\mathbf{f}} \|^2, \tag{7.37}$$

where $\widetilde{\mathbf{f}}$ is the subspace representation as in Equation 7.33. Using Equations 7.33 and 7.34, the estimator in Equation 7.37 becomes

$$\mathbf{f}_{ls} = \arg\min_{\mathbf{f}} \| \mathbf{g} - \mathbf{H}\mathbf{f} \|^2 + \lambda \| \left(\mathbf{I} - \Lambda\Lambda^{\mathrm{T}}\right)(\mathbf{f} - \mu) \|^2 . \tag{7.38}$$

7.3.6.3 SR Pixel Estimation in Local Patches

The previous two approaches work on the entire image and are limited to certain image classes. The approach presented in this section works on local image patches and can be generically applied to any image. An example of such a method is presented in Reference [38], where image patches are classified based on adaptive dynamic range coding (ADRC), and for each class, the relation between the low-resolution pixels and the corresponding high-resolution pixels is learned. Referring to Figure 7.9, suppose a low-resolution patch $[g_{p_1}(-1,-1),\ g_{p_1}(0,-1),\ \cdots,\ g_{p_1}(1,1)]$ and a corresponding high-resolution pixel $f_{p_1}(0,0)$. The low-resolution patch is binarized with ADRC as follows:

$$ADRC\left(g_{p_1}(x,y)\right) = \begin{cases} 1 & \text{if } g_{p_1}(x,y) \geq \bar{g}_{p_1}, \\ 0 & \text{otherwise}, \end{cases} \qquad (7.39)$$

where $\bar{g}_{p_1}$ is the average of the patch. Applying ADRC to the entire patch, a binary codeword is obtained. This codeword determines the class of that patch. That is, a 3×3 patch is coded with a 3×3 matrix of ones and zeroes. There are, therefore, a total of $2^9 = 512$ classes. During training, each low-resolution patch is classified; and for each class, linear regression is applied to learn the relation between the low-resolution patch and the corresponding high-resolution pixel. Assuming M low-resolution patches and corresponding high-resolution pixels are available for a particular class c, the regression parameters, $[a_{c,1}\ a_{c,2}\ \cdots\ a_{c,9}]$, are then found by solving:

$$\begin{bmatrix} f_{p_1}(0,0) \\ f_{p_2}(0,0) \\ \vdots \\ f_{p_M}(0,0) \end{bmatrix} = \begin{bmatrix} g_{p_1}(-1,-1)\ g_{p_1}(0,-1)\ \cdots\ g_{p_1}(1,1) \\ g_{p_2}(-1,-1)\ g_{p_1}(0,-1)\ \cdots\ g_{p_1}(1,1) \\ \vdots \\ g_{p_M}(-1,-1)\ g_{p_M}(0,-1)\ \cdots\ g_{p_M}(1,1) \end{bmatrix} \begin{bmatrix} a_{c,1} \\ a_{c,2} \\ \vdots \\ a_{c,9} \end{bmatrix} \qquad (7.40)$$

During testing, for each pixel, the local patch around that pixel is taken and classified. According to its class, the regression parameters are taken from a look-up table and applied to obtain the high-resolution pixels corresponding to that pixel. This single-image method can be extended to a multi-image method by including local patches from neighboring frames. This would, however, increase the dimensionality of the problem significantly and requires a much higher volume of training data. For example, if three frames were taken, the size of the ADRC would be $3 \times 3 \times 3 = 27$; and the total number of classes would be 2^{27}.

There are also other patch-based methods. In Reference [39], a feature vector from a patch is extracted through a nonlinear transformation, which is designed to form more edge classes. The feature vectors are then clustered to form classes. And, as in the case of Reference [38], a weighted sum of the pixels in the patch is taken to get a high-resolution pixel; the weights are learned during training. Reference [40] uses a Markov network to find the relation between low- and high-resolution patches. Given a test patch, the best matching low-resolution patch is determined and the corresponding high-resolution patch learned from training is obtained. In Reference [41], a multi-scale decomposition is applied to an image to obtain feature vectors, and the solution is forced to have feature vectors close to ones learned during training.

7.3.7 Other Methods and Applications

In addition to the methods already discussed in this chapter, there are other SR methods and application-specific formulations. An example application is restoration from compressed video. Most image and video compression standards are based on transforming an image from spatial domain to discrete cosine transform (DCT) domain. The DCT coefficients are then quantized and encoded. Denoting $\mathbf{T}$ as the matrix that performs block-DCT, the image acquisition process is modified as follows:

$$\tilde{\mathbf{d}}_k = Q\{\mathbf{TH}_k\mathbf{f} + \mathbf{Tn}_k\} = \mathbf{TH}_k\mathbf{f} + \mathbf{Tn}_k + \mathbf{q}_k, \tag{7.41}$$

where $Q\{\cdot\}$ is the quantization operation, $\tilde{\mathbf{d}}_k$ is the vector of quantized DCT coefficients, and $\mathbf{q}_k$ is the quantization error vector. While the exact value of the quantization error $\mathbf{q}_k$ is not known, the quantization step size is available from the data stream; therefore, the bounds on $\mathbf{q}_k$ can be obtained. In Reference [42], the additive noise term $\mathbf{Tn}_k$ is neglected, and a POCS algorithm is proposed based on the convex sets defined using the quantization bounds. In Reference [20], $\mathbf{n}_k$ and $\mathbf{q}_k$ are modeled as random variables, the probability density function (PDF) of the overall noise $\mathbf{Tn}_k + \mathbf{q}_k$ is derived, and a Bayesian algorithm is proposed. References [43] and [44] also propose a Bayesian algorithm, with regularization terms explicitly defined to penalize compression artifacts.

Other noteworthy SR methods include adaptive filtering approaches [45], [46], SR from multi-focus images [47], and combined POCS with Bayesian methods [48]. Reference [49] not only improves the resolution but also estimates albedo and height in the scene using a Bayesian framework. As an extension to SR restoration, Reference [50] investigates space-time super-resolution, where the idea is to improve both spatial and temporal resolution by combining videos captured at different frame rates or sample times. Finally, References [51] and [52] investigate the resolution enhancement limits of SR.

7.4 Other Issues

7.4.1 Image Registration

SR imaging requires subpixel accurate registration of images to improve the spatial resolution. There are several surveys on image registration, including References [53] and [54]. Image registration techniques can be categorized as *parametric* and *nonparametric*. Parametric techniques have the advantage of low computational cost; on the other hand, they cannot be directly applied to scenes where motion is not parametric. Parametric techniques can be classified as frequency domain methods and spatial domain methods.

- *Frequency domain methods.* A linear shift in spatial domain corresponds to a phase shift in frequency domain. This leads to efficient frequency domain motion estimation algorithms. References [55], [56], and [57] estimate translational motion vectors. Planar rotation and scaling can also be added to these algorithms [58], [59]. Some frequency domain methods [55], [60], [61] explicitly address the aliasing issue that may affect the results adversely.

- *Spatial domain methods.* The registration parameters may also be determined in spatial domain. Reference [62] assumes a parametric model (translation and rotation), applies Taylor series expansion to linearize the problem, and employs an iterative scheme to find the optimum parameters. Reference [63] extracts and matches features to find homography between images; the RANSAC (RANdom SAmple Consensus) technique [64] is utilized to eliminate the outliers in parameter estimation. The homography-based method can model perspective projection and be applied to a wider variety of image sequences.

Nonparametric techniques are not limited to video sequences with global geometric transformations. Their disadvantage is the high computational cost. Among the nonparametric image registration techniques are:

- *Block matching methods.* A block around the pixel in question is taken, and the best matching block in the other frame is found based a criterion, such as the mean squared error or sum of absolute difference. References [10] and [28] are among the methods that utilize the hierarchical block matching technique to estimate motion vectors. Reference [65] evaluates the performance of block matching algorithms for estimating subpixel motion vectors in noisy and aliased images. It is shown that a $(1/p)$-pixel-accurate motion estimator exhibits errors bounded within $\pm 1/(2p)$, for $p \geq 1$; however, for real data the accuracy does not increase much beyond $p > 4$.

- *Optical flow methods.* These methods assume brightness constancy along the motion path and derive the motion vector at each pixel based on a local or global smoothness model. References [31], [66], [67] and [68] are among the methods that use optical flow based motion estimation. A comparison of optical flow methods is presented in Reference [69].

It is possible that there are misregistered pixels, and these may degrade the result during restoration. Such inaccurate motion vectors can be detected and excluded from the restoration process. In References [70], two threshold values, one for regions of low local variance and the other for regions of high local variance, are applied on the motion compensated pixel residuals to determine the unreliable motion vectors, and the corresponding observed data are excluded from the POCS iterations.

Most SR algorithms, including the ones mentioned above, perform registration and restoration in two separate successive steps. There are a few SR algorithms that do joint registration and restoration. A popular approach is the Bayesian approach, where the high-resolution image $\mathbf{f}$ and the registration parameters $\mathbf{p}$ are calculated to maximize the conditional probability $P(\mathbf{f}, \mathbf{p}|\mathbf{g})$:

$$\{\mathbf{f}_{map}, \mathbf{p}_{map}\} = \arg\max_{\mathbf{f}, \mathbf{p}} P(\mathbf{f}, \mathbf{p}|\mathbf{g}) = \arg\max_{\mathbf{f}, \mathbf{p}} P(\mathbf{f}) P(\mathbf{p}|\mathbf{f}) P(\mathbf{g}|\mathbf{f}, \mathbf{p}). \qquad (7.42)$$

An example of such an algorithm is presented in Reference [71], which models both the SR image and the registration parameters as Gaussian random variables, and employs an iterative scheme to get the estimates.

7.4.2 Regularization Parameter Estimation

The regularization parameter controls the trade-off between the data fidelity and the prior information fidelity. With a bad choice of the regularization parameter, the solution might be degraded with excessive noise amplification or over-smoothing. Therefore, selection of an appropriate regularization parameter is an important part of the overall SR restoration process. The most commonly used methods for choosing the regularization parameter are reviewed below.

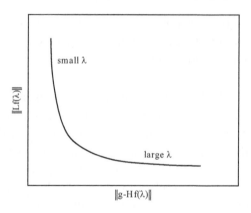

FIGURE 7.10

L-curve.

Visual inspection. When the viewer has considerable prior knowledge on the scene, it is reasonable to choose the regularization parameter through visual inspection of results with different parameter values. Obviously, this approach is not appropriate for all applications.

L-curve method [72]. Since the regularization parameter controls the trade-off between the data fidelity and prior information fidelity, it makes sense to determine the parameter by examining the behavior of these fidelity terms. Assuming that $\mathbf{f}(\lambda)$ is the solution with a particular regularization parameter λ, the data fidelity is measured as the norm of the residual $\|\mathbf{g} - \mathbf{Hf}(\lambda)\|$, and the prior information fidelity is measured as the norm of the prior term, for example, $\|\mathbf{Lf}(\lambda)\|$ in case of the Tikhonov regularization. The plot of these terms as λ is varied forms an L-shaped curve (Figure 7.10). For some values of λ, the residual changes rapidly while the prior term does not change much; this is the over-regularized region. For some other values of λ, the residual changes very little while the prior term changes significantly; this is the under-regularized region. Intuitively, the optimal λ value is the one that corresponds to the corner of the L-curve. The corner point may be defined in a number of ways, including the point of maximum curvature, and the point with slope -1. A sample application of the L-curve method in SR is presented in Reference [73].

Generalized cross-validation (GCV) method [74]. GCV is an estimator for the predictive risk $\|\mathbf{Hf} - \mathbf{Hf}(\lambda)\|^2$. The underlying idea is that the solution that is obtained using all but one observation should predict that left-out observation well if the regularization parameter is a good choice. The total error for a particular choice of the parameter is calculated by summing up the prediction errors over all observations. The optimal parameter value is the one that minimizes the total error. A search technique or an optimization method could be used to determine the optimal value.

Discrepancy principle [75]. If the variance of the noise is known, then the bound on the residual norm $\|\mathbf{g} - \mathbf{Hf}(\lambda)\|$ can be determined. Since under-regularization causes excessive noise amplification, one can choose the regularization parameter so that the residual norm is large but not larger than the bound. That is, if the bound is δ, then it is needed to find λ such that $\|\mathbf{g} - \mathbf{Hf}(\lambda)\| = \delta$.

Statistical approach. As already discussed, the statistical methods look for the image $\mathbf{f}$ by maximizing the conditional probability $p(\mathbf{g}|\mathbf{f})$ (maximum likelihood solution) or $p(\mathbf{f}|\mathbf{g})$

(maximum *a posteriori* solution). These statistical approaches can be extended by incorporating the regularization parameter into the formulation. That is, for the maximum likelihood solution, $p(\mathbf{g}|\mathbf{f}, \lambda)$ is maximized; whereas for the maximum *a posteriori* solution, $p(\mathbf{f}, \lambda|\mathbf{g})$ is maximized. The resulting optimization problems can be solved using a gradient-descent technique or the expectation-maximization (EM) technique [76]. The EM technique involves iterative application of two steps. Namely, in the expectation step, the image is restored given an estimate of the parameter. In the maximization step, a new estimate of the parameter is calculated given the restored image.

7.4.3 Blur Modeling and Identification

A number of components contribute to the overall blur of an imaging system:

- *Diffraction* depends on the shape and size of the aperture, wavelength of the incoming light, and focal length of the imaging system. For a circular aperture, the diffraction-limited PSF is the well-known Airy pattern.

- *Atmospheric turbulence blur* is especially critical in remote sensing applications. While there is a variety of factors, such as temperature, wind, and exposure time, that affect the blur, the PSF can be modeled reasonably well with a Gaussian function.

- *Out-of-focus blur* can be modeled as a disk, also known as the circle of confusion, if the aperture of the camera is circular. The diameter of the disk depends on the focal length, size of the aperture, and the distance between the object and the camera.

- *Sensor blur* is due to the spatial integration on the photosensitive region of a pixel. The shape of the PSF depends on the shape of the photosensitive region.

- *Motion blur* occurs when the exposure time is not small enough with respect to the motion in the scene. In that case, the projected imagery is smeared over the sensor according to the motion.

Overall, an imaging system may have a space- and time-variant blur. Due to the difficulty of estimating the shift-variant blur and the resulting complexity, most SR algorithms assume a linear shift-invariant PSF. Such modeling is sufficient for a majority of SR problems. Even with a shift-invariant PSF, a SR algorithm usually does not have information on the exact shape of the PSF. When the PSF estimate is not good enough, the restored image image would be either blurry or likely to suffer from ringing artifacts. Blur identification methods that have been developed for blind image deconvolution [77] over the years can be adopted for SR. Two commonly used blur identification methods are the generalized cross validation [74], [78] and maximum likelihood [79] methods. These methods assume a parametric model for the PSF, and estimate the parameters of the model. The techniques that were discussed for regularization parameter estimation can be used for blur parameter estimation as well. In References [80] and [81], the generalized cross validation technique is used for SR. The maximum likelihood approach is adopted in Reference [82]. Reference [83] takes a Bayesian approach and jointly estimates the SR image, PSF, and registration parameters.

7.5 Variations and Advances on Imaging Model

7.5.1 Motion Blur Modeling

The model in Section 7.2 assumes that the exposure time is relatively short (effectively zero) with respect to the motion in the scene. In other words, motion blur is not a part of the model. It is, of course, possible that image sequences are degraded with motion blur; and in such cases, the performance would be poor if motion blur is not modeled. Reference [10] presents an imaging model that explicitly takes exposure time into account. The model assumes a spatially and temporally continuous video signal. Given that f is the reference signal at time 0 and f_t is the signal at time t, these two signals are related to each other through a motion mapping M_t:

$$f_t(u,v) = f(M_t(u,v)). \tag{7.43}$$

During image acquisition, the signal f_t is blurred with a linear shift-invariant PSF b, which is due to optical and sensor blurs:

$$b(u,v) * f_t(u,v) = \int b((u,v) - (\xi_1, \xi_2)) f_t(\xi_1, \xi_2) d\xi_1 d\xi_2. \tag{7.44}$$

By making the change of variables $(u_r, v_r) = (M_t(\xi_1, \xi_2))$, Equation 7.44 becomes

$$b(u,v) * f_t(u,v) = \int b((u,v) - M_t^{-1}(u_r,v_r)) f(u_r,v_r) |J(M_t)|^{-1} du_r dv_r,$$

$$= \int b(u,v;u_r,v_r;t) f(u_r,v_r) du_r dv_r, \tag{7.45}$$

where $|J(M_t)|$ is the determinant of the Jacobian of M_t, M_t^{-1} is inverse motion mapping, and $b(u,v;u_r,v_r;t)$ is defined as follows:

$$b(u,v;u_r,v_r;t) = b((u,v) - M_t^{-1}(u_r,v_r)) |J(M_t)|^{-1}. \tag{7.46}$$

Note that $b(u,v;u_r,v_r;t)$ is not invariant in space or time. The video signal in Equation 7.45 is then integrated during the exposure time t_k to obtain

$$\hat{f}_k(u,v) = \frac{1}{t_k} \int_0^{t_k} \int b(u,v;u_r,v_r;t) f(u_r,v_r) du_r dv_r dt,$$

$$= \int b_k(u,v;u_r,v_r) f(u_r,v_r) du_r dv_r, \tag{7.47}$$

where $b_k(u,v;u_r,v_r)$ is the linear shift- and time-variant blur defined as follows:

$$b_k(u,v;u_r,v_r) = \frac{1}{t_k} \int_0^{t_k} b(u,v;u_r,v_r;t) dt. \tag{7.48}$$

Finally, $\hat{f}_k(u,v)$ is downsampled to obtain the kth observation $g_k(x,y)$. Discretizing this continuous model, one can write $\mathbf{g}_k = \mathbf{H}_k\mathbf{f} + \mathbf{n}_k$ and apply the techniques (that do not require linear shift-invariant PSF) as before. The only difference would be the construction of the matrix $\mathbf{H}_k$ which is no longer block circulant. In Reference [10], the POCS technique, which is very suitable for shift-variant blurs, is utilized; the algorithm is successfully demonstrated on real video sequences with pronounced motion blur.

7.5.2 Geometric and Photometric Registration

Most SR algorithms assume that input images are captured under the same photometric conditions. When there is photometric diversity among input images, these algorithms would not be able to handle them correctly. Photometric diversity in image sequences is not uncommon; exposure time, aperture size, gain, and white balance may vary during video capture since many modern cameras have automatic control units. There may also be changes in external illumination conditions. An SR algorithm, therefore, should have a photometric model and incorporate it in the restoration process. In photometric modeling, the camera response function (CRF) should also be considered. The CRF, which is the response of a sensor to incoming light, is not necessarily linear.

To address SR under photometric diversity, Reference [63] models photometric changes as global gain and offset parameters among image intensities. This affine model is successful when photometric changes are small. When photometric changes are large, nonlinearity of camera response function should be taken into consideration [3], [84].

7.5.2.1 Affine Photometric Model

Suppose that N images of a static scene are captured and these images are geometrically registered. Given that $\mathbf{q}$ denotes the irradiance of the scene and $\mathbf{g}_i$ is the ith measured image, then according to the affine model the following holds:

$$\mathbf{g}_i = a_i\mathbf{q} + b_i, \tag{7.49}$$

where the gain (a_i) and offset (b_i) parameters can model a variety of things, including global external illumination changes and camera parameters such as gain, exposure rate, aperture size, and white balancing.

The ith and the jth image are related to each other as follows:

$$\mathbf{g}_j = a_j\mathbf{q} + b_j = a_j\left(\frac{\mathbf{g}_i - b_i}{a_i}\right) + b_j = \frac{a_j}{a_i}\mathbf{g}_i + \frac{a_ib_j - a_jb_i}{a_i}. \tag{7.50}$$

Defining $\alpha_{ji} \equiv a_j/a_i$ and $\beta_{ji} \equiv (a_ib_j - a_jb_i)/a_i$, Equation 7.50 can be rewritten as follows:

$$\mathbf{g}_j = \alpha_{ji}\mathbf{g}_i + \beta_{ji}. \tag{7.51}$$

The affine relation given in Equation 7.51 is used in Reference [63] to model photometric changes among the images to be used in SR reconstruction. Namely, the images are first geometrically registered to the reference image to be enhanced. The registration method is based on feature extraction and matching; the matching criteria is normalized cross correlation which is invariant to affine changes. After geometric registration, the relative gain and offset terms with respect to the reference image are estimated in order to photometrically correct each image. This is followed by SR reconstruction.

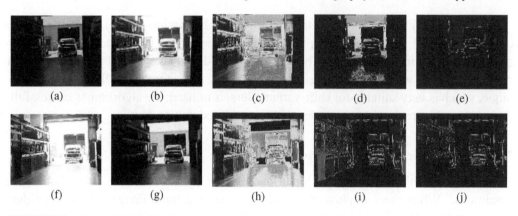

FIGURE 7.11 (See color insert.)

Comparison of the affine and nonlinear photometric conversion using differently exposed images. The images are geometrically registered: (b) is photometrically mapped on (a), and (g) is photometrically mapped on (f). Images in (c) and (h) are the corresponding residuals when the affine photometric model is used. Images in (d) and (i) are the residuals when the nonlinear photometric model is used. Images in (e) and (j) are the residuals multiplied by the weighting function.

7.5.2.2 Nonlinear Photometric Model

Although the affine transformation handles moderate photometric changes, the conversion accuracy decreases drastically in case of large changes. A typical image sensor has a nonlinear response to the amount of light it receives. Estimation of this nonlinear CRF thus becomes critical in various applications. For example, in high-dynamic range (HDR) imaging, images captured with different exposure rates are combined to produce an HDR image; this requires an accurate estimate of the CRF [85], [86], [87]. Another example is mosaicking, where border artifacts occur when CRF is not estimated accurately [88].

In nonlinear photometric modeling, the image $\mathbf{g}_i$ is related to the irradiance $\mathbf{q}$ of the scene as follows:

$$\mathbf{g}_i = \varphi\left(a_i\mathbf{q} + b_i\right), \tag{7.52}$$

where $\varphi(\cdot)$ is the camera response function (CRF), and a_i and b_i are the gain and offset parameters as in Equation 7.49. Then, two images are related to each other as follows:

$$\mathbf{g}_j = \varphi\left(\frac{a_j}{a_i}\varphi^{-1}\left(\mathbf{g}_i\right) + \frac{a_ib_j - a_jb_i}{a_i}\right) = \varphi\left(\alpha_{ji}\varphi^{-1}\left(\mathbf{g}_i\right) + \beta_{ji}\right). \tag{7.53}$$

The function $\psi_{ji} = \varphi\left(\alpha_{ji}\varphi^{-1}(\cdot) + \beta_{ji}\right)$ is known as the intensity mapping function (IMF) [89]. This nonlinear photometric modeling is adopted in Reference [84] for SR. Let $\mathbf{f}$ be the (unknown) high-resolution version of the reference image $\mathbf{g}_r$, and define $\psi_{ri}(\mathbf{g}_i)$ as the intensity mapping function that takes $\mathbf{g}_i$ and converts it to the photometric range of $\mathbf{g}_r$. Then, the cost function to be minimized is [84]:

$$C(\mathbf{f}) = \sum_i \left(\psi_{ri}(\mathbf{g}_i) - \mathbf{H}_i\mathbf{f}\right)^T \mathbf{W}_i \left(\psi_{ri}(\mathbf{g}_i) - \mathbf{H}_i\mathbf{f}\right), \tag{7.54}$$

where $\mathbf{H}_i$ includes geometric warping, blurring, and downsampling as before, and $\mathbf{W}_i$ is a diagonal matrix that is a function of $\mathbf{g}_i$. Reference [84] shows that the error in tonal

FIGURE 7.12 (See color insert.)

A subset of 22 input images showing different exposure times and camera positions.

FIGURE 7.13 (See color insert.)

Image restoration using the method in Reference [84]: (a,c) input images, and (b,d) their restored versions.

conversion increases in the lower and higher parts of the intensity range. This is mainly due to low signal-to-noise ratio in the lower parts and saturation in the higher parts. Less weight should be given to pixels (with intensities in the lower and higher parts of the range) in constructing the cost function; and the diagonal matrix $\mathbf{W}_i$ reflects that.

Figure 7.11 compares the affine and nonlinear photometric models. It is clear that the nonlinear model works better than the affine model. One may notice that the residual in Figure 7.11d is not as small as the residual in Figure 7.11i. The reason is the saturation in Figure 7.11b; some pixels in Figure 7.11a cannot be estimated in any way from Figure 7.11b. As seen in Figure 7.11e, the weighting $\mathbf{W}_i$ would suppress these pixels and prevent them from degrading the solution.

FIGURE 7.14 (See color insert.)

High-dynamic-range high-resolution image obtained using the method in Reference [3].

Figure 7.12 shows a number of input images from a dataset consisting of images captured at different exposure times and camera positions. These images are processed with the algorithm of Reference [84] to produce higher resolution versions. Two of these high-resolution images are shown in Figure 7.13. The resulting high-resolution images can then be processed with a HDR imaging algorithm to produce an image of high resolution and high dynamic range. Reference [3] combines these steps; the tone-mapped image for this dataset is given in Figure 7.14.

7.5.3   Color Filter Array Modeling

To produce color pictures, at least three spectral components are required. Some digital cameras use a beam-splitter to split the incoming light into several optical paths, and use a different spectral filter on each path to capture these color components. This approach requires three precisely aligned sensors. However, most digital cameras use a mosaic of color filters, commonly known as a *color filter array* (CFA), placed on top of the sensor. This second approach requires only one sensor and is often preferred due to its simplicity and lower cost. One issue with this approach is the need to interpolate the missing color samples. Because of the mosaic pattern of the color samples, the CFA interpolation problem is also referred to as *demosaicking*. If demosaicking is not performed well, the captured image will have visible color artifacts. A comprehensive survey of demosaicking methods is given in Reference [90].

The demosaicking research has revealed that estimation of the missing pixels can be performed much better if the correlation among different color components are exploited. This should, of course, be reflected to SR restoration since most images in real life are captured with CFA cameras. The imaging model in Section 7.2 can be updated as follows. Let $f^{(S)}$ be a color channel of a high-resolution image; there are typically three color channels, red ($f^{(R)}$), green ($f^{(G)}$), and blue ($f^{(B)}$). The kth low-resolution color channel, $g_k^{(S)}$, is obtained from this high-resolution image through spatial warping, blurring, and downsampling operations as follows:

$$g_k^{(S)} = DBM_k f^{(S)}, \text{ for } S = R, G, B, \tag{7.55}$$

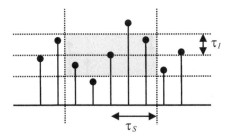

FIGURE 7.15

The spatio-intensity neighborhood of a pixel illustrated for a one-dimensional case. The gray region is the neighborhood of the pixel in the middle.

where M_k is the warping operation to account for the relative motion between observations, B is the convolution operation to account for the point spread function of the camera, and D is the downsampling operation. Note that in Section 7.2, these operations were represented as matrices $\mathbf{DBM}_k$.

The full-color image $(g_k^{(R)}, g_k^{(G)}, g_k^{(B)})$ is then converted to a mosaicked observation z_k according to a CFA sampling pattern as follows:

$$z_k(x,y) = \sum_{S=R,G,B} P_S(x,y) g_k^{(S)}(x,y), \qquad (7.56)$$

where $P_S(x,y)$ takes only one of the color samples at a pixel according to the CFA pattern. For example, at red pixel location, $[P_R(x,y), P_G(x,y), P_B(x,y)]$ is $[1,0,0]$.

There are a number of SR papers utilizing the ideas developed in demosaicking research. In Reference [91], the alternating projection method of Reference [92] is extended to multiple frames. In addition to the data fidelity constraint set, Reference [92] defines two more constraint sets; namely, the detail constraint set and the color consistency constraint set.

The *detail constraint set* is based on the observation that the high-frequency contents of color channels are similar to each other for natural images. Since the green channel is more densely sampled and therefore less likely to be aliased, the high-frequency contents of the red and blue channels are constrained to be close to the high-frequency content of the green channel. Let $\mathscr{W}_i$ be an operator that produces the ith frequency subband of an image. There are four frequency subbands ($i = LL, LH, HL, HH$) corresponding to low-pass filtering and high-pass filtering permutations along horizontal and vertical dimensions [93]. The detail constraint set, C_d, that forces the details (high-frequency components) of the red and blue channels to be similar to the details of the green channel at every pixel location (x,y), is defined as follows:

$$C_d = \left\{ g_k^{(S)}(x,y) : \left| \left(\mathscr{W}_i g_k^{(S)} \right)(x,y) - \left(\mathscr{W}_i g_k^{(G)} \right)(x,y) \right| \leq T_d(x,y) \right\}, \qquad (7.57)$$

where $i = LH, HL, HH$, and $S = R, B$. The term $T_d(x,y)$ is a nonnegative threshold that quantifies the closeness of the detail subbands to each other.

Color consistency constraint set: It is reasonable to expect pixels with similar green intensities to have similar red and blue intensities within a small spatial neighborhood.

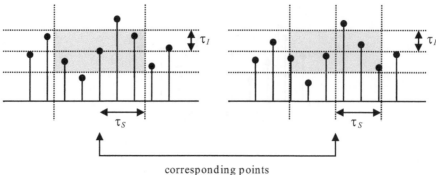

corresponding points

FIGURE 7.16

Extension of the spatio-intensity neighborhood of a pixel for multiple images. The corresponding point of a pixel is found using motion vectors; using the parameters τ_S and τ_I, the neighborhood (gray regions) is determined.

This leads to the concept of *spatio-intensity neighborhood* of a pixel. Suppose that the green channel $g_k^{(G)}$ of an image is already interpolated and the goal here is to estimate the red value at a particular pixel (x,y). Then, the spatio-intensity neighborhood of the pixel (x,y) is defined as follows:

$$\mathcal{N}(x,y) = \left\{ (u,v) : \|(u,v) - (x,y)\| \le \tau_S \text{ and } \left| g_k^{(G)}(u,v) - g_k^{(G)}(x,y) \right| \le \tau_I \right\}, \quad (7.58)$$

where τ_S and τ_I determine the extents of the spatial and intensity neighborhoods. Figure 7.15 illustrates the spatio-intensity neighborhood for a one-dimensional signal. Note that this single-frame spatio-intensity neighborhood can be extended to multiple images using motion vectors. The idea is illustrated in Figure 7.16.

The spatio-intensity neighbors of a pixel should have similar color values. One way to measure color similarity is to inspect color differences between the red and green channels and between the blue and green channels. These differences are expected to be similar within the spatio-intensity neighborhood $\mathcal{N}(x,y)$. Therefore, the color consistency constraint set can be defined as follows:

$$C_c = \left\{ g_k^{(S)}(x,y) : \left| \left(g_k^{(S)}(x,y) - g_k^{(G)}(x,y) \right) - \overline{\left(g_k^{(S)}(x,y) - g_k^{(G)}(x,y) \right)} \right| \le T_c(x,y) \right\}, \quad (7.59)$$

where $S = R, B$. The term $\overline{(\cdot)}$ denotes averaging within the neighborhood $\mathcal{N}(x,y)$, and $T_c(x,y)$ is a nonnegative threshold. It should be noted here that the spatio-intensity neighborhood concept is indeed a variant of the bilateral filter [95] with uniform box kernels instead of Gaussian kernels.

The method starts with an initial estimate and projects it onto the constraint sets defined over multiple images iteratively to obtain the missing pixels. If the blur function is set to a delta function and the downsampling operation is not included, only the missing color samples are obtained; this is called multi-frame demosaicking. Figure 7.17 provides sample results. As seen, single-frame interpolation methods do not produce satisfactory results.

(a)   (b)

(c)   (d)

FIGURE 7.17 (See color insert.)

Image demosaicking: (a) bilinear interpolation, (b) edge-directed interpolation [94], (c) multi-frame demosaicking [91], and (d) super-resolution restoration [91].

The multi-frame demosaicking can get rid of most of the color artifacts. And finally, SR restoration deblurs the image and increases the resolution further [92].

Another method combining demosaicking and super-resolution is presented in References [6] and [33]. This method is based on least squares estimation with demosaicking related regularization terms. Specifically, there are three regularization terms. The first regularization term is the bilateral total variation regularization as in Equation 7.28 applied on the luminance channel. The second regularization term is the Tikhonov regularization applied on the chrominance channels. And the third regularization term is orientation regularization, which basically forces different color channels to have similar edge orientations.

7.6 Conclusions

This chapter presented an overview of SR imaging. Basic SR methods were described, sample results were provided, and critical issues such as motion estimation and parameter estimation were discussed. Key references were provided for further reading. Two recent advances in modeling, namely, photometric modeling and color filter array modeling were discussed. While there are well-studied issues in SR imaging, there are also open problems that need further investigation, including real-time implementation and algorithm parallelization, space-variant blur identification, fast and accurate motion estimation for noisy and aliased image sequences, identifying and handling occlusion and misregistration, image prior modeling, and classification and regression in training-based methods.

As cameras are equipped with more computational power, it is becoming possible to incorporate specialized hardware with associated software and exceed the performance of traditional cameras. The jitter camera [1] is a good example; the sensor is shifted in horizontal and vertical directions during video capture, and the resulting sequence is processed with a SR algorithm to produce a higher-resolution image sequence. Imagine that the pixels in the jitter camera have a mosaic pattern of ISO gains; in that case, not only a spatial diversity but also a photometric diversity could be created. Such a camera would have the capability of producing a high-dynamic range and high-resolution image sequence. This joint hardware and software approach will, of course, bring new challenges in algorithm and hardware design.

Acknowledgment

This work was supported in part by the National Science Foundation under Grant No 0528785 and National Institutes of Health under Grant No 1R21AG032231-01.

References

[1] M. Ben-Ezra, A. Zomet, and S. Nayar, "Video super-resolution using controlled subpixel detector shifts," *IEEE Transactions on Pattern Analysis and Machine Intelligence*, vol. 27, no. 6, pp. 977–987, June 2005.

[2] N. Damera-Venkata and N.L. Chang, "Realizing super-resolution with superimposed projection," in *Proceedings of IEEE International Conference on Computer Vision and Pattern Recognition*, Minneapolis, MN, USA, June 2007, pp. 1–8.

[3] B.K. Gunturk and M. Gevrekci, "High-resolution image reconstruction from multiple differently exposed images," *IEEE Signal Processing Letters*, vol. 13, no.4, pp. 197–200, April 2006.

[4] S. Borman and R.L. Stevenson, "Super-resolution from image sequences — A review," in

Proceedings of Midwest Symposium on Circuits and Systems, Notre Dame, IN, USA, August 1998, pp. 374–378.

[5] M.G. Kang and S. Chaudhuri, "Super-resolution image reconstruction," *IEEE Signal Processing Magazine*, vol. 20, no. 3, pp. 19–20, May 2003.

[6] S. Farsiu, D. Robinson, M. Elad, and P. Milanfar, "Advances and challenges in super-resolution," in *International Journal of Imaging Systems and Technology*, vol. 14, no. 2, pp. 47–57, August 2004.

[7] S. Chaudhuri, *Super-Resolution Imaging*. Boston, MA: Kluwer Academic Publishers, January 2001.

[8] D. Capel, *Image Mosaicing and Super Resolution*. London, UK: Springer, January 2004.

[9] A. Katsaggelos, R. Molina, and J. Mateos, *Super Resolution of Images and Video*. San Rafael, CA: Morgan and Claypool Publishers, November 2006.

[10] A.J. Patti, M.I. Sezan, and A.M. Tekalp, "Superresolution video reconstruction with arbitrary sampling lattices and nonzero aperture time," *IEEE Transactions on Image Processing*, vol. 6, no. 8, pp. 1064–1076, August 1997.

[11] R.Y. Tsai and T.S. Huang, "Multiframe Image Restoration and Registration." In *Advances in Computer Vision and Image Processing*. Greenwich: JAI Press, CT, 1984.

[12] S.P. Kim and W.Y. Su, "Recursive high-resolution reconstruction of blurred multiframe images," *IEEE Transactions on Image Processing*, vol. 2, no. 4, pp. 534–539, October 1993.

[13] N.K. Bose, H.C. Kim, and H.M. Valenzuela, "Recursive total least squares algorithm for image reconstruction from noisy, undersampled frames," *Multidimensional Systems and Signal Processing*, vol. 4, no. 3, pp. 253–268, July 1993.

[14] S. Rhee and M. Kang, "Discrete cosine transform based regularized high-resolution image reconstruction algorithm," *Optical Engineering*, vol. 38, no. 8, pp. 1348–1356, April 1999.

[15] S.C. Park, M.K. Park, and M.G. Kang, "Super-resolution image reconstruction: A technical overview," *IEEE Signal Processing Magazine*, vol. 20, no. 3, pp. 21–36, May 2003.

[16] S.P. Kim and N.K. Bose, "Reconstruction of 2D bandlimited discrete signals from nonuniform samples," in *IEE Proceedings on Radar and Signal Processing*, vol. 137, no. 3, pp. 197–204, June 1990.

[17] S. Lertrattanapanich and N.K. Bose, "High resolution image formation from low resolution frames using delaunay triangulation," *IEEE Transactions on Image Processing*, vol. 11, no. 2, pp. 1427–1441, December 2002.

[18] T. Strohmer, "Computationally attractive reconstruction of bandlimited images from irregular samples," *IEEE Transactions on Image Processing*, vol. 6, no. 4, pp. 540–548, April 1997.

[19] T.Q. Pham, L.J. van Vliet, and K. Schutte, "Robust fusion of irregularly sampled data using adaptive normalized convolution," *EURASIP Journal on Applied Signal Processing*, vol. 2006, Article ID 83268, pp. 236–236, 2006.

[20] B.K. Gunturk, Y. Altunbasak, and R.M. Mersereau, "Super-resolution reconstruction of compressed video using transform-domain statistics," *IEEE Transactions on Image Processing*, vol. 13, no. 1, pp. 33–43, January 2004.

[21] M. Irani and S. Peleg, "Improving resolution by image registration," *CVGIP: Graphical Models and Image Processing*, vol. 53, no 3, pp. 231–239, May 1991.

[22] A. Zomet, A. Rav-Acha, and S. Peleg, "Robust super-resolution," in *Proceedings of IEEE International Conference on Computer Vision and Pattern Recognition*, Kauai, HI, USA, December 2001, pp. 645–650.

[23] H.W. Engl, M. Hanke, and A. Neubauer, *Regularization of Inverse Problems*. Dordrecht, Netherlands: Kluwer Academic Publishers, 1996.

[24] C. Groetsch, *Theory of Tikhonov Regularization for Fredholm Equations of the First Kind*. Boston, MA: Pittman, April 1984.

[25] L.I. Rudin, S. Osher, and E. Fatemi, "Nonlinear total variation based noise removal algorithms," *Physica D*, vol. 60, no. 1–4, pp. 259–268, November 1992.

[26] D. Geman and C. Yang, "Nonlinear image recovery with half-quadratic regularization," *IEEE Transactions on Image Processing*, vol. 4, no. 7, pp. 932–945, July 1995.

[27] P. Cheeseman, B. Kanefsky, R. Hanson, and J. Stutz, "Super-resolved surface reconstruction from multiple images," Tech. Rep. FIA-94-12, NASA Ames Research Center, December 1994.

[28] R.R. Schultz and R.L. Stevenson, "Extraction of high-resolution frames from video sequences," *IEEE Transactions on Image Processing*, vol. 5, no. 6, pp. 996–1011, June 1996.

[29] R.C. Hardie, K.J. Barnard, and E.E. Armstrong, "Joint map registration and high-resolution image estimation using a sequence of undersampled images," *IEEE Transactions on Image Processing*, vol. 6, no. 12, pp. 1621–1633, December 1997.

[30] S. Geman and D. Geman, "Stochastic relaxation, gibbs distributions, and the Bayesian restoration of images," *IEEE Transactions on Pattern Analysis and Machine Intelligence*, vol. 6, no. 6, pp. 721–741, November 1984.

[31] M.K. Ng, H. Shen, E.Y. Lam, and L. Zhang, "A total variation regularization based super-resolution reconstruction algorithm for digital video," *EURASIP Journal on Advances in Signal Processing*, vol. 2007, Article ID 74585, pp. 1–16, 2007.

[32] S.D. Babacan, R. Molina, and A.K. Katsaggelos, "Total variation super resolution using a variational approach," in *Proceedings of the IEEE International Conference on Image Processing*, San Diego, CA, USA, October 2008, pp. 641–644.

[33] S. Farsiu, D. Robinson, M. Elad, and P. Milanfar, "Fast and robust multiframe super resolution," *IEEE Transactions on Image Processing*, vol. 13, no. 10, pp. 1327–1344, October 2004.

[34] H. Stark and P. Oskoui, "High-resolution image recovery from image-plane arrays, using convex projections," *Journal of the Optical Society of America*, vol. 6, no. 11, pp. 1715–1726, November 1989.

[35] A.M. Tekalp, M.K. Ozkan, and M.I. Sezan, "High-resolution image reconstruction from lower-resolution image sequences and space-varying image restoration," in *Proceedings of the IEEE International Conference on Acoustics, Speech, and Signal Processing*, San Francisco, CA, USA, March 1992, vol. 3, pp. 169–172.

[36] D. Capel and A. Zisserman, "Super-resolution from multiple views using learnt image models," in *Proceedings of the IEEE International Conference on Computer Vision and Pattern Recognition*, Kauai, HI, USA, December 2001, vol. 2, pp. 627–634.

[37] B.K. Gunturk, A.U. Batur, Y. Altunbasak, M.H. Hayes, and R.M. Mersereau, "Eigenface-domain super-resolution for face recognition," *IEEE Transactions on Image Processing*, vol. 12, no. 5, pp. 597–606, May 2003.

[38] T. Kondo, Y. Node, T. Fujiwara, and Y.Okumura, "Picture conversion apparatus, picture conversion method, learning apparatus and learning method," U.S. Patent 6 323 905, November 2001.

[39] C.B. Atkins, C.A. Bouman, and J.P. Allebach, "Optimal image scaling using pixel classification," in *Proceedings of the IEEE International Conference on Image Processing*, Thessaloniki, Greece, October 2001, vol. 3, pp. 864–867.

[40] W.T. Freeman, T.R. Jones, and E.C. Pasztor, "Example-based super-resolution," *IEEE Computer Graphics and Applications*, vol. 22, no. 2, pp. 56–65, March/April 2002.

[41] S. Baker and T. Kanade, "Hallucinating faces," in *Proceedings of the IEEE Fourth International Conference on Automatic Face and Gesture Recognition*, Grenoble, France, March 2000, pp.83–88.

[42] Y. Altunbasak, A.J. Patti, and R.M. Mersereau, "Super-resolution still and video reconstruction from mpeg-coded video," *IEEE Transactions on Circuits and Systems for Video Technology*, vol. 12, no. 4, pp. 217–226, April 2002.

[43] C.A. Segall, R. Molina, and A.K. Katsaggelos, "High-resolution images from low-resolution compressed video," *IEEE Signal Processing Magazine*, vol. 20, no. 3, pp. 37–48, May 2003.

[44] C.A. Segall, R. Molina, and A.K. Katsaggelos, "Bayesian resolution enhancement of compressed video," *IEEE Transactions on Image Processing*, vol. 13, no. 7, pp. 898–911, July 2004.

[45] A.J. Patti, A.M. Tekalp, and M.I. Sezan, "A new motion-compensated reduced-order model kalman filter for space varying restoration of progressive and interlaced video," *IEEE Transactions on Image Processing*, vol. 7, no. 4, pp. 543–554, April 1998.

[46] M. Elad and A. Feuer, "Superresolution restoration of an image sequence: Adaptive filter approach," *IEEE Transactions on Image Processing*, vol. 8, no. 3, pp. 387–395, March 1999.

[47] D. Rajan, S. Chaudhuri, and M.V. Joshi, "Multi-objective super resolution: Concepts and examples," *IEEE Signal Processing Magazine*, vol. 20, no. 3, pp. 49–61, May 2003.

[48] M. Elad and A. Feuer, "Restoration of a single superresolution image from several blurred, noisy and undersampled measured images," *IEEE Transactions on Image Processing*, vol. 6, no. 12, pp. 1646–1658, December 1997.

[49] H. Shekarforoush, M. Berthod, J. Zerubia, and M. Werman, "Sub-pixel Bayesian estimation of albedo and height," *International Journal of Computer Vision*, vol. 19, no. 3, pp. 289–300, August 1996.

[50] E. Shechtman, Y. Caspi, and M. Irani, "Space-time super-resolution," *IEEE Transactions on Pattern Analysis and Machine Intelligence*, vol. 27, no. 4, pp. 531–545, April 2005.

[51] Z. Lin and H.Y. Shum, "Fundamental limits of reconstruction-based superresolution algorithms under local translation," *IEEE Transactions on Pattern Analysis and Machine Intelligence*, vol. 26, no. 1, pp. 83–97, January 2004.

[52] D. Robinson and P. Milanfar, "Statistical performance analysis of super-resolution," *IEEE Transactions on Image Processing*, vol. 15, no. 6, pp. 1413–1428, June 2006.

[53] L.G. Brown, "A survey of image registration techniques," *ACM Computing Survey*, vol. 24, no. 4, pp. 325–376, December 1992.

[54] B. Zitova and J. Flusser, "Image registration methods: A survey," *Image and Vision Computing*, vol. 21, no. 11, pp. 977–1000, October 2003.

[55] H.S. Stone, M.T. Orchard, E.C. Chang, and S.A. Martucci, "A fast direct Fourier-based algorithm for subpixel registration of images," *IEEE Transactions on Geoscience and Remote Sensing*, vol. 39, no. 10, pp. 2235–2243, October 2001.

[56] P. Vandewalle, S. Susstrunk, and M. Vetterli, "Double resolution from a set of aliased images," in *Proceedings of SPIE Electronic Imaging*, San Jose, CA, USA, January 2004, pp. 374–382.

[57] H. Foroosh, J.B. Zerubia, and M. Berthod, "Extension of phase correlation to subpixel registration," *IEEE Transactions on Image Processing*, vol. 11, no. 3, pp. 188–200, March 2002.

[58] B.S. Reddy and B.N. Chatterji, "An FFT-based technique for translation, rotation and scale-invariant image registration," *IEEE Transactions on Image Processing*, vol. 5, no. 8, pp. 1266–1271, August 1996.

[59] L. Lucchese and G.M. Cortelazzo, "A noise-robust frequency domain technique for estimating planar roto-translations," *IEEE Transactions on Image Processing*, vol. 48, no. 6, pp. 1769–1786, June 2000.

[60] S.P. Kim and W.Y. Su, "Subpixel accuracy image registration by spectrum cancellation," in *Proceedings of the IEEE International Conference on Acoustics, Speech, and Signal Processing*, Minneapolis, MN, USA, April 1993, vol. 5, pp. 153–156.

[61] P. Vandewalle, S. Susstrunk, and M. Vetterli, "A frequency domain approach to registration of aliased images with application to super-resolution," *EURASIP Journal on Applied Signal Processing*, vol. 2006, Article ID 71459, pp. 1–14, 2006.

[62] D. Keren, S. Peleg, and R. Brada, "Image sequence enhancement using sub-pixel displacements," in *Proceedings of the IEEE International Conference on Computer Vision and Pattern Recognition*, Ann Arbor, MI, USA, June 1988, pp. 742–746.

[63] D. Capel and A. Zisserman, "Computer vision applied to super resolution," *IEEE Signal Processing Magazine*, vol. 20, no. 3, pp. 75–86, May 2003.

[64] M.A. Fischler and R.C. Bolles, "Random sample consensus: A paradigm for model fitting with applications to image analysis and automated cartography," *Communications of the ACM*, vol. 24, no. 6, pp. 381–395, June 1981.

[65] S. Borman, M. Robertson, and R.L. Stevenson, "Block-matching sub-pixel motion estimation from noisy, under-sampled frames - An empirical performance evaluation," *SPIE Visual Communications and Image Processing*, vol. 3653, no. 2, pp. 1442–1451, January 1999.

[66] S. Baker and T. Kanade, "Super-resolution optical flow," Tech. Rep. CMU-RI-TR-99-36, The Robotics Institute, Carnegie Mellon University, October 1999.

[67] R. Fransens, C. Strecha, and L.V. Gool, "A probabilistic approach to optical flow based super-resolution," in *Proceedings of the IEEE International Conference on Computer Vision and Pattern Recognition*, Washington, DC, USA, June 2004, vol. 12, pp. 191–191.

[68] W. Zhao and H.S. Sawhney, "Is super-resolution with optical flow feasible?," in *Proceedings of the 7th European Conference on Computer Vision*, London, UK, May 2002, pp. 599–613.

[69] B. Galvin, B. McCane, K. Novins, D. Mason, and S. Mills, "Recovering motion fields: An evaluation of eight optical flow algorithms," in *Proceedings of the British Machine Vision Conference*, Southampton, UK, September 1998, pp. 195–204.

[70] P.E. Eren, M.I. Sezan, and A.M. Tekalp, "Robust, object-based high-resolution image reconstruction from low-resolution," *IEEE Transactions on Image Processing*, vol. 6, no. 6, pp. 1446–1451, October 1997.

[71] P. Cheeseman, B. Kanefsky, R. Kraft, J. Stutz, and R. Hanson, *Maximum Entropy and Bayesian Methods*, ch. Super-resolved surface reconstruction from multiple images, G.R. Heidbreder (ed.), Dordrecht, Netherlands: Kluwer Academic Publishers, 1996, pp. 293–308.

[72] P.C. Hansen, "Analysis of discrete ill-posed problems by means of the L-curve," *SIAM Review*, vol. 34, no. 4, pp. 561–580, December 1992.

[73] N.K. Bose, S. Lertrattanapanich, and J. Koo, "Advances in superresolution using L-curve," in *Proceedings of the IEEE International Symposium on Circuit and Systems*, Sydney, Australia, May 2001, vol. 2, pp. 433–436.

[74] G. Golub, M. Heath, and G. Wahba, "Generalized cross-validation as a method for choosing a good ridge parameter," *Technometrics*, vol. 21, no. 2, pp. 215–223, May 1979.

[75] V.A. Morozov, "On the solution of functional equations by the method of regularization," *Soviet Math. Dokl.*, vol. 7, pp. 414–417, 1966.

[76] A.P. Dempster, N.M. Laird, and D.B. Rubin, "Maximum likelihood from incomplete data via the EM algorithm," *Journal of the Royal Statistical Society - Series B (Methodological)*, vol. 39, no. 1, pp. 1–38, 1977.

[77] D. Kundur and D. Hatzinakos, "Blind image deconvolution," *IEEE Signal Processing Magazine*, vol. 13, no. 3, pp. 43–64, May 1996.

[78] S.J. Reeves and R.M. Mersereau, "Blur identification by the method of generalized cross-validation," *IEEE Transactions on Image Processing*, vol. 1, no. 3, pp. 301–311, July 1992.

[79] R.L. Lagendijk, A.M. Tekalp, and J. Biemond, "Maximum likelihood image and blur identification: A unifying approach," *Journal of Optical Engineering*, vol. 29, no. 5, pp. 422–435, May 1990.

[80] N. Nguyen, G. Golub, and P. Milanfar, "Blind restoration / superresolution with generalized cross-validation using Gauss-type quadrature rules," in *Conference Record of the Third-Third Asilomar Conference on Signals, Systems, and Computers*, Pacific Grove, CA, USA, October 1999, vol. 2, pp. 1257–1261.

[81] N. Nguyen, P. Milanfar, and G. Golub, "Efficient generalized cross-validation with applications to parametric image restoration and resolution enhancement," *IEEE Transactions on Image Processing*, vol. 10, no. 9, pp. 1299–1308, September 2001.

[82] J. Abad, M. Vega, R. Molina, and A.K. Katsaggelos, "Parameter estimation in super-resolution image reconstruction problems," in *Proceedings of the IEEE International Conference on Acoustics, Speech, and Signal Processing*, Hong Kong, April 2003, vol. 3, pp. 709–712.

[83] M.E. Tipping and C.M. Bishop, "Bayesian image super-resolution," in *Proceedings of the International Conference on Advances in Neural Information Processing Systems*, Vancouver, BC, Canada, December 2002, vol. 15, pp. 1279–1286.

[84] M. Gevrekci and B.K. Gunturk, "Superresolution under photometric diversity of images," *EURASIP Journal on Advances in Signal Processing*, vol. 2007, Article ID 36076, pp. 1–12, 2007.

[85] P. Debevec and J. Malik, "Recovering high dynamic range radiance maps from photographs," in *Proceedings of International Conference on Computer Graphics and Interactive Techniques*, Los Angeles, CA, USA, pp. 369–378, August 1997.

[86] S. Mann, "Comparametric equations with practical applications in quantigraphic image processing," *IEEE Transactions on Image Processing*, vol. 9, no. 8, pp. 1389–1406, August 2000.

[87] T. Mitsunaga and S. Nayar, "Radiometric self calibration," in *Proceedings of the IEEE International Conference Computer Vision and Pattern Recognition*, Fort Collins, CO, USA, June 1999, vol. 1, pp. 374–380.

[88] A. Litvinov and Y. Schechner, "Radiometric framework for image mosaicking," *Journal of the Optical Society of America A*, vol. 22, no. 5, pp. 839–848, May 2005.

[89] M.D Grossberg and S.K. Nayar, "Determining the camera response from images: What is knowable?," *IEEE Transactions on Pattern Analysis and Machine Intelligence*, vol. 25, no. 11, pp. 1455–1467, November 2003.

[90] B.K. Gunturk, J. Glotzbach, Y. Altunbasak, R.W. Schafer, and R.M. Mersereau, "Demosaicking: Color filter array interpolation in single-chip digital cameras," *IEEE Signal Processing Magazine*, vol. 22, no. 1, pp. 44–55, January 2005.

[91] M. Gevrekci, B.K. Gunturk, and Y. Altunbasak, "Restoration of Bayer-sampled image sequences," *Oxford University Press, Computer Journal*, vol. 52, no. 1, pp. 1–14, January 2009.

[92] B.K. Gunturk, Y. Altunbasak, and R.M. Mersereau, "Color plane interpolation using alternating projections," *IEEE Transactions on Image Processing*, vol. 11, no. 9, pp. 997–1013, September 2002.

[93] B.K. Gunturk, Y. Altunbasak, and R.M. Mersereau, "Multiframe resolution-enhancement methods for compressed video," *Signal Processing Letters*, vol. 9, no. 6, pp. 170–174, June 2002.

[94] C.A. Laroche and M.A. Prescott, "Apparatus and method for adaptively interpolating a full color image utilizing chrominance gradients," U.S. Patent 5 373 322, December 1994.

[95] C. Tomasi and R. Manduchi, "Bilateral filtering for gray and color images," in *Proceedings of the IEEE International Conference on Computer Vision*, Bombay, India, January 1998, pp. 839–846.

8

Image Deblurring Using Multi-Exposed Images

Seung-Won Jung and Sung-Jea Ko

8.1 Introduction

Image blur arises when a single object point spreads over several image pixels. This phenomenon is mainly caused by camera motion during exposure or a lens that is out-of-focus. The conventional approach to image deblurring is to construct an image degradation model and then solve the ill-posed inverse problem of this model. A new approach takes advantage of recent advances in image sensing technology which enable splitting or controlling the exposure time [1], [2], [3]. This approach exploits the mutually different pieces of information from multi-exposed images of the same scene to produce the deblurred image.

This chapter presents a general framework of image deblurring using multi-exposed images. The objective of this approach is to reconstruct an image that faithfully represents a real scene by using multi-exposed images of the same scene. The multi-exposed images are assumed to be captured by a single camera or multiple cameras placed at the different locations. It is often further assumed that the multiple captured images do not have any grid mismatch, as they can be aligned at the same grid by applying image registration algorithms. With these assumptions, the various applications and recent endeavors in this field are provided in detail.

This chapter is organized as follows. In Section 8.2, the characteristics of the multi-exposed images are analyzed. The typical camera architecture for capturing multi-exposed images is described in Section 8.2.1. Section 8.2.2 focuses on the characteristics of multi-exposed images captured by various methods. Section 8.3 presents various techniques for

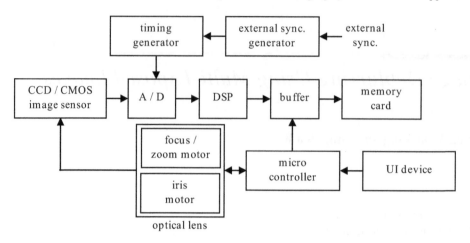

FIGURE 8.1

Block diagram of the camera system.

image deblurring using multi-exposed images. Section 8.3.1 describes the basic concept of multi-exposed image deblurring, Section 8.3.2 presents an approach that relies on the degradation model while using multi-exposed images, and Section 8.3.3 introduces a new approach which does not require a deconvolution operation. Finally, conclusions are drawn in Section 8.4.

8.2   Multi-Exposed Images

8.2.1   Multi-Exposure Camera Architectures

A charge-coupled device (CCD) or complementary metal oxide semiconductor (CMOS) sensor is used to capture real-life scenes electronically. Figure 8.1 shows a typical example of the camera system. In such a system, the digitized frames are consecutively stored in the buffer which is accessed by the digital signal processor (DSP). The exposure time, which is the main concern here, is controlled by the timing generator. Recent advances in image sensor technology have enabled high-speed capture of up to thousands of frames per second. When capturing multi-exposed images, the timing generator creates regular impulses to either produce multi-exposed images with similar characteristics, or irregular impulses to generate multi-exposed images with different characteristics.

8.2.2   Characteristics of Multi-Exposed Images

In general, the exposure time is manually controlled by the user or automatically determined by a certain algorithm. In this section, it is assumed that a suitable exposure time for capturing an image is known in advance. By taking an image with a known exposure time, it is possible to represent the brightness and the color of a real scene. However, a problem arises when the camera or the objects in the scene move during the exposure time, resulting

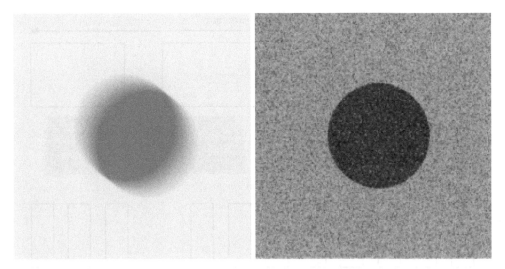

FIGURE 8.2

Example of images captured with (left) a long exposure time, and (right) a short exposure time. © 2008 IEEE

in a captured image with incomplete information about the scene. In order to extract more useful visual information in such situations, images are captured multiple times. Then, mutually different pieces of the information are merged into the final image.

To design a deblurring algorithm which can take advantage of multi-exposed images, the characteristics of these images need to be known. Multi-exposed images can be categorized as i) images taken with the same exposure time and ii) images taken with different exposure times. Before comparing these two, the basic relationship between the exposure time and image quality factors such as blur, noise, brightness, and color distortion should be discussed. Figure 8.2 shows a simple example of images taken with different exposure times. Namely, the image on the left is captured with a long exposure time and exhibits strong motion blur characteristics due to camera shake that often occurs in such scenarios. This is not the case of the image on the right which is taken with a short exposure time for which camera shake is negligible, resulting in sharp edges as well as undesired high-frequency content such as noise and brightness distortion. In short, a long exposure time is a source of image blur. Setting the exposure time short enough reduces blur while introducing noise, color artifacts, and brightness loss. Due to this relationship, the exposure time setting plays an important role in multi-exposed deblurring. Figure 8.3 illustrates three possible exposure time settings.

As shown in Figure 8.3a, using uniform long exposure intervals results in multiple blurred images. Since the speed and direction of the camera or object motion usually change during the exposure time, multiple blurred images tend to contain different information about the original scene. In other words, the point spread function (PSF) applied to the ideal image varies amongst the multi-exposed images.

Figure 8.3b shows the effect of short and uniformly split exposure intervals. As can be seen, image blur is inherently reduced and multiple noisy images are obtained instead. In this case, the deblurring problem becomes a denoising problem. Since noise is generally uncorrelated among these images, the denoising problem can be solved by accumulating

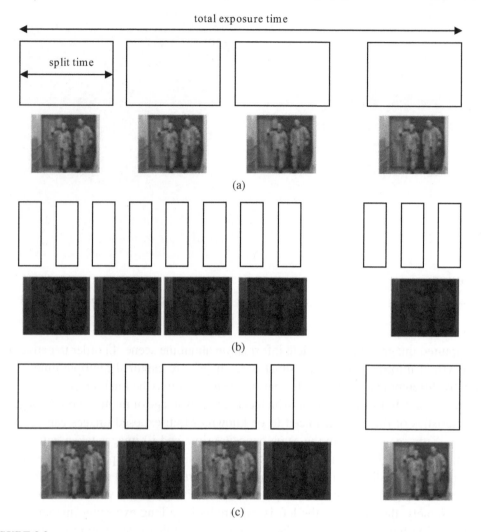

FIGURE 8.3

Three possible methods of exposure time splitting: (a) uniform long intervals, (b) uniform short intervals, and (c) nonuniform intervals.

multiple noisy images. Unfortunately, this can result in various color and brightness degradations, thus necessitating image restoration methods to produce an image with the desired visual quality.

Finally, Figure 8.3c shows the effect of nonuniform exposure intervals. Unlike the above two methods, multi-exposed images of different characteristics are taken by partitioning the exposure time into short and long intervals, thus producing both noisy and blurred images. Color and brightness information can be acquired from blurred images, whereas edge and detail information can be derived from noisy images. This image capturing method has recently gained much interest since difficult denoising or deblurring problems can be reduced into a simpler image merging problem. Moreover, the method allows focusing on the deblurring problem by using noisy images as additional information or the denoising problem by extracting brightness and color information from blurred images.

8.3　Image Deblurring Using Multi-Exposed Images

8.3.1　Basic Concept

Image deblurring is a challenging problem in the field of image and video processing. Though there has been a great deal of progress in the area in the last two decades, this problem remains unsolved. The main difficulty comes from the problem itself. In general, image degradation can be modeled as follows:

$$g = h * f + n, \tag{8.1}$$

where g, h, f, and n denote the blurred image, the PSF, the original image, and the observation noise term, respectively. The objective here is to estimate the original image f using the observed image g. The most classical approach is to estimate h and n from g, and solve f by applying the image deconvolution algorithms [4] such as Richardson-Lucy (RL) deconvolution [4]. However, the estimation of h and n is also a very demanding problem and the estimation error is unavoidable in most practical cases. To this end, joint optimization of blur identification and image restoration has been considered [5]. Though this approach performs relatively well, it is computationally complex and does not guarantee robust performance when the images are severely blurred. Since the basic difficulties of the conventional methods originate from the lack of sufficient knowledge about a real scene, the performance is inherently limited. The reader can refer to References [4], [6], [7], [8], [9], and [10] for details on image deblurring using a single image.

Recently, many researchers have devoted their attention to a new approach that utilizes multi-exposed images. Unlike the conventional approach, multiple images of the same scene are captured and used together to reconstruct an output image. Since the amount of available information about the original scene is increased by taking images multiple times, the difficulty of the original deblurring problem is significantly alleviated.

The degradation model for multi-exposed images can be expressed as follows:

$$g_i = h_i * f + n_i, \text{ for } i = 1, 2, ..., N, \tag{8.2}$$

where g_i, h_i, and n_i denote the ith blurred image, and its corresponding PSF and observation noise, respectively. The amount of the noise and blur of N multi-exposed images is dependent on the image capturing methods mentioned in Section 8.2.2. The multi-exposed deblurring techniques can be divided into two groups. The first group includes techniques which preserve the classical structure of single image deblurring and use multi-exposed images to enhance deblurring performance. The techniques in the second group convert the deblurring problem to a new formulation and use a specific solution for this new problem at hand. These two approaches are discussed in detail below.

8.3.2　Image Deblurring by Image Deconvolution

Since the classical image deconvolution algorithms and their variations are not a main concern of this chapter, only the difference between the techniques which use multi-exposed images and the techniques which use a single-exposed image will be discussed.

In particular, this section focuses on deconvolution-based multi-exposed image deblurring algorithms presented in References [11], [12], and [13].

Reference [11] describes the case where the motion blur function differs from image to image. Multi-exposed (commonly two) images are obtained by the method in Figure 8.2a, and they are assumed to be blurred by different directional motion blurs. The ith blurred image g_i is obtained as follows:

$$g_i = f *^{\theta_i} h_i, \tag{8.3}$$

where h_i is the ith directional blur PSF and θ_i is its angle. When two blurred images are considered, the following condition is satisfied due to the commutative property of convolution:

$$g_1 *^{\theta_2} h_2 = g_2 *^{\theta_1} h_1. \tag{8.4}$$

Then, h_1 and h_2 are estimated by minimizing the following error function:

$$E(h_1, h_2) = \sum_{x,y} \left[\left(g_1 *^{\theta_2} h_2 \right)(x,y) - \left(g_2 *^{\theta_1} h_1 \right)(x,y) \right]^2, \tag{8.5}$$

where x and y denote the image coordinates. By pessimistically setting the support of h_1 and h_2 to a large number, K_1 and K_2, respectively, the derivatives of Equation 8.5 provide $K_1 + K_2$ linear equations. By solving these equations, the PSFs can be obtained. Additional constraints can be included in Equation 8.5 to stabilize the solution [11].

After estimating the PSFs, the deblurred image $\widehat{f}$ is obtained by minimizing the following error function:

$$E = \sum_{i=1}^{2} \left\| g_i - \widehat{f} *^{\theta_i} h_i \right\|^2 + \lambda \left[\left(\left\| \widehat{f_x} \right\|_p \right)^p + \left(\left\| \widehat{f_y} \right\|_p \right)^p \right], \tag{8.6}$$

where λ controls the fidelity and stability of the solution. The regularization term is computed using the *p-norm* of the horizontal and vertical derivatives of $\widehat{f}$, $\widehat{f_x}$, and $\widehat{f_y}$. Then, the solution, $\widehat{f}$, is iteratively estimated as follows:

$$\widehat{f}^{(k+1)} = \widehat{f}^{(k)} + \sum_{i=1}^{2} h_i^T *^{\theta_i} \left(g_i - \widehat{f}^{(k)} *^{\theta_i} h_i \right) - \lambda \left. \frac{\partial L}{\partial f} \right|_{\widehat{f}}, \tag{8.7}$$

where $\widehat{f}^{(k)}$ is the deblurred image at the kth iteration and h_i^T denotes the flipped version of h_i. Equations 8.5, 8.6, and 8.7 are almost the same as those used in conventional single image deconvolution except for the summation term required for the two images. However, by using two blurred images, the ill-posed problem can move onto the direction of the well-posed problem and therefore this approach can improve the quality of the restored images. This straightforward generalization indicates that deblurring based on multi-exposed images can outperform its single-exposed image counterpart.

Reference [12] focuses on the fundamental relationship between exposure time and image characteristics discussed in Section 8.2. It introduces an imaging system shown in Figure 8.4 which can be seen as an alternative to the system shown in Figure 8.3c. The image obtained by the primary detector truly represents the color and brightness of the original scene but the details are blurred. On the other hand, the image taken by the secondary detector possesses the image details which allow robust motion estimation. The

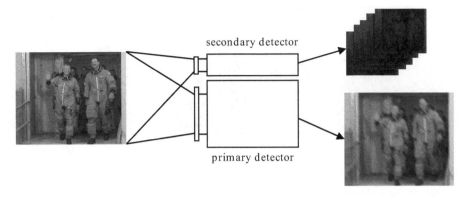

FIGURE 8.4

A conceptual design of a hybrid camera.

algorithm first estimates the global motion between successive frames by minimizing the following optical flow-based error function [14]:

$$\arg\min_{(u,v)} \sum \left(u\frac{\partial g}{\partial x} + v\frac{\partial g}{\partial y} + \frac{\partial g}{\partial t} \right)^2,$$

(8.8)

where g is the image captured by the secondary detector, $\partial g/\partial x$ and $\partial g/\partial y$ are the spatial partial derivatives, $\partial g/\partial t$ is the temporal derivative of the image, and (u, v) is the instantaneous motion at time t. Each estimated motion trajectory is then interpolated and the PSF is estimated with an assumption that the motion direction is the same as the blur direction. Then, the RL deconvolution, an iterative method that updates the estimation result at each iteration as follows:

$$\widehat{f}^{(k+1)} = \widehat{f}^{(k)} \cdot h^T * \frac{g}{h * \widehat{f}^{(k)}},$$

(8.9)

is applied to the image captured by the primary camera. In the above equation, $*$ is the convolution operator, h is the PSF estimated using the multiple images from the secondary detector, and h^T is the flipped version of h. The initial estimate, $\widehat{f}^{(0)}$, is set to g. Experimental results reveal good deblurring performance even in situations where the camera moves in an arbitrary direction [12]. Moreover, significant performance improvements can be achieved using special hardware, suggesting no need to cling to the conventional camera structures. Introducing a new or modified hardware structure allows efficiently solving the problem. Additional examples can be found in Reference [15].

Reference [13] presents a multi-exposed image deblurring algorithm that requires only two images, blurred and noisy image pairs as shown in Figure 8.5. The blurred image, g_b, is assumed to be obtained as follows:

$$g_b = f * h,$$

(8.10)

which indicates that noise in the blurred image is negligible. On the contrary, when the original image is captured within a short exposure time, the original image f can be represented as:

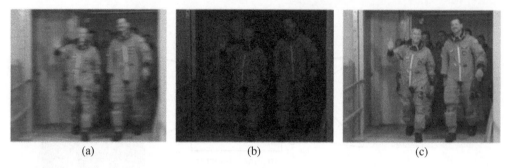

| (a) | (b) | (c) |

FIGURE 8.5 (See color insert.)

Image deblurring using a blurred and noisy image pair: (a) blurred input image, (b) noisy input image, and (c) output image obtained using two input images.

$$f = g_n + \Delta f, \qquad (8.11)$$

where g_n is the noisy image and Δf is called the residual image. It should be noted here that g_n is not the captured noisy image but the scaled version of that image. The scaling is required to compensate the exposure difference between the blurred and noisy images. Based on the above modeling, h is first estimated from Equation 8.10 using the Landweber iterative method [16] with Tikhonov regularization [17]. Unlike the conventional approaches, g_n can be used here as an initial estimate of f since g_n is a very close approximation of f compared to g_b. Therefore, the PSF estimation accuracy is significantly improved.

The estimated PSF is not directly used for the image deconvolution. Instead of recovering f, the residual image Δf is first recovered from the blurred image g_b. Combining Equations 8.5 and 8.6 results in the following:

$$\Delta g_b = g_b - g_n * h = \Delta f * h. \qquad (8.12)$$

Then, the estimated h is used to reconstruct Δf, as this method tends to produce fewer deconvolution artifacts than estimating f directly. Also, the conventional RL deconvolution is modified to suppress the ringing artifacts in smooth image regions. As can be seen in Figure 8.5, this algorithm produces a clear image without ringing artifacts.

In summary, the key advantage of multi-exposed image deconvolution is the increased accuracy of the PSF. When estimating the PSF, the noisy, but not blurred, images play an important role. Using a more accurate PSF significantly improves the output image quality compared to the classical single-image deconvolution. The three techniques described above represent the main directions in the area of multi-exposed image deconvolution. A method of capturing multi-exposed images is still an open issue, and a new exposure time setting associated with a proper deconvolution technique can further improve the performance of conventional approaches.

It should be also noted that the conventional multi-exposed image deconvolution algorithms still have some limitations. First, when the objects in a scene move, the linear shift invariant (LSI) degradation model does not work any longer. In this case, the piecewise LSI model [18], [19] or the linear shift variant degradation model (LSV) [20] should be adopted instead of the LSI model. Second, even though the accuracy of the PSF is significantly improved by using information from multi-exposed images, the estimation error is inevitable in most cases. Therefore, the deblurred output image is still distinguishable from the ground truth image. Finally, the computational complexity of the PSF estimation and image deconvolution methods is demanding, thus preventing these methods to be widely used in practical applications.

8.3.3 Image Deblurring without Image Deconvolution

The objective of multi-exposed image deconvolution in Section 8.3.2 was deblurring, that is, eliminating blur by solving the inverse problem. Instead of directly performing image deblurring, this section introduces an approach that merges the meaningful information of a scene from multi-exposed images. The four representative techniques, introduced in References [21], [22], [23], and [24], are described below.

Namely, Reference [21] presents a simple method based on the assumption that the denoising problem is much easier than the deblurring problem. From this viewpoint, the exposure time can be uniformly split and multiple noisy images can be captured as shown in Figure 8.3b. Since noise among the multi-exposed image frames is generally uncorrelated, it can be eliminated by simply averaging the multi-exposed images. Experimental results show that the the the most effective number of splits is in the range of five to ten. Due to the simplicity of the algorithm, the visual quality is unsatisfactory. On the other hand, it is guaranteed that the blur is almost completely eliminated. A similar approach can be found in Reference [25].

The method in Reference [22] requires both a blurred image and a noisy image, which is similar to Reference [13]. However, the solution is completely different. The blurred image taken with a long exposure time can faithfully represent the brightness and color of the original scene whereas the noisy underexposed image usually contains the clear object shape and detail information while having unacceptable color and brightness. The goal is to preserve the object shape and detail information of the noisy image g_l and reproduce the color and brightness of the blurred image g_h. To this end, the transformation function, F, can be defined as follows:

$$g_h = F(g_l). \tag{8.13}$$

The basic assumption behind this transformation function is that the color characteristics of the blurred image are close, if not identical, to that of the original image. Therefore, by finding a suitable transformation function, the damaged color values in the underexposed image can be recovered.

However, there are some color differences between the blurred image and the original image. As shown in Figure 8.6, the color information is similar in a global sense. However, pixels located around the edges may not find their corresponding color due to edge blur. In other words, a simple global mapping such as adaptive histogram equalization can produce annoying color artifacts. Therefore, a spatial constraint is utilized to improve the accuracy of the mapping function. The basic intuition is that the color is not severely damaged in the smooth region as shown in Figure 8.6c. Consequently, it can be constrained so that the pixels in g_l are mapped to the same locations in g_h in smooth regions. In order to realize this constraint, image segmentation is performed at g_h, followed by region matching. After finding matching pairs, the mapping function is estimated by the Bayesian framework. The mathematical derivations are omitted here and can be found in Reference [22].

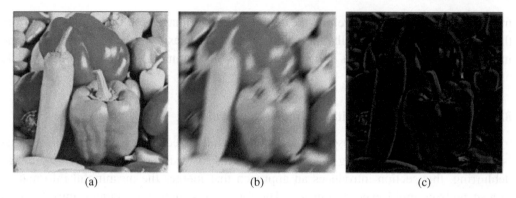

(a)   (b)   (c)

FIGURE 8.6

Color variations due to image blur: (a) original image, (b) blurred image, and (c) difference between the original image and the blurred image.

The above constitutes a novel approach that converts the deblurring problem into a color mapping problem. However, this method does not carefully address noise in an underexposed image. Since an image captured using a short exposure time tends to contain noise, the color mapping result can still be noisy. Therefore, a robust mapping function needs to be devised to improve the performance.

Reference [23] presents an image merging algorithm that requires three multi-exposed images. The first image and the third image are captured using a long exposure time to preserve brightness and color information, whereas the second image is captured with a short exposure time to preserve edges and details. The objective of this technique is to compensate the color and brightness loss of the second image using the other two images. The flow of this algorithm is shown in Figure 8.7.

The algorithm starts with global motion estimation to find the corresponding regions between successively captured images. Since the camera or objects can move during the time needed to capture three images, this step is necessary to remove a global mismatch between images. A single parametric model is adopted to estimate global motion as follows:

$$
\begin{bmatrix} x_l \\ y_l \\ s_l \end{bmatrix} = \begin{bmatrix} h_{11} & h_{12} & h_{13} \\ h_{21} & h_{22} & h_{23} \\ h_{31} & h_{32} & h_{33} \end{bmatrix} \begin{bmatrix} x_h \\ y_h \\ s_h \end{bmatrix},
\tag{8.14}
$$

where (x_l, y_l) and (x_h, y_h) are pixel positions in the underexposed image and the blurred images, respectively. A scaling parameter $s_l(s_h)$ is used to represent the position by the homogeneous coordinate. Nine unknown parameters, h_{11} to h_{33}, are then estimated by minimizing the intensity discrepancies between two images. However, since the brightness

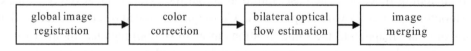

FIGURE 8.7

The block diagram of the image merging-driven deblurring algorithm.

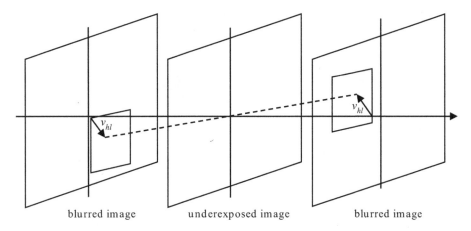

FIGURE 8.8

Bilateral optical flow estimation.

differences between the underexposed image and the two blurred images are usually significant, the direct motion estimation between differently exposed images is not desired. Therefore, the global motion is estimated using two blurred images as these are captured using the same exposure time setting. Then, the motion between the first blurred image and the underexposed image is simply assumed to be half of the estimated motion. Since optical flow estimation follows this global motion estimation, the loss of accuracy by this simplified assumption is not severe.

Since the color distribution of the underexposed image tends to occupy low levels, histogram processing is used to correct the color of the underexposed image based on the color distribution of the corresponding regions in the long-exposed (blurred) images. However, the noise in the underexposed image still remains. Therefore, for each pixel in the underexposed image, bilateral motion estimation is used to search for the correlated pixels in the blurred images. This estimation is based on the assumption of linear motion between two intervals. Unlike conventional optical flow estimation, the motion trajectory passing through a point in the intermediate frame is found by comparing a pixel at the shifted position in the first blurred image and one at the opposite position in the other blurred image as shown in Figure 8.8. Finally, the output image is obtained by averaging the pixel values of two corresponding pixels in the blurred images and the one in the underexposed image as follows:

$$\widehat{f}(\mathbf{u}) = \lambda \widehat{g_l}(\mathbf{u}) + \frac{1-\lambda}{2} \left(g_{h,1}(\mathbf{u}-\mathbf{v}) + g_{h,2}(\mathbf{u}+\mathbf{v}) \right), \tag{8.15}$$

where $\widehat{f}$ denotes the resultant image, $\widehat{g_l}$ denotes the the scaled underexposed image, and $g_{h,1}$ and $g_{h,2}$ denote the two blurred images. The terms $\mathbf{u}$ and $\mathbf{v}$ are the vector representation of the pixel position and its corresponding motion vector. The weighting coefficient λ is determined by the characteristics of the image sensor.

This algorithm, which can be seen as the first attempt to use motion estimation and compensation concepts for deblurring, produces high-quality images. Since the output image is generated by averaging two blurred images and one noisy image, most noise contributions are reduced, if not completely eliminated. Although the computational complexity for esti-

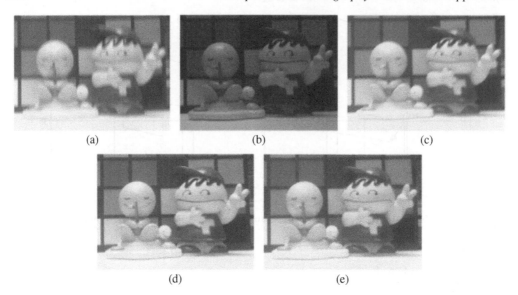

(a)   (b)   (c)

(d)   (e)

FIGURE 8.9 (See color insert.)

Comparison of two deblurring approaches: (a-c) three test images, (d) output produced using two first input images, and (e) output produced using all three input images.

mating global and local motion is demanding, this technique does not require any hardware modifications, and produces a natural output image without visually annoying artifacts.

Figure 8.9 allows some comparison of the methods in References [22] and [23]. Three test images, shown in Figures 8.9a to 8.9c, are respectively captured with a long, a short, and a long exposure time. As mentioned above, the method in Reference [22] uses two differently blurred images shown in Figures 8.9a and 8.9b, whereas the method in Reference [23] combines all three input images. As can be seen, the images output by both of these algorithms do not suffer from image blur or restoration artifacts, which is a significant advantage of this deconvolution-free approach. It can be further noted that the color in Figure 8.9d is different from that of Figure 8.9a or Figure 8.9c. This is because noise in Figure 8.9b prevents finding the optimal color mapping function.

Reference [24] approaches multi-exposed image deblurring using the assumption that the multi-exposed images are captured within a normal exposure time and these images are differently blurred. In other words, unlike previously cited techniques which rely on noisy data for image details and edge information, this method uses only blurred images. Figure 8.10 illustrates the basic concept. The original image is blurred in the diagonal, horizontal, and vertical directions, respectively. Then, the frequency spectra of these blurred images are found by applying the fast Fourier transform (FFT). Since motion blur behaves as a directional low-pass filter, each spectrum loses high frequency components depending on blur direction. In other words, each spectrum tends to have mutually different partial information of the original image spectrum. Therefore, it is possible to gather this information for compensating the loss of frequency components.

The merging operation is performed in the frequency domain via fuzzy projection onto convex sets (POCS). Based on the fact that image blur exhibits the low-pass characteristics in the frequency domain regardless of the blur type, the reliability of frequency coefficients

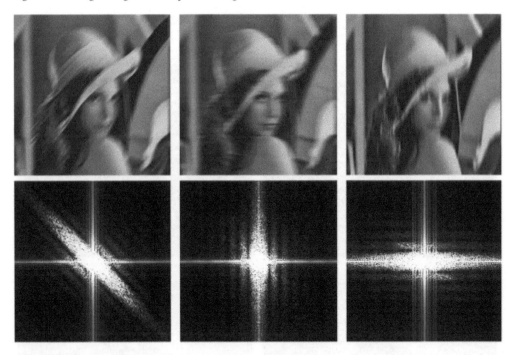

FIGURE 8.10

Frequency spectra of blurred images: (top) three differently blurred images, and (bottom) magnitude of their frequency spectra.

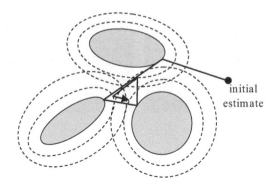

FIGURE 8.11

Procedure of the projection onto the fuzzy convex sets. © 2009 IEEE

reduces as the frequency increases. Therefore, only low-frequency regions of each spectrum are considered as convex sets. However, projection onto these sets cannot recover enough high-frequency parts. To this end, the convex sets can be expanded through fuzzification [26], [27]. The fuzzification process is illustrated in Figure 8.11. Each ellipse represents the convex set and the arrow indicates a projection operation. Both projection and fuzzification are repeated until the predefined criterion is reached. By this approach, the mutually different information about the scene can be efficiently combined. Since this method merges all the available information from multi-exposed images, the quality improvement is limited if multi-exposed images are strongly correlated to each other.

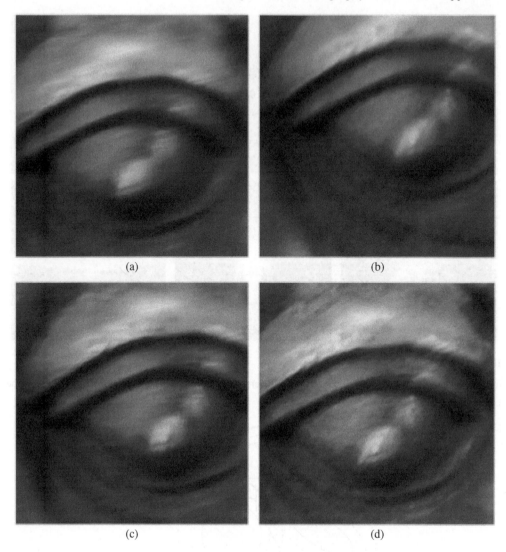

FIGURE 8.12

Deblurring using real-life blurred images: (a-c) three test images, and (d) the deblurred image.

Figure 8.12 shows the performance of the method in Reference [24] on real-life images. For the sake of visual comparison, small patches of 400×400 pixels were cropped from 1024×1024 test images. The image registration algorithm from Reference [28] is applied to all test images to match the image grids. Figure 8.12d shows the combined result when Figure 8.12a is used as a reference image for image registration. As can be seen, the output image successively merges available information from multi-blurred images.

This section explored various multi-exposed image deblurring algorithms. Since each algorithm uses its own experimental conditions, the direct comparison of these algorithms is not applicable. Instead, the conceptual differences and the characteristics of the output images need to be understood. In multi-exposed image deconvolution, the underexposed (noisy) image is used to estimate the PSF with higher accuracy. By using additional images, the parametric assumption of the PSF [9], [29] and computationally complex blind deconvolution [30], [31] are not necessary. Also, all the existing techniques developed for single-image deconvolution can be generalized for multi-exposed image deconvolution. In multi-exposed image deblurring without deconvolution, no typical degradation model is required. Therefore, this approach can be more generally used in practical applications. Compared to the first approach, it has a lot of room for improvement. For example, by changing the image capturing method and/or applying the techniques used in other fields, such as histogram mapping and motion compensated prediction, the performance can be further improved.

8.4 Conclusion

For several decades, researchers have tried to solve a troublesome deblurring problem. Due to the fundamental difficulty — the lack of available information — the performance of conventional single image deblurring methods is rather unsatisfactory. Therefore, multi-exposed image deblurring is now considered a promising breakthrough. Thanks to recent advances in image sensing technology, the amount of available information about the original scene is significantly increased. The remaining problem is how to use this information effectively. By referring to the image capturing methods and conventional multi-exposed deblurrig algorithms described in this chapter, the reader can freely develop a new solution, making no need to adhere to the classical deblurring approach any longer.

Other camera settings such as ISO and aperture can be also tuned to generate images with different characteristics. Therefore, more general image capturing methods and deblurring solutions can be designed by considering these additional factors. In addition to deblurring, multi-exposed images have another application. Based on the fact that a set of multi-exposed images can capture a wider dynamic range than a single image, a dynamic range improvement by using multi-exposed images is also a promising research topic. By jointly considering these two image processing operations, the ultimate goal of truly reproducing the real scene can be reached in the near future.

Acknowledgment

Figure 8.2 is reprinted from Reference [23] and Figure 8.11 is reprinted from Reference [24], with the permission of IEEE.

References

[1] E.R. Fossum, "Active pixel sensors: Are CCDs dinosaurs," *Proceedings of SPIE*, vol. 1900, pp. 2–14, February 1993.

[2] N. Stevanovic, M. Hillegrand, B.J. Hostica, and A. Teuner, "A CMOS image sensor for high speed imaging," in *Proceedings of the IEEE International Solid-State Circuits Conference*, San Francisco, CA, USA, February 2000, pp. 104–105.

[3] O. Yadid-Pecht and E. Fossum, "Wide intrascene dynamic range CMOS aps using dual sampling," *IEEE Transactions on Electron Devices*, vol. 44, no. 10, pp. 1721–1723, October 1997.

[4] P.A. Jansson, *Deconvolution of Image and Spectra*. New York: Academic Press, 2nd edition, October 1996.

[5] Y.L. You and M. Kaveh, "A regularization approach to joint blur identification and image restoration," *IEEE Transactions on Image Processing*, vol. 5, no. 3, pp. 416–428, February 1996.

[6] M.R. Banham and A.K. Katsaggelos, "Digital image restoration," *IEEE Signal Processing Magazine*, vol. 14, no. 2, pp. 24–41, March 1997.

[7] R.L. Lagendijk, J. Biemond, and D.E. Boekee, "Identification and restoration of noisy blurred images using the expectation-maximization algorithm," *IEEE Transactions on Acoustics, Speech, and Signal Processing*, vol. 38, no. 7, pp. 1180–1191, July 1990.

[8] A.K. Katsaggelos, "Iterative image restoration algorithms," *Optical Engineering*, vol. 28, no. 7, pp. 735–748, July 1989.

[9] G. Pavlović and A. M. Tekalp, "Maximum likelihood parametric blur identification based on a continuous spatial domain model," *IEEE Transactions on Image Processing*, vol. 1, no. 4, pp. 496–504, October 1992.

[10] D.L. Tull and A.K. Katsaggelos, "Iterative restoration of fast-moving objects in dynamic image sequences," *Optical Engineering*, vol. 35, no. 12, pp. 3460–3469, December 1996.

[11] A. Rav-Acha and S. Peleg, "Two motion-blurred images are better than one," *Pattern Recognition Letters*, vol. 26, no. 3, pp. 311–317, February 2005.

[12] M. Ben-Ezra and S.K. Nayar, "Motion-based motion deblurring," *IEEE Transactions on Pattern Analysis and Machine Intelligence*, vol. 26, no. 6, pp. 689–698, June 2004.

[13] L. Yuan, J. Sun, L. Quan, and H.-Y. Shum, "Image deblurring with blurred / noisy image pairs," *ACM Transactions on Graphics*, vol. 26, no. 3, July 2007.

[14] B.D. Lucas and T. Kanade, "An iterative image registration technique with an application to stereo vision," *Defense Advanced Research Projects Agency, DARPA81*, pp. 121–130, 1981.

[15] R. Raskar, A. Agrawal, and J. Tumblin, "Coded exposure photography: Motion deblurring using fluttered shutter," *ACM Transactions on Graphics*, vol. 25, no. 3, pp. 795–804, July 2006.

[16] H.W. Engl, M. Hanke, and A. Neubauer, *Regularization of Inverse Problems*. Dordrecht, The Netherlands: Kluwer Academic, March 2000.

[17] A. Tarantola, *Inverse Problem Theory*. Philadelphia, PA: Society for Industrial and Applied Mathematics, December 2004.

[18] J. Bardsley, S. Jefferies, J. Nagy, and R. Plemmons, "A computational method for the restoration of images with an unknown, spatially-varying blur," *Optics Express*, vol. 14, no. 5, pp. 1767–1782, March 2006.

[19] J. Bardsley, S. Jefferies, J. Nagy, and R. Plemmons, "Variational pairing of image segmentation and blind restoration," in *Proceedings of the European Conference on Computer Vision*, Prague, Czech Republic, May 2004, pp. 166–177.

[20] M.K. Ozkan, A.M. Tekalp, and M.I. Sezan, "POCS-based restoration of space-varying blurred images," *IEEE Transactions on Image Processing*, vol. 3, no. 4, pp. 450–454, July 1994.

[21] Z. Wei, Y. Cao, and A.R. Newton, "Digital image restoration by exposure-splitting and registration," in *Proceedings of the International Conference on Pattern Recognition*, Cambridge, UK, August 2004, vol. 4 pp. 657–660.

[22] J. Jia, J. Sun, C.K. Tang, and H.-Y. Shum, "Bayesian correction of image intensity with spatial consideration," in *Proceedings of the European Conference on Computer Vision*, Prague, Czech Republic, May 2004, pp. 342–354.

[23] B.D. Choi, S.W. Jung, and S.J. Ko, "Motion-blur-free camera system splitting exposure time," *IEEE Transactions on Consumer Electronincs*, vol. 54, no. 3, pp. 981–986, August 2008.

[24] S.W. Jung, T.H. Kim, and S.J. Ko, "A novel multiple image deblurring technique using fuzzy projection onto convex sets," *IEEE Signal Processing Letters*, vol. 16, no. 3, pp. 192–195, March 2009.

[25] X. Liu and A.E. Gamal, "Simultaneous image formation and motion blur restoration via multiple capture," in *Proceedings of the IEEE International Conference on Acoustics, Speech, and Signal Processing*, Salt Lake City, UT, USA, May 2001, vol. 3, pp. 1841–1844.

[26] S. Oh and R.J. Marks, "Alternating projection onto fuzzy convex sets," in *Proceedings of IEEE International Conference on Fuzzy Systems*, San Francisco, CA, USA, March 1993, pp. 1148–1155.

[27] M.R. Civanlar and H.J. Trussell, "Digital signal restoration using fuzzy sets," *IEEE Transactions on Acoustics, Speech, and Signal Processing*, vol. 34, no. 4, pp. 919–936, August 1986.

[28] B.S. Reddy and B.N. Chatterji, "An FFT-based technique for translation, rotation, and scale-invariant image registration," *IEEE Transacxtions on Image Processing*, vol. 5, no. 8, pp. 1266–1271, August 1996.

[29] D.B. Gennery, "Determination of optical transfer function by inspection of frequency-domain plot," *Optical Society of America*, vol. 63, no. 12, pp. 1571–1577, December 1973.

[30] A. Levin, "Blind motion deblurring using image statistics," in *Proceedings of the Twentieth Annual Conference on Neural Information Processing Systems*, Vancouver, BC, Canada, December 2006, pp. 841–848.

[31] G. Harikumar and Y. Bresler, "Perfect blind restoration of images blurred by multiple filters: Theory and efficient algorithms," *IEEE Transactions on Image Processing*, vol. 8, no. 2, pp. 202–219, February 1999.

9

Color High-Dynamic Range Imaging: Algorithms for Acquisition and Display

Ossi Pirinen, Alessandro Foi, and Atanas Gotchev

9.1 Introduction

Dynamic range is the term used in many fields to describe the ratio between the highest and the lowest value of a variable quantity. In imaging and especially in display technology, dynamic range is also known as the contrast ratio or, simply contrast, which denotes the brightness ratio between black and white pixels visible on the screen at the same time. For natural scenes the dynamic range is the ratio between the density of luminous intensity of the brightest sunbeam and the darkest shadow. In photography, the unit of luminance, cd/m^2 (candelas per square meter), is also known as a "nit."

The problem of high-dynamic range (HDR) imaging is two-fold: i) how to capture the true luminance and the full chromatic information of an HDR scene, possibly with a capturing device with a dynamic range smaller than the scene's, and ii) how to faithfully represent this information on a device that is capable of reproducing the actual luminances of neither the darkest nor the brightest spots, nor the true colors. The first half of the issue is commonly referred to as high-dynamic range *composition* or *recovery* and the latter as *compression* or, more commonly, *tone mapping*. In this chapter the terms used are composition for the acquisition process and tone mapping for the displaying part.

Figure 9.1 illustrates the HDR imaging problem. Namely, Figure 9.1a shows that the conventional low-dynamic range (LDR) snapshot has a very limited dynamic range, which is seen as clipped bright parts as well as underexposed dark parts. The HDR representation shown in Figure 9.1b has all the information but only a fraction of it is visible on an LDR media, such as the printout. The tone mapped HDR image shown in Figure 9.1c boasts the greatest amount of visible information.

HDR images can be acquired by capturing real scenes or by rendering three-dimensional (3D) computer graphics, using techniques like radiosity and ray tracing. HDR images of natural scenes can be acquired either by capturing multiple images of the same scene with different exposures or utilizing only recently introduced special cameras able to directly capture HDR data [1]. The multiple exposure method relies on combining the different exposures into a single image, spanning the whole dynamic range of the scene [2], [3]. The image composition step requires a preliminary calibration of the camera response [3], [4].

| (a) | (b) | (c) |

FIGURE 9.1 (See color insert.)

HDR imaging: (a) single exposure from a clearly HDR scene, (b) HDR image linearly scaled to fit the normal 8-bit display interval — the need for tone mapping is evident, and (c) tone mapped HDR image showing a superior amount of information compared with the two previous figures.

Though HDR capture and display hardware are both briefly discussed, the focus of this chapter is on the approach of multiple capture composition.

The second part of the problem is to appropriately tone map the so-obtained HDR image back to an LDR display. Tone mapping techniques range in complexity from a simple gamma-curve to sophisticated histogram equalization methods and complicated lightness perception models [5], [6], [7]. Recently, display devices equipped with extended dynamic range have also started to appear [8].

HDR imaging methods have been originally developed for RGB color models. However, techniques working in luminance-chrominance spaces seem more meaningful and preferable for a number of reasons:

- Decorrelated color spaces offer better compressibility. As a matter of fact, the near totality of image compression techniques store images in some luminance-chrominance space. When one starts with already-compressed multiple-exposure LDR images, it is more efficient to compose the HDR image directly in the same color space. The resulting HDR image is then better suited for compression and, if to be displayed, it can be mapped to sRGB during the tone mapping stage.

- Any HDR technique operating in RGB space requires post-composition white balancing since the three color channels undergo parallel transformations. While the white balancing would yield perceptually convincing colors, they might not be the true ones. For the sake of hue preservation and better compression, it is beneficial to opt for a luminance-chrominance space, even if the input data is in RGB (for instance, an uncompressed TIFF image).

- The luminance channel, being a weighted average of the R, G, and B channels, enjoys a better signal-to-noise ratio (SNR), which is crucial if the HDR imaging process takes place in noisy conditions. In this chapter, the problem of HDR imaging in a generic luminance-chrominance space is addressed, efficient algorithms for HDR image composition and tone mapping are presented, and the functionality of such an approach under various conditions is studied.

In this chapter, Section 9.2 provides general information concerning color spaces and the notation used in the chapter. Techniques essential to HDR acquisition, namely camera response calibration, are discussed in Section 9.3. A brief overview of HDR sensors is also included in Section 9.3. Section 9.4 focuses on the creation of HDR images; techniques for both monochromatic and color HDR are presented. Section 9.5 deals with displaying HDR images; display devices capable of presenting HDR content are also discussed. Section 9.6 contains examples and discussion. Conclusions are drawn in Section 9.7.

9.2 Color Spaces

Generally speaking, the retina has three types of color photo-receptor cells, known as cones. These respond to radiation with a different spectral response, that is, different cones generate the physical sensation of different colors, which is combined by the human visual system (HVS) to form the color image. From this it is intuitive to describe color with three numerical components. All the possible values of this three-component vector form a space called a color space or color model. The three components can be defined in various meaningful ways, which leads to definition of different color spaces [1], [9].

9.2.1 RGB Color Spaces

An RGB tristimulus color space can be defined as the Cartesian coordinate system-based model with red, green and blue primary spectral components [10]. All the standardized RGB (Red, Green, Blue) color spaces can be defined by giving the CIE XYZ chromaticities (x and y) of each primary color and the white reference point. The reference white point serves to define white inside the gamut of the defined color space. RGB model is the most often used representation in computer graphics and multimedia, and different RGB spaces (primary color and white point combinations) have been standardized for different applications [11]. What they all have in common is the idea to mix red, green and blue primaries in different relations to produce any given color inside the gamut of that space. All the RGB color models can be categorized under the definition *physiologically inspired color models* because the three primaries have been designed with the idea of matching the three different types of cones in the human retina. For implementation simplicity, the RGB coordinates are usually limited to $[0\ 1]^3$ (floating point) or $[0\ 255]^3$ (8-bit unsigned integer).

9.2.2 Luminance-Chrominance Color Spaces

Even though technology has steered the selection of the color model used in computer graphics toward the RGB models, that is not the *psychophysical* color interpretation. Perceptual features, namely brightness (luminance), saturation and hue, form the color sensation in the human visual system (HVS). These features can be intuitively formalized through the introduction of luminance-chrominance color models.

Consider a generic luminance-chrominance color space linearly related to the RGB space. Loosely speaking, such a space is characterized by an achromatic luminance component, which corresponds to the grayscale part of the image, and two chrominance components, which are orthogonal to gray.

Throughout the chapter, Roman letters are used to denote images in RGB and the corresponding Greek letters to denote images in luminance-chrominance space. Thus, $\mathbf{z} = \left[z^R, z^G, z^B \right]$ denotes an image in the RGB space and $\zeta = \left[\zeta^Y, \zeta^U, \zeta^V \right]$ be the same image in the luminance-chrominance space where the luminance and the two chrominance channels are denoted by Y, U, and V, respectively. Transformation of the image from RGB to luminance-chrominance is defined in matrix form as $\zeta = \mathbf{z} \mathbf{A}$ where the matrix $\mathbf{A}$ is normalized in such a way that if $\mathbf{z}(\cdot) \in [0,1]^3$ then $\zeta(\cdot) \in [0,1] \times [-0.5, 0.5]^2$. Because of this constraint, the first column of $\mathbf{A}$ has all elements positive, $a_{j,1} \geq 0$. It is further assumed that $\sum_{j=1}^{3} a_{j,1} = 1$, thus ensuring that the $[0,1]$ range of the luminance component is fully utilized. Examples of such luminance-chrominance spaces are the opponent, the YUV/YCbCr, and the YIQ color spaces [9].

Luminance-chrominance transformations become particularly significant when the image $\mathbf{z}$ is corrupted by some independent noise. In fact, because of the typical correlation among z^R, z^G, and z^B, one can observe that the luminance ζ^Y has noticeably higher signal-to-noise ratio (SNR) than the two chrominance values ζ^U and ζ^V, as well as any of the individual RGB components z^R, z^G, and z^B.

It is well known that natural color images exhibit a high correlation between the R, G, and B channels. Thus, it can be observed that the luminance Y contains most of the valuable information (edges, shades, objects, texture patterns, etc.) and the chrominances U and V contain mostly low-frequency information (considering compressed data, these channels very often come from undersampled data).

9.2.2.1 Opponent Color Space

The opponent color space, presented for example in Reference [9], is based on the 1964 color opponency theory by a German physiologist Ewald Hering. The theory suggests two pairs of opponent colors, red-green and yellow-blue, that cannot be perceived simultaneously. This theory was supported by color naming experiments where reddish-green and yellowish-blue tones were not identified. This led Hering to presume three opponent channels; red-green, yellow-blue, and black-white (achromatic luminance). In fact, Hering was one of the first to separate luminance and two chrominances in a color model. Unlike the RGB tristimulus model, driven by an intent of modeling the retinal color stimulus response, the opponent color space is based on more central mechanisms of the brain. The opponent processes are acquired through a transformation of the cone responses.

The transformation matrices for the opponent color space are

$$\mathbf{A}_{opp} = \begin{bmatrix} 1/3 & 1/2 & 1/4 \\ 1/3 & 0 & -1/2 \\ 1/3 & -1/2 & 1/4 \end{bmatrix} \text{ and } \mathbf{B}_{opp} = \mathbf{A}_{opp}^{-1} = \begin{bmatrix} 1 & 1 & 2/3 \\ 1 & 0 & -4/3 \\ 1 & -1 & 2/3 \end{bmatrix}. \tag{9.1}$$

It can be noted that the second and third columns of matrix $\mathbf{A}_{opp}$ have zero mean. This is equivalent to the inner product between the chrominance basis vectors and a vector corresponding to a gray pixel (for which $z^R = z^G = z^B$) always being zero. It means that gray is orthogonal to the chrominance components and that the inverse color transformation matrix $\mathbf{B} = \mathbf{A}^{-1}$ has the elements of its first row all equal, $b_{1,1} = b_{1,2} = b_{1,3}$. Since $1 = \sum_{j=1}^{3} a_{j,1} b_{1,j}$ and $\sum_{j=1}^{3} a_{j,1} = 1$, it is given that $b_{1,1} = b_{1,2} = b_{1,3} = 1$. This means that the luminance component ζ^Y can be directly treated as a grayscale component of the RGB image $\mathbf{z}$, because the inverse transformation of the luminance component, $\begin{bmatrix} \zeta^Y & 0 & 0 \end{bmatrix} \mathbf{B} = \mathbf{z}_{gray}(\mathbf{x})$, is a grayscale image.

Luminance-chrominance transformations can be considered as special color decorrelating transforms. In particular, up to a diagonal normalization factor, the matrix $\mathbf{A}_{opp}$ is nothing but a 3×3 DCT transform matrix. Further, it is noted that the columns of $\mathbf{A}_{opp}$ are respectively a mean filter, a finite derivative filter, and a second derivative filter.

9.2.3 YUV Color Space

The YUV color space was first designed for the purposes of television broadcasting with the intent of minimizing the bandwidth requirements. Similar to the YIQ color space which is also used in television broadcasting, it fills the definition of an opponent color space because it consists of a luminance channel and two color difference channels. As a side note it can be mentioned that the usual scheme for compressing a YUV signal is to downsample each chromatic channel with a factor of two or four so that chromatic data occupies at most half of the total video bandwidth. This can be done without apparent loss of visual quality as the human visual system is far less sensitive to spatial details in the chrominances than in the luminance. The same compression approach is also utilized among others in the well-known JPEG (Joint Photographic Experts Group) image compression standard.

The transformation matrices for YUV color space are

$$\mathbf{A}_{YUV} = \begin{bmatrix} 0.30 & -0.17 & 0.50 \\ 0.59 & -0.33 & -0.42 \\ 0.11 & 0.50 & -0.08 \end{bmatrix} \text{ and } \mathbf{B}_{YUV} = \mathbf{A}_{YUV}^{-1} = \begin{bmatrix} 1 & 0 & 1.4020 \\ 1 & -0.3441 & -0.7141 \\ 1 & 1.7720 & 0 \end{bmatrix}. \quad (9.2)$$

Because of the similar nature of the color models, the orthogonality properties given in the previous section for opponent color space hold also for the YUV space. One should note that these properties do not depend on the orthogonality of the matrix $\mathbf{A}$ (in fact the three columns of $\mathbf{A}_{YUV}$ are not orthogonal), but rather on the orthogonality between a constant vector and the second and third columns of the matrix.

9.2.3.1 HSV Color Space

The HSV color space is a representative of the class of perceptual color spaces. The name stands for the three components that form the space; hue, saturation and value (intensity, brightness). This type of color models is categorized under the umbrella of perceptual color spaces for a reason; while an untrained observer can hardly form an image of a color based on individual RGB tristimulus components, everybody can form a color based on its hue (tint, tone) and saturation.

The hue can be defined as $H = \arctan(\zeta^U / \zeta^V)$ and the saturation as $S = \sqrt{(\zeta^U)^2 + (\zeta^V)^2}$, or normalized with the value ζ^Y as $S = \sqrt{(\zeta^U)^2 + (\zeta^V)^2}/\zeta^Y$. These can be interpreted as the angular component and the length of a planar vector. Thus the triplet Y, H, and S corresponds to a luminance-chrominance representation with respect to cylindrical coordinates. It is noted that multiple definitions of both hue and saturation can be found in the literature [12] and while the presented ones are among the more frequently noted ones, they are chosen here first and foremost for the sake of simplicity.

9.3 HDR Acquisition

In this section, techniques essential for successful HDR acquisition are presented. As imaging devices directly capable of capturing HDR content are scarcely available on the consumer market, the means for capturing HDR content with a normal camera are given. In addition, a quick cross-section of the fast-evolving field of imaging sensor technologies directly capable of capturing HDR scenes is presented.

9.3.1 Camera Response Function

The most essential step in HDR image acquisition is the definition of the camera response function. This function defines the relation between the scene irradiance and the camera pixel output and can therefore be used to linearize the output data. Conversely, knowing the camera response is crucial for determining the irradiance of the source scene from the given image data.

Essentially two techniques exist for the response calibration. The approach defined in Reference [3] extends the previous work presented in Reference [2]. The basic principle of this approach is that by capturing frames of a still scene in different exposures, one is actually sampling the camera response function at each pixel. The technique is relatively simple and robust, and is described later in this section in detail.

The other calibration method is presented in Reference [4]. Instead of filling an enumerated table as in the previous approach, this method approximates polynomial coefficients for the response function. This not only enables one to find the solution of a camera response function, but it also allows exact exposure ratios which are essential in combining HDR images from source sequences whose aperture and shutter speed are not known exactly.

With the pixel exposure $e(\mathbf{x})$ defined as the product between exposure time Δt and irradiance $E(\mathbf{x})$, that is, $e(\mathbf{x}) = E(\mathbf{x})\Delta t$, the generic pixel output $\mathfrak{z}(\mathbf{x})$ is given by

$$\mathfrak{z}(\mathbf{x}) = f(e(\mathbf{x})) = f(E(\mathbf{x})\Delta t), \tag{9.3}$$

where f is the function describing the response of the output to the given exposure. Therefore, the irradiance can be obtained from the pixel output by the formal expression as follows [3]:

$$\ln E(\mathbf{x}) = g(\mathfrak{z}(\mathbf{x})) - \ln \Delta t, \tag{9.4}$$

where $g(\cdot) = \ln f^{-1}(\cdot)$ is the inverse camera response function. This function can be estimated from a set of images of a fixed scene captured with different exposure times. Images for the calibration sequence are assumed to be perfectly aligned, shot under constant illumination, and contain negligible noise. For the camera response function calibration, a much larger set of images (i.e., a denser set of exposures) than typically available for the HDR composition, is used. The camera response function is estimated (calibrated) only once for each camera. The estimated function can then be used for the linearization of the input values in all subsequent HDR compositions of the same device.

Because of underexposure (which produces dramatically low SNR) and overexposure (which results in clipping the values which would otherwise exceed the dynamic range) not all pixels from the given set of images should be used for the calibration of the camera response function, nor for the subsequent composition, with equal weights. Near the minimum of the value interval of an imaging device, the information is distorted by numerous noise components, most influential of which is the Poisson photon shot noise [13]. As the number of photons hitting a well on the sensor within a given time interval is a random process with Poisson distribution, at low light levels the variations in the number of photons become substantial in comparison to the mean levels, effectively rendering the current created in the well completely unreliable. In addition to this, components like thermal and read-out noise play some role in corrupting the low-level sensor output. In the vicinity of

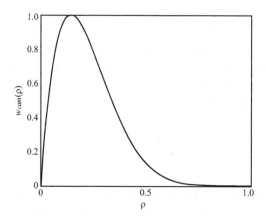

FIGURE 9.2

Weight function used for the luminance values in the camera response function definition.

the maximum of the value interval, clipping starts to play its role. Naturally a saturated pixel can only convey very limited amount of information about the scene. The described phenomena motivate the penalization of underexposed and overexposed pixels in both the camera response calibration and the HDR image composition phase. Additionally, only pixels having monotonically strictly increasing values between underexposure and over-exposure, throughout the sequence, can be considered to be valid in the camera response calibration. This is a natural idea, since the pixels used for response calibration are assumed to come from a sequence shot with increasing exposure time. As the scene is assumed to be static and the pixel irradiance constant, the only aberration from the strict monotonical increase in pixel value can come from noise. Noisy pixels should not be used for camera response calibration, as erroneous response function would induce a systematic error in the subsequent HDR compositions. Using these criteria for pixel validity, the function g can be fitted by minimizing the quadratic objective function [3]

$$\sum_{j=1}^{P}\sum_{i=1}^{N} w_{\text{cam}}(\zeta_i^Y(\mathbf{x}_j))\left[g(\zeta_i^Y(\mathbf{x}_j)) - \ln E_i - \ln \Delta t_i\right]^2 + \lambda \int_0^1 w_{\text{cam}}(\zeta)g''(\zeta)^2 d\zeta, \qquad (9.5)$$

where P is the number of pixels used from each image and w_{cam} is a weight function limiting the effect of the underexposed and overexposed pixels. The regularization term on the right-hand side uses a penalty on the second derivative of g to ensure the smoothness of the solution. In this chapter, the preference is to use a relatively high pixel count (e.g., 1000 pixels) in order to minimize the need for regularization in the data fitting. The system of equations is solved by employing singular-value decomposition, in a fashion similar to that of Reference [3]. If instead of processing directly one of the sensor's RGB outputs, a combination ζ_i^Y, (e.g., luminance) is considered, it should be noted that some pixels in this combination can include overexposed components without reaching the upper limit of the range of the combination. Such pixels have to be penalized in the processing. To ensure a more powerful attenuation of the possibly overexposed pixels, instead of a simple triangular hat function (e.g., as in Reference [3]) an asymmetric function of the form $w_{\text{cam}}(\rho) = \rho^\alpha(1-\rho)^\beta$ where $0 \le \rho \le 1$ and $1 \le \alpha < \beta$, is used. An example of such a

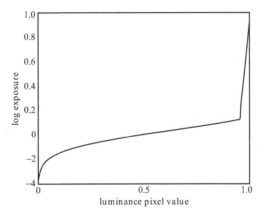

FIGURE 9.3

The estimated inverse response function g for the luminance channel of Nikon COOLPIX E4500.

weight function is given in Figure 9.2. An example of a camera response function for the luminance solved with this method is illustrated in Figure 9.3. When looking at the camera response one can observe an abrupt peak towards the right end of the plot. This is a result of the instability of the system for overexposed pixels. It becomes evident that such pixels cannot be directly included in the subsequent composition step. Instead, they have to be penalized with weights. As discussed in Reference [14] the issue of overexposure being present also in pixels whose value is lower than the maximum is not absent even for the conventional approaches dealing in the RGB space.

9.3.2 HDR Sensors

The importance and the potential of HDR imaging are further highlighted by the introduction of HDR hardware. Novel camera sensor technologies have recently been unveiled by a few companies. For the direct capture of HDR data several sensor designs exist [1], [15], [16]. Most HDR sensor designs focus on enhancing the dynamic range by adding to the maximum amount of light the sensor can capture without saturating, rather than enhancing sensitivity for low light levels. Reference [15] studies four different architectures: time-to-saturation, multiple capture, asynchronous self-reset with multiple capture, and synchronous self-reset with residue readout. Time-to-saturation and multiple capture offer extended dynamic range by various methods of varying exposure time. Varying exposure time does exactly what one would assume; it adapts the pixelwise integration time to pixel photocurrent enabling the use of long exposure times for small photocurrents and shorter exposure times for larger ones. In asynchronous self-reset the pixel voltage is sampled at regular intervals and pixels reset autonomously on saturation. Synchronous self-reset requires on-pixel counter to keep track of reset count. At the end of integration, the number of resets is combined with the residual voltage to estimate the total photocurrent.

None of the above-mentioned technologies have yet been made available to the general public. A couple of digital single-lens reflex (DSLR) cameras, however, have been introduced to the market. Fujifilm's S5 Pro and its new sensor is the first attempt of bringing direct extended dynamic range sensors to consumer, albeit professional, cameras. It improves the dynamic range of the captured photograph by including two sensor wells for each pixel on the sensor. The two wells have different sensitivities, where the less sensitive well starts reacting only when the normal well is saturated. The dynamic range improvement takes place at the bright end of the scene and the total dynamic range is according to Fuji comparable to traditional film. Another example of HDR imaging in a consumer camera is the Pentax K-7. Though not featuring a true HDR sensor, their 2009 released DSLR is probably the first commercial camera to provide in-camera HDR capture and tone mapping. An automated bracketing and composition program is made available, and the user is presented with a tone mapped HDR image.

9.4 Algorithms for HDR Composition

The history of HDR imaging is relatively short. The first articles dealing with the creation of HDR images were published in the beginning of the 1990s; see, for example, References [2] and [17]. Partially based on these previous efforts, the core of HDR imaging in RGB space was defined in Reference [3]. The vast majority of research done in the field has focused on optimizing the parameters (number of exposures, exposure times, etc.) [18] or developing different tone mapping methods [6]. Though parallel treatment of the RGB channels causes color distortion [3], with examples presented in Section 9.6, only very limited attention has been focused on addressing these problems. In one of these few studies, International Color Consortium (ICC) color profiles are suggested to be used in attempt to reach more realistic color reproduction in context with RGB HDR imaging [19]. While HDR composition techniques have practically been at a standstill, tone mapping has become a popular research topic after the introduction of Reference [3].

9.4.1 Monochromatic HDR Composition

The problem of HDR imaging and the state-of-the-art techniques proposed to solve this problem are discussed in Reference [1]. These techniques employ the same underlying concept of multiple exposure HDR. The starting point is an image sequence that fills the qualifications discussed in Section 9.3.1. With this prerequisite filled, the general line-up of the process, illustrated in the block diagram of Figure 9.4, is as follows:

- A response curve for each color channel of the camera needs to be solved. This needs to be done only once for a given imaging device.

- Based on this response curve and the exposure time, the different exposures can be linearized and the radiance map acquired utilizing a weighted average of the linearized data.

- The acquired HDR data needs to be stored in an efficient manner. This step is not necessary if the data is composed only for direct display.

- The HDR data needs to be compressed so that it can be displayed on a conventional LDR display or for example a printout. The compression of the visual range, gen-

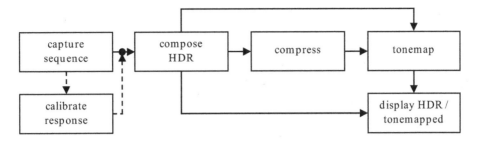

FIGURE 9.4

The block diagram of the general multiple exposure HDR imaging pipeline. Different paths indicate alternatives in the process, such as displaying a tone mapped HDR image on a normal display or alternatively displaying an HDR image on an HDR capable display.

erally known as tone mapping, can be skipped if the image is to be displayed on an HDR display device.

With the camera response function solved as described in Section 9.3.1 and the exposure times for each captured frame known, the logarithmic HDR radiance map for a monochromatic channel C can then be composed as a weighted sum of the camera output pixels values as follows:

$$\ln E_i^C = \frac{\sum_{i=1}^N w_C(z_i(\mathbf{x}_j))(g_C(z_i(\mathbf{x}_j)) - \ln \Delta t_i)}{\sum_{i=1}^N w_C(z_i(\mathbf{x}_j))}$$

In case of color images, the RGB channels are treated in parallel. This assumes that the interactions between channels are negligible, which is, as admitted in Reference [3], problematic to defend. As a result to the parallel treatment of the three channels, color distortions are in many cases introduced to the composed image. These have to be corrected by post-composition white-balancing which in turn may lead to colors that are not faithful to the original ones. Nevertheless, the approach works well enough to produce satisfactory results provided that the source sequence is very well aligned and noise is negligible.

9.4.2 Luminance-Chrominance HDR Composition

An alternative approach to parallel processing of the three RGB channels is rooted in the psychophysically inspired color models where the image is formed based on one component describing luminance and two components containing the chromatic information. Alone the luminance channel forms a grayscale representation of the scene, and as such, contains most of the information associated with an image. The two chrominance channels are nearly meaningless alone, and as such contain complementary information that is not per se crucial. This fact is also exploited in color image compression where the chromatic components are often downsampled compared to the luminance. The completely different nature of the luminance and the chrominance motivates a new approach for HDR composition. The method described here treats the luminance channel similarly to the conventional approach for a single channel of the RGB triplet and presents a more sound method for the composition of color.

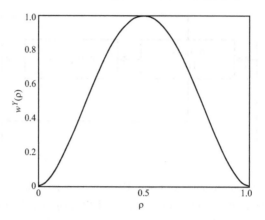

FIGURE 9.5

Weight function w^Y used for the luminance channel composition.

9.4.2.1   Problem Statement

Let $\zeta_i = [\zeta_i^Y, \zeta_i^U, \zeta_i^V]$, for $i = 1, 2, ..., N$, be a set of images in a luminance-chrominance space, captured with different exposure times Δt_i and with LDR, for $\zeta(\mathbf{x}) \in [0, 1] \times [-0.5, 0.5]^2$ where $\mathbf{x} = [x_1, x_2]$ is a pixel coordinate. The goal is to obtain a single HDR image $\tilde{\zeta} = [\tilde{\zeta}^Y, \tilde{\zeta}^U, \tilde{\zeta}^V]$ in the same color space. In the setting, the luminance and chrominance channels are treated separately. A precalibrated camera response function is used for the luminance channel, whereas a saturation-driven weighting is used for the chrominance channels.

9.4.2.2   Luminance Component Composition

The HDR luminance component is obtained by a pixelwise weighted average of the pixel log irradiance values defined according to Equation 9.4 as $\ln E_i(\mathbf{x}) = g(\zeta_i^Y(\mathbf{x})) - \ln \Delta t_i$. As observed in the previous section, pixels whose value is close to zero or unity carry little valuable information because of low SNR (underexposure) and clipping (overexposure), respectively. Such pixels are therefore penalized by employing weights during the composition. A polynomial function $w^Y(\rho) = \rho^\alpha (1 - \rho)^\beta$ with $0 \leq \rho \leq 1$, $\alpha = 2$, and $\beta = 2$ is used as a weight function, thus ensuring a smaller impact of the underexposed or overexposed pixels. An example of such a weight function for the luminance is shown in Figure 9.5. The logarithmic HDR luminance is obtained as follows:

$$\ln \tilde{\zeta}^Y(\mathbf{x}) = \frac{\sum_{i=1}^{N} w^Y(\zeta_i^Y(\mathbf{x}))(g(\zeta_i^Y(\mathbf{x})) - \ln \Delta t_i)}{\sum_{i=1}^{N} w^Y(\zeta_i^Y(\mathbf{x}))}. \tag{9.6}$$

Because of the nature of the camera response function g, the HDR luminance is obtained in logarithmic scale. After employing the natural exponential, the resulting values are positive, normally spanning $[10^{-4} \ 10^4]$, thus constituting truly a high dynamic range.

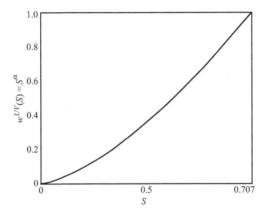

FIGURE 9.6

Weight function w^{UV} used for the composition of the chrominance channels.

9.4.2.3 Chrominance Components Composition

For the chrominance components no camera response is defined. Instead, the chrominance signals are weighted in relation to the level of color saturation. The higher the color saturation, the more the pixel contains valuable chromatic information, and thus the higher the weight. This is motivated by the fact that when a pixel is overexposed or underexposed it is always less saturated than it would be at the correct exposure. More specifically, $w^{UV}(S) = S^{\alpha}$ where $\alpha > 1$. In exhaustive experiments, $\alpha = 1.5$ has been found to be a good choice. A saturation-based chrominance weight function is illustrated in Figure 9.6. To preserve color information, the same weights are used for both chromatic components and any chromatic component $C \in \{U, V\}$ is composed as follows:

$$\tilde{\zeta}^C(\mathbf{x}) = \frac{\sum_{i=1}^{N} w^{UV}(S_i(\mathbf{x}))\zeta_i^C(\mathbf{x})}{\sum_{i=1}^{N} w^{UV}(S_i(\mathbf{x}))}, \tag{9.7}$$

where S_i denotes the saturation of ζ_i. It is pointed out that being a convex combination of the input chrominance signals, the range of $\tilde{\zeta}^C(\mathbf{x})$ is again in $[-0.5, 0.5]$. However, because of averaging, the possible number of distinct chrominance values is remarkably higher than in the original source sequence.

Another intuitively interesting approach would be to use the luminance-dependent weights w^y for both the luminance and the chrominance channels. Loosely this approach would be similar to the proposed saturation-driven weighting because of the fact that color saturation typically decreases with the luminance moving closer to the extremes. However, in experiments it was found that saturation-based weighting provides a better reproduction of color especially for saturated highly exposed or low-lit details.

9.4.2.4 HDR Saturation Control

After the composition, the luminance component, or more precisely the radiance map, has a range much higher than the original [0 1]. The newly acquired pixel irradiance values span an extended range, while the chrominance component composition, being a convex combination of the components, results in a denser sampling of the original range. This

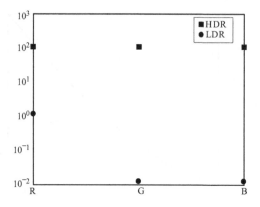

FIGURE 9.7

The illustration of the desaturating effect of the luminance-chrominance HDR composition.

presents a problem, because luminance-chrominance color spaces represent color as difference from the gray luminance component. If the HDR image were directly mapped to RGB for display, the result would be an effectively grayscale image because the differences represented by chrominance spanning $[-0.5\ 0.5]$ cannot form saturated colors in relation to a luminance component with maximum values up to 10^5. This can be practically demonstrated with the following example.

A red pixel in the RGB color space defined as $z = [1\ 0\ 0]$ is transformed into the opponent color space using the matrix A_{opp} defined in Section 9.2.2, thus providing its luminance-chrominance representation $\zeta = [1/3\ 1/2\ 1/4]$. Assuming that the HDR composition process results in a pixel irradiance of 100 and the chrominances of ζ, the HDR pixel is defined in opponent color space as $\tilde{\zeta} = [100\ 1/2\ 1/4]$. The application of the inverse transformation leads to an RGB pixel $\tilde{z} = [100\frac{2}{3}\ 99\frac{2}{3}\ 99\frac{2}{3}]$. The resulting values form a gray pixel, as the relative differences between components are not transmitted through the luminance increase. This is of course unwanted, as the correct result would be $\tilde{z} = [100\ 0\ 0]$. The phenomenon is illustrated in Figure 9.7.

Obviously, if one wants to display or process the HDR data on a device with RGB HDR input, a *saturation control* method has to be introduced. For this purpose the following approach can be used. Let $\mu(\mathbf{x})$ defined as follows:

$$\mu(\mathbf{x}) = \frac{\tilde{\zeta}^Y \sum_{i=1}^N w^{UV}(S_i(\mathbf{x}))}{\sum_{i=1}^N w^{UV}(S_i(\mathbf{x}))\, \zeta_i^Y(\mathbf{x})}. \tag{9.8}$$

be the scalar proportionality factor between the HDR luminance $\tilde{\zeta}^Y$ and the weighted average of the LDR luminances $\zeta_i^Y(\mathbf{x})$ with weights $w^{UV}(S_i(\mathbf{x}))$. In other words, $\mu(\mathbf{x})$ is a pixelwise scaling parameter defining the ratio between the weighted average of the original pixel values and the pixel irradiance values obtained through the HDR composition process. The HDR image in RGB space can now be obtained by the normalized inverse color transformation

$$\tilde{\mathbf{z}}(\mathbf{x}) = \tilde{\zeta}(\mathbf{x}) \begin{bmatrix} 1 & 0 & 0 \\ 0 & \mu(\mathbf{x}) & 0 \\ 0 & 0 & \mu(\mathbf{x}) \end{bmatrix} \mathbf{B}.$$

(a)   (b)

FIGURE 9.8 (See color insert.)

The HDR image is composed in luminance-chrominance space and transformed: (a) directly to RGB for tone mapping, (b) to RGB utilizing the presented saturation control, with subsequent tone mapping is done in RGB.

where the normalization is realized by multiplication against the diagonal matrix $diag\,(1, \mu\,(\mathbf{x}), \mu\,(\mathbf{x}))$ which scales the two chrominances $\tilde{\zeta}^U$ and $\tilde{\zeta}^V$ yielding a value of saturation which matches the full dynamic range achieved by $\tilde{\zeta}^Y$. Indeed, the weights $w^{UV}\,(S_i\,(\mathbf{x}))$ in Equations 9.7 and 9.8 are exactly the same. From the diagonal scaling matrix it can be observed that the value of $\mu\,(\mathbf{x})$ does not have any influence on the luminance of $\tilde{\mathbf{z}}\,(\mathbf{x})$. Likewise, from the definition of hue, $H = \arctan \frac{\zeta^U}{\zeta^V}$, it is clear that the hue is left intact and the only thing altered is the saturation.

Though the need for saturation control cannot be visualized on conventional displays let alone printout, in full HDR, the issue can be conveyed through the chain of RGB tone mapping. This is visualized in Figure 9.8a which shows an image composed in luminance-chrominance space and transformed to RGB for tone mapping. Due to reasons explained above, virtually all color is lost in the transformation. In Figure 9.8b, the same luminance-chrominance HDR image is transformed into RGB utilizing the saturation control described here. The colors are reproduced faithfully underlining the hue preservation properties of the luminance-chrominance method.

9.5 Algorithms for HDR Display

The dynamic range of an HDR image often spans more than five orders of magnitude, of which a conventional display is able to visualize a maximum of two orders of magnitude. This presents a problem, because while HDR images are becoming more and more available, HDR displays lag behind. The problem is then to fit the greater dynamic range of an HDR image into the limited gamut of a display device. The simplest solution is to linearly scale the data and while a simple linear scaling very rarely produces acceptable results, applying gamma curves or some more sophisticated mapping procedures will likely do better.

(a)   (b)

FIGURE 9.9 (See color insert.)

An HDR image of a stained glass window from Tampere Cathedral, Finland: (a) the image is linearly scaled for display, and (b) the tone mapping was done with the method described in Section 9.5.2.1.

The problem of tone mapping is essentially one of compression with preserved visibility. A linear scene-to-output mapping of an HDR image produces results similar to the image shown in Figure 9.9a. The tone mapped version of the same scene, using the method described in Section 9.5.2.1, is shown in Figure 9.9b. As can be seen, an HDR image is useless on an LDR display device without tone mapping.

The history of tone mapping dates back to much longer than the introduction of HDR imaging. As the dynamic range of the film has always exceeded that of the photographic paper of the era, manual control has been necessary in the development process. The darkroom was introduced in the late 19th century and since that time, a technique called dodging and burning has been used to manipulate the exposures of the photographic print.

The same idea has later been transported into controlling the visibility of digital HDR images. In this context the operation is known as tone mapping. The very basic aim of tone mapping or *tone reproduction* (the two terms are interchangeable) is to provoke the same response through the tone mapped image that would be provoked by the real scene. In other words, matching visibility. In the darkroom the procedure was of course done manually, but for digital purposes the ideal solution would fit every problem and not need any human interaction. So far the optimal tone mapping method has not been invented and one has to choose the method dependent on the problem. Many of the methods require parameter adjustment, while some are automatic. The results of the tone mapping depend highly on the chosen method as well as the parameters used. New tone mapping approaches are introduced almost monthly and the aim of the scientific community is, as usual, to develop more and more universal approaches well suited for the majority of problems. Some of the major tone mapping algorithms were reviewed in Reference [6].

9.5.1 Tone Mapping of Monochromatic HDR

Tone mapping methods can be divided in various ways. One of the most frequently used divisions is the *local* versus *global* division. Additionally, methods operating in some transform domain can be classified in their own classes. Reference [1] distinguished global, local, frequency domain, and gradient domain operators.

The global methods are the most simple class of tone mapping operators (TMOs). They all share a couple of inherent properties; the same mapping is applied for all the pixels and the tone mapping curve is always monotonic in nature. The mapping has to be monotonic in order to avoid disturbing artifacts. This imposes a great limitation on the compression / visibility preservation combination. As the usual target is to map an HDR scene into the range of standard eight-bit representation, only 256 distinct brightness values are available. Global methods excel generally in computational complexity, or the lack-thereof. As the same mapping is applied on all the pixels, the operations can be done very efficiently. On the other hand, for scenes with very high dynamic range, the compression ability of the global class may not be sufficient.

The local methods are able, to an extent, to escape the limitations met with global TMOs. In general, local methods do not rely on image-wide statistics. Instead, every pixel is compressed depending on its luminance value and the values of a local neighborhood. Often local methods try to mimic properties of the HVS; the eye is known to focus locally on an area of a scene forming an independent adaptation based on the local neighborhood contents. As a result, the cost of more flexible compression is in the computational complexity. The number of necessary computations goes up with the number of local adaptation neighborhoods. Also the higher, local compression of scene brightness may at times lead to halo-like artifacts around objects.

The transform domain operators are distinguished from global as well as local methods by the fact that they operate on the data in some domain other than the conventional spatial one. Frequency domain operators compress data, as the name suggests, utilizing a frequency-dependent scheme. The first digital tone mapping operator was already published in 1968 and it was a frequency domain one [20]. Many of the properties of modern frequency domain operators are inherited from this approach. Gradient domain operators rely on the notion that a discrimination between illuminance and reflectance is for many scenes relatively well approximated by the gradient approach. This is supported by the notion that an image area with a high dynamic range usually manifests a large gradient between neighboring pixels. The follow-up is a tone mapping operator functioning on the differentiation domain, using gradient manipulation for dynamic range reduction.

The majority of tone mapping methods work on the spatial domain and are therefore categorized under either local or global umbrella, depending on the nature of the compression. Finally it is noted that as there are numerous methods for tone mapping of HDR scenes and their basic functionality is, apart from the core idea of range compression, very different from one method to another, it is not meaningful to give a detailed description of an example implementation. All the methods are well described in the literature and the methods implemented for the luminance-chrominance approach of this thesis are described in detail in Section 9.5.2.

9.5.2 Tone Mapping of Luminance-Chrominance HDR

A number of tone mapping methods working in RGB space exist. These techniques can easily be adapted for the luminance range reduction. However, the term tone mapping would be questionable in this context, as tone is usually used in connection with color. In this chapter such a luminance range reduction operator is denoted as $\mathcal{T}$. Its output

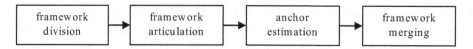

FIGURE 9.10

The block diagram of the anchoring-based tone mapping process. The blocks represent the four central stages of the procedure.

is a luminance image with range $[0, 1]$, that is, $\mathscr{T}(\tilde{\zeta}^Y)(\cdot) \in [0, 1]$. As for the chromatic channels, a simple, yet effective approach is presented.

For the compression of the luminance channel, two global luminance range reduction operators are presented. The selection is limited to global operations simply because thus far, local operators have not been able to produce results faithful to the original scene. It should be noted that the majority of tone mapping methods developed for RGB can be applied more or less directly for the compression of the luminance channel as if it were a grayscale HDR image. As such, the continuous development of tone mapping methods for RGB HDR images also benefits the compression of luminance-chrominance HDR data.

9.5.2.1 Anchoring-Based Compression

The first method is based on an anchoring theory of lightness perception and was presented for RGB in Reference [7]. The core idea of the method is to divide the luminance image into frameworks based on the theory of lightness perception [21]. This theory states that in order to relate the luminance values to perceived lightness, one or more mappings between the luminance and perceived values on the grayscale, an *anchor*, have to be defined. A block diagram of the method is depicted in Figure 9.10.

The division into frameworks is done utilizing the standard K-means algorithm and is based on the histogram of $\log_{10}$ luminance. Preliminary centroids are based within one $\log_{10}$ unit of each other. In subsequent steps the clusters based on these centroids are fused if the distance between two centroids becomes less than one and the cluster contains no pixels at all or no pixels with probability more than 0.6 of belonging to that cluster. The probability is calculated as a function of distance from the cluster centroid. When this algorithm is left to converge, the result is a division into usually no more than three frameworks with similar intensities. At the last stage of the framework division the acquired frameworks are articulated based on the dynamic range of an individual framework. A framework with a high dynamic range (above one $\log_{10}$-unit) has the maximum articulation, and the articulation goes down to zero as the dynamic range goes down to zero. The articulation is imposed on the frameworks to make sure a framework consisting of background does not play a significant role in the computation of the net lightness. At the penultimate stage the frameworks are anchored so that in an individual framework, the 95th percentile is mapped as white. Ultimately the reduced range luminance is calculated by subtracting the anchored frameworks from the original luminance.

In terms of detail visibility, the anchoring-based method is superior to some other methods. Even scenes with extremely high contrast are squeezed into the range of a consumer level liquid crystal display (LCD) display. For some scenes this may lead to results appearing slightly unnatural, as global contrast is reduced very severely. The fact that the described method allows no manual tuning serves as both a drawback and a benefit; for most images the majority of the scene is brought visible in a believable manner but for images containing extremely high dynamic range, the lack of global contrast may at times lead to results appearing slightly too flattened.

9.5.2.2 Histogram-Based Compression

Another method is based on a relatively simple histogram adjustment technique. It coarsely approximates the method described in Reference [5]. It is noted, however, that most of the sophisticated ideas introduced in Reference [5] are sacrificed in the implementation for speed and simplicity. In the histogram adjustment method the first step is to clip the HDR luminance image values to the lower limit of the HVS ($10^{-4}cd/m^2$). Then the luminance image can be downsampled with a factor of eight (the downsampling has to be preceded by a corresponding low-pass filtering to avoid aliasing). The downsampling is a simplistic approximation of the foveal fixation phenomenon encountered in adaptation of the eye. Then a luminance histogram is calculated from the downsampled image. A cumulative distribution is then defined as $P(b) = \sum_{b_i<b} f(b_i)/\sum_{b_i} f(b_i)$, where $f(b_i)$ is the frequency count for the bin number i. Now if the goal were to equalize the probability of each brightness value, this could be achieved by histogram equalization

$$B_{disp} = \log(L_{dmin}) + [\log(L_{dmax}) - \log(L_{dmin})]P(b),$$

where L_{dmax} and L_{dmin} are, respectively, the maximum and minimum display luminance values and B_{disp} is the output display brightness. However, while histogram equalization compresses dynamic range in sparsely populated regions of the histogram, it also expands contrast in highly populated zones resulting in exaggerated contrast. This can be avoided by imposing a linear ceiling on the contrast produced by the method. The linear ceiling can be defined as follows:

$$f(b) \leq \frac{\sum_{b_i} fb_i \Delta b}{\log(L_{dmax}) - \log(L_{dmin})},$$

where Δb is the histogram bin stepsize. This ceiling has to be imposed on the histogram equalization in an iterative manner, because when values exceeding the ceiling are truncated, the histogram sample count $\sum_{b_i} fb_i$ is altered thus altering also the ceiling. The display mapping is then obtained by using the newly acquired luminance histogram (or its cumulative distribution) in the histogram equalization formula defined above.

9.5.2.3 Chrominance Component Compression

The sRGB gamut does not allow the rendition of very dark or very bright vivid and saturated colors which exist in real scenes and which are captured in HDR images. Therefore, there exists a need for chromatic tone mapping. In this approach, in order to get faithful colors that fit into the sRGB gamut the hue is kept intact by sacrificing saturation. Introducing a scaling factor δ for the two chrominance values will then not change the hue, but will scale down the saturation. The scheme that is used to guarantee legal sRGB values is embedded in the color space transformation itself and described below.

Let $\mathbf{B} = \mathbf{A}^{-1}$ be the luminance-chrominance to RGB transformation matrix. Let us also define the gray (achromatic) image and its chromatic complement image in RGB space as follows:

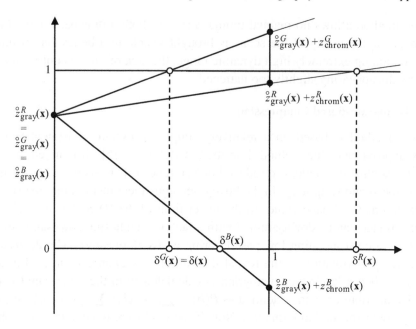

FIGURE 9.11

Illustration of the definition of the chromatic tone mapping parameter δ.

$$\mathring{\mathbf{z}}_{\mathsf{gray}}(\mathbf{x}) = \left[\mathring{z}^R_{\mathsf{gray}}(\mathbf{x}) \; \mathring{z}^G_{\mathsf{gray}}(\mathbf{x}) \; \mathring{z}^B_{\mathsf{gray}}(\mathbf{x}) \right] = \left[\begin{array}{c} \mathscr{T}\left(\tilde{\zeta}^Y \right)(\mathbf{x}) \\ 0 \\ 0 \end{array} \right]^T \mathbf{B}, \qquad (9.9)$$

$$\mathbf{z}_{\mathsf{chrom}}(\mathbf{x}) = \left[z^R_{\mathsf{chrom}}(\mathbf{x}) \; z^G_{\mathsf{chrom}}(\mathbf{x}) \; z^B_{\mathsf{chrom}}(\mathbf{x}) \right] = \left[\begin{array}{c} 0 \\ \tilde{\zeta}^U(\mathbf{x}) \\ \tilde{\zeta}^V(\mathbf{x}) \end{array} \right]^T \mathbf{B}. \qquad (9.10)$$

It can be noted that $\mathring{\mathbf{z}}_{\mathsf{gray}}(\mathbf{x})$ is truly a gray image because in RGB to luminance-chrominance transforms $b_{1,1} = b_{1,2} = b_{1,3}$. Then a map $\delta \geq 0$ is needed, such that

$$\mathring{\mathbf{z}}(\mathbf{x}) = \mathring{\mathbf{z}}_{\mathsf{gray}}(\mathbf{x}) + \delta(\mathbf{x}) \mathbf{z}_{\mathsf{chrom}}(\mathbf{x}) \in [0, 1]^3. \qquad (9.11)$$

It can be defined by $\delta(\mathbf{x}) = \min \left\{ 1, \delta^R(\mathbf{x}), \delta^G(\mathbf{x}), \delta^B(\mathbf{x}) \right\}$, where

$$\delta^R(\mathbf{x}) = \begin{cases} \mathring{z}^R_{\mathsf{gray}}(\mathbf{x}) / -z^R_{\mathsf{chrom}}(\mathbf{x}) & \text{if } z^R_{\mathsf{chrom}}(\mathbf{x}) < 0, \\ (1 - \mathring{z}^R_{\mathsf{gray}}(\mathbf{x})) / z^R_{\mathsf{chrom}}(\mathbf{x}) & \text{if } z^R_{\mathsf{chrom}}(\mathbf{x}) > 0, \\ 1 & \text{if } z^R_{\mathsf{chrom}}(\mathbf{x}) = 0, \end{cases} \qquad (9.12)$$

and δ^G and δ^B are defined analogously. Thus, $\delta(\mathbf{x})$ is the largest scalar smaller or equal to one, which allows the condition in Equation 9.11 to hold. Figure 9.11 illustrates the definition of $\delta(\mathbf{x})$. From the figure it is easy to realize that the hue, that is, the angle of the vector, of $\mathring{\mathbf{z}}(\mathbf{x})$ is not influenced by δ, whereas the saturation is scaled proportionally to it. Roughly speaking, the low-dynamic range image $\mathring{\mathbf{z}}(\mathbf{x})$ has colors which have the same hue as those in the HDR image $\tilde{\zeta}$ and which are desaturated as little as is needed to fit within the sRGB gamut.

It is now obvious that the tone mapped LDR image can be defined in luminance-chrominance space as follows:

$$\mathring{\zeta}(\mathbf{x}) = \left[\mathscr{T}\left(\tilde{\zeta}^Y\right)(\mathbf{x}) \quad \delta(\mathbf{x})\,\tilde{\zeta}^U(\mathbf{x}) \quad \delta(\mathbf{x})\,\tilde{\zeta}^V(\mathbf{x}) \right]. \tag{9.13}$$

The tone mapped luminance-chrominance image $\mathring{\zeta}$ can be compressed and stored directly with an arbitrary method (for example, DCT-based compression, as in JPEG), and for display transformed into RGB using the matrix **B**. It is demonstrated that this approach yields lively, realistic colors.

9.5.3 HDR Display Hardware

Obviously even the most elaborate tone mapping algorithms cannot match the sensation of directly viewing a HDR scene. The limitations in dynamic range of modern displays relate to the limited maximum presentable intensity and more importantly the minimum presentable intensity. Conventional LCDs and even plasma screens both suffer from the phenomenon of some light being present in pixels that should in fact be completely black. For LCDs this is due to the LCD screen not being able to completely block the backlight from reaching the observer. For plasma screens grayish blacks are caused by the precharge induced in the plasma needed to achieve an acceptable response time.

In recent years some advances have been made in display technology. Reference [8] presents the fundamentals of a conceptual adaptive backlight LCD display. In this display, the backlight is implemented with an light emitting diode (LED) matrix. Such backlight can then be locally dimmed and areas of the backlight can even be completely turned off to achieve true black in some image areas while simultaneously displaying maximum intensity in others, yielding a theoretical infinite contrast ratio. The adaptive backlight was later formulated into two technology packages by Dolby. Dolby Contrast and Dolby Vision offer different levels of backlight control and accuracy and are at the time of writing waiting to be licensed by the first display manufacturer. Meanwhile numerous display manufacturers have introduced LED-based LCD screens of their own, significantly extending the dynamic range (contrast ratio) of the conventional LCD television. Many of the caveats of current display technologies will be antiquated by the maturation of organic light-emitting diode (OLED) technology and similar technologies where all pixels control their own intensity and can be turned on or off independently of neighboring image elements.

FIGURE 9.12 (See color insert.)

Four exposures from the real LDR sequence used to compose an HDR image. The exposure times of the frames are 0.01, 0.0667, 0.5, and 5 seconds.

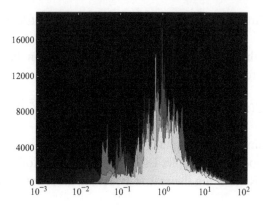

FIGURE 9.13 (See color insert.)

The histograms of the HDR image composed in luminance-chrominance space and transformed into RGB. Histogram is displayed in logarithmic scale.

9.6   Examples

This section presents examples of HDR scenes imaged and visualized with methods described in this chapter. In the experiments, both real and synthetic images are used. Special attention is focused on the effects of noise and misalignment, both realistic components of distortion when HDR scenes are imaged with off-the-shelf components. Extensive performance comparisons against state-of-the-art RGB methods can be found in Reference [14].

9.6.1   HDR from Real LDR Data

To display the basics of luminance-chrominance HDR composition, a sequence of 28 frames was captured with a remotely operated Canon EOS 40D set on a tripod. The images were captured with ISO 100, thus having very low noise. The image sequence is composed into HDR in opponent color space with the methods described in this chapter. First a response function is solved for the camera as described in Section 9.3.1. Using the solved response function an HDR image in luminance-chrominance space is acquired as described in Section 9.4.2. The exposure times for the captured sequence range from 0.01s to 5.00s; Figure 9.12 shows extracts of this sequence. The histogram for the R, G, and B channels of the HDR image is shown on logarithmic axis in Figure 9.13. Figures 9.14a and 9.14b show the tone mapping results achieved using the two methods described in this chapter. Figure 9.14c shows an adaptive logarithmic tone mapping result of the RGB transformed HDR image achieved with the method described in Reference [22]. Note that this tone mapping method was implemented as in Reference [23], with exposure adjustment, bias, shadow luminance, and contrast parameters set to 0.64, 0.77, 1.00, and 0.00, respectively.

9.6.2   HDR from Synthetic LDR Data

The HDR image acquired as described in Section 9.6.1 can also be transformed into RGB space using the approach discussed in Section 9.4.2.4. The HDR image transformed into

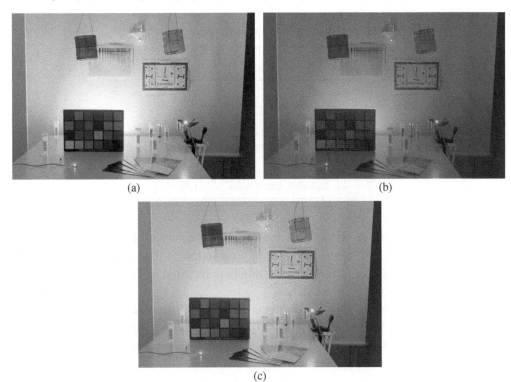

(a)

(b)

(c)

FIGURE 9.14 (See color insert.)

The HDR image composed in luminance-chrominance space from real data and tone mapped using: (a) the histogram adjustment technique presented in Section 9.5.2.2, (b) the anchoring technique presented in Section 9.5.2.1, and (c) the adaptive logarithmic technique presented in Reference [22] applied in RGB.

RGB space can be used as ground truth for subsequent synthetic composition examples. The HDR image is assumed to represent the true scene irradiance E, with an HDR pixel value $E(\mathbf{x}) = 1$ equal to a scene irradiance of $1\ W/m^2$. Where comparison with RGB techniques is provided, the tone mapping method described in Reference [22] is used to process the images composed both in RGB and in luminance-chrominance to allow for relevant visual comparison of the results.

9.6.2.1 Generating Synthetic LDR Data

In practice, the LDR frames are captured sequentially, one after the other, each time with a different exposure time. Particularly with a hand-held camera, this results in some misalignment between the frames. Unless the scene is perfectly static, misalignment can occur even when the camera is held on a tripod, simply because of minor changes in the scene content (movement of the subjects, etc.). In simulations, to create realistic misalignment of the LDR frames, the HDR image has been sampled using linear interpolation after a set of geometrical deformations. These include rotation, translation, as well as parabolic stretch, faithfully mimicking the apparent nonrigid deformation produced by the optical elements of the camera. Although using random deformation parameters to generate each frame, the actual misalignment is quantified by tracking the position of the three light-emitting diodes

TABLE 9.1
Average LED position variances σ_{LED}^2 for the source image sequences of Figures 9.16 and 9.18.

Sequence	σ_{LED}^2
Reference	0
1st degraded, no noise	0.2289
2nd degraded, added noise	0.2833
Only added noise	0
3rd degraded, no noise	2.4859
4th degraded, added noise	1.5339

(LEDs, used as markers in the HDR scene) over the sequence of frames. In particular, for each sequence of LDR frames, the standard deviation of the position of the center of each of the three LEDs was computed. The numbers reported in Table 9.1 are the average variances for position of the three LEDs.

A synthetic LDR image sequence is obtained by simulating acquisition of the above HDR image into LDR frames. The LDR frames are produced using Equations 9.3 and 9.4 applied separately to the R, G, and B channels. The camera response function f used in the equations is defined as follows:

$$f\left(e\left(\mathbf{x}\right)\right) = \max\left\{0, \min\left\{1, \left\lfloor\left(2^8 - 1\right)\kappa e\left(\mathbf{x}\right)\right\rceil\left(2^8 - 1\right)^{-1}\right\}\right\}. \tag{9.14}$$

where $\kappa = 1$ is a fixed factor representing the acquisition range of the device (full well) and the $\lfloor\cdot\rceil$ brackets denote the rounding to the nearest integer, thus expressing 8-bit quantization. This is a simplified model which, modulo the quantization operation, corresponds to a linear response of the LDR acquisition device.

Noise is also introduced to the obtained LDR images using a signal-dependent noise model of the sensor [24], [25]. More precisely, noise corrupts the term $\kappa e\left(\mathbf{x}\right)$ in the acquisition formula, Equation 9.14, which thus becomes

$$f\left(e\left(\mathbf{x}\right)\right) = \max\left\{0, \min\left\{1, \left\lfloor\left(2^8 - 1\right)\varepsilon\left(\mathbf{x}\right)\right\rceil\left(2^8 - 1\right)^{-1}\right\}\right\}, \tag{9.15}$$

where $\varepsilon\left(\mathbf{x}\right) = \kappa e\left(\mathbf{x}\right) + \sqrt{a\sigma\left(\kappa e\left(\mathbf{x}\right)\right) + b}\,\eta\left(\mathbf{x}\right)$ and η is standard Gaussian noise, $\eta\left(\mathbf{x}\right) \sim \mathcal{N}\left(0, 1\right)$. In the experiment, parameters a=0.004 and b=0.02^2 are used; this setting corresponds to the noise model of a Fujifilm FinePix S9600 digital camera at ISO1600 [25]. Note that the selected noise level is relatively high and intentionally selected to clearly visualize the impact of noise on the HDR compositions.

9.6.2.2 Composition from Synthetic LDR Frames

Generating the LDR frames synthetically provides exact knowledge of the camera response. Thus, the logarithmic inverse g of the camera response function f used in the composition is the logarithm $g(x) = \ln x$. In the case of synthetic LDR frames, when processing in RGB, this response is used for all channels, processed in parallel, to compose the original scene. When processing synthetic LDR frames in a luminance-chrominance space, the

FIGURE 9.15

The four synthetic LDR frames used to create the HDR images shown tone mapped in the first row of Figure 9.16. The exposure times used in creating the frames are 0.064, 0.256, 1.024, and 4.096 seconds.

same inverse response function is used for the luminance channel, while chrominance channels are processed as described in Section 9.4.2.3. Since RGB composition is essentially an inverse of the acquisition process followed to obtain the synthetic LDR frames, the RGB composition result is, apart from some minute quantization-induced differences, perfect. For luminance-chrominance this cannot be expected, since the luminance channel is always obtained as a combination of the RGB channels, which due to clipping reduces the accuracy of the response. As will be shown, however, the reduced accuracy does not compromise the actual quality of the composed HDR image. Instead, the luminance-chrominance approach leads to much more accurate composition when the frames are degraded by noise or misaligned.

Because of noisy image data, the normalized saturation definition given in Section 9.2.3.1 modified with a regularization term τ is employed. The regularized saturation is then obtained as $S = \sqrt{(\zeta^U)^2 + (\zeta^V)^2}/\sqrt{(\zeta^Y)^2 + \tau^2}$ with $\tau = 0.1$. Without regularization, such saturation would become unstable at low luminance values, eventually resulting in miscalculation of the weights for the composition of the chrominance channels.

9.6.2.3 Experiments and Discussion

The trivial case of HDR composition from perfect synthetic frames shown in Figure 9.15 is provided for reference. Further, two different levels of misalignment are presented with and without added noise. Also, a perfectly aligned case is considered with added noise. For every case, images composed in both LCR and RGB space and tone mapped with the method of Reference [22] (exposure adjustment 0.64, bias 0.77, shadow luminance 1.00, contrast 0.00) in RGB space are presented. This demonstrates the ability of luminance-chrominance composition to function under imperfect conditions.

As can be seen in Figures 9.16a and 9.16b, the HDR compositions obtained using uncorrupted frames are, except a slight overall tonal difference, identical. Both LCR and RGB perform very well with synthetic frames with no degradation. Figures 9.16c and 9.16d show tone mapped HDR images composed using frames subjected to simulated camera shake. The average variance of the LED positions used to measure sequence stability is 0.2289 pixels, as shown in Table 9.1. The luminance-chrominance composed frame is again visually pleasing whereas in its RGB counterpart some significant color artifacts, such as reddish and blueish artifacts on the upper edge of the color chart, can be seen. This is further illustrated in Figure 9.17, which displays magnified fragments of Figure 9.16. Similar artifacts can also be seen on the upper-right edge of the white-screen. These artifacts are caused by parallel processing of the RGB channels. Figures 9.16e and 9.16f show composition results

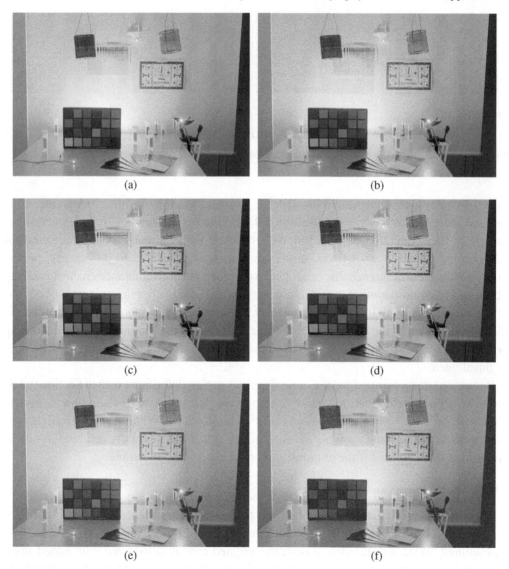

(a)　　　　　　　　　　　　　　　(b)

(c)　　　　　　　　　　　　　　　(d)

(e)　　　　　　　　　　　　　　　(f)

FIGURE 9.16

Experiments with synthetic data: (a,b) luminance-chrominance and RGB images composed using synthetic LDR data subject to no degradation, (c,d) images composed using sequence subject to misalignment with measured average LED position variance of 0.2289, and (e,f) images composed using sequence subject to misalignment with measured average LED position variance of 0.2833 as well as noise.

for images degraded by misalignment (average LED position variance 0.2833 pixels) and noise. Though masked somewhat by the noise, color artifacts can again be witnessed, for example in the upper right corner of the color chart and in the vertical shelf edge found in the left side of the scene. It is also evident that both these compositions suffer from noise, which is particularly obvious in dark regions.

Figure 9.18 shows similar behavior as discussed above. Figures 9.18a and 9.18b show composition results for a synthetic sequence degraded by noise; again, the comparison

FIGURE 9.17

Magnified details extracted from the images of Figure 9.16: (a,b) luminance-chrominance and RGB images composed using synthetic LDR data subject to no degradation, (c,d) images composed using sequence subject to misalignment with measured average LED position variance of 0.2289, and (e,f) images composed using sequence subject to misalignment with measured average LED position variance of 0.2833 as well as noise.

favors the luminance-chrominance composed image. Figures 9.18c and 9.18d show composed images obtained using the source sequence degraded by camera shake with the average variance of 2.6858 pixels. The luminance-chrominance composed image, apart from slight blurriness, seems visually acceptable whereas the RGB image suffer from color artifacts present at almost all of the edges. Among the worst are the greenish artifacts at the upper-right edge of the white-screen and red distortion on the wires of the LED at the upper part of the scene, as further illustrated in Figure 9.19 which displays magnified fragments of

FIGURE 9.18

Experiments with synthetic data: (a,b) luminance-chrominance and RGB images composed using synthetic LDR data subject to noise, (c,d) images composed using sequence subject to misalignment with measured average LED position variance of 2.4859, and (e,f) images composed using sequence subject to misalignment with measured average LED position variance of 1.5339 as well as noise.

Figure 9.18. Figures 9.18e and 9.18f show the composition results for a sequence degraded by noise and camera shake with average variance of 1.5339 pixels. Again, both noise and color artifacts are very much present in the RGB composed image, whereas the luminance-chrominance composed image handles the imperfect conditions visibly significantly better.

It is interesting to comment about the blueish colored noise visible on the darker parts of the HDR images produced by the RGB composition of noisy LDR frames. First, as can be seen in Figure 9.15, these are areas which remain rather dark even in the frame

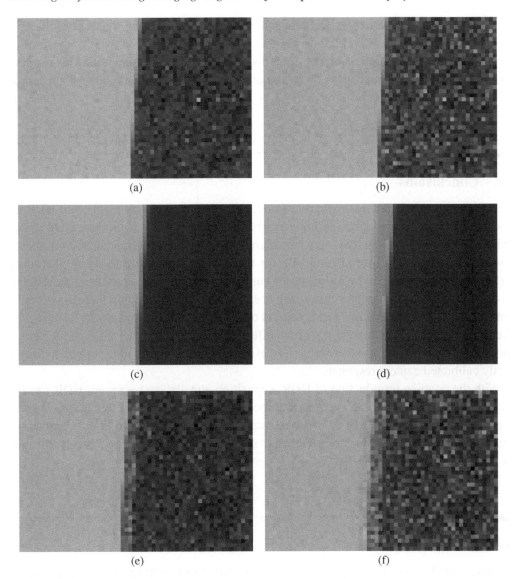

FIGURE 9.19

Magnified details extracted from the images of Figure 9.18: (a,b) luminance-chrominance and RGB images composed using synthetic LDR data subject to noise, (c,d) images composed using sequence subject to misalignment with measured average LED position variance of 2.4859, and (e,f) images composed using sequence subject to misalignment with measured average LED position variance of 1.5339 as well as noise.

with the longest exposure ($\Delta t = 4.096$ s). Second, as can be observed in Figure 9.14c, the scene is dominated by a cream yellow cast, mainly due to the tone of the lamp used for lighting. Thus, in each LDR frame, in these areas the blue component is the one with both the lowest intensity and the poorest signal-to-noise ratio. Because of the way weights are defined for the composition of these dark areas, only the longest exposed frame contributes with significant weights. This situation is quite different from that of the other parts of the image, which are instead produced as an average of two or more frames, one of which is properly exposed. In particular, for the darkest components, it is the right tail of the noise distribution which is awarded larger weights (see Figure 9.5). Moreover, because of the clipping at zero in Equation 9.15, the noise distribution itself is asymmetric. This results in a positive bias in the composition of the darker parts, which causes the blueish appearance of the noise over those parts.

9.7 Conclusions

This chapter presented methods for capture, composition, and display of color HDR images. In particular, composition techniques which effectively allow to produce HDR images using LDR acquisition devices were considered. Composition in luminance-chrominance space is shown to be especially suitable for the realistic imaging case where the source LDR frames are corrupted by noise and/or misalignment. In addition to being more robust to degradations, the luminance-chrominance approach to HDR imaging focuses special attention on the faithful treatment of color. As opposed to traditional methods working with the RGB channels, this method does not suffer from systematic color artifacts in cases of misaligned data nor color balancing errors usually induced among other things by imperfectly calibrated camera response.

With the ongoing introduction of HDR acquisition and display hardware, HDR imaging techniques are set to gain an even more important role in all parts of computational photography and image processing. With new applications, both industrial and commercial, introduced nearly daily, it is not farfetched to say that HDR will in more than one meaning be "the new color television."

Acknowledgment

This work was in part supported by OptoFidelity Ltd (www.optofidelity.com) and in part by the Academy of Finland (project no. 213462, Finnish Programme for Centres of Excellence in Research 2006-2011, project no. 118312, Finland Distinguished Professor Programme 2007-2010, and project no. 129118, Postdoctoral Researcher's Project 2009-2011).

Figures 9.2, 9.3, 9.5, 9.6, and 9.11 are reprinted from Reference [14], with the permission of John Wiley and Sons.

References

[1] E. Reinhard, G. Ward, S. Pattanaik, and P. Debevec, *High Dynamic Range Imaging*. San Francisco, CA, USA: Morgan Kaufmann Publishers, November 2005.

[2] S. Mann and R. Picard, "Being 'undigital' with digital cameras: Extending dynamic range by combining differently exposed pictures," in *Proceedings of the IS&T 46th Annual Conference*, Boston, MA, USA, May 1995, pp. 422–428.

[3] P. Debevec and J. Malik, "Recovering high dynamic range radiance maps from photographs," in *Proceedings of the 24th Conference on Computer Graphics and Interactive Techniques*, New York, USA, August 1997, pp. 369–378.

[4] T. Mitsunaga and S. Nayar, "Radiometric self calibration," in *Proceedings of the IEEE Conference on Computer Vision and Pattern Recognition*, Fort Collins, CO, USA, June 1999, pp. 1374–1380.

[5] G. Larson, H. Rushmeier, and C. Piatko, "A visibility matching tone reproduction operator for high dynamic range scenes," *IEEE Transactions Visual and Computer Graphics*, vol. 3, no. 4, pp. 291–306, October 1997.

[6] K. Devlin, A. Chalmers, A. Wilkie, and W. Purgathofer, "Tone reproduction and physically based spectral rendering," in *Proceedings of EUROGRAPHICS*, Saarbrucken, Germany, September 2002, pp. 101–123.

[7] G. Krawczyk, K. Myszkowski, and H. Seidel, "Lightness perception in tone reproduction for high dynamic range images," *Computer Graphics Forum*, vol. 24, no. 3, pp. 635–645, October 2005.

[8] H. Seetzen, W. Heidrich, W. Stuerzlinger, G. Ward, L. Whitehead, M. Trentacoste, A. Ghosh, and A. Vorozcovs, "High dynamic range display systems," *ACM Transactions on Graphics*, vol. 23, no. 3, pp. 760–768, August 2004.

[9] K.N. Plataniotis and A.N. Venetsanopoulos, *Color Image Processing and Applications*, New York, USA: Springer-Verlag, July 2000.

[10] R.C. Gonzales and R.E. Woods, *Digital Image Processing*, 2nd Edition, Upper Saddle River, NJ: Prentice-Hall, January 2002.

[11] S. Susstrunk, R. Buckley, and S. Swen, "Standard RGB color spaces," in *Proceedings of IS&T/SID 7th Color Imaging Conference*, Scottsdale, AZ, USA, November 1999, pp. 127–134.

[12] A. Ford and A. Roberts, "Colour space conversions." Available online: http://herakles.fav.zcu.cz/research/night_road/westminster.pdf.

[13] A. Blanksby, M. Loinaz, D. Inglis, and B. Ackland, "Noise performance of a color CMOS photogate image sensor," Electron Devices Meeting, Technical Digest, Washington, DC, USA, December 1997.

[14] O. Pirinen, A. Foi, and A. Gotchev, "Color high dynamic range imaging: The luminance-chrominance approach," *International Journal of Imaging Systems and Technology*, vol. 17, no. 3, pp. 152–162, October 2007.

[15] S. Kavusi and A.E. Gamal, "Quantitative study of high-dynamic-range image sensor architectures," in *Proceedings of Sensors and Camera Systems for Scientific, Industrial, and Digital Photography Applications*, San Jose, CA, USA, January 2004, pp. 264–275.

[16] J. Burghartz, H.G. Graf, C. Harendt, W. Klingler, H. Richter, and M. Strobel, "HDR CMOS imagers and their applications," in *Proceedings of the 8th International Conference on Solid-State and Integrated Circuit Technology*, Shanghai, China, October 2006, pp. 528–531.

[17] B.C. Madden, "Extended intensity range imaging," Technical Report, GRASP Laboratory, University of Pennsylvania, 1993.

[18] S. O'Malley, "A simple, effective system for automated capture of high dynamic range images," in *Proceedings of the IEEE International Conference on Computer Vision Systems*, New York, USA, January 2006, p. 15.

[19] M. Goesele, W. Heidrich, and H.P. Seidel, "Color calibrated high dynamic range imaging with icc profiles," in *Proceedings of the 9th Color Imaging Conference*, Scottsdale, AZ, USA, November 2001, pp. 286–290.

[20] A.V. Oppenheim, R. Schafer, and T. Stockham, "Nonlinear filtering of multiplied and convolved signals," *Proceedings of the IEEE*, vol. 56, no. 8, pp. 1264–1291, August 1968.

[21] A. Gilchrist, C. Kossyfidis, F. Bonato, T. Agostini, J. Cataliotti, X. Li, B. Spehar, V. Annan, and E. Economou, "An anchoring theory of lightness perception," *Psychological Review*, vol. 106, no. 4, pp. 795–834, October 1999.

[22] F. Drago, K. Myszkowski, T. Annen, and N. Chiba, "Adaptive logarithmic mapping for displaying high contrast scenes," *Computer Graphics Forum*, vol. 22, no. 3, pp. 419–426, July 2003.

[23] "Picturenaut 3.0." Available online: http://www.hdrlabs.com/picturenaut/index.html.

[24] A. Foi, M. Trimeche, V. Katkovnik, and K. Egiazarian, "Practical Poissonian-Gaussian noise modeling and fitting for single-image raw-data," *IEEE Transactions on Image Processing*, vol. 17, no. 10, pp. 1737–1754, October 2008.

[25] A. Foi, "Clipped noisy images: Heteroskedastic modeling and practical denoising," *Signal Processing*, vol. 89, no. 12, pp. 2609–2629, December 2009.

10

High-Dynamic Range Imaging for Dynamic Scenes

Celine Loscos and Katrien Jacobs

10.1 Introduction

Computational photography makes it possible to enhance traditional photographs digitally [1]. One of its branches is high-dynamic range imaging, which enables access to a wider range of color values than traditional digital photography. Typically, a high-dynamic range (HDR) image stores RGB color values as floating point numbers, enlarging the conventional discretized RGB format (8-bit per color channel) used for low-dynamic range (LDR) images [2]. It is possible to visualize the HDR images through a specifically-built HDR display [3], [4] or to perceptually adapt them for LDR displays using tone mapping [5], [6], [7].

Example of a dynamic scene illustrated on a sequence of LDR images taken at different exposures. On the quay, pedestrians stroll. The boat, floating on the water, oscillates with the water movement. The water and the clouds are subjected to the wind. Therefore water wrinkles change from one image to another.

HDR imaging revolutionized digital imaging. Its usage became popular to the professional and amateur photographer after its inclusion in software such as HDRShop [8], Photogenics [9] and Cinepaint [10]. The autobracketing functionality nowadays available on many mainstream digital cameras makes the HDR capture less cumbersome and hence more attractive. Its dissemination is becoming common in press articles on photography. Besides becoming a must-have in digital photography, it is also used for scientific activities such as computer graphics, image processing and virtual reality. With its larger range of luminance values it provides a much more detailed support for imagery. Showing none or fewer saturated areas ensures more accurate calculations and more realistic simulations.

While HDR imaging becomes more mature in its use, its development is still at an early stage. This chapter will show that most HDR capture methods only work well when scenes are static, meaning that no movement or changes in the scene content is allowed during the capture process. The real world, however, is made of dynamic scenes (Figure 10.1), with objects in movement during the capture process. In indoor scenes this can be people moving, doors or windows opening, and objects being manipulated. In outdoor scenes this dynamic character is even more prominently present as pedestrians walk through the scene, leaves and trees move due to wind, and cloud movement or reflection on water change the lighting in the scene. The camera itself may also be in motion. Thus, in order to consider HDR imaging for a wider range of applications, motion and dynamic scenes need to be integrated in HDR technology targeting both HDR photographs and movies.

In this chapter, Section 10.2 presents the definition of HDR images. Section 10.3 describes how to create an HDR image from multiple exposures. Section 10.4 explores new approaches that extend the types of possible scenes. Section 10.5 presents a vision on future HDR imaging development for generic scenes. Conclusions are drawn in Section 10.6.

10.2 High-Dynamic Range Images: Definition

In conventional photography saturation occurs when the light quantity reaching the sensors exceeds the range of allowed values. In order to capture higher illumination values without saturation effects one often reduces the shutter speed or the aperture of the camera. A similar saturation effect occurs with dark areas in a scene. In Figure 10.1a, parts of the boat and the harbor are underexposed but brighter parts such as the water or the sky are represented by the colors in the appropriate range. In Figure 10.1d, the sky is overexposed and its color representation is saturated. However, color information is well represented in the areas of the harbor and the boat.

Dark parts of a scene are typically captured using a decreased shutter speed or larger aperture of the camera, which is the exact opposite of what is used when capturing a bright scene. It seems impossible to capture bright and dark areas in a scene simultaneously using conventional methods. Furthermore when storing the image digitally, illumination values are clamped to fit within a fixed range. Conventional photographs are therefore called *low-dynamic range* images. A *high-dynamic range* image stores a wider range of color values, enabling more precision and preventing the risk of overexposed or underexposed areas.

Each pixel of an image contains color information that can be viewed as the amount of light reaching the camera's sensors. With conventional cameras, this color is encoded in a restricted range of values and color space, and processed initially by the capture sensors of the equipment, then to represent pleasing observed colors and to have a controlled storage space. Typically, color contrast and white balance are adjusted and color values are tone mapped. Color processing may present a problem when one wants full control over the image. Recently, manufacturers started offering the possibility of storing images in a RAW format which directly reflects sensor data, in which pixel values contain supposedly no compression and no color postprocessing.

In an HDR image, the stored data corresponds to a physical quantity of light called radiance. This quantity can be viewed as the amount of light reaching an observer's eye or leaving a surface in a certain direction. It is directly linked to the color intensity found in video or photographs. An HDR image therefore stores radiance information in an RGB extended format. Several formats exist, such as HDR or OpenEXR, usable within different software and presenting different compression algorithms. Often, HDR values are stored between 16 bits to 32 bits per channel.

10.3 HDR Image Creation from Multiple Exposures

Although some HDR cameras [11] or other hardware technology [12] exist, HDR images are usually created by combining information coming from a set of LDR images (minimum of three) that capture the scene using different exposures [2]. The idea is that some of the LDR images contain clear information on the darkest parts of the scene, while others contain information on the brightest parts. This ensures that information from darker areas as well as brighter ones can be recovered. Recently, some methods work on extending the high

TABLE 10.1

A regular interval for exposures values (EV) with
a fixed aperture ($N = 1.0$) and varying time t.

t	4s	2s	1s	1/2s	1/4s
EV	-2	-1	0	1	2

dynamic range of one LDR image only [13], [14]. These methods extrapolate information
in underexposed or overexposed areas by either using user intervention or neighboring parts
of the image. Although this approach can produce reasonable results, it is generally more
accurate to recover information using several exposures of the same scene.

10.3.1 Capture with Still Cameras

The LDR pictures are usually taken sequentially at regular EV (exposure value) intervals
that link the aperture of the camera to the exposure time. In order to better control the
quality of the acquired LDR images, it is preferable to have a manual camera on which one
can set the white balance and the focus. In addition, the use of a tripod to minimize the
movement between each picture is highly encouraged.

To compute EV, the following equation is used:

$$EV = log_2(N^2/t),\qquad(10.1)$$

where N is the f-number (aperture value) and t is the exposure time (shutter speed). It is
usually preferred to fix the aperture and vary the exposure time to prevent traditional photographic effects associated with aperture changes, such as depth of field variation. Table 10.1
lists regular or linear intervals of EV for $N = 1.0$ and the required shutter speed. Having
regular intervals for EV is not compulsory but makes it easier for future computations.

The auto-bracketing function of existing cameras is very useful to capture a set of LDR
images at regular EV intervals. The camera chooses the best EV_0 exposure value by calculating what combination of aperture and shutter speed introduces the least saturation
in the image. Once this EV_0 is established, darker and brighter pictures are taken by respectively decreasing and increasing the shutter speed while keeping the aperture width
fixed. As an example, if $2n + 1$ is the number of pictures taken, EV varies along the range
$EV_{-n}, ...EV_{-1}, EV_0, EV_1, ..., EV_n$. The number of pictures taken varies with the camera,
three, five or nine pictures ($n = 1, 2, 4$) being the standard. An example of a series of five
pictures is shown in Figure 10.2.

FIGURE 10.2 (See color insert.)

A series of five LDR images taken at different exposures.

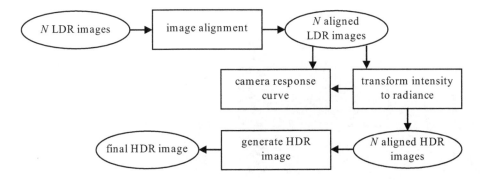

FIGURE 10.3

The different steps needed to create an HDR image from several LDR images in traditional methods.

10.3.2 Processing the Data

Several steps are needed in order to combine the LDR images into one HDR image. The process is described in the diagram in Figure 10.3.

Alignment: Even when using a tripod, there may be misalignments between the input images. The image alignment could be done by an automatic procedure but is not always easy to achieve. Typically, transformations are recovered for translational and rotational movements in the image plane. However, some images may require non linear transformation such as warping or scaling in case of rotations or translations of the camera during the capture. Image features are used in order to perform the alignment. They can be found using a medium bitmap transform [15], [16], [17] or the scale invariant feature transform [18]. For video content, warping is also used [19] or a combination between feature matching and optical flow [20].

Recovery of the inverse camera response function: The image saved by the camera is very different from the actual scene in terms of its colors. The manufacturer's choice of sensors as well as postprocessing have a large influence on the final image. It is said that each camera has its own response curve that transforms radiance values into image pixel values. In order to retrieve the radiance value it is compulsory to retrieve the inverse camera function. Considering the irradiance E reaching the camera sensor, there is a linear transform

$$E = Lt \tag{10.2}$$

that links linearly the irradiance with the scene radiance L, often with the exposure time t. The brightness or intensity value M stored in the picture corresponds to the transformation f, such as $M = f(E)$. To retrieve the irradiance, and therefore the radiance, we need to find the inverse response function $g = f^{-1}$. In general, the procedure is to select corresponding pixels in the LDR images captured with different exposures, plot the various brightness values for each pixel, and deduce the best fitting curve separately for each channel R, G, and B. Several methods exist to retrieve this function, often making an assumption on the curve. Firstly, it is reasonable to make the assumption that f is monotonic and increasing (ensuring that it can be inverted). Also, assumptions are made on its shape: log shape [21], polynomial shape [22], or a gamma shape [23]. The function can also be reconstructed

(a)   (b)

FIGURE 10.4 (See color insert.)

The HDR image obtained using traditional methods, (a) from the images shown in Figure 10.2 after alignment, (b) from the image sequence shown in Figure 10.1. (Images built using HDRShop [8].)

from other known functions [24]. The procedure is very sensitive to the chosen pixels and to the input images. If the camera parameters do not change, the same curve is often reused for other sets of LDR images.

Converting the intensity values of the LDR images to radiance values: The inverse response function f is applied to each pixel of the LDR images to convert the RGB color value into an HDR radiance value.

Generating the HDR image: The radiance values stored in the LDR images are combined using a weighted average function into a single value for each pixel that will form the final HDR image. The weights are used to eliminate saturated values, or misaligned pixels. Examples of a resulting image are shown in Figure 10.4; namely, Figure 10.4a shows the output image obtained using the input images in Figure 10.2 after alignment (translation and rotation) whereas Figure 10.4b shows the image resulting from the aligned image sequence shown in Figure 10.1. Traditional methods calculate the final radiance $E(i, j)$ of an HDR image for pixel (i,j) such as:

$$E(i, j) = \frac{\sum_{r=1}^{R} w(M_r(i, j)) \left(\frac{g^{-1}(M_r(i,j))}{\Delta t_r} \right)}{\sum_{r=1}^{R} w(M_r(i, j))}, \tag{10.3}$$

where $M_r(i, j)$ is the intensity of pixel (i, j) for the image r of R images, w its associated weight and Δt_r the exposure time. In traditional combining methods, w often only takes into account overexposure or underexposure. A hat function can be used [25], illustrated in Figure 10.5, to select well-exposed values.

10.3.3 Limitations

Limitations occur at different stages of the construction of an HDR image from several LDR images. The existing alignment procedures often used in HDR generation methods rarely consider nonlinear transformations (such as zooming or warping) which are much more computationally expensive and difficult to recover. Moreover, the alignment can easily be perturbed when object movement is present in the scene.

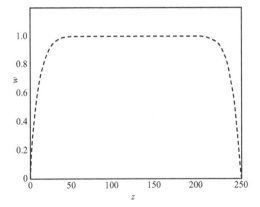

FIGURE 10.5

The hat function used to clamp values underexposed or overexposed in Reference [25].

The currently known algorithms that retrieve the camera response function work reasonably well for many cases, but there is no guarantee that these will always work. Firstly, the chosen shape (e.g., log, gamma or polynomial) may not fit the original curve; secondly, it is very dependent on the chosen pixels; thirdly a large dataset of curves is required to ensure a reasonable success in the procedure used in Reference [24]. In practice, since the response curve estimation is unstable, it is often preferable to calculate it once for a certain setting of the camera, and use this retrieved curve for other sets of LDR images.

The reconstruction of an HDR image from aligned LDR images works only if the scene is static. Ghosting effects or other types of wrongly recovered radiance values will be present for dynamic scenes. Such effects are illustrated in Figure 10.6. The time to take an HDR image is at the minimum the sum of the exposure times. For certain scenes, such as exterior scenes or populated scenes, it is almost impossible to guarantee complete stillness during this time. An ignored problem in the literature is the changes in illumination or covering objects in movement during the shoot. This happens for example on a cloudy day which may lead to rapid changes in illumination. It may also happen that an object moves fast

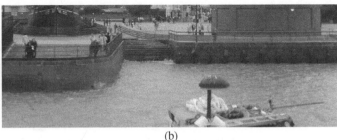

(a)   (b)

FIGURE 10.6

Example of ghosting effects in the HDR image due to movement in the input LDR image sequence: (a) a zoomed portion of the HDR image shown in Figure 10.4 with hosting effects due to people walking near the bridge, (b) a zoomed portion of the HDR reconstructed image from the sequence shown in Figure 10.1, with ghosting effects present in the zone with pedestrians and on the platform floating on the water.

and changes position reaching both shaded and illuminated areas. This leads to radiance incoherence in the LDR input images.

10.4 High-Dynamic Range Images for Dynamic Scenes

Dynamic scenes are actually the most encountered scenes where movement is not controlled. During the capture of the multiple exposures objects may move in the scene. As shown in Section 10.3.3, this leads to difficulties and errors at different stages of the reconstruction procedure: during the alignment procedure, the inverse camera curve retrieval, and the combining of the LDR image sequence. To compensate for this, effort is made to choose static pixels for the alignment procedure and the inverse curve retrieval. Also, regions showing movement can be detected before being treated during the final combining. Some existing methods are discussed below with respect to their approach to detecting and managing movement in images.

10.4.1 Movement Detection and Feature Insertion in the Final Image

Reference [16] presents two methods used to detect movement in a sequence of LDR exposures. The first method, movement removal using variance, uses the irradiance variation across these exposures as an indicator. Similar methods can also be found in Reference [2]. The second method measures uncertainty using entropy as an indicator of potential movements. Both methods are described below.

10.4.1.1 Movement Removal Using Variance

Using the camera response curve an HDR image called E_i is calculated for each of the $2n + 1$ LDR exposures I_i. The pixels affected by movement will show a large irradiance variation over the different E_i. Therefore, the variance of a pixel over the different E_i's can be used as a likelihood measure for movement. The movement cluster is derived from a variance image (VI), which is created by storing the variance of a pixel over the different exposures in a matrix with the same dimensions as the LDR images.

Some pixels in the I_i's can be saturated, others underexposed. Compared to their counterparts in the other exposures, such pixels do not contain any reliable irradiance information. When calculating the variance of a pixel over a set of images it is important to ignore the variance introduced by saturated or underexposed pixels. This can be achieved by calculating the variance $VI(.)$ of a pixel (k, l) as a weighted variance [2]

$$VI(k,l) = \frac{\sum_{i=0}^{N} W_i(k,l)E_i(k,l)^2 / \sum_{i=0}^{N} W_i(k,l)}{(\sum_{i=0}^{N} W_i(k,l)E_i(k,l))^2 / (\sum_{i=0}^{N} W_i(k,l))^2} - 1. \tag{10.4}$$

The weights $W_i(k,l)$ are the same as those used during the HDR image generation as described in Reference [2]. The variance image can be calculated for one color channel or as the maximum of the variance over three color channels.

FIGURE 10.7 (See color insert.)

Movement removal using variance and uncertainty [16]: (a) sequence of LDR images captured with different exposure times; several people walk through the viewing window, (b) variance image VI, (c) uncertainty image UI, (d) HDR image after object movement removal using the variance image, and (e) HDR image after movement removal using the uncertainty image. © 2008 IEEE

The assumption is made that in general, moving objects cover a wide range of adjacent pixel clusters, called movement clusters, rather than affecting isolated pixels solely. The movement clusters are derived by applying a threshold T_{VI} on VI, resulting in a binary image VI_T. For well-defined and closed movement clusters, the morphological operations erosion and dilation are applied to the binary image VI_T. In Reference [16] a suitable threshold value for T_{VI} is stated to be 0.18. The HDR reconstruction is done using the weighted sum as in Equation 10.3 except for the identified movement clusters. In those regions, pixels are replaced by the ones of the best exposed LDR radiance image.

An example is shown in Figure 10.7. Namely, Figure 10.7a shows a set of four exposures in which people walk through the viewing window. Figure 10.7b presents the variance image used to detect this movement. The HDR image generated using this variance image is shown in Figure 10.7d.

This method defines that highly variant pixels in VI indicate movement. Other influences exist, besides remaining camera misalignments, that might result in a highly variant VI value:

- *Camera curve*: The camera curve might fail to convert the intensity values to irradiance values correctly. This influences the variance between corresponding pixels in the LDR images and might compromise the applicability of the threshold to retrieve movement clusters.

- *Weighting factors*: Saturation and underexposure of pixels in an LDR image can result in incorrect irradiance values after transformation to irradiance values using the camera curve. Defining the weighting factors is not straightforward and various different methods exist to define the weights [2].

- *Inaccuracies in exposure speed and aperture width used*: In combination with the camera curve this produces incorrect irradiance values after transformation. Changing the aperture width causes change in depth of field, which influences the quality of the irradiance values.

10.4.1.2 Movement Removal Using Entropy

The second method described in Reference [16] detects movement across a sequence of LDR exposures using a statistical, contrast-independent measure based on the concept of entropy. In information theory, entropy is a scalar statistical measure defined for a statistical process. It defines the uncertainty that remains about a system, after having taken into account the observable properties. Let X be a random variable with probability function $p(x) = P(X = x)$, where x ranges over a certain interval. The entropy $H(X)$ of a variable X is given by

$$H(X) = -\sum_x P(X = x) \log(P(X = x)). \tag{10.5}$$

To derive the entropy of an image I, written as $H(I)$, the intensity of a pixel in an image is thought of as a statistical process. In other words, X is the intensity value of a pixel, and $p(x) = P(X = x)$ is the probability that a pixel has intensity x. The probability function $p(x) = P(X = x)$ is the normalized histogram of the image. Note that the pixel intensities range over a discrete interval, usually defined as the integers in $[0, 255]$, but the *number of bins M* of the histogram used to calculate the entropy can be less than 256.

The entropy of an image provides some useful information and the following remarks can be made:

- The entropy of an image has a positive value between $[0, \log(M)]$. The lower the entropy, the less different intensity values are present in the image; the higher the entropy, the more different intensity values there are in the image. However, the actual intensity values do not have an influence on the entropy.

- The actual *order* or *organization* of the pixel intensities in an image does not influence the entropy. As an example, consider two images with equal amounts of black and white organized in squares as in a checkerboard. If the first image has only 4 large squares and the second image consists of 100 smaller squares they still contain the same amount of entropy.

- Applying a scaling factor on the intensity values of an image does not change its entropy, if the intensity values do not saturate. In fact, the entropy of an image does not change if an injective function is applied to the intensity values. An injective function associates distinct arguments to distinct values, examples are the logarithm, exponential, scaling, etc.

- The entropy of an image gives a measure of the uncertainty of the pixels in the image. If all intensity values are equal, the entropy is zero and there is no uncertainty about the intensity value a randomly chosen pixel can have. If all intensity values are different, the entropy is high and there is a lot of uncertainty about the intensity value of any particular pixel.

The movement detection method discussed in this section shares some common elements with the one presented in References [26] and [27]. Both methods detect movement in a sequence of images, but restrict this sequence to be captured under the same conditions (illumination and exposure settings). The method presented here can be applied to a sequence of images captured under different exposure settings. It starts by creating an uncertainty image UI, which has a similar interpretation as the variance image VI used in Section 10.4.1.1; pixels with a high UI value indicate movement. The following explains how the calculation of UI proceeds.

For each pixel with coordinates (k,l) in each LDR image I_i the local entropy is calculated from the histograms constructed from the pixels that fall within a two-dimensional window V with size $(2v+1) \times (2v+1)$ around (k,l). Each image I_i therefore defines an entropy image H_i, where the pixel value $H_i(k,l)$ is calculated as follows:

$$H_i(k,l) = -\sum_{x=0}^{M-1} P(X=x)\log(P(X=x)), \qquad (10.6)$$

where the probability function $P(X=x)$ is derived from the normalized histogram constructed from the intensity values of the pixels within the two-dimensional window V, or over all pixels p in

$$\{p \in I_i(k-v:k+v,l-v:l+v)\}. \qquad (10.7)$$

From these entropy images a final uncertainty image UI is defined as the local weighted entropy difference:

$$UI(k,l) = \sum_{i=0}^{N-1}\sum_{j=0}^{j<i} \frac{w_{ij}}{\sum_{i=0}^{N-1}\sum_{j=0}^{j<i} v_{ij}} h_{ij}(k,l), \qquad (10.8)$$

$$h_{ij}(k,l) = |H_i(k,l) - H_j(k,l)|, \qquad (10.9)$$

$$w_{ij} = \min(W_i(k,l), W_j(k,l)). \qquad (10.10)$$

It is important that the weights $W_i(k,l)$ and $W_j(k,l)$ remove any form of underexposure or saturation to ensure the transformation between the different exposures is an injective function. Therefore they are slightly different from those used during the HDR generation. In Reference [16] a relatively small hat function with lower and upper thresholds equal to 0.05 and 0.95 for normalized pixel intensities is used. The weight w_{ij} is created as the minimum of $W_i(k,l)$ and $W_j(k,l)$, which further reflects the idea that underexposed and saturate pixels do not yield any entropic information.

The reasoning behind this uncertainty measure follows from the edge enhancement that the entropy images H_i provide. The local entropy is high in areas with edges and details. These high entropic areas do not change between the images in the exposure sequence, except when corrupted by a moving object or saturation. The difference between the entropy images therefore provides a measure for the difference in features, such as intensity edges, between the exposures. Entropy does this without the need to search for edges and corners which can be difficult in low contrast areas. In fact, the entropy images are invariant to the local contrast in the areas around these features. If two image regions share the exact same structure, but with a different intensity, the local entropy images will fail to detect this change. This can be considered a drawback of the entropic movement detector as it also implies that when one homogenous colored object moves against another homogeneously colored object, the uncertainty measure would only detect the boundaries of the moving objects of having changed. Fortunately, real-world objects usually show some spatial variety, which is sufficient for the uncertainty detector to detect movement. Therefore the indifference to local contrast is only an advantage, particularly in comparison to the variance detector discussed previously in this section.

The difference in local entropy between two images induced by the moving object, depends on the difference in entropy of the moving object and the background environment. Though the uncertainty measure is invariant to the contrast of these two, it is not invariant to the entropic similarity of the two. For instance, if the local window is relatively large, the moving object is small relative to this window, and the background consists of many static objects that are small and similar, then the entropic difference defined in Equation 10.8 might not be large. Decreasing the size of the local window will result in an increased entropic difference, but a too small local window might be subject to noise and outliers. It was found in Reference [16] that a window size of 5×5 pixels returned good results.

Similarly to the variance-based method [2], [16], the movement clusters are now defined by applying a threshold T_{UI} on UI, resulting in a binary image UI_T. For well-defined, closed, movement clusters, the morphological operations erosion and dilation are applied to UI_T. A threshold T_{UI} equal to 0.7 for $M = 200$ provides satisfactory results, although it does not seem to be as robust as the threshold for the variance detector. As for the variance-based method, for the HDR reconstruction, pixels in a detected movement area are replaced by the pixels of only one LDR image chosen to have least saturation in this area.

An example is shown in Figure 10.7. Namely, Figure 10.7a shows a set of four exposures that indicate object movement. Figure 10.7c presents the uncertainty image. The resulting HDR image after movement removal using this uncertainty image is shown in Figure 10.7e.

The creation of UI is independent from the camera curve calibration. As mentioned earlier, this has as an extra advantage that the detection of movement clusters could potentially be used in the camera calibration phase.

(a)   (b)   (c)

FIGURE 10.8

HDR reconstruction using the background estimation method of Reference [28]: (a) input LDR images, (b) computed labeling, and (c) reconstructed HDR image. Courtesy of Granados et al. [28].

10.4.2 HDR Reconstruction of Background Information

Background reconstruction is related to the HDR reconstruction of dynamic scenes and is the complementary action to movement detection.

10.4.2.1 Background Estimation

Reference [28] proposes a solution to background estimation from a non-time sequence of images, meaning that a sequence of images from a same viewpoint can be taken at different times and therefore contain different objects while maintaining the same background. The input images are LDR, taken under the same camera setting and lighting conditions. Background pixels are determined by minimizing a cost function, built on ideas presented in References [29] and [30]. The method works under the assumption that background regions constitute the majority of the image, have a high occurrence count in the set of images, and are motionless in the set of images.

The cost function is represented by an energy function associated to each pixel summing a data term with a smoothness term and a hard constraint. The data term represents a weighted sum of the *likelihood* of the pixel to belong to the background and its *stationariness*. The weight is assigned using an entropy measure to preserve the most reliable information. The smoothness term is used to select regions in images that are labeled differently but present similarities in intensities, to minimize discontinuities at the different labeled regions. It also leads to fewer labeled regions. The hard constraint term helps to keep image semantic consistency by keeping whole parts of objects. This method can be applied to HDR reconstruction with a set of aligned LDR images with radiance values and a selection of valid pixels, that is, not presenting overexposed and underexposed values.

10.4.2.2 Probabilistic Computation of Background Estimation

Reference [25] proposes to reconstruct HDR images directly from the LDR images without motion estimation, movement detection or other motion representation. Instead, the final HDR image is directly built from the input LDR images. Weights are calculated through an iterative process to represent the chance of each pixel to belong to the static part of the scene together with its chance of being correctly exposed. A specific weight w, that takes into account not only exposure quality but also the chance of being static, is obtained. Calculations are made in the $L\alpha\beta$ color space.

An iterative process is set to calculate the weights w_{pqs} of a pixel (p,q) in image s with an intensity Z. All weights are first initialized to the average over the color space of a hat function which is low for values close to the extremes 0 and 255. This hat function is represented in Figure 10.5 and defined as follows:

$$w(Z) = 1 - (2 \cdot \frac{Z}{255} - 1)^{12}.$$
(10.11)

Reference [25] uses the set N that contains neighboring pixels $\mathbf{y}_{pqs}$ of pixel $\mathbf{x}_{ijr}$ with $(p,q) \neq (i,j)$, where $\mathbf{x}$ and $\mathbf{y}$ denote vectors in $\mathbb{R}^5$ representing the $L\alpha\beta$ color space and the two-dimensional position. In practice, the neighboring pixels are located inside a 3×3 window. The neighboring pixels help to evaluate the likelihood of the pixel to belong to the background. New weights are calculated at iteration $t+1$ as follows:

$$w_{pqs,t+1} = w(Z_s(p,q))P(\mathbf{x}_{pqs} \mid F).$$
(10.12)

The probability function $P(\cdot)$ is defined as

$$P(\mathbf{x}_{ijr} \mid F) = \frac{\sum_{pqs \in N(x_{ijr})} w_{pqs} K_{\mathbf{H}}(\mathbf{x}_{ijr} - \mathbf{y}_{pqs})}{\sum_{pqs \in N(x_{ijr})} w_{pqs}},$$
(10.13)

where

$$K_{\mathbf{H}}(\mathbf{x}) = |\mathbf{H}|^{\frac{1}{2}}(2\Pi)^{\frac{d}{2}} exp(-\frac{1}{2}\mathbf{x}^T\mathbf{H}^{-1}\mathbf{x})$$
(10.14)

for a d-variate Gaussian, with $\mathbf{H}$ being a symmetric, positive definite, $d \times d$ bandwidth matrix. For example, $\mathbf{H}$ can be an identity matrix.

By performing 10 to 15 iterations, the method can effectively remove ghosting effects; it works particularly well if the background is predominant in the image [25]. When there is some overlap in the region in movement in most images, the object in movement is still present in the reconstructed image. This is shown in Figure 10.9.

This method is extended in Reference [31]. The color values of the input images are calibrated through a histogram matching; however, a more robust radiometric alignment could be used instead. The matrix $\mathbf{H}$ used in Equation 10.13 is set to

$$\mathbf{H} = \begin{pmatrix} \tilde{\sigma}_L(N_{i,j}) & 0 & 0 \\ 0 & \tilde{\sigma}_\alpha(N_{i,j}) & 0 \\ 0 & 0 & \tilde{\sigma}_\beta(N_{i,j}) \end{pmatrix},$$
(10.15)

with the weighted standard deviation $\tilde{\sigma}$ calculated for L, α, and β. This matrix is used instead of the identity matrix chosen in Reference [25]. To prevent objects from being segmented in the final reconstruction, a weight propagation algorithm is proposed to cover regions that are likely to belong to the same object in each image. This algorithm requires fewer iterations than the algorithm described in Reference [25], using often only one iteration for improved results. An example is shown in Figure 10.10.

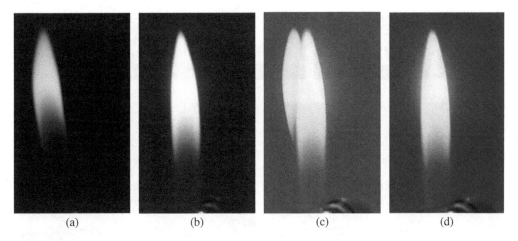

(a)   (b)   (c)   (d)

FIGURE 10.9

HDR reconstruction using the background estimation method of Reference [25]: (a,b) two of the input LDR images, (c) HDR reconstruction with traditional methods, and (d) HDR reconstruction after 10 iterations. Courtesy of Khan et al. [25]. © 2006 IEEE

(a)   (b)   (c)

FIGURE 10.10

Comparison of various HDR image reconstruction methods: (a) HDR image reconstructed without any movement identification, (b) results obtained using the method in Reference [25] after six iterations, and (c) results obtained using the method in Reference [31] using only one iteration. Courtesy of Pedone et al. [31].

10.4.3 Calculation of Errors to Detect Moving Regions

Reference [32] proposes an approach to manage object movement in images using the mathematical relation between images. It uses the linear relation between the exposure value and the radiance with the exposure time (see Equation 10.2) to identify inconsistent pixels. For an exposure value X of pixel p_i in the ith image, and with a relative exposure ev_{ij} between the ith and the jth images, the exposure value of p_j should satisfy

$$X(p_j) = X(p_i)ev_{ij}. \tag{10.16}$$

FIGURE 10.11

(a-e) LDR sequence of images showing walking pedestrians, (f) HDR reconstruction using traditional methods, and (g) HDR reconstruction after ghost removal [32]. Image courtesy of Nokia, copyright 2009. © 2009 IEEE

To identify inconsistent pixels, the algorithm actually looks at the offset of the logarithmic relation

$$ln(X(p_2)) = ln(X(p_1)) + ln(ev_{12}) \tag{10.17}$$

for two patches, 1 and 2, of two different exposures. A measure of outliers gives the percentage of ghosting in each patch. This measure is used when combining the exposures together to stitch the final HDR image. Patches found inconsistent are not considered in the stitching. A significant advantage of the approach is that if the movement occurs only in parts of the images, the radiance of a moving object can still be reconstructed from input images where this object was still. Only the largest set $\mathcal{I}$ of patches with consistent values is considered. Reconstruction of the radiance around a moving object is done so that no seams will be visible. A reconstruction based on a Poisson equation is used to create valid values around the moving area from the identified set $\mathcal{I}$ of consistent images for each area. An example of the method is shown in Figure 10.11.

A similar approach is employed in Reference [17] which identifies errors in pixels between images using the computation of a predicted color from Equation 10.2 as follows:

$$\tilde{z}_{i,k} = f\left(\frac{\Delta t_k}{\Delta t_j} f^{-1}(z_{i,j})\right), \tag{10.18}$$

where Δt_j and Δt_k denote the exposures, i denote the pixel, and $f(\cdot)$ is the camera response function, with $f^{-1}(z_{i,.j}) = X_{i,j} = E_i . \Delta t_j$. Errors are computed by comparing $\tilde{z}_{i,k}$ with $z_{i,k}$. If an error is identified, the pixel is not used in the final HDR combination. The final image is built using a selected region showing least saturation in the region of detected invalid pixels. Unlike Reference [32], this method aims to have all moving objects appearing in images, although this requires the user to interact manually with the picture.

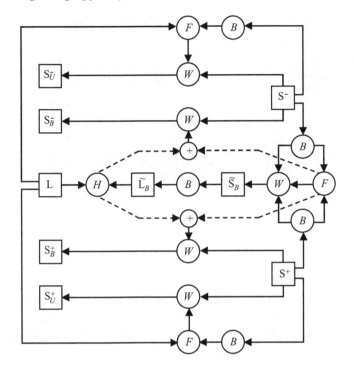

FIGURE 10.12

HDR stitching procedure of Reference [19] for an image sequence S^-, L, and S^+.

10.4.4 Motion Detection Using Optical Flow

It is possible to address HDR video using optical flow to detect motion. Two approaches were developed that retrieve HDR content for video data. In those two cases, optical flow alone is shown to be insufficient due to the complexity of the different motion paths and scene details. The two optical flow-driven approaches are described below; namely, one based on hierarchical homography and warping [19] and the other one based on feature point matching [20].

10.4.4.1 HDR Video Capture, Reconstruction, and Display

Reference [19] proposes an early solution for HDR video. The video is captured using a Lady Bug Firewired point gray camera, that is programmed to store alternatively a low exposure value with a high exposure value. The ratio between exposures can range from 1 to a maximum set by the user (16 in Reference [19]). The exposure settings vary automatically depending on the scene content. Since frame content varies temporally and spatially, both need to be treated for HDR reconstruction. Neighboring frames are used to reconstruct intermediate exposure content as well as the scene and object motion. The process of transferring radiance information from neighboring frames is called HDR stitching.

Figure 10.12 summarizes the process for a sequence of images with short (S^-), long (L), and short (S^+) exposures. The process for L^-, S, and L^+ is shown to be similar. It is considered that S^-, L, and S^+ have different scene content due to motion. In order to reconstruct the radiance map of L as HDR content, the information contained in S^- and S^+

FIGURE 10.13

HDR video reconstruction: (top) input photographs of an video sequence, and (bottom) HDR reconstruction. Courtesy of Kang et al. [19]. © 2003 ACM

needs to be used, but is incoherent with what is contained in L. New images are created, S_U^-, S_U^+ and S_B^-, S_B^+, that match the content of L for the HDR reconstruction. All images are converted to new images with radiance values $(\hat{S}_U^-, \hat{S}_U^+, \hat{S}_B^-, \hat{S}_B^+, \hat{L})$ using a retrieved inverse camera curve, as proposed in Reference [22]. The reconstructed HDR radiance image $\hat{L}_{HDR}$ contains pixels with the following values:

- If the considered pixel in $\hat{L}$ is not saturated, a weighted average is computed using values in $\hat{S}_B^-$, $\hat{S}_B^+$ and L, with the weights being low for overexposed or underexposed values and representing a plausibility map (Hermite cubic).

- If the considered pixel in $\hat{L}$ is saturated, the value in $\hat{S}_B^-$ is used if it is not overexposed or underexposed, and $\hat{S}_B^+$ otherwise.

The main difficulty relies on computing S_U^-, S_U^+, S_B^-, and S_B^+. Motion estimation between S^-, L, and S^+ is calculated combing optical flow [33], [34] with hierarchical homography for more accuracy. Residual flow vectors are computed between two warped images from the source images and are accumulated in a pyramid. A global affine flow is also added. This estimate is used to generate (S_U^-, S_U^+) for a unidirectional warping (unidirectional flow field) and (S_B^-, S_B^+) for a bidirectional warping. To display the final video, a temporal tone-mapper is proposed. Figure 10.13 show some results. This technique can also be applied to produce a still HDR photograph. One extremely interesting aspect of this method is that camera motion and scene content movements are treated together with the same method.

10.4.4.2 Optical Flow Used for Video Matching

There is a need for image alignment for sequences taken at different moments which, however, share spatial and temporal content where differences may be found in object positions or scene illumination. Reference [20] proposes a method to image alignment using optical flow-based warping and feature point matching to allow robust image registration.

A weight w_i is associated to the ith correspondence between two pixels, which is the product of two computed values; a pixel matching probability P_i and a motion consistency probability M_i. The computation of P_i is inspired by a technique which allows small spatial variations, including changes in scale, rotations and skew in an image region [35]. The algorithm runs on each pair of images to match. If images contain substantial intensity differences (illumination or exposure), images are normalized. For each pixel of an image, the algorithm analyzes the surrounding of its corresponding pixel in its pair image using a 3×3 window. A score is defined based on the maximum and minimum intensities found. If the pixel value is outside this range, it receives a penalty. A pixel intensity dissimilarity is calculated averaging the scores over a region, which then serves to compute a probability on pixel matching. The computation of M_i uses weights that measure a correspondence field between matched feature points, vector fields using locally weighted linear regression, and a predicted locally weighted regression.

This technique is shown to have an application to HDR video for aligning several image sequences taken at different exposures before the HDR reconstruction is done, for example, with the method of Reference [19].

10.4.5 Limitations

The methods presented in this section show that it is possible to address the problem of dynamic scenes in HDR imaging. They all seem to work well for some typical scenes, where some small parts are in movement, but cannot be considered generic for all types of scenes. In particular, most of the methods require a well-defined background region with foreground moving objects. Only the entropy-based method [16] can differentiate well moving objects to a similar background. Most methods assume that the moving objects are located in a cluster that is small compared to the static background in the image. This is a strong limitation, since scenes with a predominant moving area cannot be treated.

Certain methods only work for background reconstruction [17], [25], [28], [32]. As a consequence, objects in movement disappear from the reconstructed HDR image or can be cut [25], [32] during the reconstruction process. When the moving areas of the image are conserved in the final HDR reconstruction, they are often represented by the associated region of a chosen LDR radiance image [16], [17] or their HDR representation is limited by the amount of valid data [25], [31]. Depending on the algorithm, this may imply duplication of some objects and overexposed or underexposed areas still present in the final HDR reconstruction. Warping, as presented in References [19] and [20], can solve the reconstruction problem particularly for nonrigid movement but it is difficult to have optical flow-based methods to be robust. In particular, the method in Reference [19] relies strongly on proximity of the moving object in the image sequence.

An interesting point is that some methods solve directly for image alignment and movement detection [19], [20]. Finally, none of the above methods propose solutions to the retrieval of the inverse camera response function when scenes are dynamic. All assume that the camera curve was previously precomputed.

10.5 A Vision of Future Developments

It can be considered for now that it becomes less important to have a good calibration procedure to retrieve the inverse curve of the camera. In fact, most digital cameras now propose to store the captured image directly in a RAW format that most of the time stores pixel values without any processing. The RAW format also uses more bits, typically between 12 to 14 bits against the 8 bits per color channel for more traditional formats, to store the intensity information increasing thus the range of intensities. The RAW format stores all the necessary information to convert its uncompressed stored image into the final image. While the RAW format helps to acquire more precise information for HDR photography, its storage size makes it impractical for HDR video. Therefore, more research needs to be undertaken to calculate the inverse camera curves for camera acquired dynamic scenes, so that a natural extension could be made to HDR video capture.

For a more accurate and robust camera inverse curve retrieval, it is crucial to develop methods that make no assumption on the camera curve shape; as mentioned in Reference [24], the shape of the curve varies a lot from one camera to another. A function built with no assumption on the shape will be more robust and should lead to more accurate reconstructed radiance values. Further improvements can be achieved by identifying well viewpoint / object movement; it is vital to develop methods that take movement into account rather than ignoring pixels in motion, as it may occur in some pictures that most pixels in the image are in movement.

The reconstruction of the radiance in moving areas is important when the required reconstructed HDR image needs to be faithful to the input scene so that all objects of the scene and only those are present in the final image and are represented only with valid radiance values. Methods based on HDR reconstruction from only one LDR image may be used to fill in missing areas of the reconstructed HDR image.

Another issue is to guarantee the accuracy of the captured radiance. Reference [36] presents a method to characterize color accuracy in HDR images. Up to now, methods have rather focused on plausible radiance results. Some applications may require physically accurate or perceptually accurate radiance values, and it is important to develop HDR methods that take this into account.

Finally, hardware implementation can reduce the need of postprocessing [19], [12]. Multiple camera systems could be used, although image registration and synchronization will be a crucial determinant in the HDR reconstruction. It is also expected that manufacturers will develop new sensors that will capture higher ranges of values as it is currently the trend. New storage and compression formats will probably be designed to fit with HDR video requirements.

10.6 Conclusion

HDR imaging is a growing field already popular in its use. However, up to now, it was mostly restricted to static scenes. This chapter identified the issues linked to dynamic scenes and described the methods recently presented in the literature that contribute to a significant step in solving HDR imaging for dynamic scenes. Nevertheless, some work is still needed to address some of their limitations and improve their robustness to more generic scenes. There is also a need for postprocessing approaches for image calibration and HDR reconstruction in moving scenes as well as hardware development of new HDR capture systems and sensors to capture a higher range in a shorter time.

The research field in HDR imaging is extremely active, both in computer vision and computer graphics communities, and important advances should continue to appear in the near future.

Acknowledgment

This work is funded by a grant given by the Spanish ministry of education and science (MEC) for the project TIN2008-02046-E/TIN "High-Dynamic Range Imaging for Uncontrollable Environments." It is also supported by the Ramon y Cajal program of the Spanish government. We would like to thank authors of References [19], [25], [28], [31], and [32] for allowing us to use their images for illustration. We would also like to thank Florent Duguet (Altimesh) for his valuable advice.

Figure 10.7 is reprinted from Reference [16], Figure 10.9 is reprinted from Reference [25], Figure 10.11 is reprinted from Reference [32], with the permission of IEEE. Figure 10.13 is reprinted from Reference [19], with the permission of ACM.

References

[1] R. Raskar and J. Tumblin, *Computational Photography: Mastering New Techniques for Lenses, Lighting, and Sensors*. A K Peters Ltd., December 2009.

[2] E. Reinhard, G. Ward, S. Pattanaik, and P. Debevec, *High Dynamic Range Imaging: Acquisition, Display, and Image-Based Lighting*. San Francisco, CA: Morgan Kaufmann Publishers, August 2005.

[3] BrightSide Technologies, "High dynamic range displays." www.brightsidetech.com.

[4] P. Ledda, A. Chalmers, and H. Seetzen, "HDR displays: A validation against reality," in *Proceedings of IEEE International Conference on Systems, Man and Cybernetics*, The Hague, Netherlands, October 2004, vol. 3, pp. 2777–2782.

[5] P. Ledda, A. Chalmers, T. Troscianko, and H. Seetzen, "Evaluation of tone mapping operators using a high dynamic range display," *ACM Transactions on Graphics*, vol. 24, no. 3, pp. 640–648, August 2005.

[6] M. Čadík, M. Wimmer, L. Neumann, and A. Artusi, "Evaluation of HDR tone mapping methods using essential perceptual attributes," *Computers & Graphics*, vol. 32, no. 3, pp. 330–349, June 2008.

[7] A. Yoshida, V. Blanz, K. Myszkowski, and H.-P. Seidel, "Perceptual evaluation of tone mapping operators with real-world scenes," *Proceedings of SPIE*, vol. 5666, pp. 192–203, January 2005.

[8] HDRshop 2.0. www.hdrshop.com.

[9] Idruna software, "Photogenics HDR." www.idruna.com.

[10] T. C. project, "Cinepaint software." www.cinepaint.org.

[11] Spheron, "Spherocamhdr." www.spheron.com.

[12] S. Nayar and T. Mitsunaga, "High dynamic range imaging: Spatially varying pixel exposures," in *Proceedings of IEEE Conference on Computer Vision and Pattern Recognition*, Hilton Head, SC, USA, June 2000, pp. 472–479.

[13] L. Wang, L.-Y. Wei, K. Zhou, B. Guo, and H.-Y. Shum, "High dynamic range image hallucination," in *Proceedings of Eurographics Symposium on Rendering*, Grenoble, France, June 2007, pp. 321–326.

[14] F. Banterle, K. Debattista, A. Artusi, S. Pattanaik, K. Myszkowski, P. Ledda, M. Bloj, and A. Chalmers, "High Dynamic Range Imaging and Low Dynamic Range Expansion for Generating HDR Content," in *Proceeding of the 30th Annual Conference of the European Association for Computer Graphics*, Munich, Germany, April 2009, pp. 17–44.

[15] G. Ward, "Fast, robust image registration for compositing high dynamic range photographs from handheld exposures," *Journal of Graphics Tools*, vol. 8, no. 2, pp. 17–30, May 2003.

[16] K. Jacobs, C. Loscos, and G. Ward, "Automatic high-dynamic range generation for dynamic scenes," *IEEE Computer Graphics and Applications*, vol. 28, no. 2, pp. 24–33, March-April 2008.

[17] T. Grosch, "Fast and robust high dynamic range image generation with camera and object movement," in *Proceedings of International Workshop on Vision, Modeling and Visualization*, Aachen, Germany, November 2006, pp. 277–284.

[18] A. Tomaszewska and R. Mantiuk, "Image registration for multi-exposure high dynamic range image acquisition," in *Proceedings of the International Conference on Computer Graphics, Visualization and Computer Vision*, Plzen, Czech Republic, January 2007.

[19] S.B. Kang, M. Uyttendaele, S. Winder, and R. Szeliski, "High dynamic range video," *ACM Transactions on Graphics*, vol. 22, no. 3, pp. 319–325, July 2003.

[20] P. Sand and S. Teller, "Video matching," *ACM Transactions on Graphics*, vol. 22, no. 3, pp. 592–599, August 2004.

[21] P.E. Debevec and J. Malik, "Recovering high dynamic range radiance maps from photographs," in *Proceedings of the 24th ACM Annual Conference on Computer Graphics and Interactive Techniques*, Los Angeles, CA, USA, August 1997, pp. 369–378.

[22] T. Mitsunaga and S.K. Nayar, "Radiometric self calibration," in *Proceedings of IEEE Conference on Computer Vision and Pattern Recognition*, Fort Collins, CO, USA, June 1999, pp. 374–380.

[23] S. Mann and R.W. Picard, "Being 'undigital' with digital cameras: Extending dynamic range by combining differently exposed pictures," in *Proceedings of the IS&T 46th Annual Conference*, Scottsdale, AZ, USA, May 1995, pp. 422–428.

[24] M.D. Grossberg and S.K. Nayar, "Modeling the space of camera response functions," *IEEE Transactions on Pattern Analysis and Machine Intelligence*, vol. 26, no. 10, pp. 1272–1282, October 2004.

[25] E.A. Khan, A.O. Akyz, and E. Reinhard, "Ghost removal in high dynamic range images," in *Proceedings of the IEEE International Conference on Image Processing*, Atlanta, GA, USA, October 2006, pp. 2005–2008.

[26] Y.F. Ma and H.J. Zhang, "Detecting motion object by spatiotemporal entropy," in *Proceedings of the IEEE International Conference on Multimedia and Expo*, Tokyo, Japan, August 2001, pp. 265–268.

[27] G. Jing, C.E. Siong, and D. Rajan, "Foreground motion detection by difference-based spatial temporal entropy image," in *Proceedings of the IEEE Region Ten Conference*, Chiang Mai, Thailand, November 2004, pp. 379–382.

[28] M. Granados, H.P. Seidel, and H. Lensch, "Background estimation from non-time sequence images," in *Proceedings of the Graphics Interface Conference*, Windsor, ON, Canada, May 2008, pp. 33–40.

[29] A. Agarwala, M. Dontcheva, M. Agrawala, S. Drucker, A. Colburn, B. Curless, D. Salesin, and M. Cohen, "Interactive digital photomontage," *ACM Transactions on Graphics*, vol. 23, no. 3, pp. 294–302, August 2004.

[30] S. Cohen, "Background estimation as a labeling problem," in *Proceedings of the IEEE International Conference on Computer Vision*, Beijing, China, October 2005, pp. 1034–1041.

[31] M. Pedone and J. Heikkilä, "Constrain propagation for ghost removal in high dynamic range images," in *Proceedings of the International Conference on Computer Vision Theory and Applications*, Funchal, Portugal, January 2008, pp. 36–41.

[32] O. Gallo, N. Gelfand, W. Chen, M. Tico, and K. Pulli, "Artifact-free high dynamic range imaging," in *Proceedings of the IEEE International Conference on Computational Photography*, San Francisco, CA, USA, April 2009.

[33] J.R. Bergen, P. Anandan, K.J. Hanna, and R. Hingorani, "Hierarchical model-based motion estimation," in *Proceedings of the Second European Conference on Computer Vision*, Santa Margherita Ligure, Italy, May 1992, pp. 237–252.

[34] B.D. Lucas and T. Kanade, "An iterative image registration technique with an application to stereo vision," in *Proceedings of the Seventh International Joint Conference on Artificial Intelligence*, Vancouver, BC, Canada, August 1981, pp. 674–679.

[35] S. Birchfield and C. Tomasi, "A pixel dissimilarity measure that is insensitive to image sampling," *IEEE Transactions on Pattern Analysis and Machine Intelligence*, vol. 20, no. 4, pp. 401–406, April 1998.

[36] M.H. Kim and J. Kautz, "Characterization for high dynamic range imaging," *Computer Graphics Forum*, vol. 27, no. 2, pp. 691–697, April 2008.

11

Shadow Detection in Digital Images and Videos

Csaba Benedek and Tamás Szirányi

11.1 Introduction

Shadow detection is an important preprocessing task and a hot topic in computer vision. There exist numerous applications which vary in their motivations to address shadows in acquired digital images and video. For example, in *video surveillance* [1], [2], *aerial exploitation* [3], and *traffic monitoring* [4] shadows are usually mentioned as harmful effects, because they make it difficult to separate and track moving objects via background subtraction (Figure 11.1). In *remote sensing*, shadows may reduce the performance of change detection techniques [5]. Similarly, in *scene reconstruction* it is a fundamental problem to distinguish surface edges from illumination differences [6]. It should also be noted that shadow-free images are commonly required for purely visual purposes [6].

FIGURE 11.1 (See color insert.)

Results of background subtraction using the algorithm of Reference [7]. Object silhouettes are strongly corrupted, and multiple moving objects cannot be separated due to cast shadows.

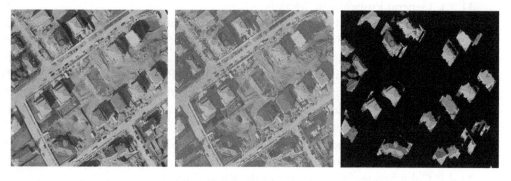

FIGURE 11.2 (See color insert.)

Built-in area extraction using cast shadows: (left) input image, (middle) output of a color-based shadow filter with red areas indicating detected shadows, and (right) built-in areas identified as neighboring image regions of the shadows' blobs considering the sun direction.

On the other hand, shadows may be helpful phenomena in many situations. The so-called *shape from shading* [8] methods derive the three-dimensional (3D) parameters of objects based on estimated shadowing effects. Shadows also provide general descriptors for the illumination conditions in scenes, which can be used for image and video indexing or event analysis [9]. For example, the darkness of a shadow indicates whether an outdoor shot was taken in sunlit or overcast weather; meanwhile the size and orientation of the shadow blobs are related to the time and date of frame capture. If multiple shadows are observable with different darkness, several light sources in the scene can be expected. Object extraction in still images can also be facilitated by shadow detection. In aerial image analysis, it is often necessary to detect static scene objects, such as buildings [10], [11] and trees [12], which constitute a challenging pattern recognition problem. Note that even a noisy shadow map is a valuable information source, because the object candidate regions can be estimated as image areas lying next to the shadow blobs in the sun direction as demonstrated in Figure 11.2.

As suggested above, shadow detection is a wide concept; different classes of approaches should be separated depending on the environmental conditions and the exact goals of the systems. This chapter focuses on the video surveillance problem; demonstrating some challenges and solutions related to shadow detection in digital video. In surveillance video streams, foreground areas usually contain the regions of interest, moreover, an accurate object-silhouette mask can directly provide useful information for several applications, for instance, people detection [13], [14], [15], vehicle detection [4], tracking [16], [17], biometrical identification through gait recognition [18], [19], and activity analysis [7]. However, moving cast shadows on the background make it difficult to estimate shape [20] or behavior [14] of moving objects, because they can be erroneously classified as part of the foreground mask. Considering that under some illumination conditions more than half of the nonbackground image areas may belong to cast shadows, their filtering has a crucial role in scene analysis.

This chapter will build upon a few assumptions for the scene and the input data. First, the camera is fixed and has no significant ego-motion. Expected are static background objects (for example, there is no running river or flickering object in the background); therefore, all motions are caused either by moving objects or by shadows. Moreover, a topically valid image is expected in each moment; this can be obtained by the conventional Gaussian mixture method [7]. There is one emissive light source in the scene (the sun or an artificial source), but the presence of additional effects (for example, reflection), is considered; such effects may change the spectrum of illumination locally. It is assumed that the estimated background values of the pixels correspond to the illuminated surface points.

On the other hand, several properties of real situations are considered. The background may change over time, due to varying lighting conditions and changes of static objects. Crowded and empty scenarios may alternate, and background or shadow colored object parts are expected. Due to the daily changes of the sun position and weather, shadow properties may strongly alter as well.

11.2 Shadow Detection in Video Surveillance: An Overview

The shadow filtering problem has been handled in various ways in the literature. *Geometry-based* approaches estimate the spatial transform between the objects and their cast shadows in the projected image plane [21], [22]. However, these methods are highly restricted to specific conditions and object types; *color filtering* techniques are therefore more commonly used. Methods can also be distinguished based on the requirement for their input. Methods for still images [6], [23], which attempt to find and remove shadows in the single frames independently, are usually used for high quality photos where the background has a uniform color or texture pattern. On the other hand, these methods are less efficient in video surveillance, where images with poor quality and resolution [6] are often expected and the computational complexity should be kept low to allow real-time processing [23]. Some approaches focus on the discrimination of shadow edges and edges due to object boundaries [24], [25]. However, it may be difficult to extract connected foreground regions from a ragged edge map of a noisy video frame [24]. Complex scenes containing several small objects or shadow parts may be disadvantageous for these methods.

Considering the above discussion, this chapter focuses on video and region-based shadow modeling techniques. These techniques can be categorized with respect to the description of the shadow-background color transform, which can be *nonparametric* and *parametric* [26]. Nonparametric techniques are often referred to as shadow invariant approaches, since instead of detecting the shadows they remove them by converting the pixel values into an illuminant invariant feature space. Usually a conventional color space transformation is applied to fulfill this task; the normalized RGB (or rg) [27], [28] and $C_1C_2C_3$ spaces [29] purely contain chrominance color components which are less dependent on luminance. Similar constancy of the hue channel in HSV space is exploited in Reference [30]. However, as Reference [29] points out, illumination invariant approaches have several limitations regarding reflective surfaces and the lighting conditions of the scenes. Outdoors, shadows will have a blue color cast (due to the sky), while lit regions have a yellow cast (sunlight), hence the chrominance color values corresponding to the same surface point in shadow and sunlight [25] may differ significantly. It was also found that the shadow invariant methods often fail in outdoor scenes and are more usable in indoor scenes. Moreover, by ignoring the luminance components of the color, these models become sensitive to noise.

Consequently, *parametric* models will be of interest in this chapter. First, the mean background values of the individual pixels are estimated using a statistical background model [7], then feature vectors from the actual and the estimated background values of the pixels are extracted in order to model the feature domain of shadows in a probabilistic way. Parametric shadow models may be categorized as *local* or *global*.

In a local shadow model [31], independent shadow processes are proposed for each pixel. The local shadow parameters are trained using a second mixture model similar to background-based training [7]. This way, the differences in the light absorption / reflection properties of the scene points can be taken into account. However, each pixel should be shadowed several times till its estimated parameters converge under unchanged illumination conditions; a hypothesis often not satisfied in surveillance videos.

In this chapter, Section 11.3 introduces a novel statistical shadow model. This model follows an approach which characterizes shadows in an image using *global* parameters; this approach describes the relation of the corresponding background and shadow color values. Since this transformation is considered here as a random transformation affected by a perturbation, illumination artifacts are taken into consideration. On the other hand, the shadow parameters are derived from global image statistics; therefore, the model performance is reasonable also on image regions where motion is rare.

Color space choice is a key issue in a number of methods; this problem will be intensively studied in Section 11.4. The initial model presented in Section 11.3 will be extended to the CIE L*u*v* space which allows measuring the perceptual distance between colors using Euclidean distance [32] and in which the color components are approximately uncorrelated with respect to camera noise and changes in illumination [33]. Since the model parameters will be derived statistically, there is no need for accurate color calibration and the CIE D65 standard can be used. It is also not critical to consider an exact physical meaning of the color components, which is usually environment-dependent [29], [34]; only an approximate interpretation of the L, u, v components will be used and the validity of the model will be shown via experiments.

TABLE 11.1

Comparison of various methods.

Reference	Shadow detection	Adaptive	Scenes
[35]	global, constant ratio	no	outdoor
[27]	illumination invariant	no	indoor
[29]	illumination invariant	no	both
[31]	local process	yes	indoor
[36]	no	—	both
[37]	global, constant ratio	no	indoor
[1]	global, probabilistic	yes	both

Section 11.4 presents a detailed qualitative and quantitative study of the color space selection problem of shadow detection. In this sense, this section can be considered both as the premise and generalization of Section 11.3. The choice of the CIE L*u*v* space will be justified here, but on the other hand, experiments will refer to the previously introduced model elements, extending their validity to various color spaces. The reason for dedicating an independent part to this issue is that statistical feature modeling and color space analysis are two different and, in themselves, composite aspects of shadow detection. Although interaction between the two approaches will be emphasized several times, a separate discussion can help the clarity of presentation. Due to the various experiments, the consequences of Section 11.4 may be more generally usable than in the context of the proposed statistical model framework.

For validation, both real surveillance video shots and test sequences from a well-known benchmark set [26] will be used. Table 11.1 summarizes the different goals and tools regarding some of the above mentioned state-of-the-art methods and the proposed model. For a detailed comparison see also Section 11.3.6.

11.3 A Bayesian Approach for Modeling Shadows in Video Scenes

The shadow detection problem will be solved here using a Bayesian image segmentation framework which separates foreground, background and shadow regions in the video frames. Assuming the two dimensional pixel grid S and using a first ordered neighborhood system on S, the procedure assigns a label $\omega(s)$ to each pixel $s \in S$ to form the label-set $\Phi = \{fg,bg,sh\}$ which corresponds to three possible classes: foreground (fg), background (bg), and shadow (sh). The label field is modeled by a Markov random field (MRF) [38]; the segmentation is equivalent to a global labeling $\underline{\omega} = \{[s, \omega(s)] \mid s \in S\}$ and the probability of a given $\underline{\omega} \in \Omega$ follows Gibbs distribution [38].

The observation at pixel s is the three dimensional color vector, expressed here in the CIE L*u*v* space as $\overline{o}(s) = [o_L(s), o_u(s), o_v(s)]^T$. Set $\mathcal{O} = \{\overline{o}(s) \mid s \in S\}$ refers to the global image data. The key point in the model is to define the conditional density functions $p_\phi(s) = P(\overline{o}(s) \mid \omega(s) = \phi)$, for all $\phi \in \Phi$ and $s \in S$. For example, $p_{bg}(s)$ is the probability that the background process generates the observed feature value $\overline{o}(s)$ at s.

Note that foreground modeling is not addressed in this chapter, since it has been extensively covered in the literature. The simplest approach is using uniform foreground distributions $p_{fg} = u$ [35] which is equivalent to outlier detection. More sophisticated models are based on temporal foreground descriptions [36] or pixel state transition probabilities [37]. The model described below uses the spatial foreground calculus [1] which is insensitive to the frame rate parameter of a video stream, thus ensuring robustness in surveillance environments.

The background's and shadow's conditional density functions are defined in Sections 11.3.1 to 11.3.4, and the segmentation procedure will be presented in detail in Section 11.3.6. Note that minimization will be done using the minus-logarithm of the global probability term. Therefore, $\varepsilon_\phi(s) = -\log p_\phi(s)$ will be used to denote the local energy terms in order to simplify the notation.

11.3.1 General Probabilistic Models

The distribution of feature values in the background and in the shadow is modeled here by Gaussian density functions, similarly to References [26], [37], and [39]. For simplicity, the joint distribution of the color components is approximated by a three dimensional Gaussian density function with diagonal covariance matrix $\overline{\overline{\Sigma}}_k(s) = \mathrm{diag}\{\sigma_{k,L}^2(s), \sigma_{k,u}^2(s), \sigma_{k,v}^2(s)\}$ for $k \in \{\mathrm{bg}, \mathrm{sh}\}$. Accordingly, the distribution parameters are the mean vectors $\overline{\mu}_k(s) = [\mu_{k,L}(s), \mu_{k,u}(s), \mu_{k,v}(s)]^T$ and the standard deviation vectors $\overline{\sigma}_k(s) = [\sigma_{k,L}(s), \sigma_{k,u}(s), \sigma_{k,v}(s)]^T$. Using this diagonal model avoids matrix inversion and determinant recovering during the calculation of the probabilities, and the term $\varepsilon_k(s) = -\log p_k(s)$ can be derived directly from the one dimensional marginal probabilities as follows:

$$\varepsilon_k(s) = C + \sum_{i=\{L,u,v\}} \log \sigma_{k,i}(s) + \frac{1}{2}\left(\frac{o_i(s) - \mu_{k,i}(s)}{\sigma_{k,i}(s)}\right)^2, \qquad (11.1)$$

with $C = 2\log 2\pi$. According to Equation 11.1, each feature contributes with its own additional term to the energy calculus. Therefore, the model is modular; the one dimensional model parameters, $[\mu_{k,i}(s), \sigma_{k,i}^2(s)]$, can be estimated separately.

The use of a Gaussian distribution to model the observed color of a single background pixel is well established in the literature, the corresponding parameter estimation procedures can be found in References [7] and [40]. The components of the background parameters $[\overline{\mu}_{bg}(s), \overline{\sigma}_{bg}(s)]$ can be trained in a similar manner to the conventional online K-means algorithm [7]. The vector $[\mu_{bg,L}(s), \mu_{bg,u}(s), \mu_{bg,v}(s)]^T$ estimates the mean background color of pixel s measured over the recent frames, whereas $\overline{\sigma}_{bg}(s)$ is an adaptive noise parameter. An efficient outlier filtering technique [7] excludes most of the nonbackground pixel values from the parameter estimation process, which works without user interaction.

As stated in Section 11.2, shadows are characterized by describing the background-shadow color value transformation in the images. The shadow calculus is based on the illumination-reflection model [41] introduced in Section 11.3.2. This model assumes constant lighting, along with flat and Lambertian reflecting surfaces; however, video surveillance scenes do not usually fulfill these requirements. Therefore, a probabilistic approach

FIGURE 11.3

Illustration of two illumination artifacts using the frame extracted from the *Entrance am* test sequence: (left) input frame with "1" indicating the dark shadow part between the legs (more object parts change the reflected light) and "2" indicating penumbra artifact near the edge of the shadow, (middle) segmented image using the constant ratio model which causes errors, (right) segmented image using the proposed model which is more robust.

presented in Section 11.3.3 will be used to describe the deviation of the scene from the ideal surface assumptions in order to obtain more robust shadow detection (Figure 11.3).

11.3.2 Shadow Description by Lambertian Color Features

According to the illumination model [41] the response $g(s)$ of a given image sensor placed at pixel s can be written as

$$g(s) = \int e(\lambda, s)\rho(\lambda, s)v(\lambda)d\lambda, \qquad (11.2)$$

where $e(\lambda, s)$ is the illumination function at a given wavelength λ, the term $\rho(s)$ depends on the surface albedo and geometry, and $v(\lambda)$ denotes the sensor sensitivity. Accordingly, the difference between the shadowed and illuminated background values of a given surface point is caused only by the different local value of $e(\lambda, s)$. In outdoor scenes, the illumination function observed in sunlight is the composition of the direct component (sun), the Rayleigh scattering (sky), resulting in a blue tinge to ambient light [42], and residual light components reflected from other objects. On the other hand, the effect of the direct component is missing in the shadow.

Although the validity of Equation 11.2 is already limited by several scene assumptions [41], it is in general still too difficult to exploit appropriate information about the corresponding background-shadow values since the components of the illumination function are unknown. Therefore, further strong simplifications are used in the applications. According to Reference [6], the camera sensors must be exact Dirac delta functions $v(\lambda) = q_0 \cdot \delta(\lambda - \lambda_0)$ and the illumination must be Planckian [43]. In this case, Equation 11.2 implies the well-known constant ratio rule. Namely, the ratio of the shadowed $g_{sh}(s)$ and illuminated value $g_{bg}(s)$ of a given surface point is considered to be constant over the image, that is, $g_{sh}(s)/g_{bg}(s) = A$.

The constant ratio rule has been used in several applications [35], [37], [39]. Here the shadow and background Gaussian terms corresponding to the same pixel are related via a globally constant linear density transform. In this way, the results may be reasonable when all the direct, diffused and reflected light can be considered constant over the scene.

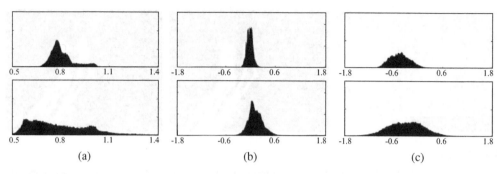

FIGURE 11.4

Histograms of (a) ψ_L, (b) ψ_u, and (c) ψ_v values for (top) shadowed and (bottom) foreground points collected over a 100-frame period of the video sequence *Entrance pm* (frame rate 1 fps).

However, the reflected light may vary over the image in case of several static or moving objects, and the reflecting properties of the surfaces may differ significantly from the Lambertian model (see Figure 11.3). The efficiency of the constant ratio model is also restricted by several practical reasons, like quantization errors of the sensor values, saturation of the sensors, imprecise estimation of $g_{bg}(s)$ and A, or video compression artifacts. Based on the experiments presented in Section 11.3.6, these inaccuracies cause poor detection rates in some outdoor scenes.

11.3.3 Proposed Shadow Model

The previous section suggests that the ratio of the shadowed and background luminance values of the pixels may be useful, but not powerful enough as a descriptor of the shadow process. Instead of constructing a more difficult illumination model, for example in 3D with two cameras, the problems can be overcome using a statistical model. Each pixel s can be associated with the variable $\psi_L(s)$ defined as:

$$\psi_L(s) = \frac{o_L(s)}{\mu_{\mathrm{bg,L}}(s)}, \tag{11.3}$$

where, as defined earlier, $o_L(s)$ is the observed luminance value at s, and $\mu_{\mathrm{bg},L}(s)$ is the mean value of the local Gaussian background term estimated over the previous frames [7].

Thus, if the $\psi_L(s)$ value is close to the estimated shadow darkening factor, s is more likely to be a shadowed point. More precisely, in a given video sequence, the distribution of the shadowed ψ_L values can be estimated globally in the video parts. Based on experiments with manually generated shadow masks, a Gaussian approximation seems to be reasonable regarding the distribution of shadowed ψ_L values. Figure 11.4 shows the global ψ statistics regarding a 100-frame period of one outdoor test sequence. For comparison, this figure also shows the statistics for the foreground points which follow a significantly different, more uniform distribution.

Due to the spectral differences between the direct and ambient illumination, cast shadows may also change the u and v color components [25]. An offset between the shadowed and background u values of the pixels can be efficiently modeled by a global Gaussian term in a given scene (similarly for v component). Hence, $\psi_u(s)$ (and $\psi_v(s)$) can be defined as:

FIGURE 11.5 (See color insert.)

Different parts of the day on *Entrance* sequence with the corresponding segmentation results: (a) *morning am*, (b) *noon*, (c) *afternoon pm*, and (d) *wet weather*.

$$\psi_u(s) = o_u(s) - \mu_{\text{bg,u}}(s). \qquad (11.4)$$

Note that as shown in Figure 11.4, the shadowed $\psi_u(s)$ and $\psi_v(s)$ values follow approximately normal distributions.

Consequently, the shadow color process is characterized by a three dimensional Gaussian random variable:

$$\forall s \in S : \overline{\psi}(s) = [\psi_L(s), \psi_u(s), \psi_v(s)]^T \sim N[\overline{\mu}_\psi, \overline{\sigma}_\psi]. \qquad (11.5)$$

Using Equations 11.3 and 11.4, the color values in the shadow at a given pixel position are also generated by a Gaussian distribution, that is,

$$[o_L(s), o_u(s), o_v(s)]^T \sim N[\overline{\mu}_{\text{sh}}(s), \overline{\sigma}_{\text{sh}}(s)] \qquad (11.6)$$

with the following parameters:

$$\mu_{\text{sh},L}(s) = \mu_{\psi,L} \cdot \mu_{\text{bg},L}(s), \qquad (11.7)$$

$$\sigma_{\text{sh},L}^2(s) = \sigma_{\psi,L}^2 \cdot \mu_{\text{bg},L}^2(s). \qquad (11.8)$$

For the u (and similarly v) component, the following can be written:

$$\mu_{\text{sh},u}(s) = \mu_{\psi,u} + \mu_{\text{bg},u}(s), \quad \sigma_{\text{sh},u}^2(s) = \sigma_{\psi,u}^2. \qquad (11.9)$$

11.3.4 Parameter Settings

The proposed method, built into a 24-hour surveillance system of a university campus (Figure 11.5), works with scene-dependent and condition-dependent parameters. *Scene-dependent* parameters can be considered constant in a specific field, and are influenced by, for example, camera settings and prior knowledge about the appearing objects or reflection properties. Strategies on how to set these parameters if a surveillance environment is given will be provided later. *Condition-dependent* parameters vary in time in a scene; therefore, adaptive algorithms should be used in this case. Note that as discussed in Section 11.3.1, only the one dimensional marginal distribution parameters should be estimated for the background and shadow processes. The *background* parameter estimation and update procedure are automated, based on the work in Reference [7] which presents reasonable results and it is computationally more effective than the standard expectation-maximization algorithm.

As shown in Figure 11.5, changes in global illumination significantly alter the *shadow* properties. Moreover, changes can occur rapidly; in indoor scenes due to switching on/off different light sources and in outdoor scenes due to the appearance of clouds. Regarding the shadow parameter settings, parameter initialization and re-estimation are discriminated. From a practical point of view, initialization may be supervised by marking shadowed regions in a few video frames by hand, once after switching on the system. Maximum likelihood estimates of the shadow parameters can be calculated based on the training data. On the other hand, there is usually no opportunity for continuous user interaction in an automated surveillance environment; thus, the system must adapt to illumination changes by initiating a claim to an automatic re-estimation procedure. Therefore, supervised initialization is used here. The parameter adaptation process will be described later.

According to Section 11.3.3, the shadow process has six parameters, stored in three-component vectors $\overline{\mu}_{\psi}$ and $\overline{\sigma}_{\psi}$. Figure 11.6a shows the one-dimensional histograms for the ψ_L, ψ_u and ψ_v values of shadowed points for each video shot. It can be observed that while the variation of parameters $\overline{\sigma}_{\psi}$, $\mu_{\psi,u}$ and $\mu_{\psi,v}$ are low, $\mu_{\psi,L}$ varies in time significantly. Therefore, the parameters should be updated in two different ways.

11.3.4.1 Re-Estimation of the Chrominance Parameters

The update procedure for parameters $[\mu_{\psi,u}, \sigma_{\psi,u}]$ and $[\mu_{\psi,v}, \sigma_{\psi,v}]$ is similar to the one presented in Reference [44]. The procedure will be shown for the u component only, as the v component is updated in the same way.

The parameters are re-estimated at fixed time-intervals $\mathscr{T}$, set as $\mathscr{T} = 60$ sec. Parameters at time t are denoted here by $\mu_{\psi,u}[t], \sigma_{\psi,u}[t]$. The term

$$W_{t_2} = \{\psi_u^{[t]}(s) | t = t_1, \dots, t_2 - 1, \ \omega^{[t]}(s) = \text{sh}, \ s \in S\} \tag{11.10}$$

denotes the set containing the observed ψ_u values collected over the pixels detected as shadows between time $t_1 = t_2 - \mathscr{T}$ and t_2. In the above equation, the upper index $[t]$ refers to time, $|W_{t_2}|$ is the number of the elements in W_{t_2}, and M_{t_2} and D_{t_2} are the empirical mean and the standard deviation values of W_{t_2}. The parameters are updated as follows:

$$\mu_{\psi,u}[t_2] = (1 - \xi^{[t_2]}) \cdot \mu_{\psi,u}[t_1] + \xi^{[t_2]} \cdot M_{t_2}, \tag{11.11}$$

$$\sigma_{\psi,u}^2[t_2] = (1 - \xi^{[t_2]}) \cdot \sigma_{\psi,u}^2[t_1] + \xi^{[t_2]} \cdot D_{t_2}^2. \tag{11.12}$$

Parameter $\xi^{[t]}$, for $0 \le \xi^{[t]} \le 1$, is a weighting term depending on $|W_t|$. Namely, $\xi^{[t]}$ and the influence of the M_t and D_t^2 terms increase with the number of detected shadow points.

11.3.4.2 Re-Estimation of the Luminance Parameters

The parameter $\mu_{\psi,L}$ corresponds to the average background luminance darkening factor of the shadow. Except from windowless rooms with constant lighting, $\mu_{\psi,L}$ is strongly

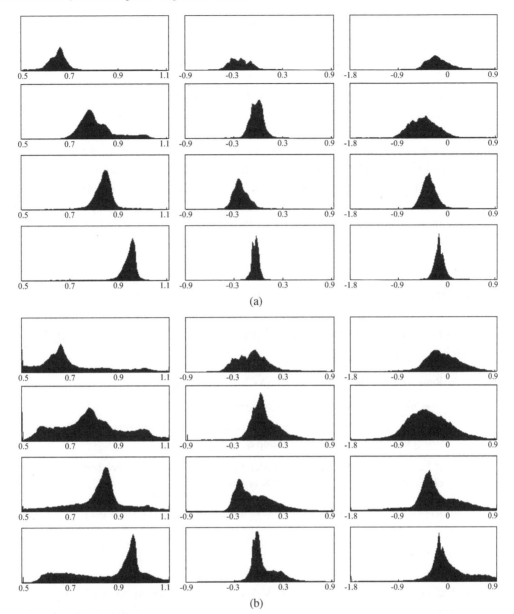

FIGURE 11.6

Extracted $\overline{\psi}$ statistics from four sequences recorded by the entrance camera of the university campus: (a) shadow statistics, (b) nonbackground statistics, (left) ψ_L, (middle) ψ_u, and (right) ψ_v. Rows correspond to video shots from different parts of the day.

condition dependent. In outdoor scenes, it can vary between 0.6 in direct sunlight and 0.95 in overcast weather. The simple re-estimation from the previous section does not work in this case, since the illumination properties between time t and $t + \mathcal{T}$ may change a lot, which would result in absolutely false detected shadow values in set W_t presenting false M_t and D_t parameters for the re-estimation procedure.

For this reason, the actual $\mu_{\psi,L}$ value is obtained using the statistics of all nonbackground ψ_L values (with background filtering performed using the Stauffer-Grimson algorithm). Figure 11.6b shows that the peaks of the nonbackground ψ_L-histograms are approximately in the same location as they were in Figure 11.6a. The videos of the first and second rows were recorded around noon, where the shadows were relatively small, but the peak is still in the right location.

The previous experiments encourage identifying $\mu_{\psi,L}$ with the location of the peak on the nonbackground ψ_L-histograms for the scene. The $\mu_{\psi,L}$ value is updated as depicted in Algorithm 11.1. Namely, using a data structure $[\psi_L, t]$ which contains a ψ_L value with its timestamp, the latest occurring $[\psi_L, t]$ pairs of the nonbackground points in a set $\mathscr{Q}$ are stored, and the histogram h_L of the ψ_L values in $\mathscr{Q}$ are continuously updated. The key point is the management of set $\mathscr{Q}$; therefore, two parameters, MAX and MIN, are defined to control the size of $\mathscr{Q}$.

ALGORITHM 11.1 Algorithm for updating the $\mu_{\psi,L}$ shadow parameter.

1. For each frame t determine $\Psi_t = \{ [\psi_L^{[t]}(s), t] \mid s \in S, \omega^{[t]}(s) \neq \text{bg}\}$.

2. Append Ψ_t to $\mathscr{Q}$.

3. Remove elements from $\mathscr{Q}$ as follows:
   - if $|\mathscr{Q}| < \text{MIN}$, keep all the elements,
   - if $|\mathscr{Q}| \geq \text{MIN}$, find the oldest timestamp t_e in $\mathscr{Q}$ and remove all the elements from $\mathscr{Q}$ with time stamp t_e.

4. If $|\mathscr{Q}| > \text{MAX}$ after step 3: in order of their timestamp remove further "old" elements from $|\mathscr{Q}|$ till $|\mathscr{Q}| \leq \text{MAX}$ is reached.

5. Update the histogram h_L regarding $\mathscr{Q}$ and apply $\mu_{\psi,L}^{[t+1]} = \text{argmax}\{h_L\}$

Consequently, $\mathscr{Q}$ always contains the latest available ψ_L values. The algorithm keeps the size of $\mathscr{Q}$ between prescribed bounds MAX and MIN ensuring the topicality and relevancy of the data contained. The actual size of $\mathscr{Q}$ is around MAX in the case of cluttered scenarios. In the case of few or no motions in the scene, the size of $\mathscr{Q}$ decreases until it reaches MIN. This increases the influence of the forthcoming elements, and causes quicker adaptation, since it is faster to modify the shape of a smaller histogram.

The parameter $\sigma_{\psi,L}$ is updated similarly to $\sigma_{\psi,u}$ but only in the time periods when $\mu_{\psi,L}$ does not change significantly. Note that the above update process may fail in shadow-free scenarios. However, that case occurs mostly under artificial illumination conditions, where the shadow detector can be switched off.

11.3.5 MRF Optimization

The MAP estimator is realized by combining a conditional independent random field of signals and an unconditional Potts model [45]. The optimal segmentation corresponds to

TABLE 11.2

Computational efficiency comparison of various methods (using published frame-rates).

Criterion	Method / Reference			
	[31]	[36]	[37]	Proposed
Classes	3	2	3	3
MRF Opt	—	graph cut	ICM	ICM
Frame-rate	10 fps	11 fps	1-2 fps	3 fps

the global labeling, $\widehat{\omega}$, defined as follows:

$$\widehat{\omega} = \arg\min_{\omega \in \Omega} \left\{ \sum_{s \in S} \underbrace{-\log P\left(\overline{o}(s) \mid \omega(s)\right)}_{\varepsilon_{\omega(s)}(s)} + \sum_{r,s \in S} \Theta\left(\omega(r), \omega(s)\right) \right\}, \qquad (11.13)$$

where the minimum is searched over all the possible segmentations (Ω) of a given input frame. The first part of Equation 11.13 contains the sum of the local class-energy terms regarding the pixels of the image (see Equation 11.1). The second part is responsible for the smooth segmentation; with $\Theta\left(\omega(r), \omega(s)\right) = 0$ if s and r are not neighboring pixels, and

$$\Theta\left(\omega(r), \omega(s)\right) = \begin{cases} -\delta & \text{if } \omega(r) = \omega(s), \\ +\delta & \text{if } \omega(r) \neq \omega(s). \end{cases} \qquad (11.14)$$

As for optimization, the deterministic modified Metropolis (MMD) [46] relaxation method was found similarly efficient but significantly faster for this task than the original stochastic algorithm [47]. Namely, it runs about 1 fps when processing 320×240 images whereas the running speed of the ICM method [48] with the proposed model is 3 fps in exchange for some degradation in the segmentation results. For comparison, frame-rates of three latest reference methods are shown in Table 11.2. It can be observed that the proposed model has approximately the same complexity as [37]. Although the processing speed of methods in References [31] and [36] is notably higher, one should consider that the method in Reference [31] does not use any spatial smoothing (like MRF), thus requiring a separate noise filter in the postprocessing phase. On the other hand, the method in Reference [36] performs only a two-class segmentation (background and foreground). That simplification enables using the quick graph-cut based MRF optimization techniques, which is not the case for three classes [49].

11.3.6 Experimental Results

The goal of this section is to qualitatively and quantitatively demonstrate the benefit of using the novel shadow model introduced in this chapter. The proposed method was validated on several test sequences; here, the results are shown for following videos:

- *Laboratory* test sequence from the ATON benchmark set [26]. This shot contains a simple environment where previous methods [37] have produced already accurate results.

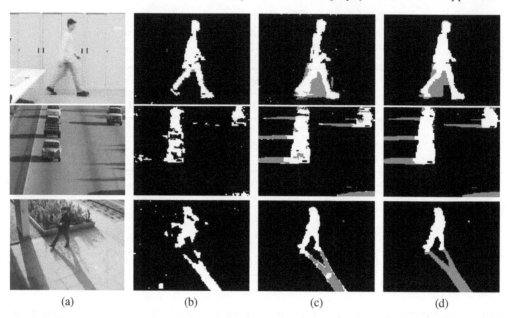

FIGURE 11.7

Shadow model validation: (a) video image, (b) $C_1C_2C_3$ space-based illumination invariants [29], (c) constant ratio model of Reference [35] without object-based postprocessing, and (d) proposed statistical shadow model. Test image sequences: (top) *Laboratory*, (middle) *Highway*, and (bottom) *Entrance am*.

- *Highway* video (ATON benchmark set). This sequence contains dark shadows but homogenous background without illumination artifacts. In contrast to Reference [35], the proposed method reaches the appropriate results without postprocessing, which is strongly environment-dependent.

- *Corridor* indoor surveillance video. Although it is a simple office environment the bright objects and background elements often saturate the image sensors and it is hard to accurately separate the white shirts of the people from the white walls in the background.

- Four surveillance video sequences captured by the entrance (outdoor) camera of the university campus in different lighting conditions (Figure 11.5: *Entrance am, Entrance noon, Entrance pm,* and *Entrance overcast*). These sequences contain difficult illumination and reflection effects and suffer from sensor saturation (dark objects and shadows). Here, the presented model improves the segmentation results significantly versus previous methods.

Figure 11.7 shows the results of different shadow detectors. For the sake of comparison, an illumination invariant method based on Reference [29] and a constant ratio model [35] were implemented in the same framework. It was observed that the performance differences between the previous and the proposed methods increase with surveillance scene complexity. In the *Laboratory* sequence, the constant ratio and the proposed method are similarly accurate. For the *Highway* video, the illumination invariant and constant ratio methods approximately find objects without shadows; these results are, however, much

TABLE 11.3
Overview of the evaluation parameters.

Test video	Frames*	fre**	Duration (min) ***
Laboratory	205	2-4 fre†	1:28
Entrance am	160	2 fre	1:20
Entrance pm	75	1 fre	1:15
Entrance noon	251	1 fre	4:21
Highway	170	5-8 fre†	0:29

*The number of frames in the ground truth set, **fre – the number of frames with ground-truth within one second of the video, ***Length of the evaluated video part, †fre was higher in busy scenarios.

noisier compared to that of the proposed model. The illumination invariant method fails completely on the *Entrance am* surveillance video, as shadows are not removed while the foreground component is noisy due to the lack of using luminance features in the model. The constant ratio model also produces poor results; due to the long shadows and various field objects the constant ratio model becomes inaccurate. The proposed model handles these artifacts in a robust way.

The quantitative evaluations are done through manually generated ground-truth sequences. Since the application's goal is foreground detection, the crossover between shadow and background does not account for errors. Denoting the number of correctly identified foreground pixels by TP, misclassified background pixels by FP, and misclassified foreground pixels of evaluation images by FN, the evaluation metrics consisting of the *Recall* rate, Rc, and the *Precision* of the detection, Pr, can be defined as follows:

$$\text{Recall} = \frac{TP}{TP+FN}, \quad \text{Precision} = \frac{TP}{TP+FP}. \tag{11.15}$$

In addition to these measures, the so-called F-measure (FM) [50]

$$FM = \frac{2 \cdot Rc \cdot Pr}{Rc + Pr}. \tag{11.16}$$

which combines Rc and Pr in a single efficiency measure (it is the harmonic mean of Rc and Pr) will be also used. It should be noted that while Rc and Pr have to be used jointly to characterize a given algorithm, FM constitutes a stand-alone evaluation metrics.

TABLE 11.4
Quantitative evaluation results.

Dataset	Recall Rc		Precision Pr		FM-measure	
	CR	SS	CR	SS	CR	SS
Laboratory	0.950	0.941	0.883	0.929	0.915	0.935
Highway	0.886	0.890	0.644	0.805	0.746	0.845
Entrance am	0.946	0.968	0.596	0.774	0.731	0.861
Entrance noon	0.980	0.963	0.742	0.833	0.845	0.894
Entrance pm	0.972	0.961	0.621	0.830	0.756	0.891

CR – constant ratio model, SS – proposed statistical shadow model.

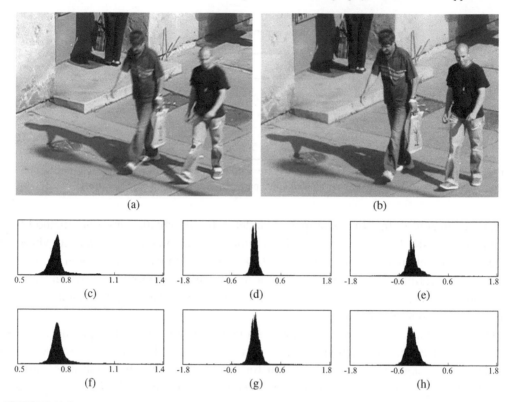

FIGURE 11.8

Distribution of the shadowed $\overline{\psi}$ values in simultaneous sequences from a street scenario recorded by different CCD cameras: (a,c-e) three-sensor camera recorder, (b,f-h) digital camera with the Bayer CFA; (c,f) L component, (d,g) u component, and (e,h) u component.

For numerical validation, 861 frames were used from the *Laboratory*, *Highway*, *Entrance am*, *Entrance noon*, and *Entrance pm* sequences. Table 11.3 lists some details about these test sets. Table 11.4 compares the detection results of the proposed method and the constant ratio model, showing that the proposed shadow calculus improves the precision rate as it significantly decreases the number of false negative shadow pixels and simultaneously preserves the high foreground recall rate. Consequently, the proposed model outperforms the constant ratio method on all test sequences in terms of the *FM*-measure.

11.3.6.1 Influence of CCD Selection on the Shadow Domain

A statistical shadow model was introduced without any knowledge about the technical details and embedded control of the different cameras. However, to explore the influence of various camera technologies on the performance of shadow detection methods, test videos were simultaneously recorded for a street scenario using a three charge-coupled device (CCD) digital video camcorder and a conventional digital camera equipped with a Bayer color filter array (CFA). By examining the corresponding shadow domains in Figure 11.8, it can be observed that the distributions of the shadowed $\overline{\psi}$ values are very similar. However, stronger noise in the Bayer CFA camera results in higher variance parameters associated with the u and v components.

11.4 Color Space Selection in Cast Shadow Detection

This section focuses on a particular aspect of shadow detection, that is, improving segmentation performance through color space selection. Note that for practical purposes, the number of free parameters of the method should be kept low. Experimentation will be conducted to obtain some knowledge about: i) the performance improvement due to using color images instead of grayscale images, ii) the performance gain resulting from using uncorrelated spaces instead of the standard RGB space, iii) efficiency of chrominance (illumination invariant), luminance, or mixed spaces, and iv) the relation between performance of the methods and the type of visual scenes. In the experiment, color-based clustering of the individual pixels and Bayesian foreground-background-shadow segmentation through the generalization of the model introduced in Section 11.3 will be considered. It will be shown that that CIE L*u*v* color space is the most effective choice in both cases.

11.4.1 Color Spaces: Significance of the Right Choice

Appropriate color space selection is a crucial step for many image processing problems [25], [54], [55]. Since the shadow model proposed in Section 11.3 is primarily based on describing the shadow's color domain, issues of color spaces should also be investigated in this case. Although shadow detection is a well examined problem and some comparative works [26], [56] have also been published on this topic, previous reviews classify and compare the existing methods based on their *model structures*. It is noted in Reference [26] that the methods work in different color spaces, like RGB [35] and HSV [51]. However, it remains open-ended, how important is the appropriate color space selection, and which color space is the most effective regarding shadow detection. Reference [39] uses only gray levels for shadow segmentation, whereas other approaches deal with CIE L*u*v* [31] and CIE L*a*b* [52]. An overview is provided in Table 11.5. Note that experimental evalua-

TABLE 11.5
Color space selection in the state-of-the-art methods.

Reference	Color space	PPCC
[28]	rg	invariant
[29]	$C_1C_2C_3$	invariant
[27]	rg	invariant
[35]	RGB	1
[39]	grayscale	2
[37]	grayscale	2
[51]	HSV	1.33
[31]	CIE L*u*v*	2
[52]	CIE L*a*b*/HSV	—
[53]	RGB	— ‡
[2]	all from above	2

PPCC – the average number of shadow parameters for one color channel in parametric methods; ‡ proportional to the number of support vectors after training.

tion of color spaces for shadow edge classification can be found in Reference [25] and that this chapter addresses detection of the shadowed and foreground regions, which is a fairly different problem.

For the above reasons, this section aims to experimentally compare different color models for the purpose of cast shadow detection in digital video. Since the validity of such experiments is limited to the examined model structures, it is important to make the comparison in a relevant framework. Taking a general approach, the task is considered as a classification problem in the space of the extracted features, describing the different cluster domains with relatively few free parameters. Note that most models in Table 11.5 use two parameters for each color channel; drawbacks of methods which use fewer parameters were discussed in Section 11.3.

Popular models can be categorized as *deterministic* (per pixel) [51] or *statistical* (probabilistic) [35]. Up to now, this chapter has only dealt with statistical models, since these models have proved to be advantageous considering the whole segmentation process. A deterministic method will be introduced here; the pixels are classified independently before the rate of the correct pixel-classification is investigated. In this way, a relevant quantitative comparison of the different color spaces can be achieved, because the decision for each pixel depends only on the corresponding local color-feature value. Note that postprocessing and prior effects whose efficiency may be environment-dependent will not be considered. A probabilistic interpretation of this model will be given and used in the MRF framework which was introduced in Section 11.3. The results after MRF optimization will be compared both qualitatively and quantitatively.

11.4.2 Generalized Feature Vector of Shadow Separation

Feature extraction is done similarly to Section 11.3, although here a generalization of the $\overline{\psi}$ shadow features will be used to handle different color spaces. It should be remembered that the constant ratio model introduced in Section 11.3.2 assumes that the ratio of the shadowed and illuminated sensor values is nearly constant over the entire image(s). To handle various artifacts, one can prescribe a *domain* [53] or a *distribution* (see Section 11.3.3) instead of a single value for the ratios, which results in a powerful detector.

The following examines how one can use this approach in different color systems. It is assumed here that the camera output the frames in the RGB space, and for the different color space conversions the formulas presented in Reference [57] will be applied. The ITU D65 standard is used again for the calibration of the CIE L*u*v* and L*a*b* spaces.

Following the strategy used in the CIE L*u*v*-based model, the color components directly related to the brightness of the pixels will be handled separately from the remaining ones which correspond to chrominances of the observed colors. In this way, the color spaces can be classified as chrominance spaces (e.g., the normalized rg and $C_1C_2C_3$ spaces), luminance spaces (e.g., grayscale and RGB), and mixed spaces (e.g., HSV, CIE L*u*v* and L*a*b*). Table 11.6 shows classification of channels for different color spaces.

The shadow descriptor is derived in an analogous manner to the approach of Section 11.3.3. Namely, the probabilistic ratio method is used for the luminance components, while the offsets between the shadowed and illuminated chrominance values of the pixels are modeled by a Gaussian additive term. For a pixel expressed in a given color space

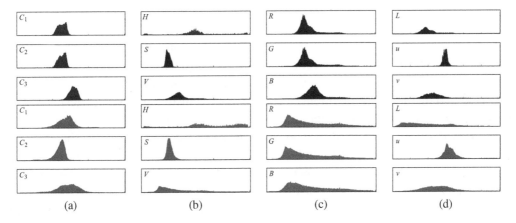

FIGURE 11.9

One dimensional projection of histograms of $\overline{\psi}$ values in the *Entrance pm* test sequence: (top) shadow, (bottom) foreground; (a) $C_1C_2C_3$, (b) HSV, (c) RGB, and (d) L*u*v*.

as a three-component vector $[o_0, o_1, o_2]$ and its estimated (illuminated) background value expressed as $[\mu_{bg,0}, \mu_{bg,1}, \mu_{bg,2}]$, the shadow descriptor $\overline{\psi} = [\psi_0, \psi_1, \psi_2]$ can be defined as

$$\psi_i(s) = \frac{o_i(s)}{\mu_{bg,i}(s)}, \quad \text{for } i = 0, 1, 2, \tag{11.17}$$

if i being the index of a luminance component, and

$$\psi_i(s) = o_i(s) - \mu_{bg,i}(s), \quad \text{for } i = 0, 1, 2, \tag{11.18}$$

if i is the index of a chrominance component. The descriptor in grayscale and in the rg space are defined similarly to Equations 11.17 and 11.18 considering that $\overline{\psi}$ will be a scalar and a two-dimensional vector, respectively.

The efficiency of the proposed feature selection regarding three color spaces is demonstrated in Figure 11.9 on the plots of one-dimensional marginal histograms of the ψ_0, ψ_1, and ψ_2 values for manually marked shadowed and foreground points of a 75-frame long outdoor surveillance video sequence, *Entrance pm*. Apart from some outliers, the shadowed ψ_i values lie for each color space and each color component in a short interval, while the difference between the upper and lower bounds of the foreground values is usually greater.

TABLE 11.6

Luminance and chrominance channels in different color spaces.

| Channel | \multicolumn{7}{c}{Color space} |
	gray	rg	$C_1C_2C_3$	HSV	RGB	L*a*b*	L*u*v*
Luminance	g	—	—	H	R,G,B	L*	L*
Chrominance	—	r,g	C_1,C_2,C_3	S,V	—	a*,b*	u*,v*

11.4.3 Quantitative Comparison through a Deterministic Classifier

In this section, the MRF concept will temporarily be put aside. Using a deterministic approach instead, the shadow detection problem is considered as a simple classification task in the $\overline{\psi}$-feature space. As shown in Figure 11.9, while the $\overline{\psi}$ statistics characterize the scene and illumination conditions, the foreground $\overline{\psi}$ histograms only correspond to the occurring foreground objects in the evaluated sequence. On the other hand, an efficient shadow model is also expected to work with differently colored objects. Therefore, the upcoming discrimination process will follow a one-class-classification approach. Namely, the pixel s will be classified as a shadowed point, if its $\overline{\psi}(s)$ value lies in the estimated *shadow domain*, and the *outlier points* will be labeled as foreground. As usual, the shadow domain is defined by a manifold having a prescribed number of free parameters which fit the model to a given scene/situation. For grayscale images the shadowed ψ features should be included by an interval [37], while for color scenes different domain models can be used. These models include a three-dimensional rectangular bin [51] (ratio/difference values for each channel lie between defined threshold), an ellipsoid [35], or the domain may have a general shape [53]. In the latter case a support vector domain description is proposed in the RGB color ratio space.

For each domain selection, an overlap between the classes should be considered; for example, foreground points may appear whose feature values lie in the shadow domain. Therefore, the optimal domain should be as narrow as possible while containing almost all the feature values corresponding to the shadowed points. Accordingly, if one only prescribes that a shadow descriptor should be accurate, the most general domain shape seems to be the most appropriate. However, in practice, issues related to parameter estimation and adaptation have to be considered (see Section 11.3.4). Therefore, the domains with relatively few free parameters, for which an automatic update strategy can be constructed, are generally preferred.

It can be observed that according to Figure 11.9, the shadowed ψ_0, ψ_1, and ψ_2 values follow approximately normal distributions. Therefore, a 3D joint normal representation of the $\overline{\psi}$ features in shadows is straightforward (similar to Section 11.3). Since the equipotential surfaces of the 3D Gaussian density functions are ellipsoids, a natural choice is to use an elliptical shadow domain boundary. Here, the equation of a standard ellipsoid body having *parallel axes* with the coordinate axes in the $\psi_0 - \psi_1 - \psi_2$ Cartesian coordinate system will be used:

$$\text{Pixel } s \text{ is shadowed} \Leftrightarrow \sum_{i=0}^{2} \left(\frac{\psi_i(s) - a_i}{b_i} \right)^2 \leq 1, \tag{11.19}$$

where $[a_0, a_1, a_2]$ is the coordinate of the ellipsoid center and (b_0, b_1, b_2) are the semi-axis lengths. In other words, $[a_0, a_1, a_2]$ is equivalent to the mean $\overline{\psi}(s)$ value of shadowed pixels in a given scene, while b_0, b_1, and b_2 depend on the spatiotemporal variance of the $\overline{\psi}(s)$ measurements under shadows. It will be shown later that the similarity to the $\overline{\mu}_\psi$ and $\overline{\sigma}_\psi$ parameters from Section 11.3 is not by chance; thus, parameter adaptation can also be done in a similar manner.

Note that with the SVM method [53], the number of free parameters is related to the number of support vectors, which can be much greater than the six scalars of the proposed model. Moreover, for each situation, a novel SVM should be trained. Also note that one

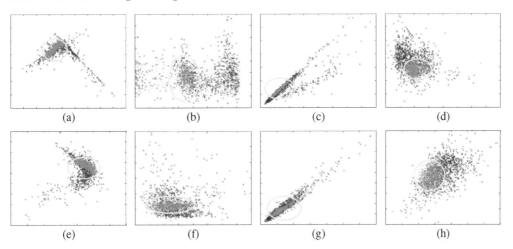

FIGURE 11.10 (See color insert.)

Two dimensional projection of foreground (dark) and shadow (bright) $\overline{\psi}$ values in the *Entrance pm* test sequence: (a) $C_1 - C_2$, (b) $H - S$, (c) $R - G$, (d) $L - u$, (e) $C_2 - C_3$, (f) $S - V$, (g) $G - B$, and (h) $u - v$. The ellipse denotes the projection of the optimized shadow boundary.

could use an arbitrarily oriented ellipsoid, but compared to Equation 11.19, it is more difficult to define since it needs the accurate estimation of nine parameters. The domain defined by Equation 11.19 becomes an interval for grayscale images and a two dimensional ellipse for the rg space.

Figure 11.10 shows the two dimensional scatter plots about the foreground and shadow $\overline{\psi}$ values. It can be observed that the components of vector $\overline{\psi}$ are strongly correlated in RGB space (and also in $C_1 C_2 C_3$) and that the previously defined ellipse cannot present a narrow boundary. In HSV space, the shadowed values are not within a convex hull, even if the hue component is considered periodic (hue $= 2k\pi$ means the same color for each $k = 0, 1, \ldots$). Based on the above facts, the CIE L*u*v* space seems to be a good choice. In the following, this statement will be supported by numerical results.

11.4.3.1 Evaluation of the Deterministic Model

The evaluations were done using manually generated ground-truth versions of the test video *Laboratory*, *Highway*, *Entrance am*, *Entrance noon* and *Entrance pm*, with the same test parameters as before (Table 11.3).

This section will show the tentative limits of the elliptical shadow domain defined by Equation 11.19. The goal of these experiments is to compare the foreground-shadow discriminating ability of the different color spaces purely based on the extracted per pixel $\overline{\psi}$ features. Therefore, the parameters are set here manually and do not take into consideration local connectivity or postprocessing. In the experiments, two sets of $\overline{\psi}$ values corresponding respectively to manually marked foreground and shadowed pixels are collected for each test sequence.

Figure 11.11 shows the plots of the *Precision* (*Pr*) and *Recall* (*Rc*) values achieved for some optimized ellipse parameters using the *Laboratory* and *Entrance pm* test sequences. It can be observed that the CIE L*a*b* and L*u*v* spaces produce the best results (highest

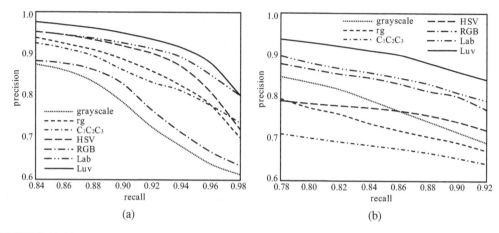

FIGURE 11.11

Evaluation of the deterministic model using recall-precision curves corresponding to different parameter-settings for (a) *Laboratory* and (b) *Entrance pm* sequences.

Pr/Rc curves) in both cases. However, the relative performance of the other color systems strongly varies for these two test videos. In the indoor scene, the grayscale and RGB segmentation procedures are less efficient than the other ones, whereas for the outdoor scenes the performance of the chrominance spaces is prominently poor.

Table 11.7 shows the achieved *FM* rates. Also here, it can be seen that the CIE L*a*b* and L*u*v* spaces are the most efficient. As for the other color systems, in sequences containing dark shadows (*Entrance pm*, *Highway*), the chrominance spaces produce poor results, while the luminance and mixed spaces are similarly effective. Performance of the chrominance spaces is reasonable for brighter shadows (*Entrance am*, *Laboratory*), whereas the luminance spaces are relatively poor. In the latter case, the color constancy of the chrominance channels seems to be more relevant than the luminance-darkening domain. It was also observed that the hue coordinate in HSV is very sensitive to illumination artifacts (see also Figure 11.9), thus the HSV space is more efficient in the case of light shadows. Table 11.8 summarizes the relationship between the darkness of shadows and the performance of various color spaces. In this table, darkness is characterized by the mean of the grayscale ψ_0 values of shadowed points.

TABLE 11.7

Evaluation of the deterministic model (Equation 11.16) using the *FM* measure.

Test video	Color space						
	gray	rg	$C_1C_2C_3$	HSV	RGB	Lab	Luv
Laboratory	0.860	0.889	0.883	0.912	0.863	0.922	0.933
Highway	0.849	0.766	0.761	0.835	0.850	0.851	0.855
Entrance am	0.849	0.920	0.926	0.931	0.861	0.954	0.956
Entrance noon	0.916	0.898	0.890	0.936	0.941	0.946	0.953
Entrance pm	0.817	0.774	0.758	0.814	0.845	0.862	0.882

TABLE 11.8

Indication of the most and least successful color spaces for various test sequences based on the experiments in Section 11.4.3.1. For numerical evaluation refer to Figure 11.11 and Table 11.7.

Test video	Scene	Dark†	Worst	Best
Laboratory	indoor	0.73	gray, RGB	Luv, Lab
Entrance am	outdoor	0.50	gray, RGB	Luv, Lab
Entrance pm	outdoor	0.39	$C_1C_2C_3$, rg	Luv, Lab
Entrance noon	outdoor	0.35	$C_1C_2C_3$, rg	Luv, Lab
Highway	outdoor	0.23	$C_1C_2C_3$, rg	Luv, Lab

†The mean darkening factor of shadows in grayscale.

11.4.4 Segmentation with Different Color Spaces

The results in the previous section show that when using the elliptical shadow domain defined by Equation 11.19, the CIE L*u*v* color space gives the most efficient separation of shadowed and foreground pixels. However, those experiments needed manually evaluated training data to set the parameters. In the following, the above model will be suited to the adaptive Bayesian model-framework (Section 11.3) to demonstrate that the advantage of using the appropriate color space can be also measured directly in the applications.

First, a probabilistic interpretation is added to the shadow classification step defined in Section 11.4.3. By rewriting Equation 11.19, the current $\overline{\psi}(s)$ value of pixel s is matched to a probability density function $f\left(\overline{\psi}(s)\right)$. Its class is decided as follows:

$$\text{pixel } s \text{ is shadowed} \Leftrightarrow f\left(\overline{\psi}(s)\right) \geq t. \tag{11.20}$$

Based on the one-dimensional marginal histograms in Figure 11.9, $f\left(\overline{\psi}(s)\right)$ is modeled by a multivariate Gaussian density function, similar to the CIE L*u*v* case introduced in Section 11.3. To keep the six-parameter shadow model, a diagonal covariance matrix will be used, requiring the definition of a three element-mean value vector and the three diagonal components of the covariance matrix. In this way, the variety of $\overline{\psi}$ values observed in shadows is modeled. This variety is caused by camera noise, fine alterations in illumination, and differences in albedo and geometry of the different surface points. However, the changes in the different color components are considered to be independent, exploiting the fact that many color spaces, for example, CIE L*u*v*, CIE L*a*b*, and HSV have approximately uncorrelated basis [33]. As for the RGB space, this diagonal approach is less accurate. However, it will be shown that for most of the sequences the performance of this oversimplified RGB model is reasonable.

Note that as shown in Reference [2], if f is a Gaussian density function (η), the domains defined by Equations 11.19 and 11.20 are equivalent:

$$f\left(\overline{\psi}(s)\right) = \eta\left(\overline{\psi}(s), \overline{\mu}_\psi, \overline{\overline{\Sigma}}_\psi\right) \tag{11.21}$$

$$= \frac{1}{(2\pi)^{\frac{3}{2}}\sqrt{\det\overline{\overline{\Sigma}}_\psi}} \exp\left[-\frac{1}{2}(\overline{\psi}(s) - \overline{\mu}_\psi)^T \overline{\overline{\Sigma}}_\psi^{-1}(\overline{\psi}(s) - \overline{\mu}_\psi)\right], \tag{11.22}$$

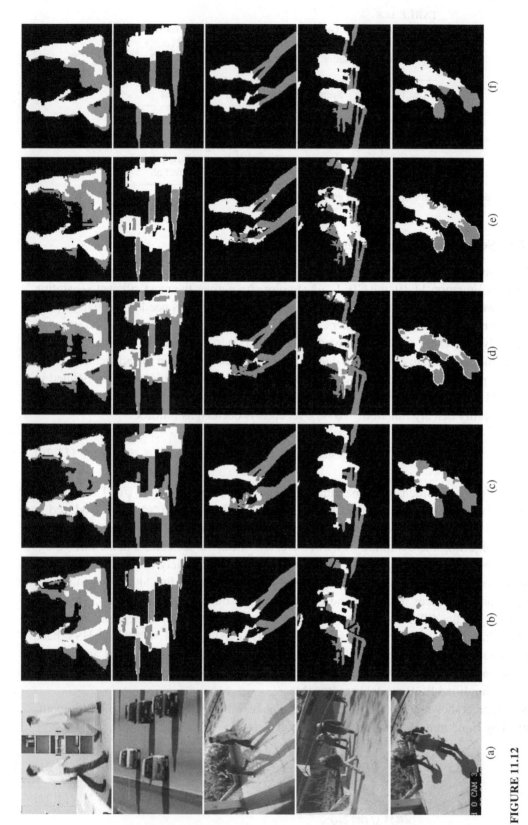

FIGURE 11.12

MRF segmentation results with different color models: (a) input video frame, (b) grayscale, (c) $C_1C_2C_3$, (d) HSV, (e) RGB, and (f) CIE L*u*v*. Test sequences from top to bottom: *Laboratory*, *Highway*, *Entrance am*, *Entrance pm*, and *Entrance noon*.

where

$$\overline{\mu}_{\psi} = [a_0, a_1, a_2]^T, \quad \overline{\overline{\Sigma}}_{\psi} = \text{diag}\{b_0^2, b_1^2, b_2^2\}, \quad t = (2\pi)^{-\frac{3}{2}}(b_0 b_1 b_2)^{-1} e^{-\frac{1}{2}}. \qquad (11.23)$$

In the following, the previously defined probability density functions will be used in the MRF model in a straightforward way, as $p_{\text{sh}}(s) = f(\overline{\psi}(s))$. The flexibility of this MRF model comes from the fact that $\overline{\psi}(s)$ shadow descriptors are defined for different color spaces differently (see Section 11.4.2).

11.4.4.1 Test Results

Figure 11.12 shows the MRF segmentation results of two frames from each test sequence using five color spaces: grayscale, $C_1 C_2 C_3$, HSV, RGB and CIE L*u*v*. Note that since the results of the CIE L*a*b* space were very similar to the L*u*v* outputs while *rg* worked similarly to $C_1 C_2 C_3$, they are skipped in this comparison. It can be observed that the CIE L*u*v* space outperforms the other ones significantly and that the largest errors were produced using $C_1 C_2 C_3$, especially in the case of sharp shadows. A typical problem of the HSV and RGB spaces is the foreground halo effect that may appear around some dark shadowed parts due to the penumbra of cast shadows [29] and video compression. These erroneous areas correspond to shadows, but they are lighter than the central areas, thus they lie outside of the shadow domain in the feature space. On the other hand, the proposed color space-based probabilistic model removes these artifacts.

Hereinafter, quantitative evaluations are performed using the MRF model. Section 11.4.3.1 measured the ability to discriminate foreground and shadowed pixels. Since the present model uses three classes and the goal here is accurate foreground detection, the confusion rate between foreground and background should be also considered. However, similarly to Section 11.3.6, the crossover between shadow and background does not count for errors (both of them are nonforeground areas).

Table 11.9 shows the clear superiority of the CIE L*u*v* space. However, the relative performance of the color spaces does not show exactly the same tendencies as measured in Section 11.4.3.1. The differences between Tables 11.7 and 11.9 are caused by effects of the composite foreground model, MRF neighborhood conditions and errors in parameter estimation, since the artifacts may appear different in the different sequences. Therefore, the numerical results from Section 11.4.3.1 can be considered to be more relevant to characterizing the capabilities of the color spaces for shadow separation. However, the experiments

TABLE 11.9
Evaluation of the MRF model using F^* coefficients.

Test video	Color space				
	gray	$C_1 C_2 C_3$	HSV	RGB	Luv
Laboratory	0.855	0.844	0.918	0.893	0.930
Highway	0.685	0.688	0.712	0.768	0.832
Entrance am	0.845	0.834	0.813	0.849	0.861
Entrance noon	0.825	0.843	0.689	0.827	0.892
Entrance pm	0.770	0.610	0.740	0.790	0.895

described in this section confirm that appropriate color space selection is also crucial in the applications, and the CIE L*u*v* space is preferred for this task.

11.5 Conclusion

This chapter examined the color modeling problem of cast shadows, focusing on video surveillance applications. A novel adaptive model for shadow segmentation without strong restrictions on *a priori* probabilities, image quality, objects' shapes and processing speed was introduced. The proposed modeling framework was generalized for different color spaces and used to compare these color spaces in detail. It was observed that the appropriate color space selection is an important issue in classification, and the CIE L*u*v* space is the most efficient both color-based clustering of the individual pixels and in the case of Bayesian foreground-background-shadow segmentation. The proposed method was validated on various video shots, including well-known benchmark videos and real-life surveillance sequences, indoor and outdoor shots, which contain both dark and light shadows. Experimental results showed the advantages of the proposed statistical approach over earlier methods.

References

[1] C. Benedek and T. Szirányi, "Bayesian foreground and shadow detection in uncertain frame rate surveillance videos," *IEEE Transactions on Image Processing*, vol. 17, no. 4, pp. 608–621, April 2008.

[2] C. Benedek and T. Szirányi, "Study on color space selection for detecting cast shadows in video surveillance," *International Journal of Imaging Systems and Technology*, vol. 17, no. 3, pp. 190–201, June 2007.

[3] C. Benedek, T. Szirányi, Z. Kato, and J. Zerubia, "Detection of object motion regions in aerial image pairs with a multi-layer Markovian model," *IEEE Transactions on Image Processing*, vol. 18, no. 10, pp. 2303–2315, October 2009.

[4] J. Kato, T. Watanabe, S. Joga, L. Ying, and H. Hase, "An HMM/MRF-based stochastic framework for robust vehicle tracking," *IEEE Transactions on Intelligent Transportation Systems*, vol. 5, no. 3, pp. 142–154, September 2004.

[5] C. Benedek and T. Szirányi, "Change detection in optical aerial images by a multi-layer conditional mixed Markov model," *IEEE Transactions on Geoscience and Remote Sensing*, vol. 47, no. 10, pp. 3416–3430, October 2009.

[6] D. Finlayson, S.D. Hordley, C. Lu, and M.S. Drew, "On the removal of shadows from images," *IEEE Transactions on Pattern Analysis and Machine Intelligence*, vol. 28, no. 1, pp. 59–68, January 2006.

[7] C. Stauffer and W.E.L. Grimson, "Learning patterns of activity using real-time tracking," *IEEE Transactions on Pattern Analysis and Machine Intelligence*, vol. 22, no. 8, pp. 747–757, August 2000.

[8] R.Z. Ping-Sing, R. Zhang, P. Sing Tsai, J.E. Cryer, and M. Shah, "Shape from shading: A survey," *IEEE Transactions on Pattern Analysis and Machine Intelligence*, vol. 21, no. 8, pp. 690–706, August 1999.

[9] Z. Szlávik, L. Kovács, L. Havasi, C. Benedek, I. Petrás, A. Utasi, A. Licsár, L. Czúni, and T. Szirányi, "Behavior and event detection for annotation and surveillance," in *Proceedings of the International Workshop on Content-Based Multimedia Indexing*, London, UK, June 2008, pp. 117–124.

[10] A. Katartzis and H. Sahli, "A stochastic framework for the identification of building rooftops using a single remote sensing image," *IEEE Transactions on Geoscience and Remote Sensing*, vol. 46, no. 1, pp. 259–271, January 2008.

[11] B. Sirmacek and C. Unsalan, "Building detection from aerial imagery using invariant color features and shadow information," in *Proceedings of the International Symposium on Computer and Information Sciences*, Istanbul, Turkey, October 2008, pp. 1–5.

[12] G. Perrin, X. Descombes, and J. Zerubia, "2D and 3D vegetation resource parameters assessment using marked point processes," in *Proceedings of the International Conference on Pattern Recognition*, Hong-Kong, August 2006, pp. 1–4.

[13] R. Cutler and L.S. Davis, "Robust real-time periodic motion detection, analysis, and applications," *IEEE Transactions on Pattern Analysis and Machine Intelligence*, vol. 22, no. 8, pp. 781–796, August 2000.

[14] L. Havasi, Z. Szlávik, and T. Szirányi, "Higher order symmetry for non-linear classification of human walk detection," *Pattern Recognition Letters*, vol. 27, no. 7, pp. 822–829, May 2006.

[15] L. Havasi, Z. Szlávik, and T. Szirányi, "Detection of gait characteristics for scene registration in video surveillance system," *IEEE Transactions on Image Processing*, vol. 16, no. 2, pp. 503–510, February 2007.

[16] L. Czúni and T. Szirányi, "Motion segmentation and tracking with edge relaxation and optimization using fully parallel methods in the cellular nonlinear network architecture," *Real-Time Imaging*, vol. 7, no. 1, pp. 77–95, February 2001.

[17] Z. Zivkovic, *Motion Detection and Object Tracking in Image Sequences*, PhD thesis, PhD thesis, University of Twente, 2003.

[18] J.B. Hayfron-Acquah, M.S. Nixon, and J.N. Carter, "Human identification by spatio-temporal symmetry," in *Proceedings of the International Conference on Pattern Recognition*, Washington, DC, USA, August 2002, pp. 632–635.

[19] L. Wang, T. Tan, H. Ning, and W. Hu, "Silhouette analysis-based gait recognition for human identification," *IEEE Transactions on Pattern Analysis and Machine Intelligence*, vol. 25, no. 12, pp. 1505–1518, December 2003.

[20] S.C. Zhu and A.L. Yuille, "A flexible object recognition and modeling system," *International Journal of Computer Vision*, vol. 20, no. 3, pp. 187–212, October 1996.

[21] L. Havasi and T. Szirányi, "Estimation of vanishing point in camera-mirror scenes using video," *Optics Letters*, vol. 31, no. 10, pp. 1411–1413, May 2006.

[22] A. Yoneyama, C.H. Yeh, and C.C.J. Kuo, "Moving cast shadow elimination for robust vehicle extraction based on 2D joint vehicle/shadow models," in *Proceedings of the IEEE Conference on Advanced Video and Signal Based Surveillance*, Miami, FL, USA, July 2003, p. 229.

[23] C. Fredembach and G.D. Finlayson, "Hamiltonian path based shadow removal," in *Proceedings of the British Machine Vision Conference*, Oxford, UK, September 2005, pp. 970–980.

[24] T. Gevers and H. Stokman, "Classifying color edges in video into shadow-geometry, highlight, or material transitions," *IEEE Transactions on Multimedia*, vol. 5, no. 2, pp. 237–243, June 2003.

[25] E.A. Khan and E. Reinhard, "Evaluation of color spaces for edge classification in outdoor scenes," in *Proceedings of the International Conference on Image Processing*, Genoa, Italy, September 2005, pp. 952–955.

[26] A. Prati, I. Mikic, M.M. Trivedi, and R. Cucchiara, "Detecting moving shadows: Algorithms and evaluation," *IEEE Transactions on Pattern Analysis and Machine Intelligence*, vol. 25, no. 7, pp. 918–923, July 2003.

[27] N. Paragios and V. Ramesh, "A MRF-based real-time approach for subway monitoring," in *Proceedings of the IEEE Conference on Computer Vision and Pattern Recognition*, Hawaii, USA, December 2001, pp. 1034–1040.

[28] A. Cavallaro, E. Salvador, and T. Ebrahimi, "Detecting shadows in image sequences," in *Proceedings of the of European Conference on Visual Media Production*, London, UK, March 2004, pp. 167–174.

[29] E. Salvador, A. Cavallaro, and T. Ebrahimi, "Cast shadow segmentation using invariant color features," *Computer Vision and Image Understanding*, vol. 95, no. 2, pp. 238–259, August 2004.

[30] F. Porikli and J. Thornton, "Shadow flow: A recursive method to learn moving cast shadows," in *Proceedings of the IEEE International Conference on Computer Vision*, Beijing, China, October 2005, pp. 891–898

[31] N. Martel-Brisson and A. Zaccarin, "Moving cast shadow detection from a Gaussian mixture shadow model," in *Proceedings of the IEEE Computer Society Conference on Computer Vision and Pattern Recognition*, San Diego, CA, USA, June 2005, pp. 643–648.

[32] Y. Haeghen, J. Naeyaert, I. Lemahieu, and W. Philips, "An imaging system with calibrated color image acquisition for use in dermatology," *IEEE Transactions on Medical Imaging*, vol. 19, no. 7, pp. 722–730, July 2000.

[33] M.G.A. Thomson, R.J. Paltridge, T. Yates, and S. Westland, "Color spaces for discrimination and categorization in natural scenes," in *Proceedings of Congress of the International Colour Association*, Rochester, NY, USA, June 2002, pp. 877–880.

[34] T. Gevers and A.W. Smeulders, "Color based object recognition," *Pattern Recognition*, vol. 32, no. 3, pp. 453–464, March 1999.

[35] I. Mikic, P. Cosman, G. Kogut, and M.M. Trivedi, "Moving shadow and object detection in traffic scenes," in *Proceedings of the International Conference on Pattern Recognition*, Barcelona, Spain, September 2000, pp. 321–324.

[36] Y. Sheikh and M. Shah, "Bayesian modeling of dynamic scenes for object detection," *IEEE Transactions on Pattern Analysis and Machine Intelligence*, vol. 27, no. 11, pp. 1778–1792, November 2005.

[37] Y. Wang, K.F. Loe, and J.K. Wu, "A dynamic conditional random field model for foreground and shadow segmentation," *IEEE Transactions on Pattern Analysis and Machine Intelligence*, vol. 28, no. 2, pp. 279–289, February 2006.

[38] S. Geman and D. Geman, "Stochastic relaxation, Gibbs distributions and the Bayesian restoration of images," *IEEE Transactions on Pattern Analysis and Machine Intelligence*, vol. 6, no. 6, pp. 721–741, June 1984.

[39] J. Rittscher, J. Kato, S. Joga, and A. Blake, "An HMM-based segmentation method for traffic monitoring," *IEEE Transactions on Pattern Analysis and Machine Intelligence*, vol. 24, no. 9, pp. 1291–1296, September 2002.

[40] D.S. Lee, "Effective Gaussian mixture learning for video background subtraction," *IEEE Transactions on Pattern Analysis and Machine Intelligence*, vol. 27, no. 5, pp. 827–832, May 2005.

[41] D.A. Forsyth, "A novel algorithm for color constancy," *International Journal of Computer Vision*, vol. 5, no. 1, pp. 5–36, January 1990.

[42] D.K. Lynch and W. Livingstone, *Color and Light in Nature*, UK: Cambridge University Press, 1955.

[43] G. Wyszecki and W. Stiles, *Color Science: Concepts and Methods, Quantitative Data and Formulas*, 2nd Edition, USA: John Wiley & Sons, 1982.

[44] Y. Wang and T. Tan, "Adaptive foreground and shadow detection in image sequences," in *Proceedings of the International Conference on Pattern Recognition*, Quebec, Canada, August 2002, pp. 983–986.

[45] R. Potts, "Some generalized order-disorder transformation," *Proceedings of the Cambridge Philosophical Society*, vol. 24, no. 1, p. 106, January 1952.

[46] Z. Kato, J. Zerubia, and M. Berthod, "Satellite image classification using a modified Metropolis dynamics," in *Proceedings of the International Conference on Acoustics, Speech and Signal Processing*, San Francisco, CA, USA, March 1992, pp. 573–576.

[47] N. Metropolis, A. Rosenbluth, M. Rosenbluth, A. Teller, and E. Teller, "Equation of state calculations by fast computing machines," *Journal of Chemical Physics*, vol. 21, no. 6, pp. 1087–1092, June 1953.

[48] J. Besag, "On the statistical analysis of dirty images," *Journal of Royal Statistics Society*, vol. 48, no. 3 pp. 259–302, March 1986.

[49] Y. Boykov, O. Veksler, and R. Zabih, "Fast approximate energy minimization via graph cuts," *IEEE Transactions on Pattern Analysis and Machine Intelligence*, vol. 23, no. 11, pp. 1222–1239, November 2001.

[50] C.J.V. Rijsbergen, *Information Retrieval*, 2nd Edition, London, UK: Butterworths, 1979.

[51] R. Cucchiara, C. Grana, G. Neri, M. Piccardi, and A. Prati, "The Sakbot system for moving object detection and tracking," in *Video-Based Surveillance Systems-Computer Vision and Distributed Processing*, Boston, MA, USA, November 2001, pp. 145–157.

[52] M. Rautiainen, T. Ojala, and H. Kauniskangas, "Detecting perceptual color changes from sequential images for scene surveillance," *IEICE Transactions on Information and Systems*, vol. 84, no. 12, pp. 1676–1683, December 2001.

[53] K. Siala, M. Chakchouk, F. Chaieb, and O. Besbes, "Moving shadow detection with support vector domain description in the color ratios space," in *Proceedings of the International Conference on Pattern Recognition*, Cambridge, UK, August 2004, pp. 384–387.

[54] V. Meas-Yedid, E. Glory, E. Morelon, C. Pinset, G. Stamon, and J.C. Olivo-Marin, "Automatic color space selection for biological image segmentation," in *Proceedings of the International Conference on Pattern Recognition*, Washington, DC, USA, August 2004, pp. 514–517.

[55] P. Guo and M.R. Lyu, "A study on color space selection for determining image segmentation region number," in *Proceedings of the International Conference on Artificial Intelligence*, Las Vegas, NV, USA, June 2000, pp. 1127–1132.

[56] A. Prati, I. Mikic, C. Grana, and M. Trivedi, "Shadow detection algorithms for traffic flow analysis: A comparative study," in *Proceedings of the IEEE Intelligent Transportation Systems Conference*, Oakland, CA, USA, August 2001, pp. 340–345.

[57] M. Tkalcic and J. Tasic, "Colour spaces - perceptual, historical and applicational background," in *Proceedings of Eurocon*, Ljubljana, Slovenia, September 2003, pp. 304–308.

12

Document Image Rectification Using Single-View or Two-View Camera Input

Hyung Il Koo and Nam Ik Cho

12.1   Introduction

Digital cameras have several advantages, for instance, portability and fast response, over flatbed scanners. Therefore, there have been a number of attempts to replace flatbed scanners with digital cameras. Unfortunately, camera captured images often suffer from perspective distortions due to oblique shot angle, geometric distortions caused by curved book surfaces, specular reflections, and unevenness of brightness due to uncontrolled illumination and vignetting. Hence their visual quality is usually inferior to flatbed scanned images, and the optical character recognition (OCR) rate is also low. Camera captured document images thus need to be enhanced to alleviate these problems and to widen the area of valuable text processing tools (for example, OCR and text-to-speech (TTS) for the visually impaired, automatic translation of books, and easy digitization of printed material) for the camera captured inputs. This chapter focuses on removing perspective and geometric distortions in captured document images, operations that are referred to as document dewarping or document rectification.

(a)   (b)

(c)

FIGURE 12.1

Document image rectification using a stereo-pair by an algorithm presented in Section 12.2: (a,b) input stereo pair, and (c) rectified result with specular reflection removal. © 2009 IEEE

12.1.1   Overview

To rectify document images, many methods directly estimate a three-dimensional (3D) structure. A straightforward method is to use depth-measuring hardware, such as structured light or a laser scanner. Since the inferred surface may not be isometric to the plane, several methods for modifying the surface have been proposed [1], [2], [3]. Although these approaches can be used in a wide range of paper material including old documents damaged by aging or water, it is burdensome to use depth-measuring equipment.

To overcome this problem, 3D structure can be estimated from multiple images without using depth measuring devices. For example, a specialized stereo vision system is proposed in Reference [4]. However, this method still needs hardware, although it is much simpler than the depth acquisition hardware. A stereo vision method presented in Reference [5] has a disadvantage in that it requires reference points. A more recent document dewarping algorithm presented in Reference [6] alleviates the problems by using video sequences. However, the algorithm requires hundreds of input images which are the results of scanning the entire book surface carefully. Section 12.2 presents a method [7] which rectifies documents using a stereo pair. The method needs no special hardware; as shown in Figure 12.1, it uses just two images captured from different viewpoints.

Although using multiple images has several desirable properties, such as content independence and the ability to remove specular reflection, such methods suffer from computational complexity in 3D reconstruction. Therefore, a number of single-view methods that do not require 3D reconstruction have also been proposed [8], [9], [10], [11]. Most of these methods avoid the 3D reconstruction problem by exploiting the clues from the two-dimensional text line structure with some additional assumptions. These methods are usually computationally efficient and easy-to-use; however, they are limited to the rectification of text regions due to their dependencies on text lines.

Unlike the text region, the rectification of figures using a single image requires determining the boundaries of distorted figures. Given the distorted boundaries, the rectification can be done by using applicable surface assumptions [12] or simple boundary interpolation [13], [14], [15]. However, it is burdensome to extract the boundaries. Section 12.3 presents a method that segments figures from a single view using a bonding box interface [16], which substantially facilitates the segmentation process. A new boundary interpolation method that can improve the visual quality of the output image (Figure 12.2) is also presented. The overall process is very efficient, so that a rectified result is obtained within a few seconds, whereas the stereo methods require almost a minute.

12.2   Document Rectification from a Stereo Pair

This section presents a document rectification algorithm which uses two images taken from two different views [7]. In the method, the surface of a book is reconstructed from the corresponding points in two images, $\mathscr{I}_1$ and $\mathscr{I}_2$, and the geometric correction is performed using the reconstructed book surface. Finally, the geometrically corrected images are stitched for a visually better composite.

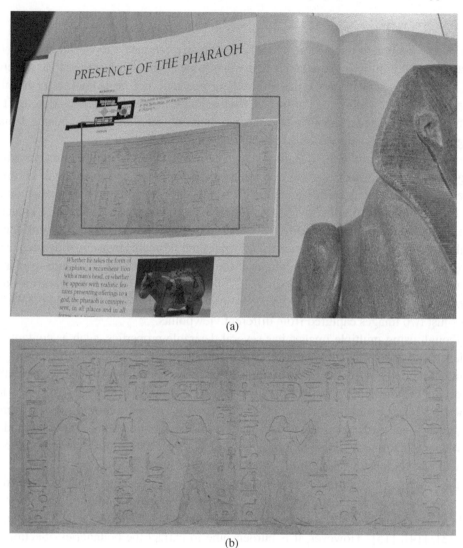

(a)

(b)

FIGURE 12.2

Document image rectification using a single image by an algorithm presented in Section 12.3: (a) input image with a user-provided bounding box, (b) segmented and rectified result.

12.2.1   Assumptions

The framework operates under three assumptions. The first is that book surfaces satisfy the cylindrical surface model (CSM) assumption, which is known to be sufficient for many kinds of document surfaces including the unfolded books [17]. The second assumption is that the intrinsic matrix of a camera is known. More precisely, a standard pin-hole camera (image coordinates are Euclidean coordinates having equal scales in both directions and the principal point is at the center of the image) is assumed, and it is also assumed that the estimated focal length is available from Exchangeable Image File Format (EXIF) tags of image files; most current digital cameras satisfy these assumptions. The third assumption

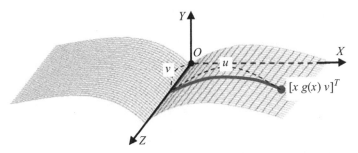

FIGURE 12.3

A book coordinate system and a modeled book surface. © 2009 IEEE

is that the contents of a book are quite distinctive compared to the background (that is, most of correspondences are found on the book surface), which can be easily achieved by placing the document on the relatively homogeneous background or capturing a book as large as possible.

12.2.2 Book Surface Model

Suppose a *uv*-plane on which the imaginary flat book surface lies. By assuming that the book's binding lies on the *v* axis, the unfolded (curved) book surface can be considered as the warping of this flat imaginary plane. This is illustrated in Figure 12.3, where a mapping from (u, v) to a point in the world coordinate is defined as

$$S(u, v) = [x \ \ g(x) \ \ v]^T \tag{12.1}$$

using the CSM assumption. That is, a point (u, v) on a flat surface goes to $[x \ \ g(x) \ \ v]$ in the world coordinate space when the surface is warped, where $g(x)$ is the height of the book surface from the *uv* plane. The relation between u, x and $g(x)$ is given by

$$u = \int_0^x \sqrt{1 + \left(\frac{dg}{dt}\right)^2}\, dt, \tag{12.2}$$

because u is the arc length of the curve $g(x)$. Although the center O of the world coordinate is located at the top of the book as shown in Figure 12.3, its position is not important as long as it is located on the book's binding.

12.2.3 Cost Function

If the surface function g and the camera matrix $\mathbf{P}_1$ and $\mathbf{P}_2 \in \mathfrak{R}^{3\times4}$ of two images $\mathscr{I}_1$ and $\mathscr{I}_2$ are given, the flattened book surface on the *uv*-plane can be obtained by using back-projection. Hence, document dewarping is equivalent to finding $\mathbf{P}_1$, $\mathbf{P}_2$, and g from $\mathscr{I}_1$ and $\mathscr{I}_2$. For this purpose, the corresponding points between two images are detected using scale invariant feature transform (SIFT) [18], denoted as $\{x_1^i \leftrightarrow x_2^i\}_{i=1}^N$ where x_j^i is the *i*-th corresponding point in the *j*-th image ($j = 1, 2$). Then, $\mathbf{P}_1$, $\mathbf{P}_2$, and 3D points $\{X_i\}_{i=1}^N$ can be reconstructed from the corresponding points up to similarity when the intrinsic matrix of a camera is known [19]. In other words, $\mathbf{P}_1'$, $\mathbf{P}_2'$, and $\{X_i'\}$ can be computed from the corresponding points which satisfy the similarity relation with P_1, P_2, and $\{X_i\}$ for some similarity transformation $H \in \mathfrak{R}^{4 \times 4}$, as follows:

$$P_1' = P_1 H^{-1}, \tag{12.3}$$

$$P_2' = P_2 H^{-1}, \tag{12.4}$$

$$\tilde{X}_i' = H \tilde{X}_i, \tag{12.5}$$

where $\tilde{X}_i \in \mathfrak{R}^4$ and $\tilde{X}_i' \in \mathfrak{R}^4$ are homogeneous representation of X_i and X_i', respectively. The similarity transform H can be represented as

$$H = \begin{bmatrix} sR & t \\ 0_{1 \times 3} & 1 \end{bmatrix}, \tag{12.6}$$

where s is a scale factor, $R = [r_1 \ r_2 \ r_3]$ is a 3×3 rotation matrix, and t is a 3×1 translation vector. By setting $s = 1$, Equation 12.5 reduces to

$$X_i' = R X_i + t. \tag{12.7}$$

Note that X_i represents a point on the book surface in the book (world) coordinate system shown in Figure 12.3 and X_i' means a point on the book surface in the camera coordinate system. By applying proper rotation and translation, $\{X_i'\}$ are transformed to $\{X_i\}$, and a surface function $g(x)$ can be estimated from them. The cost function for finding R, t and g can be expressed as follows:

$$(R, t, g) = \arg\min_{R, t, g} \sum_{i=1}^{N} d^2(S(g), X_i) = \arg\min_{R, t, g} \sum_{i=1}^{N} d^2(S(g), R^T(X_i' - t)), \tag{12.8}$$

where $S(g)$ is the surface induced from a function g, and $d(\cdot, \cdot)$ is the distance between a surface and a point. Intuitively, this cost function can be seen as finding the best surface that fits 3D points $\{X_i\}$.

Although a surface induced from any $g(x)$ is isometric to the plane (flat surface), the class of $g(x)$ can be restricted using *a priori* knowledge that the book surface is smooth except at the binding. Specifically, two polynomials are used for the modeling of $g(x)$ as follows:

$$g(x) = \begin{cases} g_+(x) & \text{if } x > 0, \\ 0 & \text{if } x = 0, \\ g_-(x) & \text{if } x < 0, \end{cases} \tag{12.9}$$

where $g_+(x)$ is for the right side and $g_-(x)$ is for the left side of book binding. Although a small number of parameters for $g(x)$ should be estimated, the direct minimization of Equation 12.8 is intractable due to highly nonlinear nature of the function and the presence of outliers.

12.2.4 Shape Reconstruction

This section presents the method that minimizes Equation 12.8. The method consists of two steps; one is the initial parameter estimation with outlier rejection and the other is the minimization of Equation 12.8 using initial estimation. More precisely, R is estimated from the general property of books and the projection analysis, and then g and t are estimated from the estimated R.

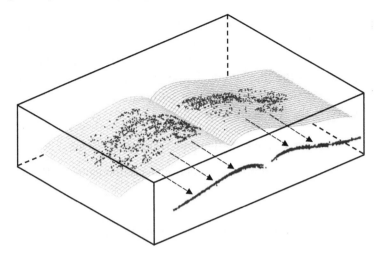

FIGURE 12.4

Points on the book surface and their projection in the direction of r_3. © 2009 IEEE

12.2.4.1 Estimation of Rotation Matrix

The initial estimate of $\mathbf{R} = [r_1 \; r_2 \; r_3]$ comes from the geometric properties of book surfaces. Because the variance along the Y axis (see Figure 12.3) is the smallest among all possible one-dimensional variances of $\{X_i\}$, the value of r_2 indicates the minimum variance direction of $\{X_i'\}$. This direction can be computed by the eigenvalue decomposition. When the eigenvectors of $\mathbf{C}' = \frac{1}{N}\sum_{i=1}^{N} X_i' X_i'^T$ are denoted as v_1, v_2, and v_3 in the increasing order of their corresponding eigenvalues, $r_2 = v_1$. Moreover, since v_2, v_3, and r_3 are placed on the same plane, r_3 can be represented as a linear combination of v_2 and v_3 for some $\hat{\theta}$, as follows:

$$r_3 = \frac{v_2}{|v_2|}\cos\hat{\theta} + \frac{v_3}{|v_3|}\sin\hat{\theta}. \tag{12.10}$$

The value of $\hat{\theta}$ is estimated using projection analysis. As can be seen in Figure 12.4, the projection of a surface onto the XY-plane forms a curve $y = g(x)$, and the projection of the points $\{X_i'\}$ onto the plane whose normal vector is r_3 (see Figure 12.4) provides points on the surface curve. Therefore

$$\hat{\theta} = \arg\min_{\theta} \mu(\{P_{r(\theta)}(X_i')\}), \tag{12.11}$$

where

$$r(\theta) = \frac{v_2}{|v_2|}\cos\theta + \frac{v_3}{|v_3|}\sin\theta. \tag{12.12}$$

The term $P_r(X_i')$ denotes the projected point of X_i' on the plane whose normal vector is r, and $\mu(\cdot)$ is a measure of area that the distributed points occupy. Since a finite number of noisy points are available, the area measure $\mu(\cdot)$ is approximated to a discrete function. By changing θ with predefined steps, $\hat{\theta}$ that minimizes $\mu(\{P_r(X_i')\})$ is found and r_3 is computed from Equation 12.10. Finally, r_1 can be obtained from $r_1 = r_2 \times r_3$. In implementation, a coarse-to-fine approach is adopted for the efficiency and the estimation of $\mathbf{R}$ is refined by a local two-dimensional search around the current estimate.

Although the 3D projection analysis for r_3 is robust to outliers, the estimate of r_2 using eigenvalue decomposition may fail in the presence of a single significant outlier. For the rejection of such outliers, samples whose distances from the center of mass are less than 3σ are only used, where σ is the standard deviation of distances from the center to the points. In this outlier rejection step, only significant outliers are rejected, as a refined rejection process is addressed in the following stage.

12.2.4.2 Estimation of the Curve Equation and Translation Vector

Because the surface is isometric, Z-directional components of $\mathbf{R}^T(X_i' - t)$ in Equation 12.8 can be ignored:

$$(\mathbf{R}, t, g) = \arg\min_{\mathbf{R}, t, g} \sum_{i=1}^{N} d^2(S(g), \mathbf{R}^T(X_i' - t)), \tag{12.13}$$

$$= \arg\min_{\mathbf{R}, t, g} \sum_{i=1}^{N} d^2(g, (x_i - a, y_i - b)), \tag{12.14}$$

where $(x_i, y_i) = (r_1^T X_i', r_2^T X_i')$ and $(a, b) = (r_1^T t, r_2^T t)$. The term $d(\cdot, \cdot)$ in Equation 12.13 is a distance function between a point and a surface, whereas $d(\cdot, \cdot)$ in Equation 12.14 is a distance function between a point and a curve. Since $(x_i - a, y_i - b)$ should be on the curve g, the parameters of g, that is, coefficients of a polynomial, can be estimated using the least squares method. In this case, the problem in Equation 12.8 reduces to the following minimization problem:

$$\sum_{i=1}^{M} (g(x_i - a) - (y_i - b))^2 \tag{12.15}$$

where M is the number of points. The curve in the left side ($x < a$) and the right side ($x > a$) are represented by different polynomials as $g_-(x-a)+b$ and $g_+(x-a)+b$, where $g_+(x) = \sum_{k=1}^{K} p_k x^k$, $g_-(x) = \sum_{k=1}^{K} q_k x^k$, and K is the order of polynomials which is empirically set to four [7].

In order to find p_k, q_k, and (a, b), a set of candidates for a is determined from the histogram of $\{x_i\}_{i=1}^{M}$. A point is chosen as a candidate if its corresponding bin is the local minimum in the histogram, which is based on the fact that there is a relatively small number of features around the book's binding. Then for each a, a total of m samples are randomly selected on both sides and an overdetermined system is solved to get $2K + 1$ unknowns ($p_1, p_2, \cdots, p_k, q_1, q_2, \cdots, q_K$, and b). Differing from the conventional random sample consensus (RANSAC), the criterion used here is to find the minimum of

$$\sum_{i=1}^{M} \phi_T(g(x_i - a) - (y_i - b)), \tag{12.16}$$

where

$$\phi_T(x) = \begin{cases} x^2 & \text{if } x < T, \\ T^2 & \text{otherwise}, \end{cases} \tag{12.17}$$

is a truncated square function and T is set to ten percent of the range of $\{x_i\}$, which is equivalent to MSAC (M-estimator sample consensus) in Reference [20]. After iteration, the hypothesis that minimizes Equation 12.16 is selected, and a rough estimate of a and inliers are obtained, where the inlier criterion is $|g(x_i - a) - (y_i - b)| < T$. Finally, two polynomials $g_+(x)$ and $g_-(x)$ on each side are estimated using a standard curve fitting method and their intersection point is determined as (a, b).

12.2.5 Image Stitching and Blending

As described in the previous section, two geometrically rectified images $\mathscr{J}_1$ and $\mathscr{J}_2$ are obtained from $\mathscr{I}_1$ and $\mathscr{I}_2$, respectively. In the case that each image contains sufficient information, either of $\mathscr{J}_1$ and $\mathscr{J}_2$ is sufficient for OCR and other purposes. However, in general, there may be out-of-focus blur and specular reflection that deteriorate image quality. Although these effects are often ignored in the literature, an image stitching process is needed for high quality output generation, especially for nontextual items such as natural photos. However, since the image alignment is imperfect, a simple blending (average) approach would result in double images. Moreover, the geometry of the stitching boundary cannot be determined in advance because the camera position is not fixed. In order to handle the challenges, a photomontage approach presented in Reference [21] is adopted as discussed below.

12.2.5.1 Image Stitching Based on Energy Minimization

Due to the asymmetry of the amount of information, better parts from each of the images are selected and stitched into a single image. The stitching is formulated as a labeling problem that assigns a label $L(p) \in \{0, 1\}$ for each pixel $p \in \mathscr{P}$, where $\mathscr{P}$ is a set of sites. Precisely, $L(p) = j$ means that the pixel at the site p comes from $\mathscr{J}_j$, for $j = 1, 2$.

In this formulation, two requirements are encoded in the cost function; one is to select more informative pixel and the other is to create a seamless mosaic. The cost function is defined as follows:

$$C(L) = \sum_{p \in \mathscr{P}} C_d(p, L(p)) + \eta \sum_{p \in \mathscr{P}} \sum_{q \in \mathscr{N}_p} C_i(p, q, L(p), L(q)), \qquad (12.18)$$

where $\mathscr{N}_p$ is the set of first order neighborhoods at the site p and the *data penalty* C_d represents the sharpness of the pixel p in $\mathscr{J}_{L(p)}$, which will be explained in the next subsection. The *interaction penalty* term C_i is expressed as follows:

$$C_i(p, q, L(p), L(q)) = \|\mathscr{J}_{L(p)}(p) - \mathscr{J}_{L(q)}(p)\| + \|\mathscr{J}_{L(p)}(q) - \mathscr{J}_{L(q)}(q)\|, \qquad (12.19)$$

where $\mathscr{J}_j(p)$ means the pixel value of $\mathscr{J}_j$ at p. Note that this cost function is the same as that of Reference [21]. The optimal labeling can be found using the graph-cut technique [22].

12.2.5.2 Data Penalty

In the case of specular-free documents, it is ideal to define the *data penalty* to reflect the sharpness, for instance, using a measure of the high-frequency components. Thus, the *data penalty* term can be expressed as

$$C_d^{(1)}(p, L(p)) = -\cos \theta(p, L(p)), \qquad (12.20)$$

where $\theta(p,L(p))$ is the angle between the line of sight and the surface normal at the point p in $\mathscr{I}_{L(p)}$. Because a small value of $\theta(p,L(p))$ means that the tangential surface at point p is closer to the perpendicular surface of the line of sight, it is approximately proportional to sharpness and naturally handles the problems caused by a little misalignment.

Since the surface and camera matrices were estimated during the 3D reconstruction process, the algorithm can handle the specular reflection without any device, whereas the existing algorithm dealt with it using some hardware [23]. Assuming that the distance from the flash to the camera center is small compared to the distance between the camera and a document, the possible glare regions can be determined. It comes from the specular distribution function model that this follows $\cos^n \phi$, where ϕ is the angle between the line of sight and principal reflected ray [23]. Hence, specular reflection can be removed by discarding the region with small ϕ. This can be done by modifying Equation 12.20 as follows:

$$C_d^{(2)}(p,L(p)) = \begin{cases} -\cos\theta(p,L(p)), & \text{if } \theta(p,L(p)) \geq \theta_0, \\ B & \text{otherwise}, \end{cases} \qquad (12.21)$$

where $\theta_0 = 10°$ and B is set to be a sufficiently large number so that specular-suspected regions (points with small ϕ) are rejected. If both $\theta(p,0) < \theta_0$ and $\theta(p,1) < \theta_0$ hold for some p, the specular reflection at p cannot be removed. Therefore, the directions and positions of cameras should be different for specular reflection removal, that is, glare spots should be placed separately.

12.2.5.3 Switching Criterion

In order to select a proper data penalty function between Equations 12.20 and 12.21, the presence of specular reflection should be determined. This can be resolved by counting the number of saturated pixels in the region that satisfies $\theta(p,L(p)) < \theta_0$, which is the region where specular reflections can occur. Namely, $S = \max\{n(S_1), n(S_2)\}$ is used for this purpose, with

$$S_j = \{p \,|\, J_j(p) > I_{thres}, \theta(p,j) < \theta_0\}, \quad \text{for } j = 1,2, \qquad (12.22)$$

where $n(\cdot)$ is the number of elements in a set. It is experimentally found for images with dimensions 1600×1200 pixels that $S > 1000$ in the presence of noticeable specular reflection, otherwise $S < 100$.

12.2.5.4 Image Blending

For a seamless result, image blending along the stitch boundary is indispensable. For blending, a multiresolution blending technique presented in Reference [24] is used. Experiments show that the blending algorithm works well for the complex boundaries generated by Equation 12.21 as well as for the boundaries found using Equation 12.20.

12.2.6 Experimental Results

Two digital cameras, a Canon PowerShot A630 and a Canon EOS 30D, were used to acquire images. The software tool OmniPage Pro 14 was used for OCR with no preprocessing or postprocessing for input images of 1600×1200 pixels with 150 dpi.

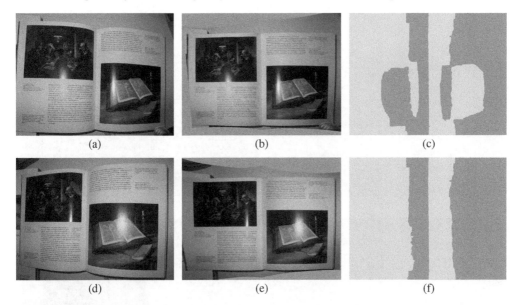

FIGURE 12.5 (See color insert.)

Document rectification from a stereo pair: (a,d) input stereo pair, (b,e) rectified images, and (c,f) stitching boundary using Equations 12.20 and 12.21. © 2009 IEEE

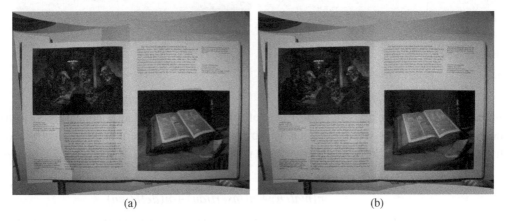

FIGURE 12.6 (See color insert.)

Composite image generation: (a) without blending, and (b) with blending (final result). © 2009 IEEE

Figures 12.5 and 12.6 show experimental results for the glossy papers with flash illumination. The stitching boundaries resulting from two *data penalty* terms (Equations 12.20 and 12.21) are shown in Figures 12.5c and 12.5f. As expected, the boundaries in Figure 12.5c are more complex than those present in Figure 12.5f. Hence, the text misalignment is sometimes observed, which seldom occurs in the case of using Equation 12.20 due to its simple boundary. However, specular reflections and illumination inconsistency are successfully removed as shown in Figure 12.6b.

The character recognition rate (CRR) is commonly measured by counting the number of correctly recognized characters or words. However, it is not an accurate measure in the

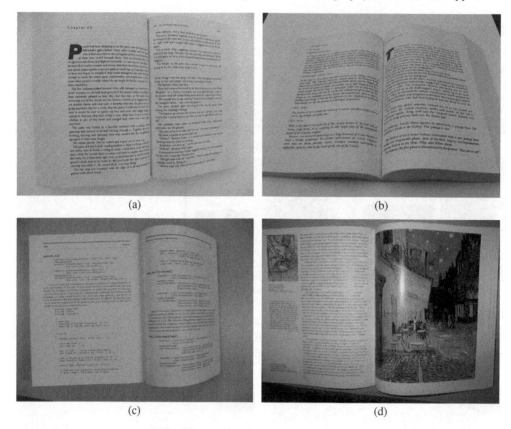

FIGURE 12.7

Images used in Table 12.1. © 2009 IEEE

sense that this number does not faithfully reflect the amount of hand-labor needed to fix incorrect characters or words after OCR. To faithfully reflect the amount of hand-labor, that is, deleting and/or inserting characters, a new CRR measure is defined here as follows:

$$CRR = 100 \times \frac{n(match)}{n(match) + n(insertion) + n(deletion)}. \quad (12.23)$$

For example, if "*Hello, World!*" is recognized as "*Helmo, Wold!!*", there are 9 matches. Also note that two deletions ("*m*" and "*!*") and two insertions ("*l*" and "*r*") are required for the correction of recognition result. Hence, the CRR of that text is $CRR = 100 \times 9/(9+2+2) \simeq 69\%$. In the computation of CRR, dynamic programming is used to find correspondences between recognized text and ground truth.

Because OCR performance depends on the line of sight and the types of content and material, quantitative evaluation is actually not a simple task. Thus, experiments in four different situations (Figure 12.7) are conducted, with results summarized in Table 12.1. Namely, Figure 12.7a represents the situation when the images are taken rather perpendicularly to the book surface; in this case the CRRs of $\mathcal{I}_1$ and $\mathcal{I}_2$ are relatively high and the CRR of rectified images is close to that of an image from flatbed scanner. Figure 12.7b represents the situation when images are captured obliquely. Although the OCR improvement is very high, the final recognition rate for Figure 12.7b is less than that of

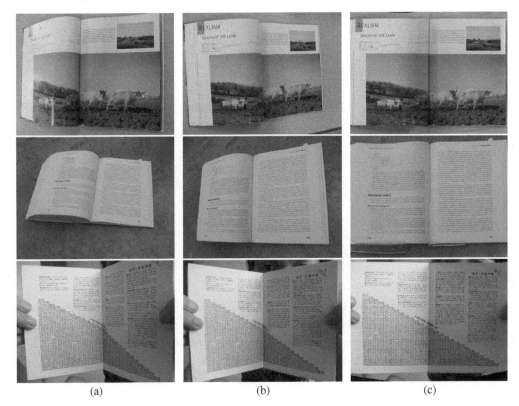

(a)   (b)   (c)

FIGURE 12.8 (See color insert.)

Additional experimental results: (a,b) input stereo pairs, and (c) final results. © 2009 IEEE

TABLE 12.1

CRR for several image pairs ($\mathscr{I}_1$ and $\mathscr{I}_2$ are image pairs before dewarping, $\mathscr{J}_1$ and $\mathscr{J}_2$ are image pairs after dewarping, "composite" denotes a stitched image obtained from $\mathscr{J}_1$ and $\mathscr{J}_2$, and "scan" denotes an image from a flatbed scanner).

pair	$\mathscr{I}_1$	$\mathscr{I}_2$	$\mathscr{J}_1$	$\mathscr{J}_2$	composite	scan
Figure 12.7a	85.92	81.54	99.57	99.67	99.57	99.85
Figure 12.7b	43.35	31.51	93.01	92.73	94.26	97.69
Figure 12.7c	78.78	73.78	90.75	87.82	94.56	96.36
Figure 12.7d	78.43	88.93	95.76	95.54	99.38	99.74

the image pair shown in Figure 12.7a, due to out-of-focus blur. Figures 12.7c and 12.7d correspond to the cases where each image does not have complete information due to blur and specular reflection. The problem is alleviated only after combining the information of two images. Figure 12.8 shows more results, additional examples can be found at http://ispl.snu.ac.kr/~hikoo/documents.html.

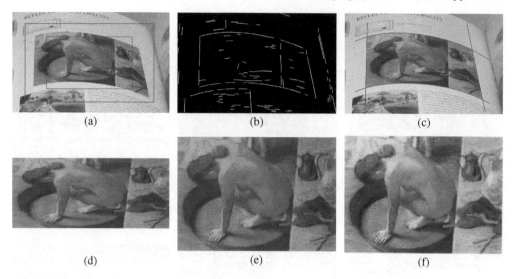

FIGURE 12.9 (See color insert.)

Single-view rectification procedure: (a) user interaction, the inner box is displayed just for the intuitive understanding of the system and is ignored after user interaction, (b) feature extraction, one of three feature maps
is shown, (c) result of the presented segmentation method, (d) result of References [13] and [27] using the
segmented image, aspect ratio 2.32, (e) result of the presented rectification method using the segmented image,
aspect ratio 1.43, and (f) figure scanned by a flatbed scanner for comparison, aspect ratio 1.39.

12.3   Figure Rectification from a Single View

Although the method in the previous section provides a unified approach to document
rectification, it requires long processing times. Moreover, it suffers from skew and curved
boundaries due to various imperfections in 3D reconstruction caused, for instance, by radial
distortion, incorrect focal length, and model error. These imperfections do not usually
constitute a severe problem for OCR; however, they become critical when users try to
digitize figures and photos in documents.

This section introduces a method which provides a high quality rectified image from a
single-view image. In this method, a user draws a bounding box on a target figure by drag-
ging the mouse (Figure 12.9a). Then, the algorithm automatically segments the figure and
generates a rectified image (Figure 12.9e). The method tolerates roughly placed bounding boxes, and it provides an easy-to-use interface for the digitization of printed figures
compared to conventional methods. Along with the new scenario that simplifies user interaction, another main feature of the algorithm is its fast response. Unlike conventional
methods, including the one presented in Section 12.2, the algorithm proposed here can segment and rectify megapixel size figures within one to two seconds, depending on the user's
bounding box. This efficiency is achieved by developing a segmentation algorithm that
fully exploits the properties of printed figures. Namely, low curvature edge maps are extracted from an input image, and a small number of line segments are found from the maps
(Figure 12.9b). Since the curvature of the figure boundaries is not large, the extracted line
segments can be used to construct a set of candidate boundaries. Then, the optimal one in

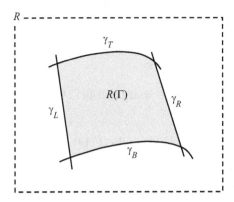

FIGURE 12.10

Bounding box definition: $\mathscr{R}$ denotes the bounding box given by the user, Γ represents four bounding curves $(\gamma_L, \gamma_T, \gamma_R$ and $\gamma_B)$, and $\mathscr{R}(\Gamma)$ indicates the region corresponding to the printed figure.

this set is searched by using an alternating optimization scheme. After boundary extraction, the extracted figure is rectified. The rectification method is the combination of the metric rectification method used for planar documents [25], [26] and the boundary interpolation methods used for curved documents [13], [27].

12.3.1 Assumptions and Notations

The framework operates under three assumptions. The first is that images are acquired so that two side curves γ_L and γ_R are roughly parallel with y-axis, as depicted in Figure 12.10. The second assumption is that the surfaces of documents on which the figure lies satisfy CSM conditions [13], [17], so that two side curves are modeled as straight lines and the rectification method based on boundary interpolation is justified [13]. The third assumption is that γ_T and γ_B are low curvature curves. The low curvature condition means that the curves can be approximated by piecewise linear segments. Although EXIF information improves the rectification performance, its availability is not an essential requirement.

From the first assumption, top and bottom curves are represented as $y = \gamma_T(x)$ and $y = \gamma_B(x)$, where γ_T and γ_B are fourth-order polynomials which can be justified by the third assumption. Two side curves are represented as $x = \gamma_L(y)$ and $x = \gamma_R(y)$, where γ_L and γ_R are first-order polynomials from the second assumption. These constraints will be relaxed to second-order polynomials in a refinement step for the compensation of nonlinearities in the imaging system such as radial distortions. Let Ω denote a domain where a given image is defined, and $\mathscr{R}$ be a bounding box given by the user. Also, let $\partial\mathscr{R}$ denote the boundary of $\mathscr{R}$, and Γ be used to represent four curves, that is, $\Gamma = (\gamma_L, \gamma_R, \gamma_T, \gamma_B)$. The region enclosed by Γ is denoted as $\mathscr{R}(\Gamma)$. Figure 12.10 shows a graphical representation of this scenario.

12.3.2 Figure Segmentation

In this section, the figure segmentation is formulated as an energy minimization problem. The energy function is the sum of the regional and boundary terms, which can be seen as the analogies of likelihood and smoothness terms in conventional MRF-based methods [16].

12.3.2.1 Energy Formulation

Since the region of a figure is described via Γ as four bounding curves, the segmentation problem can be formulated as follows:

$$\hat{\Gamma} = \arg\min \Phi(\Gamma, \Theta), \tag{12.24}$$

where

$$\Phi(\Gamma, \Theta) = \Phi_L(\Gamma, \Theta) + \Phi_S(\Gamma) \tag{12.25}$$

is defined as the sum of two terms; the first one encodes data fidelity and the second one encodes the energy of the boundaries. Finally, Θ represents the parameters of a probabilistic model that is explained later in this chapter.

12.3.2.2 Encoding Data Fidelity

The term $\Phi_L(\Gamma, \Theta)$ is similar to the data term in conventional MRF-based methods [16]. It is defined as follows:

$$\Phi_L(\Gamma, \Theta) = \sum_{(x,y) \in \mathscr{R}(\Gamma)} V_F(x, y; \Theta) + \sum_{(x,y) \in \mathscr{R} - \mathscr{R}(\Gamma)} V_B(x, y; \Theta). \tag{12.26}$$

To design $V_F(x, y; \Theta)$ and $V_B(x, y; \Theta)$, a Gaussian mixture model (GMM) of color distributions [16] is used. Therefore, Θ represents the GMM parameters of two distributions and the pixelwise energies are defined as the negative log of probabilities which are given by GMM.

The estimate of Θ is obtained from user-provided seed pixels. Although the method of user interaction is the same as that of Reference [16], there is a subtle difference in the estimation process. In the conventional object segmentation problem, a given image usually consists of an object and background, and it is reasonable to consider the pixels in a box ($\mathscr{R}$) as seeds for the object, and the pixels outside the box ($\Omega \backslash \mathscr{R}$) as seeds for the background. However, typical documents contain several figures, and it is very likely that other figures (especially those having similar color distributions) exist outside the box as can be seen in Figure 12.9a. Hence the initialization method that considers all pixels in $\Omega \backslash \mathscr{R}$ as seeds for the background is often problematic. Rather, the pixels around the outer box (∂R) are used as seeds for the background, and the pixels inside the box ($\mathscr{R} \backslash \partial R$) are used as seeds for the figure. The modified initialization method not only provides better performance but also improves efficiency due to a small number of seeds. This method will be used in the *Grabcut* algorithm [16] for fair comparison in the experimental section.

12.3.2.3 Encoding the Energy of the Boundaries

For the explanation of $\Phi_S(\Gamma)$, it is assumed that two binary random fields, $\mathscr{E}_H$ for horizontal and $\mathscr{E}_V$ for vertical edges are given. Namely, $\mathscr{E}_H(x, y) = 1$ when a low slope and low curvature curve passes through (x, y), and $\mathscr{E}_H(x, y) = 0$ otherwise. The term $\mathscr{E}_V$ is similarly defined. The construction of $\mathscr{E}_H$ and $\mathscr{E}_V$ is explained in the next section.

Using the estimated $\mathscr{E}_H$ and $\mathscr{E}_V$, the term $\Phi_S(\Gamma)$ can be expressed as follows:

$$\Phi_S(\Gamma) = \lambda \times \begin{cases} \psi_H(\gamma_T) + \psi_H(\gamma_B) + \psi_V(\gamma_L) + \psi_V(\gamma_R) & \text{if all curves are plausible,} \\ \infty & \text{otherwise,} \end{cases} \tag{12.27}$$

FIGURE 12.11

Support areas used to obtain the statistics for low curvature edge extraction.

where

$$\psi_H(\gamma) = \sum_{\mathscr{E}_H(x,y)=1} \phi_T(y - \gamma(x)). \tag{12.28}$$

Here $\phi_T(\cdot)$ is a truncated square function defined in Equation 12.17. Note that $\psi_V(\cdot)$ is similarly defined. Intuitively, both functions are minimized when the curve passes as many as all the possible edge points. In all experiments, $\lambda = 100$ and $T = 3$.

Plausible curves in Equation 12.27 are the ones that satisfy several hard constraints. Hard constraints are imposed on the slope of curves, the intersection point of γ_L and γ_R (that is, the position of the vanishing point), and so on. However, these restrictions are not critical because any Γ violating these conditions is likely to have high energy. These conditions are instead used to reject unlikely candidates at the early stage.

12.3.2.4 Edge Extraction Using Separability between Two Regions

Given a point, the presence of a horizontal edge can be determined by analyzing the statistics of three supports in Figure 12.11. In Reference [28], the edge is detected based on the separability of two regions, which is defined as follows:

$$\mu = \frac{(\hat{m}_1 - \hat{m})^2 + (\hat{m}_2 - \hat{m})^2}{2\,\hat{\sigma}^2}, \tag{12.29}$$

where $\hat{m}_1$ and $\hat{m}_2$ are the empirical means of pixels in upper and lower supports, respectively, The terms $\hat{m}$ and $\hat{\sigma}$ denote the empirical mean and standard deviations of pixels on the support.

Low curvature edges can be detected in a scale invariant manner by evaluating Equation 12.29 for several types of supports. The scale-invariant separability at (x,y) is defined as

$$\mu(x,y) = \max(\mu_1(x,y), \mu_2(x,y), \ldots, \mu_l(x,y)), \tag{12.30}$$

where $\mu_i(x,y)$ is the separability evaluated by Equation 12.29 using the i-th support. The horizontal edge field $\mathscr{E}_H$ is defined as follows

$$\mathscr{E}_H(x,y) = \begin{cases} 1 & \text{if } \mu(x,y) > T \text{ and } \mu(x,y) > \mu(x,y+k), \\ 0 & \text{otherwise,} \end{cases} \tag{12.31}$$

where $k = \pm1, \pm2, \ldots, \pm m$ and T is the threshold.

Direct implementation of the multiscale scheme is computationally expensive because support size and image size are usually large in this application. However, this scheme can be efficiently implemented by using integral images [29], [30].

12.3.2.5   Optimization

Since minimizing the term in Equation 12.25 is not a simple task, a new optimization method is presented. This method consists of four steps: i) clustering edges ($\mathscr{E}_H$ and $\mathscr{E}_V$) into line segments $\mathscr{A}$, ii) constructing a candidate boundary set $\mathscr{C}$ from $\mathscr{A}$, iii) finding a coarse solution in the candidate set $\mathscr{C}$ as $\hat{\Gamma} = \arg\min_{\Gamma \in \mathscr{C}} \Phi(\Gamma, \Theta)$, and iv) refining the coarse estimate $\hat{\Gamma}$. In this section, the width and height of $\mathscr{R}$ are respectively denoted as W and H.

12.3.2.6   Clustering Edge Points into Line Segments

The points in $\mathscr{E}_H$ and $\mathscr{E}_V$ are clustered into line segments. Although a number of robust line segment extraction algorithms have been proposed, a simple progressive probabilistic Hough transform (PPHT) is sufficient for this purpose [31]. The sets of horizontal and vertical line segments are denoted as $\mathscr{A}_H$ and $\mathscr{A}_V$, which are extracted from $\mathscr{E}_H$ and $\mathscr{E}_V$, respectively. Another set, $\mathscr{A}_V^L \subset \mathscr{A}_V$, consists of vertical line segments which are placed on the left half of $\mathscr{R}$ (they are used as proposals for the left boundary) whereas the set of line segments on the right half is denoted as $\mathscr{A}_V^R \subset \mathscr{A}_V$ (for the right boundary). Similarly, $\mathscr{A}_H^T \subset \mathscr{A}_H$ and $\mathscr{A}_H^B \subset \mathscr{A}_H$ imply horizontal line segments on the upper half and lower half of $\mathscr{R}$ respectively.

12.3.2.7   Candidate Set Construction

Since the two side curves are straight lines, it is relatively easy to find the candidates for them; $u \in \mathscr{A}_V^L$ becomes a candidate for γ_L and $u \in \mathscr{A}_V^R$ becomes a candidate for γ_R. However, the problem becomes complicated when finding the candidates for γ_T and γ_B because these are not straight lines in general. In order to handle this problem, $\mathscr{A}_H^T \times \mathscr{A}_H^T$ is considered as a candidate set for γ_T and similarly $\mathscr{A}_H^B \times \mathscr{A}_H^B$ as a candidate set for γ_B. Specifically, $(u,v) \in \mathscr{A}_H^T \times \mathscr{A}_H^T$ becomes a straight line when $u = v$ as illustrated in Figure 12.12a. Otherwise, (u,v) is considered as a curve passing u and v. Precisely, the curve that tries to pass the leftmost, center, and the rightmost points of u and v is modeled as a third-order polynomial and an overdetermined system is solved to get four unknowns of the curve hypothesis. As illustrated in Figures 12.12b and 12.12c, the curve becomes very close to the figure boundaries where both u and v are line segments on the curve; otherwise it severely deviates from the curve. In summary, the candidate set for Γ is $\mathscr{C} = \mathscr{A}_V^L \times \mathscr{A}_V^R \times (\mathscr{A}_H^T \times \mathscr{A}_H^T) \times (\mathscr{A}_H^B \times \mathscr{A}_H^B)$. Even if the size of each set is not large (for example, $|\mathscr{A}| \simeq 10^2$), the set of candidates can be a very large set since $|\mathscr{C}| \propto |\mathscr{A}|^6$.

12.3.2.8   Finding a Coarse Solution in the Candidate Set

A coarse solution $\hat{\Gamma}$ can be found by minimizing Equation 12.25 over $\mathscr{C}$ as follows:

$$\hat{\Gamma} = \arg\min_{\Gamma \in \mathscr{C}} \left(\Phi_L(\Gamma, \Theta) + \Phi_S(\Gamma) \right). \tag{12.32}$$

Unfortunately, the direct minimization process requires $O(|\mathscr{A}|^6 \times W \times H)$ operations. Since the computational load of direct evaluations is prohibitively large, an efficient minimization method based on the alternating optimization scheme is developed. Namely, from the initial estimates for top and bottom curves ($\gamma_T(x)$ and $\gamma_B(x)$ are, respectively, top and

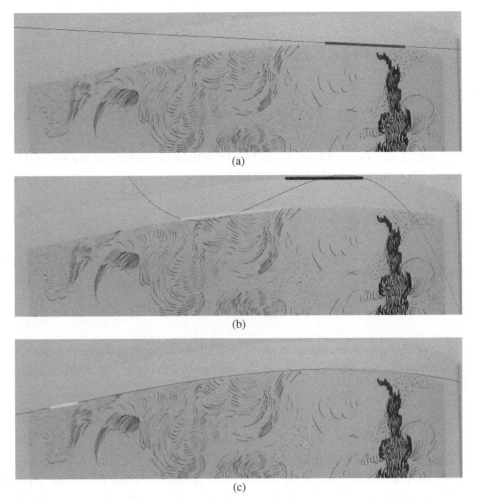

(a)

(b)

(c)

FIGURE 12.12

Three choices of $(u,v) \in \mathscr{A}_H^T \times \mathscr{A}_H^T$.

bottom sides of $\mathscr{R}$), two side curves are estimated. Then, these estimated two side curves are used to estimate top and bottom curves. Thus, the problem in Equation 12.32 reduces to two subproblems, namely

$$(\hat{\gamma}_L, \hat{\gamma}_R) = \arg \min_{(\gamma_L, \gamma_R)} (\Phi_L(\Gamma, \Theta) + \Phi_S(\Gamma)) \tag{12.33}$$

fixing (γ_T, γ_B), and

$$(\hat{\gamma}_T, \hat{\gamma}_B) = \arg \min_{(\gamma_T, \gamma_B)} (\Phi_L(\Gamma, \Theta) + \Phi_S(\Gamma)) \tag{12.34}$$

fixing (γ_L, γ_R). Since the minimization methods for Equations 12.33 and 12.34 are similar, only the method for Equation 12.33 is presented. When γ_T and γ_B are fixed, $\Phi_L(\Gamma, \Theta)$ can be represented as follows:

$$\Phi_L(\Gamma, \Theta) = \eta_L(\gamma_L) + \eta_R(\gamma_R) + \text{constant}, \tag{12.35}$$

where

$$\eta_L(\gamma_L) = \sum_{y=0}^{H-1} \left(\sum_{x=0}^{\gamma_L(y)} V_B'(x,y) + \sum_{x=\gamma_L(y)}^{W-1} V_F'(x,y) \right) \qquad (12.36)$$

and

$$\eta_R(\gamma_R) = \sum_{y=0}^{H-1} \sum_{x=\gamma_R(y)}^{W-1} \left(V_B'(x,y) - V_F'(x,y) \right). \qquad (12.37)$$

Here, V_F' and V_B' are obtained from V_F and V_B by setting the outside of the top and bottom curves as 0. By constructing the following tables:

$$T_1(y,z) = \sum_{x=0}^{z} V_B'(x,y), \qquad (12.38)$$

$$T_2(y,z) = \sum_{x=z}^{W-1} V_F'(x,y), \qquad (12.39)$$

$$T_3(y,z) = \sum_{x=z}^{W-1} \left(V_B'(x,y) - V_F'(x,y) \right), \qquad (12.40)$$

the term $\Phi_L(\Gamma, \Theta)$ can be evaluated in $O(H)$ operations using the tables. Moreover, $\Phi_S(\Gamma)$ is also represented via $\psi_V(\gamma_L) + \psi_V(\gamma_R) + \text{constant}$. Putting it all together into Equation 12.33 provides

$$(\hat{\gamma}_L, \hat{\gamma}_R) = \arg \min_{\gamma_L \in \mathscr{A}_V^L, \gamma_R \in \mathscr{A}_V^R} (\eta_L(\gamma_L) + \psi_V(\gamma_L) + \eta_R(\gamma_R) + \psi_V(\gamma_R)). \qquad (12.41)$$

The computational cost for this scheme can be summarized as i) $W \times H$ operations required for the construction of tables, ii) $|\mathscr{A}| \times H$ operations required for the precomputations of $\eta_L(\gamma_L)$ and $\psi_V(\gamma_L)$ for all $\gamma_L \in \mathscr{A}_V^L$, iii) $|\mathscr{A}| \times H$ operations required for the precomputations of $\eta_R(\gamma_R)$ and $\psi_V(\gamma_R)$ for all $\gamma_R \in \mathscr{A}_V^R$, and iv) $|\mathscr{A}|^2$ operations used for the test of hard constraints and the minimization of Equation 12.41. This results in total $O(W \times H + |\mathscr{A}| \times H + |\mathscr{A}|^2)$ computations. The computational cost of Equation 12.34 can be reduced in a similar manner. Experiments show that Equations 12.33 and 12.34 converge to their optimal solutions very quickly; they are repeated only twice.

12.3.2.9   Coarse Solution Refinement

Starting from the coarse solution $\hat{\Gamma}$, the boundary is refined using randomly generated proposals. In the refinement process, a fourth-order polynomial model is adopted for top and bottom curves, and second-order model is used for two side curves so that the nonlinearity in the imaging system can be compensated.

12.3.3   Rectification

This section introduces a rectification algorithm which improves conventional boundary interpolation methods. Using the boundaries in Figures 12.9c and 12.13a the conventional methods yield distorted results, as shown in Figures 12.9d and 12.13c. In order to alleviate this distortion, a new rectification method is presented. The method consists of two steps.

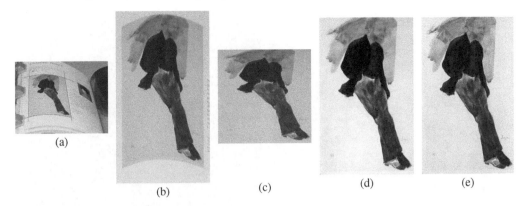

FIGURE 12.13

Figure rectification: (a) segmentation result, (b) transformed result, note that the four corners of the figure compose a rectangle, (c) rectification of the segmentation result using boundary interpolation, aspect ratio 1.02, (d) rectification of the the transformed result using boundary interpolation, aspect ratio 0.614, and (e) ground truth, scanned image with aspect ratio 0.606.

The first is the rectification process for an imaginary rectangle consisting of four corners of a figure, which is the same as metric rectification methods for planar documents used in References [25] and [26] except that the rectangle is an imaginary one. Boundary interpolation is then applied to the transformed image [13]. Figure 12.13 illustrates this process. In the proposed method, boundary interpolation is applied to Figure 12.13b to produce the result shown in Figure 12.13d. For completeness, conventional methods are applied to Figure 12.13a to produce the result shown in Figure 12.13c. As can be seen, the proposed method can largely remove distortions. In metric rectification of an imaginary rectangle, the focal length is obtained from EXIF (if available) [7].

12.3.4 Experimental Results

The performance of the proposed segmentation algorithm is compared with that of the *Grabcut* method because it is an interactive image segmentation system that uses the same interface [16]. As can be seen in Figure 12.14, the *Grabcut* method does not work well because there are a number of high-contrast edges in an object and the data fidelity term has limited power in discriminating figures from the background. However, this is not the case when the proposed method is used, as successfully segmented results are produced. In rectification, the proposed method is based on Coons patch [27]. As shown in the examples presented in Figures 12.9 and 12.13, simple modification of the rectification method substantially improves subjective quality. The robustness and efficiency of the system are also demonstrated at http://ispl.snu.ac.kr/~hikoo/Research.htm.

For comparison purposes, Figures 12.15a and 12.15b are used as an input stereo pair to the method presented in Section 12.2. Figure 12.16 shows the rectification results of the stereo algorithm, the results achieved using the method presented in this section, and the results produced through image scanning. In subjective comparison, the visual quality is similar when the image is captured in a perpendicular direction as illustrated in Figure 12.16a to 12.16c. However, camera-based methods show degraded results when the

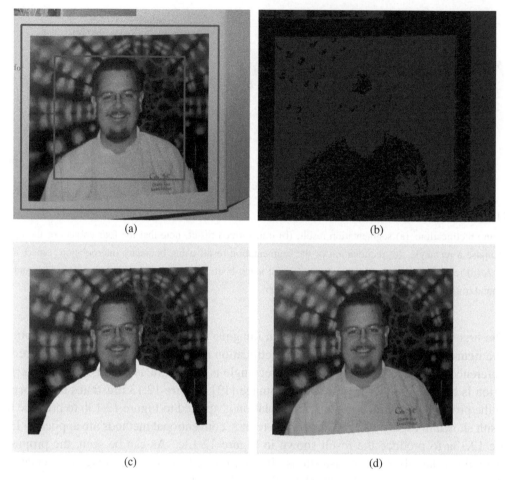

(a)   (b)

(c)   (d)

FIGURE 12.14

Comparison of the Grabcut algorithm and the proposed segmentation method: (a) user interaction-based input, (b) light pixels stand for figures and dark pixels stand for blank in feature space, (c) Grabcut output, and (d) presented method output.

scene is obliquely captured as illustrated in the bottom row in Figure 12.16. In such cases, although both camera-driven methods suffer from blur caused by perspective contraction and shallow depths of field, the stereo method is less affected by geometric distortions than the method presented in this section. On the other hand, the latter method has advantages on the boundaries because the algorithm forces the boundaries of restored images to be straight, while skews and boundary fluctuations are usually observed in the method presented in Section 12.2 (an image was manually deskewed and cropped in order to obtain Figures 12.16a and 12.16d).

In terms of computational complexity, the stereo system usually requires about one minute to produce a 1600×1200 output image. The method presented in this section takes 4.8 seconds in feature extraction, 1.3 seconds for segmentation, and 0.5 seconds in rectification for handling a 3216×2136 image. By taking about five seconds in user interaction, this method produces an output within two seconds even for seven Megapixel images.

FIGURE 12.15

Evaluation of the method in Section 12.2: (a,b) stereo pair, and (c,d) magnified and cropped images with user interactions.

12.4 Conclusion

This chapter presented two camera-driven methods for geometric rectification of documents. One is a stereo-based method using explicit 3D reconstruction. The method works irrespective of contents on documents and provides several advantages, such as specular reflection removal. Therefore, this method can be used for OCR and digitization of figures and pictures in indoor environment. The other one is a single-view method which rectifies a figure from a user-provided bounding box. This method is shown to be efficient, robust, and easy-to-use. It should be noted that the camera captured images often suffer from photometric and geometric distortion. Therefore, removal of uneven illumination and motion/out-of-focus blur are also essential in enhancing camera captured document images, although these operations are not discussed in this chapter. Nevertheless, as demonstrated in this chapter, digital camera-driven systems for document image acquisition, analysis, and processing have the potential to replace flatbed scanners.

FIGURE 12.16 (See color insert.)

Performance comparison of the two approaches: (a,d) results of the method presented in Section 12.2 with aspect ratio 1.52 and 1.50, (b,e) results of the method presented in Section 12.3 with aspect ratio 1.52 and 1.53, and (c,f) scanned images with aspect ratio 1.54 and 1.51.

Acknowledgment

Figures 12.1 and 12.3 to 12.8 are reprinted from Reference [7], with the permission of IEEE.

References

[1] M.S. Brown and C.J. Pisula, "Conformal deskewing of non-planar documents," in *Proceedings of IEEE Conference on Computer Vision and Pattern Recognition*, San Diego, CA, USA, June 2005, pp. 998–1004.

[2] M.S. Brown and W.B. Seales, "Document restortion using 3D shape: A general deskewing algorithm for arbitrarily warped documents," in *Proceedings of International Conference on Computer Vision*, Vancouver, BC, Canada, July 2001, pp. 367–374.

[3] M. Pilu, "Undoing paper curl distortion using applicable surfaces," in *Proceedings of IEEE Conference on Computer Vision and Pattern Recognition*, Kauai, HI, USA, December 2001, pp. 67–72.

[4] A. Yamashita, A. Kawarago, T. Kaneko, and K.T. Miura, "Shape reconstruction and image restoration for non-flat surfaces of documents with a stereo vision system," in *Proceedings of International Conference on Pattern Recognition*, Cambridge, UK, August 2004, pp. 482–485.

[5] A. Ulges, C.H. Lampert, and T. Breuel, "Document capture using stereo vision," in *Proceed-

ings of ACM symposium on Document Engineering*, Milwaukee, Wi, USA, October 2004, pp. 198–200.

[6] A. Iketani, T. Sato, S. Ikeda, M. Kanbara, N. Nakajima, and N. Yokoya, "Video mosaicing based on structure from motion for distortion-free document digitization," in *Proceedings of Asian Conference on Computer Vision*, Tokyo, Japan, November 2007, pp. 73–84.

[7] H.I. Koo, J. Kim, and N.I. Cho, "Composition of a dewarped and enhanced document image from two view images," *IEEE Transactions on Image Processing*, vol. 18, no. 7, pp. 1551–1562, July 2009.

[8] J. Liang, D. DeMenthon, and D. Doermann, "Flattening curved documents in images," in *Proceedings of IEEE Conference on Computer Vision and Pattern Recognition*, San Diego, CA, USA, June 2005, pp. 338–345.

[9] F. Shafait and T.M. Breuel, "Document image dewarping contest," in *Proceedings of 2nd International Workshop on Camera-Based Document Analysis and Recognition*, Curitiba, Brazil, September 2007, pp. 181–188.

[10] N. Stamatopoulos, B. Gatos, I. Pratikakis, and S. Perantonis, "A two-step dewarping of camera document images," in *Proceedings of International Workshop on Document Analysis Systems*, Nara, Japan, September 2008, pp. 209–216.

[11] S.S. Bukhari, F. Shafait, and T.M. Breuel, "Coupled snakelet model for curled textline segmentation of camera-captured document images," in *Proceedings of International Conference on Document Analysis and Recognition*, Barcelona, Spain, July 2009, pp. 61–65.

[12] N.A. Gumerov, A. Zandifar, R. Duraiswami, and L.S. Davis, "3D structure recovery and unwarping of surfaces applicable to planes," *International Journal of Computer Vision*, vol. 66, no. 3, pp. 261–281, March 2006.

[13] Y.C. Tsoi and M.S. Brown, "Geometric and shading correction for images of printed materials: A unified approach using boundary," in *Proceedings of IEEE Conference on Computer Vision and Pattern Recognition*, Washington, DC, June 2004, pp. 240–246.

[14] Y.C. Tsoi and M.S. Brown, "Multi-view document rectification using boundary," in *Proceedings of IEEE Conference on Computer Vision and Pattern Recognition*, Chicago, IL, USA, June 2007, pp. 1–8.

[15] M.S. Brown, M. Sun, R. Yang, L. Yung, and W.B. Seales, "Restoring 2D content from distorted documents," *IEEE Transactions on Pattern Analysis and Machine Intelligence*, vol. 29, no. 11, pp. 1904–1916, November 2007.

[16] C. Rother, V. Kolmogorov, and A. Blake, "Grabcut: Interactive foreground extraction using iterated graph cuts," *ACM Transactions on Graphics*, vol. 23, no. 3, pp. 309–314, August 2004.

[17] H. Cao, X. Ding, and C. Liu, "A cylindrical surface model to rectify the bound document," in *Proceedings of International Conference on Computer Vision*, Nice, France, October 2003, pp. 228–233.

[18] D.G. Lowe, "Distinctive image features from scale-invariant keypoints," *International Journal of Computer Vision*, vol. 60, no. 2, pp. 91–110, November 2004.

[19] R.I. Hartley and A. Zisserman, *Multiple view geometry in computer vision*. Cambridge UK: Cambridge University Press, 2nd edition, April 2004.

[20] P.H.S. Torr and A. Zisserman, "MLESAC: A new robust estimator with application to estimating image geometry," *Computer Vision and Image Understanding*, vol. 78, no. 1, pp. 138–156, April 2000.

[21] A. Agarwala, M. Dontcheva, S. Drucker, A. Colburn, B. Curless, D. Salesin, and M. Cohen, "Interactive digital photomontage," *ACM Transactions on Graphics*, vol. 23, no. 3, pp. 294–302, August 2004.

[22] R. Szeliski, R. Zabih, D. Scharstein, O. Veksler, V. Kolmogorov, A. Agarwala, M. Tappen, and C. Rother, "A comparative study of energy minimization methods for markov random fields with smoothness-based priors," *IEEE Transactions on Pattern Analysis and Machine Intelligence*, vol. 30, no. 6, pp. 1068–1080, June 2008.

[23] S. Pollard and M. Pilu, "Building cameras for capturing documents," *International Journal on Document Analysis and Recognition*, vol. 7, no. 2-3, pp. 123–137, July 2005.

[24] P.J. Burt and E.H. Adelson, "A multiresolution spline with applications to image mosaics," *ACM Transaction on Graphics*, vol .2, no .4, pp. 217–236, October 1983.

[25] P. Clark and M. Mirmehdi, "Estimating the orientation and recovery of text planes in a single image," in *Proceedings of the 12th British Machine Vision Conference*, Manchester, UK, pp. 421–430, September 2001.

[26] M. Pilu, "Extraction of illusory linear clues in perspectively skewed documents," in *IEEE Conference on Computer Vision and Pattern Recognition*, Kauai, HI, USA, December 2001, pp. 363–368.

[27] S.A. Coons, "Surfaces for computer-aided design of space forms," Technical Report, Cambridge, MA, USA, 1967.

[28] K. Fukui, "Edge extraction method based on separability of image features," *IEICE transactions on information and systems*, vol. 78, no. 12, pp. 1533–1538, December 1995.

[29] P. Viola and M. Jones, "Rapid object detection using a boosted cascade of simple features," in *IEEE Conference on Computer Vision and Pattern Recognition*, Kauai, HI, USA, December 2001, pp. 511–518.

[30] O. Tuzel, F. Porikli, and P. Meer, "Region covariance: A fast descriptor for detection and classification," in *Proceedings of European Conference on Computer Vision*, Graz, Austria, May 2006, pp. 589–600.

[31] J. Matas, C. Galambos, and J. Kittler, "Robust detection of lines using the progressive probabilistic hough transform," *Computer Vision and Image Understanding*, vol. 78, no. 1, pp. 119–137, April 2000.

13

Bilateral Filter: Theory and Applications

Bahadir K. Gunturk

13.1 Introduction

The bilateral filter is a nonlinear weighted averaging filter, where the weights depend on both the spatial distance and the intensity distance with respect to the center pixel. The main feature of the bilateral filter is its ability to preserve edges while doing spatial smoothing. The term *bilateral filter* was introduced in Reference [1]; the same filter was earlier called the SUSAN (Smallest Univalue Segment Assimilating Nucleus) filter [2]. The variants of the bilateral filter have been published even earlier as the sigma filter [3] and the neighborhood filter [4].

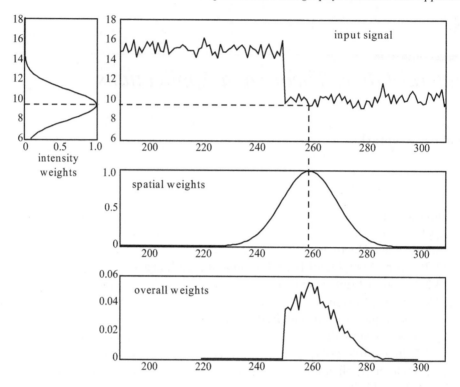

FIGURE 13.1

Illustrative application of the bilateral filter. The range kernel and the spatial kernel are placed at $I(x = 260)$. The product of $K_r(\cdot)$ and $K_d(\cdot)$ determines the weights of the pixels in a local neighborhood. As seen in the overall weights subplot, pixels on the left side of the edge have zero weights in getting the output at $x = 260$ even though they are spatially close.

At a pixel location $\mathbf{x} = (x_1, x_2)$, the output of the bilateral filter is calculated as follows:

$$\hat{I}(\mathbf{x}) = \frac{1}{C(\mathbf{x})} \sum_{\mathbf{y} \in \mathcal{N}(\mathbf{x})} K_d \left(\|\mathbf{y} - \mathbf{x}\| \right) K_r \left(|I(\mathbf{y}) - I(\mathbf{x})| \right) I(\mathbf{y}), \qquad (13.1)$$

where $K_d(\cdot)$ is the spatial domain kernel, $K_r(\cdot)$ is the intensity range kernel, $\mathcal{N}(\mathbf{x})$ is a spatial neighborhood of $\mathbf{x}$, and $C(\mathbf{x})$ is the normalization constant expressed as

$$C(\mathbf{x}) = \sum_{\mathbf{y} \in \mathcal{N}(\mathbf{x})} K_d \left(\|\mathbf{y} - \mathbf{x}\| \right) K_r \left(|I(\mathbf{y}) - I(\mathbf{x})| \right). \qquad (13.2)$$

The kernels $K_d(\cdot)$ and $K_r(\cdot)$ determine how the spatial and intensity differences are treated. The most commonly used kernel is the Gaussian kernel[1] defined as follows:

$$K_d \left(\|\mathbf{y} - \mathbf{x}\| \right) = \exp \left(\frac{-\|\mathbf{y} - \mathbf{x}\|^2}{2\sigma_d^2} \right), \qquad (13.3)$$

---

[1]In the text, the Gaussian kernel is the default kernel unless otherwise stated.

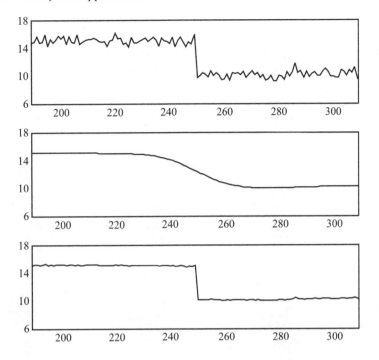

FIGURE 13.2

Top: Input signal. Middle: Output of the Gaussian filter with $\sigma_d = 10$. Bottom: Output of the bilateral filter with $\sigma_d = 10$ and $\sigma_r = 1.5$.

$$K_r\left(|I(\mathbf{y}) - I(\mathbf{x})|\right) = \exp\left(\frac{-|I(\mathbf{y}) - I(\mathbf{x})|^2}{2\sigma_r^2}\right). \tag{13.4}$$

The contribution (weight) of a pixel $I(\mathbf{y})$ is determined by the product of $K_d(\cdot)$ and $K_r(\cdot)$. As illustrated in Figure 13.1, the range kernel pulls down the weights of the pixels that are not close in intensity to the center pixel even if they are in close spatial proximity. This leads to the preservation of edges. Figure 13.2 demonstrates this property of the bilateral filter and compares it with the Gaussian low-pass filter which blurs the edge.

While the Gaussian kernel is the choice for both $K_d(\cdot)$ and $K_r(\cdot)$ in References [1] and [2], the sigma filter [3] and the neighborhood filter [4] use different kernels. The sigma filter [3] calculates the local standard deviation σ around $I(\mathbf{x})$ and uses a thresholded uniform kernel

$$K_r\left(|I(\mathbf{y}) - I(\mathbf{x})|\right) = \begin{cases} 1 & \text{if } |I(\mathbf{y}) - I(\mathbf{x})| \leq 2\sigma \\ 0 & \text{otherwise} \end{cases} \tag{13.5}$$

This range kernel essentially eliminates the use of outliers in calculating the spatial average. The spatial kernel of the sigma filter, on the other hand, is a uniform box kernel with a rectangular or a circular support. For a circular support with radius ρ_d, the spatial kernel is defined as follows:

$$K_d\left(\|\mathbf{y} - \mathbf{x}\|\right) = \begin{cases} 1 & \text{if } \|\mathbf{y} - \mathbf{x}\| \leq \rho_d \\ 0 & \text{otherwise} \end{cases} \tag{13.6}$$

In case of the neighborhood filter [4], the range kernel is Gaussian as in Equation 13.3 and the spatial kernel is a uniform box as in Equation 13.6. Among these kernel options,

FIGURE 13.3

Output of the bilateral filter for different values of σ_d and σ_r: (top) $\sigma_d = 1$, (middle) $\sigma_d = 3$, (bottom) $\sigma_d = 9$; (a) $\sigma_r = 10$, (b) $\sigma_r = 30$, (c) $\sigma_r = 90$, and (d) $\sigma_r \to \infty$.

the Gaussian kernel is the most popular choice for both the range and spatial kernels, as it gives an intuitive and simple control of the behavior of the filter with two parameters.

The Gaussian kernel parameters σ_d and σ_r control the decay of the weights in space and intensity. Figure 13.3 demonstrates the behavior of the bilateral filter for different combinations of σ_d and σ_r. It can be seen that the edges are preserved better for small values of σ_r. In fact, an image is hardly changed as $\sigma_r \to 0$. As $\sigma_r \to \infty$, $K_r(\cdot)$ approaches to 1 and the bilateral filter becomes a Gaussian low-pass filter. On the other hand, σ_d controls the spatial extent of pixel contribution. As $\sigma_d \to 0$, the filter acts on a single pixel. As $\sigma_d \to \infty$, the spatial extent of the filter will increase, and eventually, the bilateral filter will act only on intensities regardless of position, in other words, on histograms.

Denoting $H(\cdot)$ as the histogram over the entire spatial domain, the filter becomes:

$$\hat{I}(\mathbf{x}) = \frac{1}{C(\mathbf{x})} \sum_{\mathbf{y}} \exp\left(\frac{-|I(\mathbf{y}) - I(\mathbf{x})|^2}{2\sigma_r^2}\right) I(\mathbf{y})$$

$$= \frac{1}{C(\mathbf{x})} \sum_{i=0}^{255} iH(i) \exp\left(\frac{-|i - I(\mathbf{x})|^2}{2\sigma_r^2}\right) \tag{13.7}$$

where the normalization constant is defined as

FIGURE 13.4

Iterative application of the bilateral filter with $\sigma_d = 12$ $\sigma_r = 30$. Left: Input image. Middle: Result of the first iteration. Right: Result of the third iteration.

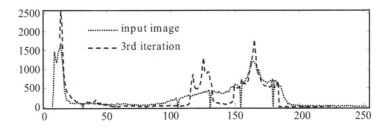

FIGURE 13.5

Histograms of the input image and the output image after the third iteration of the bilateral filter.

$$C(\mathbf{x}) = \sum_{\mathbf{y}} \exp\left(\frac{-|I(\mathbf{y}) - I(\mathbf{x})|^2}{2\sigma_r^2}\right)$$
$$= \sum_{i=0}^{255} H(i) \exp\left(\frac{-|i - I(\mathbf{x})|^2}{2\sigma_r^2}\right). \tag{13.8}$$

Considering the histogram $H(i)$ as the probability density function (pdf) of intensities, $H(i)\exp(-|i - I(\mathbf{x})|^2/(2\sigma_r^2))$ is the smoothed pdf, and Equation 13.7 can be interpreted as finding the *expected* value of the pixel intensities. When σ_r also goes to infinity, the bilateral filter returns the expected (average) value of all intensities.

From the histogram perspective, the bilateral filter can be interpreted as a *local mode* filter [5], returning the expected value of local histograms. This effect is demonstrated through iterative application of the bilateral filter in Figures 13.4 and 13.5. As seen, the filtered image approaches the modes of the distribution through the iterations. The output histogram has certain peaks, and in-between values are reduced.

Using the bilateral filter, an image can be decomposed into its large-scale (base) and small-scale (detail) components. The large-scale component is a smoothed version of the input image with main edges preserved, and the small-scale component is interpreted as having the texture details or noise, depending on the application and parameter selection. The small-scale component is obtained by subtracting the filtered image from the original image. Figure 13.6 shows the effect of the σ_r value on extracting detail components.

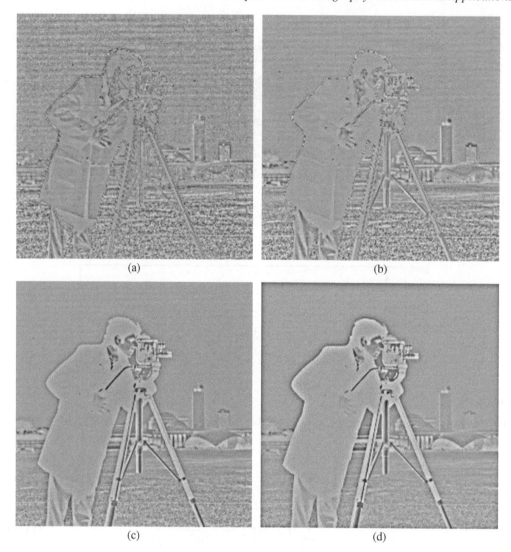

FIGURE 13.6

A detail component is obtained by subtracting the filtered image from the original image. In this figure, $\sigma_d = 3$ and (a) $\sigma_r = 10$, (b) $\sigma_r = 30$, (c) $\sigma_r = 90$, and (d) $\sigma_r \to \infty$ in raster scan order.

13.2 Applications

The bilateral filter has found a number of applications in image processing and computer vision. This section reviews some popular examples of using bilateral filter in practice.

13.2.1 Image Denoising

The immediate application of the bilateral filter is image denoising, because this filter can do spatial averaging without blurring edges. The critical question is how to adjust

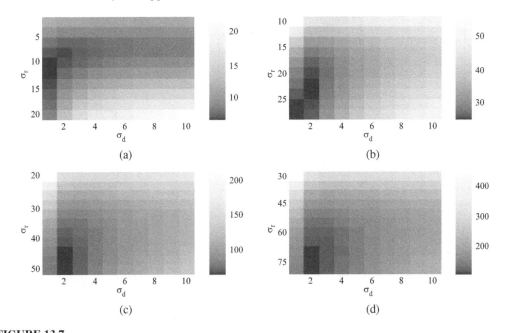

FIGURE 13.7

Average MSE values between original images and denoised images for different values of σ_d, σ_r, and the noise standard deviation σ_n: (a) $\sigma = 5$, (b) $\sigma = 10$, (c) $\sigma = 15$, and (d) $\sigma = 20$.

the parameters of the bilateral filter as a function of noise or local texture. Reference [6] presents an empirical study on optimal parameter selection. To understand the relation among σ_d, σ_r, and the noise standard deviation σ_n, zero-mean white Gaussian noise is added to some test images and the bilateral filter is applied with different values of the parameters σ_d and σ_r. The experiment is repeated for different noise variances and the mean squared error (MSE) values are recorded. The average MSE values are given in Figure 13.7. These MSE plots indicate that the optimal σ_d value is relatively insensitive to noise variance compared to the optimal σ_r value. It appears that σ_d could be chosen around two regardless of the noise power; on the other hand, the optimal σ_r value changes significantly as the noise standard deviation σ_n changes. This is an expected result because if σ_r is smaller than σ_n, noisy data could remain isolated and untouched, as in the case of the salt-and-pepper noise problem of the bilateral filter [1]. That is, σ_r should be sufficiently large with respect to σ_n.

To see the relation between σ_n and the optimal σ_r, σ_d is set to some constant values, and the optimal σ_r values (minimizing MSE) are determined as a function of σ_n. The experiments are again repeated for a set of images; the average values and the standard deviations are displayed in Figure 13.8. It can be observed that the optimal σ_r is linearly proportional to σ_n. There is obviously no single value for σ_r/σ_n that is optimal for all images and σ_d values; and in fact, future research should investigate spatially adaptive parameter selection to take local texture characteristics into account. On the other hand, these experiments give us some guidelines in selection of the parameters.

Reference [6] further suggests a multiresolution framework for the bilateral filter. In this way, different noise components (fine-grain and coarse-grain noise) can be determined

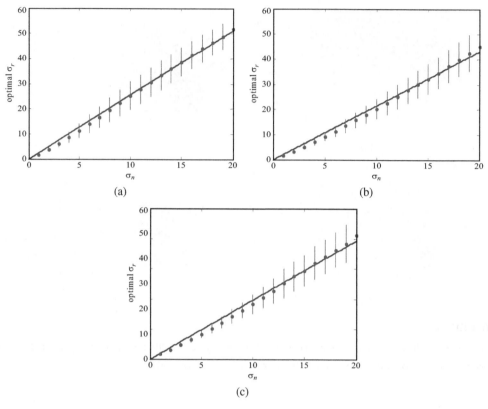

FIGURE 13.8

The optimal σ_r values plotted as a function of the noise standard deviation σ_n based on the experiments with a number of test images [6]: (a) $\sigma_d = 1.5$, (b) $\sigma_d = 3.0$, and (c) $\sigma_d = 5.0$. The data points are the mean of optimal σ_r values that produce the smallest MSE for each σ_n value. The vertical lines denote the standard deviation of the optimal σ_r for the test images. The least squares fits to the means of the optimal σ_r/σ_n data are plotted as diagonal lines. The slopes of these lines are, from left to right, 2.56, 2.16, and 1.97.

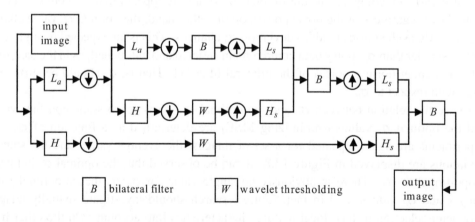

FIGURE 13.9

Illustration of the multiresolution denoising framework [6]. The analysis and synthesis filters (L_a, H_a, L_s, and H_s) form a perfect reconstruction filter bank.

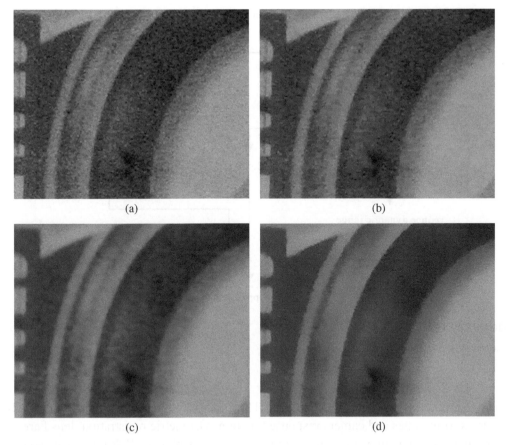

FIGURE 13.10

Image denoising using the bilateral filter: (a) input image, (b) bilateral filtering [1] with $\sigma_d = 1.8$ and $\sigma_r = 3 \times \sigma_n$, (c) bilateral filtering [1] with $\sigma_d = 5.0$ and $\sigma_r = 20 \times \sigma_n$, and (d) multiresolution bilateral filtering [6] with $\sigma_d = 1.8$ and $\sigma_r = 3 \times \sigma_n$ at each resolution level.

and eliminated at different resolution levels for better results. The proposed framework is illustrated in Figure 13.9. A signal is decomposed into its frequency subbands with wavelet decomposition; as the signal is reconstructed back, bilateral filtering is applied to the approximation subbands and wavelet thresholding is applied to the detail subbands. At each level, the noise standard deviation σ_n is estimated, and the bilateral filter parameter σ_r is set accordingly. Unlike the standard single-level bilateral filter [1], this multiresolution bilateral filter has the potential of eliminating coarse-grain noise components. This is demonstrated in Figure 13.10.

13.2.2 Tone Mapping of High-Dynamic Range Images

Because a real scene can have much wider dynamic range than a camera can capture, images often suffer from saturation. By changing the exposure rate, it is possible to get information from different parts of the same scene. In high-dynamic range (HDR) imaging, multiple low-dynamic range (LDR) images captured with different exposure rates are com-

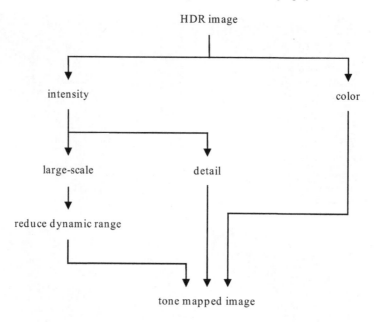

FIGURE 13.11

The tone-mapping method of Reference [10].

bined to produce a HDR image [7], [8], [9]. This process requires estimation or knowledge
of the exposure rates and camera response function. Geometric registration, lens flare and
ghost removal, vignetting correction, compression and display of HDR images are some of
the other challenges in HDR imaging.

After an HDR image is generated, it has to be tone-mapped to display on a screen, which
typically has less dynamic range than the HDR image. The bilateral filter has been successfully used for this purpose [10]. As illustrated in Figure 13.11, the intensity and color
channels of a HDR image are first extracted. The intensity channel is then decomposed
into its large-scale and detail components using the bilateral filter. The dynamic range of
the large-scale component is reduced (using, for instance, linear or logarithmic scaling) to
fit into the dynamic range of the display; it is then combined with the detail component
to form the tone-mapped intensity, which is finally combined with the color channel to
form the final image. The detail component preserves the high frequency content of the
image. Since bilateral filtering is used to obtain the large-scale component, the edges are
not blurred and the so-called halo artifacts are avoided. Figure 13.12 demonstrates this
framework in a practical situation.

13.2.3 Contrast Enhancement

A commonly used model in image formation is such that an intensity $S = LR$ is the
product of an illumination component L and a reflectance component R. Retinex methods try to remove the illumination component based on the assumption that illumination changes slowly compared to reflectance. Taking the logarithm of $S = LR$ results in
$s = \log(S) = \log(L) + \log(R) = l + r$. By low-pass filtering s, the illumination component

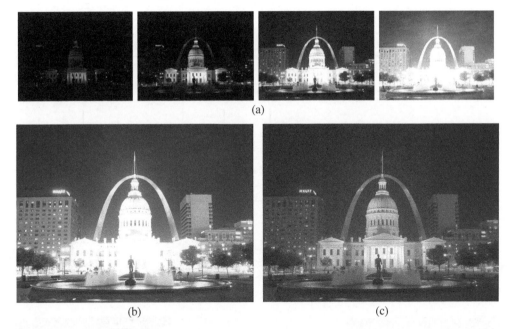

FIGURE 13.12 (See color insert.)

High definition range imaging using the bilateral filter: (a) a set of four input images, (b) linearly scaled HDR image, and (c) image obtained using the tone-mapping method of Reference [10].

l can be estimated.

Reference [11] uses two bilateral filters, one for extracting the illumination component and the other for denoising the reflectance component. Since the reflectance R is in the range $[0,1]$, it holds that $l \leq s$. Therefore, in calculating the illumination component l, only the pixels with value larger than the value of the center pixel are included in the bilateral filter. Once l is calculated, $s - l$ gives the reflectance component r. Reference [11] uses a second bilateral filter to remove noise from the reflectance component. As the noise is more pronounced in the darker regions, the bilateral filter is adapted spatially through the range parameter as $\sigma_r(\mathbf{x}) = (c_1 s(\mathbf{x})^{c_2} + c_3)^{-1}$, where c_1, c_2, and c_3 are some constants. With this adaptation, larger σ_r (therefore, stronger filtering) is applied for smaller s.

Another contrast enhancement algorithm where bilateral filtering is utilized is presented in Reference [12]. Similar to Reference [10], an image is decomposed into its large-scale and detail components using the bilateral filter. The large-scale component is modified with a histogram specification; the detail component is modified according to a *textureness* measure, which quantifies the degree of local texture. The textureness measure T_I is obtained by *cross (or joint) bilateral filtering* the high-pass filtered image H_I as follows:

$$T_I(\mathbf{x}) = \frac{1}{C(\mathbf{x})} \sum_{\mathbf{y} \in \mathcal{N}(\mathbf{x})} K_d\left(\|\mathbf{y} - \mathbf{x}\|\right) K_r\left(|I(\mathbf{y}) - I(\mathbf{x})|\right) |H_I(\mathbf{y})|, \qquad (13.9)$$

where $|\cdot|$ returns the absolute values, and the cross bilateral filter smooths $|H_I|$ without blurring edges. The term cross (or joint) bilateral filter [13], [14] is used because input to the kernel $K_r(\cdot)$ is I, but not $|H_I|$. In other words, the edge information comes from I while $|H_I|$ is filtered.

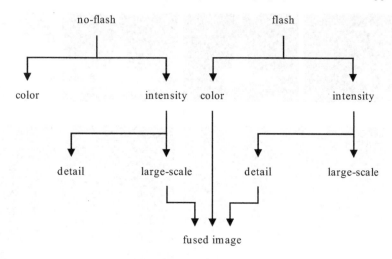

FIGURE 13.13

Flowchart for the fusion of flash and no-flash images.

FIGURE 13.14 (See color insert.)

Image fusion using the bilateral filter: (a) no-flash image, (b) flash image, and (c) fusion of the flash and no-flash images.

13.2.4 Image Fusion

Another application of a bilateral filter is the fusion of different image modalities, such as flash and no-flash images [13], [14], and visible spectrum and infrared spectrum images [15]. A flash image has high signal-to-noise ratio; however, it has unpleasing direct flash lighting. The no-flash version of the same scene, on the other hand, suffers from low signal-to-noise ratio. As illustrated in Figure 13.13, the color and detail components of the flash image and the large-scale component of the no-flash image are combined. The resulting image has the sharpness of the flash image and the tonal characteristics of the no-flash image. A sample result is given in Figure 13.14. One potential problem in combining flash

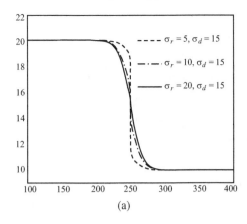

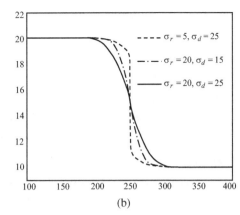

(a)   (b)

FIGURE 13.15

Bilateral filtering is applied on a step edge signal to illustrate the effects of the filter parameters σ_r and σ_d on blocking artifacts.

and no-flash images is the flash shadows. In Reference [13], shadow regions are detected and excluded from bilateral filtering in extracting the detail layer. The same work also proposes the use of the cross bilateral filter in obtaining the large-scale component of the no-flash image when it is too dark and thus suffers from a low signal-to-noise ratio:

$$\hat{I}_{no-flash}(\mathbf{x}) = \frac{1}{C(\mathbf{x})} \sum_{\mathbf{y} \in \mathcal{N}(\mathbf{x})} K_d\left(\|\mathbf{y} - \mathbf{x}\|\right) K_r\left(\left|I_{flash}(\mathbf{y}) - I_{flash}(\mathbf{x})\right|\right) I_{no-flash}(\mathbf{y}) \quad (13.10)$$

Similarly, in Reference [15], the large-scale component of a (noisy) visible-spectrum image is combined with the detail component of the corresponding infrared image. To obtain the large-scale component of the visible-spectrum image, the infrared image is utilized:

$$\hat{I}_S(\mathbf{x}) = \frac{1}{C(\mathbf{x})} \sum_{\mathbf{y} \in \mathcal{N}(\mathbf{x})} K_d\left(\|\mathbf{y} - \mathbf{x}\|\right) K_{r_1}\left(\|\mathbf{I}_{RGB}(\mathbf{y}) - \mathbf{I}_{RGB}(\mathbf{x})\|\right) K_{r_2}\left(\left|I_{IR}(\mathbf{y}) - I_{IR}(\mathbf{x})\right|\right) I_S(\mathbf{y}),$$

$$(13.11)$$

where K_{r_1} and K_{r_2} are range kernels for the visible-spectrum and the infrared images, $I_{IR}(\mathbf{x})$ is the infrared image, $I_S(\cdot)$ denotes a color channel with $S \in \{red, green, blue\}$, and $\mathbf{I}_{RGB}(\mathbf{x}) = [I_{red}(\mathbf{x}), I_{green}(\mathbf{x}), I_{blue}(\mathbf{x})]$. Reference [15] argues that the *dual* kernel $K_{r_1}(\cdot)K_{r_2}(\cdot)$ detects edges better because it is sufficient if an edge appears in the *RGB* image or the *IR* image.

13.2.5 Compression Artifact Reduction

Block-based discrete cosine transform (DCT) is adopted by various image and video compression standards, such as JPEG, MPEG, and H.26x. One problem associated with the block-based processing is the blocking artifacts, the discontinuities along the block boundaries caused by the coarse quantization of the DCT coefficients. The blocking artifacts and other compression artifacts, such as the mosquito or ringing artifacts, become more severe with higher compression rates.

Reference [16] presents a spatially adaptive version of the bilateral filter to reduce block discontinues and ringing artifacts effectively while avoiding over-smoothing of texture re-

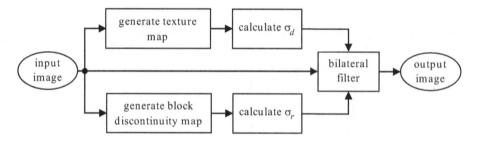

FIGURE 13.16

The block diagram of the method in Reference [16]. Discontinuity and texture detection modules produce space varying maps that are used to compute the range and domain parameters of the bilateral filter. The bilateral filter is then applied to the image based on these parameters.

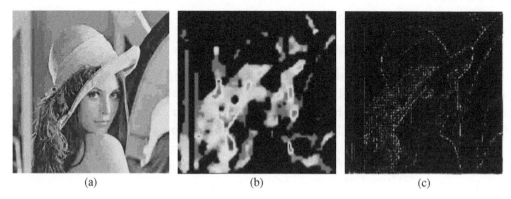

            (a)   (b)   (c)

FIGURE 13.17

Discontinuity and texture map generation: (a) input compressed image, (b) texture map, and (c) block discontinuity map produced by the method of Reference [16].

gions. The parameters of the bilateral filter should be carefully chosen for this purpose. As illustrated in Figure 13.15, when the σ_r value is less than the discontinuity amount, the filter is basically useless for eliminating the discontinuity. When σ_r is larger than the discontinuity amount, the discontinuity can be eliminated. The extent of the smoothing can be controlled by the σ_d value. The larger the σ_d value, the wider the extent of smoothing is. On the other hand, if σ_r value is less than the discontinuity amount, elimination of the discontinuity is impossible no matter the value of σ_d.

    Figure 13.16 shows the flowchart of this method. The block discontinuity amounts are detected at the block boundaries, and then spatially interpolated to obtain a discontinuity map. The σ_r value at each pixel is adjusted accordingly; specifically, σ_r at a pixel is linearly proportional to the discontinuity map value. On the other hand, the σ_d value is adjusted according to the local texture to avoid over-smoothing. A texture map is obtained by calculating the local standard deviation at every pixel; the σ_d value is set inversely proportional to the texture map value at each pixel. Figure 13.17 shows discontinuity and texture maps for a compressed image. Figure 13.18 compares the results of the standard bilateral filter and the spatially adaptive bilateral filter.

(a) (b)

(c) (d)

FIGURE 13.18

Compression artifact reduction using the bilateral filter: (a) original image, (b) compressed image, (c) result of the standard bilateral filter with $\sigma_r = 20$ and $\sigma_d = 3$, and (d) result of the adaptive bilateral filter [16].

13.2.6 Mesh Smoothing

Since three-dimensional (3D) scanning devices may produce noisy observations, denoising of 3D meshes is necessary in many computer graphics applications. In Reference [17], each vertex point of a mesh is denoised along its normal direction; this would change the geometric shape information but not any other parametric information of the mesh. Let S be the noise-free surface, and M be the noisy observed mesh. A vertex point $\mathbf{x} = (x_1, x_2, x_3) \in M$ is updated along the surface normal $\mathbf{n_x}$ as follows [17]:

$$\hat{\mathbf{x}} = \mathbf{x} + d\mathbf{n_x}. \tag{13.12}$$

The update amount d is calculated through the application of the bilateral filter within a local neighborhood:

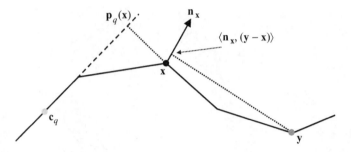

FIGURE 13.19

Illustration of the mesh denoising methods of References [17] and [18].

$$d = \frac{1}{C(\mathbf{x})} \sum_{\mathbf{y}} K_d\left(\|\mathbf{y} - \mathbf{x}\|\right) K_r\left(\langle \mathbf{n_x}, (\mathbf{y} - \mathbf{x})\rangle\right) \langle \mathbf{n_x}, (\mathbf{y} - \mathbf{x})\rangle, \qquad (13.13)$$

where the inner product $\langle \mathbf{n_x}, (\mathbf{y} - \mathbf{x})\rangle$ gives the projection of the difference $(\mathbf{y} - \mathbf{x})$ onto the surface normal $\mathbf{n_x}$, and thus the bilateral filter smooths the projections within a local space and updates the vertex as a weighted sum of the projections.

Reference [18] takes a different approach to mesh smoothing. Suppose that q is a surface within a neighborhood of the vertex $\mathbf{x}$, $\mathbf{c}_q$ is the centroid of the surface, and $\mathbf{a}_q$ is the area of the surface. The prediction $\mathbf{p}_q(\mathbf{x})$ of the vertex $\mathbf{x}$ based on the surface q is the projection of $\mathbf{x}$ to the plane tangent to the surface q. Then the vertex $\mathbf{x}$ is updated as follows:

$$\hat{\mathbf{x}} = \frac{1}{C(\mathbf{x})} \sum_{q} \mathbf{a}_q K_d\left(\|\mathbf{c}_q - \mathbf{x}\|\right) K_r\left(\|\mathbf{p}_q(\mathbf{x}) - \mathbf{x}\|\right) \mathbf{p}_q(\mathbf{x}). \qquad (13.14)$$

The inclusion of the area $\mathbf{a}_q$ is to give more weight to predictions coming from larger surfaces. Figure 13.19 illustrates the methods of References [17] and [18].

13.2.7 Image Interpolation

Since the bilateral filter adapts its filter coefficients to preserve edges, it can be used in image interpolation. A very good fit is the demosaicking problem. In single-sensor digital cameras, a color filter array is placed on the sensor to capture one spectral component at each pixel. The mosaic of color samples is then interpolated to obtain full-color channels. The interpolation process is known as demosaicking [19]. The popular Bayer pattern repeats a green-red-green-blue pattern to form the color filter array; the number of green samples is twice the number of red/blue samples. Therefore, the edge information coming from the green channel can be utilized to interpolate the red and blue channels.

In Reference [20], the red channel I_R is updated using the cross bilateral filter as follows:

$$\hat{I}_R(\mathbf{x}) = \frac{1}{C(\mathbf{x})} \sum_{\mathbf{y}} M_R(\mathbf{y}) K_d\left(\|\mathbf{y} - \mathbf{x}\|\right) K_r\left(|I_G(\mathbf{y}) - I_G(\mathbf{x})|\right) I_R(\mathbf{y}), \qquad (13.15)$$

where M_R is a mask of ones and zeros, indicating the locations of red samples, and $K_r\left(|I_G(\mathbf{y}) - I_G(\mathbf{x})|\right)$ captures the edge information from the green channel. The blue channel is updated similarly.

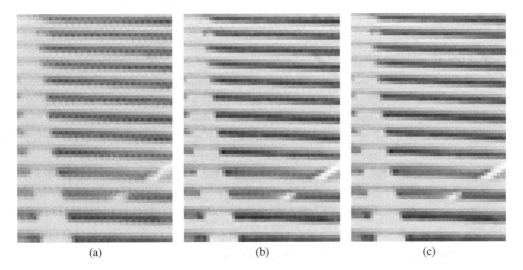

(a)   (b)   (c)

FIGURE 13.20

Demosaicking using the bilateral filter: (a) bilinear interpolation of a Bayer sampled data, (b) the standard POCS interpolation method of Reference [22], (c) the POCS method with the addition of the bilateral constraint set [21].

In Reference [21], the green channel is again used as a reference to interpolate the red and blue channels. Instead of applying the bilateral filter directly, the interpolation problem is formulated as an optimization problem, and solved using the projections onto convex sets (POCS) technique [22]. The POCS technique starts with an initial estimate and updates it iteratively by projecting onto constraint sets. Denoting $BF(\cdot)$ as the bilateral filter, the constraint set S_R on the red (or blue) channel to limit the deviation from the green channel is defined as follows:

$$S_R = \{I_R(\mathbf{x}) : |(I_R(\mathbf{x}) - I_G(\mathbf{x})) - BF(I_R(\mathbf{x}) - I_G(\mathbf{x}))| \leq T\}, \qquad (13.16)$$

where T is a positive threshold. For an object in the scene, the difference $I_R - I_G$ should be constant or change smoothly. This constraint set guarantees that $I_R - I_G$ changes smoothly in space; and the bilateral filter prevents crossing edges. Reference [21] defines additional constraint sets, including data fidelity and frequency similarity constraint sets, and uses the POCS technique to perform the interpolation. A sample result is shown in Figure 13.20.

13.2.8   Other Applications

In addition to the applications mentioned so far, the bilateral filter has been utilized in various other applications, including optical flow estimation [23], depth map estimation [24], video stylization [25], texture and illumination separation [26], orientation smoothing [27], medical imaging [28], and video enhancement [29]. The main underlying idea in all these applications is the use of multiple kernels, where spatial distance as well as range distance are taken into account to preserve discontinuities in the signal. The specific choice of kernels and parameters depends on the application.

13.3 Fast Bilateral Filter

The direct implementation of the bilateral filter is presented in Algorithm 13.1 which shows the flow of operations performed for each pixel $\mathbf{x} \in \mathscr{S}$, where $\mathscr{S}$ is the set of all pixels in the image. The computational complexity of this implementation is $\mathscr{O}(|\mathscr{S}||\mathscr{N}|)$, where $|\mathscr{S}|$ is the number of pixels in the entire image and $|\mathscr{N}|$ is the number of pixels in the neighborhood $\mathscr{N}$. The local neighborhood is typically chosen such that $\|\mathbf{y} - \mathbf{x}\| \leq 3\sigma_d$; therefore the neighborhood size $|\mathscr{N}|$ is proportional to σ_d^2. While the overall complexity $\mathscr{O}(|\mathscr{S}|\sigma_d^2)$ is bearable for small σ_d, it quickly becomes restrictive with increasing σ_d. To address this issue, a number of fast implementation/approximation methods have been proposed.

> **ALGORITHM 13.1** Bilateral filter implementation.
>
> 1. Initialize $\hat{I}(\mathbf{x}) = 0$ and $C(\mathbf{x}) = 0$.
> 2. For each $\mathbf{y}$ in the local neighborhood $\mathscr{N}$ of $\mathbf{x}$:
>   - Calculate the weight $w = K_d(\|\mathbf{y} - \mathbf{x}\|) K_r(|I(\mathbf{y}) - I(\mathbf{x})|)$.
>   - Update $\hat{I}(\mathbf{x}) = \hat{I}(\mathbf{x}) + w * I(\mathbf{y})$.
>   - Update $C(\mathbf{x}) = C(\mathbf{x}) + w$.
> 3. Normalize $\hat{I}(\mathbf{x}) = \hat{I}(\mathbf{x})/C(\mathbf{x})$.

13.3.1 Kernel Separation

One method of speeding up the bilateral filter is to separate the two-dimensional (2D) filter kernel into two one-dimensional (1D) kernels: first, the rows of an image are filtered, the result is then filtered along the columns [30]. This reduces the complexity to $\mathscr{O}(|\mathscr{S}|\sigma_d)$. Although its performance is good in smooth regions and horizontal/vertical edges, the algorithm does not perform satisfactorily on texture regions and slanted edges.

13.3.2 Bilateral Grid

Another fast bilateral filter algorithm is obtained through representing an image in a 3D space, where the signal intensity is added to the spatial domain as the third dimension [31]. First, note that the bilateral filter can be represented in a vector form:

$$\begin{pmatrix} C(\mathbf{x})\hat{I}(\mathbf{x}) \\ C(\mathbf{x}) \end{pmatrix} = \sum_{\mathbf{y} \in \mathscr{N}(\mathbf{x})} K_d(\|\mathbf{y} - \mathbf{x}\|) K_r(|I(\mathbf{y}) - I(\mathbf{x})|) \begin{pmatrix} I(\mathbf{y}) \\ 1 \end{pmatrix}. \tag{13.17}$$

This vector representation can be used to interpret the bilateral filter as linear filtering of the entries of a vector-valued image separately, followed by division of the first entry by the second.

More explicitly, the bilateral filter is implemented by defining the 3D grids, Γ_1 and Γ_2, of a 2D image I as follows:

$$\Gamma_1(x_1, x_2, r) = \begin{cases} I(x_1, x_2) & \text{if } r = I(x_1, x_2), \\ 0 & \text{otherwise.} \end{cases} \tag{13.18}$$

$$\Gamma_2(x_1, x_2, r) = \begin{cases} 1 & \text{if } r = I(x_1, x_2), \\ 0 & \text{otherwise.} \end{cases} \tag{13.19}$$

These grids are then convolved with a 3D Gaussian, $K_{d,r}$, whose standard deviation is σ_d in the 2D spatial domain and σ_r in the 1D intensity domain, providing the following:

$$\hat{\Gamma}_1(x_1, x_2, r) = K_{d,r} * \Gamma_1(x_1, x_2, r), \tag{13.20}$$

$$\hat{\Gamma}_2(x_1, x_2, r) = K_{d,r} * \Gamma_2(x_1, x_2, r). \tag{13.21}$$

Finally, the result of the bilateral filter at position (x_1, x_2) with input intensity $I(x_1, x_2)$ is obtained as follows:

$$\hat{I}(x_1, x_2) = \frac{\hat{\Gamma}_1(x_1, x_2, I(x_1, x_2))}{\hat{\Gamma}_2(x_1, x_2, I(x_1, x_2))}. \tag{13.22}$$

Since $\hat{\Gamma}_1$ and $\hat{\Gamma}_2$ are obtained through lowpass filtering, they are bandlimited and can be represented well with their low-frequency components. Therefore, the grids Γ_1 and Γ_2 can be downsampled without losing much performance to speed up the algorithm. Reference [31] proposes downsampling of the spatial domain $\mathscr{S}$ by σ_d and the intensity range $\mathscr{R}$ by σ_r. The complexity of the algorithm then becomes $\mathcal{O}(|\mathscr{S}| + |\mathscr{S}||\mathscr{R}|/(\sigma_d \sigma_r))$.

13.3.3 Local Histogram-Based Bilateral Filter with Uniform Spatial Kernel

With the uniform box kernel, the bilateral filter can be written in terms of local histograms:

$$\hat{I}(\mathbf{x}) = \frac{1}{C(\mathbf{x})} \sum_{\mathbf{y} \in \mathcal{N}(\mathbf{x})} K_r(|I(\mathbf{y}) - I(\mathbf{x})|) I(\mathbf{y})$$

$$= \frac{1}{C(\mathbf{x})} \sum_{i=0}^{255} i H_{\mathbf{x}}(i) K_r(|i - I(\mathbf{x})|), \tag{13.23}$$

where

$$C(\mathbf{x}) = \sum_{\mathbf{y} \in \mathcal{N}(\mathbf{x})} K_r(I(\mathbf{y}) - I(\mathbf{x}))$$

$$= \sum_{i=0}^{255} H_{\mathbf{x}}(i) K_r(i - I(\mathbf{x})) \tag{13.24}$$

and $H_{\mathbf{x}}$ is the histogram in a local neighborhood of $\mathbf{x}$.

There are few advantages of this formulation [32]. Namely, $K_r(|i - I(\mathbf{x})|)$ can be calculated for all values of i and at all locations $\mathbf{x}$ independently and therefore in parallel. Similarly, $H_{\mathbf{x}}(i)$ and $i H_{\mathbf{x}}(i)$ can be calculated independent of $K_r(|i - I(\mathbf{x})|)$. The term $H_{\mathbf{x}}(i)$ can be calculated in constant time using the integral histogram technique [33]. Finally, the algorithm can be further speeded up by quantizing the histogram. As a result, the bilateral filter can be implemented in constant time, that is, independent of the image size or the kernel size. A similar histogram-based method was proposed in Reference [34] although the histogram computation is not as efficient as the one presented in Reference [32].

13.3.4 Polynomial Representation of Range Filter

In addition to the histogram-based approach, Reference [32] presents another approach, where there is no restriction on the domain filter, but the range kernel is approximated with a polynomial. Doing a first-order Taylor series expansion on the Gaussian range filter, the following is obtained:

$$
\begin{aligned}
\hat{I}(\mathbf{x}) &= \frac{1}{C(\mathbf{x})} \sum_{\mathbf{y} \in \mathcal{N}(\mathbf{x})} K_d\left(\|\mathbf{y} - \mathbf{x}\|\right) e^{\frac{-|I(\mathbf{y}) - I(\mathbf{x})|^2}{2\sigma_r^2}} I(\mathbf{y}) \\
&\simeq \frac{1}{C(\mathbf{x})} \sum_{\mathbf{y} \in \mathcal{N}(\mathbf{x})} K_d\left(\|\mathbf{y} - \mathbf{x}\|\right) \left(1 - \frac{1}{2\sigma_r^2}|I(\mathbf{y}) - I(\mathbf{x})|^2\right) I(\mathbf{y}) \\
&= \frac{1}{C(\mathbf{x})} \sum_{\mathbf{y} \in \mathcal{N}(\mathbf{x})} K_d\left(\|\mathbf{y} - \mathbf{x}\|\right) \left(1 - \frac{1}{2\sigma_r^2}I^2(\mathbf{y}) + \frac{2}{2\sigma_r^2}I(\mathbf{y})I(\mathbf{x}) - \frac{1}{2\sigma_r^2}I^2(\mathbf{x})\right) I(\mathbf{y}) \\
&= \frac{1}{C(\mathbf{x})} \left[\sum_{\mathbf{y} \in \mathcal{N}(\mathbf{x})} K_d\left(\|\mathbf{y} - \mathbf{x}\|\right) I(\mathbf{y}) - \frac{1}{2\sigma_r^2} \sum_{\mathbf{y} \in \mathcal{N}(\mathbf{x})} K_d\left(\|\mathbf{y} - \mathbf{x}\|\right) I^3(\mathbf{y}) \right. \\
&\quad \left. + \frac{I(\mathbf{x})}{\sigma_r^2} \sum_{\mathbf{y} \in \mathcal{N}(\mathbf{x})} K_d\left(\|\mathbf{y} - \mathbf{x}\|\right) I^2(\mathbf{y}) - \frac{I^2(\mathbf{x})}{2\sigma_r^2} \sum_{\mathbf{y} \in \mathcal{N}(\mathbf{x})} K_d\left(\|\mathbf{y} - \mathbf{x}\|\right) I(\mathbf{y}) \right].
\end{aligned}
\tag{13.25}
$$

This equation reveals that the bilateral filter could be approximated from spatially filtered I, I^2, and I^3. Defining z_n as the convolution of $K_d(\cdot)$ with I^n:

$$
z_n(\mathbf{x}) = \sum_{\mathbf{y} \in \mathcal{N}(\mathbf{x})} K_d\left(\|\mathbf{y} - \mathbf{x}\|\right) I^n(\mathbf{y}),
\tag{13.26}
$$

the bilateral filtered image is

$$
\hat{I} = \frac{1}{C} \left[\left(1 - \frac{1}{2\sigma_r^2}I^2\right) z_1 + \left(\frac{1}{\sigma_r^2}I\right) z_2 - \left(\frac{1}{2\sigma_r^2}\right) z_3 \right],
\tag{13.27}
$$

where

$$
C = \left(1 - \frac{1}{2\sigma_r^2}I^2\right) + \left(\frac{1}{\sigma_r^2}I\right) z_1 - \left(\frac{1}{2\sigma_r^2}\right) z_2.
\tag{13.28}
$$

In other words, the bilateral filter is implemented through linear filtering and element-by-element multiplication / division of pixel intensities. The performance of this algorithm is good for small σ_r, but it degrades quickly for large σ_r since the polynomial representation does not approximate the Gaussian well. The use of higher-order polynomials should improve the results.

13.4 Theoretical Foundations

Although the bilateral filter was first proposed as an intuitive tool, recent research has pointed out the connections with some well-established techniques.

13.4.1 Relation to Robust Estimation and Weighted Least Squares

In an inverse problem (e.g., denoising, restoration, interpolation), one can impose smoothness or any other prior information on the solution by adding a regularization term to the cost function. In robust estimation, the regularization term includes a robust function to reduce the effects of outliers. Consider the following regularization term:

$$\Phi\left(I(\mathbf{x})\right) = \sum_{\mathbf{y}} K_d\left(\|\mathbf{y} - \mathbf{x}\|\right) \rho\left(I(\mathbf{y}) - I(\mathbf{x})\right), \tag{13.29}$$

where $\rho(\cdot)$ is a robust function that penalizes the difference between $I(\mathbf{x})$ and $I(\mathbf{y})$. The regularization term also includes $K_d\left(\|\mathbf{y} - \mathbf{x}\|\right)$ to give more weight to the pixels that are close to $\mathbf{x}$. If $\rho(\cdot)$ is differentiable, the solution that minimizes the cost function can be found iteratively using a gradient descent technique. For instance, an iteration of the steepest descent algorithm is

$$\hat{I}(\mathbf{x}) = I(\mathbf{x}) - \mu \frac{\partial \Phi\left(I(\mathbf{x})\right)}{\partial I(\mathbf{x})}$$

$$= I(\mathbf{x}) + \mu \sum_{\mathbf{y}} K_d\left(\|\mathbf{y} - \mathbf{x}\|\right) \rho'\left(I(\mathbf{y}) - I(\mathbf{x})\right), \tag{13.30}$$

where μ is the step size.

By expressing $\rho'(\cdot)$ as $\rho'\left(I(\mathbf{y}) - I(\mathbf{x})\right) = K_r\left(|I(\mathbf{y}) - I(\mathbf{x})|\right)\left(I(\mathbf{y}) - I(\mathbf{x})\right)$, the iteration in Equation 13.30 becomes equivalent to the bilateral filter:

$$\hat{I}(\mathbf{x}) = I(\mathbf{x}) + \mu \sum_{\mathbf{y}} K_d\left(\|\mathbf{y} - \mathbf{x}\|\right) K_r\left(|I(\mathbf{y}) - I(\mathbf{x})|\right)\left(I(\mathbf{y}) - I(\mathbf{x})\right)$$

$$= I(\mathbf{x}) - \mu \sum_{\mathbf{y}} K_d\left(\|\mathbf{y} - \mathbf{x}\|\right) K_r\left(|I(\mathbf{y}) - I(\mathbf{x})|\right) I(\mathbf{x})$$

$$+ \mu \sum_{\mathbf{y}} K_d\left(\|\mathbf{y} - \mathbf{x}\|\right) K_r\left(|I(\mathbf{y}) - I(\mathbf{x})|\right) I(\mathbf{y})$$

$$= \frac{1}{C(\mathbf{x})} \sum_{\mathbf{y}} K_d\left(\|\mathbf{y} - \mathbf{x}\|\right) K_r\left(|I(\mathbf{y}) - I(\mathbf{x})|\right) I(\mathbf{y}), \tag{13.31}$$

where μ is set to $1/C(\mathbf{x})$. This means that different versions of the bilateral filter can be defined based on robust estimation with the range kernel $K_r(\alpha) = \rho'(\alpha)/\alpha$. For example, using $\rho(\alpha) = 1 - exp(\alpha^2/(2\sigma^2))$ provides the standard Gaussian kernel. Other possible choices include the Tukey function

$$\rho(\alpha) = \begin{cases} (x/\sigma)^2 - (x/\sigma)^4 + \frac{1}{3}(x/\sigma)^6 & \text{if } |x| \leq \sigma, \\ \frac{1}{3} & \text{otherwise,} \end{cases} \tag{13.32}$$

and the Huber function

$$\rho(\alpha) = \begin{cases} x^2/2\sigma + \sigma/2 & \text{if } |x| \leq \sigma, \\ |x| & \text{otherwise.} \end{cases} \tag{13.33}$$

Also note that, as seen in Equation 13.30, the contribution of an input to the update is proportional to $\rho'(\cdot)$, the so-called *influence function* [10]. The influence function, in other

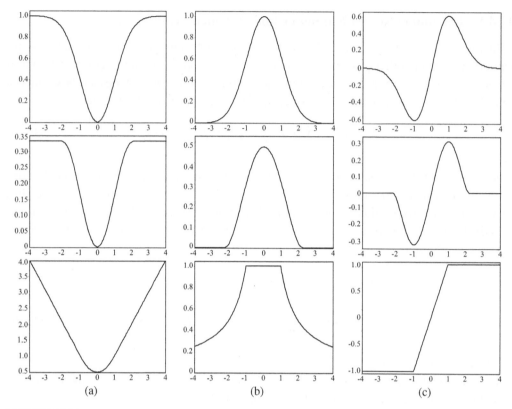

FIGURE 13.21

Example robust functions with their corresponding influence functions and kernels: (top) Gaussian function, (middle) Tukey function, (bottom) Huber function; (a) ρ, (b) K_r, and (c) ρ'.

words, can be used to analyze the response of the filter to outliers. The Gaussian, Tukey, and Huber robust functions, with corresponding influence functions and kernels are given in Figure 13.21.

The bilateral filter is also related to the weighted least squares estimation [35]. Defining $\mathbf{v}$ as the vectorized version of the image I, the regularization term in weighted least squares estimation as given by

$$\Phi\left(\mathbf{v}\right) = \sum_{m}\left(\mathbf{v} - \mathbf{D}^{m}\mathbf{v}\right)^{T}\mathbf{W}\left(\mathbf{v} - \mathbf{D}^{m}\mathbf{v}\right), \tag{13.34}$$

where $\mathbf{D}^{m}$ operator shifts a signal by m samples, and $\mathbf{W}$ is a weighting matrix. With proper choice of weighting matrix (in particular, choosing the weighting matrix such that it penalizes the pixel intensity differences and shift amounts with Gaussian functions), an iteration of the weighted least squares estimation becomes equivalent to the bilateral filter [35].

13.4.2 Relation to Partial Differential Equations

The behavior of the bilateral filter can be analyzed from the perspective of partial differential equations [36], [37]. Reference [37] examines the asymptotical behavior of the bilateral filter with a uniform spatial kernel. Specifically, the continuous version of the bilateral filter:

$$\hat{I}(\mathbf{x}) = \frac{1}{C(\mathbf{x})} \int\limits_{\mathbf{x}-\sigma_d}^{\mathbf{x}+\sigma_d} e^{\frac{-|I(\mathbf{y})-I(\mathbf{x})|^2}{2\sigma_r^2}} I(\mathbf{y})d\mathbf{y}, \tag{13.35}$$

is considered to show that the evolution (or temporal derivative) of the signal $I_t(\mathbf{x}) \equiv \hat{I}(\mathbf{x}) - I(\mathbf{x})$ is proportional to its second derivatives $I_{\eta\eta}(\mathbf{x})$ in the gradient direction and $I_{\xi\xi}(\mathbf{x})$ in the orthogonal direction of the gradient:

$$I_t(\mathbf{x}) \cong a_1 I_{\xi\xi}(\mathbf{x}) + a_2 I_{\eta\eta}(\mathbf{x}), \tag{13.36}$$

where a_1 and a_2 are functions of σ_d, σ_r, and the gradient of $I(\mathbf{x})$. The signs of a_1 and a_2 determine the behavior of the filter.

Namely, if $\sigma_r \gg \sigma_d$, both a_1 and a_2 become equal positive constants, the sum $a_1 I_{\xi\xi}(\mathbf{x}) + a_2 I_{\eta\eta}$ becomes the Laplacian of the signal and the bilateral filter becomes a lowpass filter. If σ_r and σ_d have the same order of magnitude, the filter behaves like a Perona-Malik model [38] and shows smoothing and enhancing characteristics. Since a_1 is positive and decreasing, there is always diffusion in the tangent direction $I_{\xi\xi}$. On the other hand, a_2 is positive when the gradient is less than a threshold (which is proportional to σ_d/σ_r), and the filter is smoothing in the normal direction $I_{\eta\eta}$. The term a_2 is negative when the gradient is larger than the threshold; in this case, the filter shows an enhancing behavior. Finally, if $\sigma_r \ll \sigma_d$, then a_1 and a_2 both tend to zero, and the signal is hardly altered.

Reference [37] also shows that, when σ_r and σ_d have the same order of magnitude, the bilateral filter and the Perona-Malik filter can be decomposed in the same way as in Equation 13.36, and produce visually very similar results even though their weights are not identical. Similar to the Perona-Malik filter, the bilateral filter can create contouring artifacts, also known as *shock* or *staircase* effects. These shock effects occur at signal locations where the convex and concave parts of the signal meet, in other words, at inflection points where the second derivative is zero (see Figure 13.22). Reference [37] proposes to do linear regression to avoid these artifacts. With linear regression, the weights a_1 and a_2 in Equation 13.36 become positive in smooth regions, and no contours or flat zones are created.

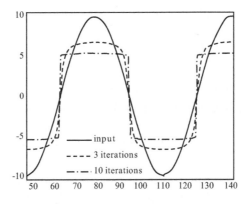

FIGURE 13.22

The shock effect is illustrated through iterative application of the bilateral filter with $\sigma_r = 4$ and $\sigma_d = 10$.

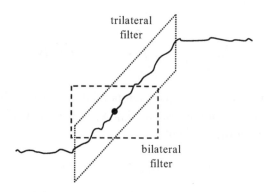

FIGURE 13.23

The trilateral filter [39] adapts to local slope.

13.5   Extensions of the Bilateral Filter

There are various extensions of the bilateral filter. One possible extension is the cross (or joint) bilateral filter, where the filter kernel uses one image to filter another. This is helpful when the image used in the kernel provides better information on large-scale features than the image to be filtered. Sample applications are fusion of flash and no-flash images, and multispectral images as discussed in Section 13.2. Other extensions of the bilateral filter include trilateral, temporal, and non-local means filtering.

The trilateral filter adapts to local slope and extends the effective range of the bilateral filter [39]. As illustrated in Figure 13.23, this is achieved by tilting the filter according to the local slope. Reference [39] also proposes to adapt the local neighborhood by extending the neighborhood up to where the local slope differences are within a threshold.

The bilateral filter can be extended to process video sequences. A video frame I_{t_0} is updated using the weighted sum of pixels from both the actual and the neighboring frames:

$$\hat{I}_{t_0}(\mathbf{x}) = \frac{1}{C(\mathbf{x})} \sum_{\mathbf{y} \in \mathcal{N}(\mathbf{x})} \sum_t K_d \left(\|\mathbf{y} - \mathbf{x}\| \right) K_r \left(|I(\mathbf{y}) - I(\mathbf{x})| \right) K_t \left(|t - t_0| \right) I_t(\mathbf{y}), \qquad (13.37)$$

where the temporal kernel could be chosen as a Gaussian:

$$K_t(t - t_0) = \exp\left(\frac{-|t - t_0|^2}{2\sigma_t^2} \right) \qquad (13.38)$$

The non-local means filter [40] uses the norm of the residual between $\underline{I}(\mathbf{x})$, which is a vectorized version of a local region around $I(\mathbf{x})$, and $\underline{I}(\mathbf{y})$ instead of $|I(\mathbf{x}) - I(\mathbf{y})|$:

$$\hat{I}(\mathbf{x}) = \frac{1}{C(\mathbf{x})} \sum_{\mathbf{y} \in \mathcal{N}(\mathbf{x})} K_d \left(\|\mathbf{y} - \mathbf{x}\| \right) K_r \left(\|\underline{I}(\mathbf{y}) - \underline{I}(\mathbf{x})\| \right) I(\mathbf{y}). \qquad (13.39)$$

Apparently, the bilateral filter is a specific case of the non-local means filter. The non-local means filter has been shown to perform better than the bilateral filter since it can be used with a larger spatial support σ_d as it robustly finds similar regions through $K_r \left(\|\underline{I}(\mathbf{y}) - \underline{I}(\mathbf{x})\| \right)$. The disadvantage is, however, its high computational cost.

In addition to the kernels that have mentioned so far (e.g., uniform box kernel, Gaussian kernel, and kernels derived from robust functions in Section 13.4.1), it is possible to use other kernel types and modifications. For example, salt-and-pepper noise cannot be eliminated effectively with the standard bilateral filter because a noisy pixel is likely to be significantly different from its surrounding, resulting in $K_r(\cdot)$ to be zero. By using the median value $I_{med}(\mathbf{x})$ of the local neighborhood around $\mathbf{x}$, the impulse noise can be eliminated:

$$K_r(|I(\mathbf{y}) - I_{med}(\mathbf{x})|) = \exp\left(\frac{-|I(\mathbf{y}) - I_{med}(\mathbf{x})|^2}{2\sigma_r^2}\right). \tag{13.40}$$

A different extension is presented in Reference [37]. Instead taking a weighted average within a neighborhood, a plane is fit to the weighted pixels in the neighborhood, and the center value of the plane is assigned as the output. This regression filter was shown to reduce the staircase artifacts.

13.6 Conclusions

This chapter surveyed the bilateral filter-driven methods and their applications in image processing and computer vision. The theoretical foundations of the filter were provided; in particular, the connections with robust estimation, weighted least squares estimation, and partial differential equations were pointed out. A number of extensions and variations of the filter were discussed. Since the filter is nonlinear, its fast implementation is critical for practical applications; therefore, the main implementation approaches for fast bilateral filter were discussed as well.

The bilateral filter has started receiving attention very recently, and there are open problems and room for improvement. Future research topics include optimal kernel and parameter selection specific to applications, fast and accurate implementations for multidimensional signals, efficient hardware implementations, spatial adaptation, and modifications to avoid staircase artifacts.

Acknowledgment

This work was supported in part by the National Science Foundation under Grant No. 0528785 and National Institutes of Health under Grant No. 1R21AG032231-01.

References

[1] C. Tomasi and R. Manduchi, "Bilateral filtering for gray and color images," in *Proceedings of the IEEE International Conference on Computer Vision*, Bombay, India, January 1998, pp. 839–846.

[2] S.M. Smith and J.M. Brady, "Susan - A new approach to low level image processing," *International Journal of Computer Vision*, vol. 23, no. 1, pp. 45–78, May 1997.

[3] J.S. Lee, "Digital image smoothing and the sigma filter," *Graphical Models and Image Processing*, vol. 24, no. 2, pp. 255–269, November 1983.

[4] L. Yaroslavsky, *Digital Picture Processing - An Introduction*. Berlin, Germany: Springer-Verlag, December 1985.

[5] J. van de Weijer and R. van den Boomgaard, "Local mode filtering," in *Proceedings of IEEE Conference on Computer Vision and Pattern Recognition*, Kauai, HI, USA, December 2001, vol. 2, pp. 428–433.

[6] M. Zhang and B.K. Gunturk, "Multiresolution bilateral filtering for image denoising," *IEEE Transactions on Image Processing*, vol. 17, no. 12, pp. 2324–2333, December 2008.

[7] E. Reinhard, G. Ward, S. Pattanaik, and P. Debevec, *High Dynamic Range Imaging: Acquisition, Display, and Image-Based Lighting*. San Francisco, CA: Morgan Kaufmann, August 2005.

[8] P. Debevec and J.Malik, "Recovering high dynamic range radiance maps from photographs," in *Proceedings of International Conference on Computer Graphics and Interactive Techniques*, San Diego, CA, USA, August 1997, pp. 369–378.

[9] M.A. Robertson, S. Borman, and R.L. Stevenson, "Dynamic range improvement through multiple exposures," in *Proceedings of the IEEE International Conference on Image Processing*, Kobe, Japan, October 1999, vol. 3, pp. 159–163.

[10] F. Durand and J. Dorsey, "Fast bilateral filtering for the display of high-dynamic-range images," *ACM Transactions on Graphics*, vol. 21, no. 3, pp. 257–266, July 2002.

[11] M. Elad, "Retinex by two bilateral filters," in *Proceedings of Scale-Space and PDE Methods in Computer Vision*, vol. 3459, Hofgeismar, Germany, April 2005, pp. 217–229.

[12] S. Bae, S. Paris, and F. Durand, "Two-scale tone management for photographic look," *ACM Transactions on Graphics*, vol. 25, no. 3, pp. 637–645, July 2006.

[13] E. Eisemann and F. Durand, "Flash photography enhancement via intrinsic relighting," *ACM Transactions on Graphics*, vol. 25, no. 3, pp. 673–678, August 2004.

[14] G. Petschnigg, M. Agrawala, H. Hoppe, R. Szeliski, M. Cohen, and K. Toyama, "Digital photography with flash and no-flash image pairs," *ACM Transactions on Graphics*, vol. 25, no. 3, pp. 664–672, August 2004.

[15] E.P. Bennett, J.L. Mason, and L. McMillan, "Multispectral video fusion," in *Proceedings of International Conference on Computer Graphics and Interactive Techniques*, Boston, MA, USA, July 2006, p. 123.

[16] M. Zhang and B.K. Gunturk, "Compression artifact reduction with adaptive bilateral filtering," *Proceedings of SPIE*, vol. 7257, p. 72571A, January 2009.

[17] S. Fleishman, I. Drori, and D. Cohen-Or, "Bilateral mesh denoising," *ACM Transactions on Graphics*, vol. 22, no. 3, pp. 950 – 953, July 2003.

[18] T.R. Jones, F. Durand, and M. Desbrun, "Non-iterative, feature-preserving mesh smoothing," *ACM Transactions on Graphics*, vol. 22, no. 3, pp. 943 – 949, July 2003.

[19] B.K. Gunturk, J. Glotzbach, Y. Altunbasak, R.W. Schafer, and R.M. Mersereau, "Demosaicking: Color filter array interpolation in single-chip digital cameras," *IEEE Signal Processing Magazine*, vol. 22, no. 1, pp. 44–55, January 2005.

[20] R. Ramanath and W.E. Snyder, "Adaptive demosaicking," *Journal of Electronic Imaging*, vol. 12, no. 4, pp. 633–642, October 2003.

[21] M. Gevrekci, B.K. Gunturk, and Y. Altunbasak, "Restoration of Bayer-sampled image sequences," *Computer Journal*, vol. 52, no. 1, pp. 1–14, January 2009.

[22] B.K. Gunturk, Y. Altunbasak, and R.M. Mersereau, "Color plane interpolation using alternating projections," *IEEE Transactions on Image Processing*, vol. 11, no. 9, pp. 997–1013, September 2002.

[23] J. Xiao, H. Cheng, H. Sawhney, C. Rao, and M. Isnardi, "Bilateral filtering-based optical flow estimation with occlusion detection," in *Proceedings of European Conference on Computer Vision*, Graz, Austria, May 2006, vol. 1, pp. 211–224.

[24] E.A. Khan, E. Reinhard, R. Fleming, and H. Buelthoff, "Image-based material editing," *ACM Transactions on Graphics*, vol. 25, no. 3, pp. 654–663, July 2006.

[25] H. Winnemoller, S.C. Olsen, and B. Gooch, "Real-time video abstraction," *ACM Transactions on Graphics*, vol. 25, no. 3, pp. 1221–1226, July 2006.

[26] B.M. Oh, M. Chen, J. Dorsey, and F. Durand, "Image-based modeling and photo editing," in *Proceedings of ACM Annual conference on Computer Graphics and Interactive Techniques*, Los Angeles, CA, USA, August 2001, pp. 433–442.

[27] S. Paris, H. Briceno, and F. Sillion, "Capture of hair geometry from multiple images," *ACM Transactions on Graphics*, vol. 23, no. 3, pp. 712—719, July 2004.

[28] W.C.K. Wong, A.C.S. Chung, and S.C.H. Yu, "Trilateral filtering for biomedical images," in *Proceedings of IEEE International Symposium on Biomedical Imaging*, Arlington, VA, USA, April 2004, pp. 820–823.

[29] E.P. Bennett and L. McMillan, "Video enhancement using per-pixel virtual exposures," *ACM Transactions on Graphics*, vol. 24, no. 3, pp. 845–852, July 2005.

[30] T.Q. Pham and L.J. Vliet, "Separable bilateral filtering for fast video preprocessing," in *Proceedings of the IEEE International Conference on Multimedia and Expo*, Amsterdam, Netherlands, July 2005, pp. 1–4.

[31] S. Paris and F. Durand, "A fast approximation of the bilateral filter using a signal processing approach," in *Proceedings of European Conference on Computer Vision*, Graz, Austria, May 2006, pp. 568–580.

[32] F. Porikli, "Constant time O(1) bilateral filtering," in *Proceedings of International Conference on Computer Vision and Pattern Recognition*, Anchorage, AK, USA, June 2008, pp. 1–8.

[33] F. Porikli, "Integral histogram: A fast way to extract histograms in Cartesian spaces," in *Proceedings of International Conference on Computer Vision and Pattern Recognition*, San Diego, CA, USA, June 2005, pp. 829–836.

[34] B. Weiss, "Fast median and bilateral filtering," *ACM Transactions on Graphics*, vol. 25, no. 3, pp. 519–526, July 2006.

[35] M. Elad, "On the origin of the bilateral filter and ways to improve it," *IEEE Transactions on Image Processing*, vol. 11, no. 10, pp. 1141–1151, October 2002.

[36] D. Barash, "Fundamental relationship between bilateral filtering, adaptive smoothing, and the nonlinear diffusion equation," *IEEE Transactions on Pattern Analysis and Machine Intelligence*, vol. 24, no. 6, pp. 844–847, June 2002.

[37] A. Buades, B. Coll, and J. Morel, "Neighborhood filters and PDE's," *Numerische Mathematik*, vol. 105, no. 1, pp. 1–34, October 2006.

[38] P. Perona and J. Malik, "Scale-space and edge detection using anistropic diffusion," *IEEE Transactions on Pattern Analysis and Machine Intelligence*, vol. 12, no. 7, pp. 629–639, July 1990.

[39] P. Choudhury and J.E. Tumblin, "The trilateral filter for high contrast images and meshes," in *Proceedings of Eurographics Symposium on Rendering*, Leuven, Belgium, June 2003, vol. 44, pp. 186–196.

[40] A. Buades, B. Coll, and J. Morel, "On image denoising methods," Technical Report 2004-15, CMLA, 2004.

14

Painterly Rendering

Giuseppe Papari and Nicolai Petkov

14.1 Introduction

For centuries, artists have been developing different tools and various styles to produce artistic images, which are usually in a way more interesting than mere representations of scenes from the real world. While classical tools, such as brushes, ink pens or pencils, require skills, effort and talent, science and technology can make more advanced tools that can be used by all people, not only artists, to produce artistic images with little effort.

Recently, scientists have been showing increasing interest in visual arts. On one side, psychologists and neurophysiologists attempt to understand the relation between the way artists produce their works and the function of the visual system of the brain. Examples of such studies include the use of principles of gestalt psychology to understand and describe art [1], [2], [3], deriving spatial organization principles and composition rules of artwork from neural principles [1], [4], or understanding the biological basis of aesthetic experiences [5], [6], [7]. A recent overview of these findings is presented in Reference [8]. On the other side, computer scientists and engineers are developing painterly rendering algorithms which imitate painting styles. There is a large variety of such algorithms, both unsupervised [9], [10], [11], [12] and interactive [13], [14], [15], [16]. Much effort has been made to model different painterly styles [17], [18], [19], [20] and techniques [21], [22], [23], [24], [25], [26], especially watercolor [27], [21], [28], [29], [30], [31], [32],

[33], [34], and to design efficient interactive user interfaces [14], [35], [36], some of which deploy special-purpose hardware [35]. An overview of these techniques can be found in Reference [37]. The importance of such algorithms is two-fold; computers have the potential to help the non-specialist to produce their own art and the artist to develop new forms of art such as impressionistic movies [9], [38], [39] or stereoscopic paintings [40], [41].

This chapter focuses on unsupervised painterly rendering, that is, fully automatic algorithms which convert an input image into a painterly image in a given style. This problem has been faced in different ways; for example, artistic images can be generated by simulating the process of putting paint on paper or canvas. A synthetic painting is represented as a list of brush strokes which are rendered on a white or canvas textured background. Several mathematical models of a brush stroke are proposed, and special algorithms are developed to automatically extract brush stroke attributes from the input image. Another approach suggests abstracting from the classical tools that have been used by artists and focusing on the visual properties, such as sharp edges or absence of natural texture, which distinguish painting from photographic images. These two classes of algorithms will be explored and discussed in the next sections. The existence of major areas of painterly rendering, in which the input is not a photographic image and whose treatment goes beyond the purpose of this chapter, will also be acknowledged. Important examples are painterly rendering on video sequences and the generation of artistic images from three-dimensional models of a real scene.

The chapter is organized as follows. Section 14.2 describes brush stroke oriented painterly rendering algorithms, including physical models of the interaction of a fluid pigment with paper or canvas. More specifically, Section 14.2.1 focuses on imitating the appearance of a single brush stroke in a given technique, such as watercolor, impasto, or oil-painting, when all its attributes are given. Section 14.2.2 presents algorithms for extracting the brush stroke attributes from an input image. Section 14.3 describes methods which aim at simulating the visual properties of a painting regardless of the process that artists perform. Conclusions are drawn in Section 14.4.

14.2 Painterly Rendering Based on Brush Stroke Simulation

This section surveys brush stroke oriented approaches which generate synthetic painterly images by attempting to mimic classical tools and the painting process performed by artists. In these approaches, a painting is represented as a list L of brush strokes to be rendered in a given order on a white or canvas-textured background. Each brush stroke is characterized by some attributes, such as position, shape, size, orientation, color, and texture, which completely define its appearance in the painting. These parameters are listed and briefly described in Table 14.1. The general scheme of this class of algorithms is shown in Figure 14.1; the image is first analyzed to extract the values of the brush stroke attributes and to generate L, each brush stroke in the list L is then rendered based on the values of its descriptive attributes. An example of a result obtained with one of these techniques is shown in Figure 14.2.

TABLE 14.1

Most important brush stroke attributes encountered in the literature.

Attribute	Meaning
Position $\mathbf{r} = (x, y)$	Position of the center of the stroke
Shape	Rectangle, rectangloid, or more general shape
Size (length, width)	The length l and width w of the brush stroke
Orientation	Single valued for rectangular strokes, multivalued for curved strokes
Color	Average color of the stroke
Transparency	Determines the superposition of more brush strokes on the same point
Texture	The texture pattern of the stroke

14.2.1 Brush Stroke Rendering

Algorithms to render a brush stroke can be divided into two classes. The first class includes approaches based on a physical simulation of the process of transferring ink from a brush to a physical mean such as paper or canvas. The second class includes techniques which try to imitate the appearance of a brush stroke while ignoring the physics behind it.

Reference [42] presents a simple and intuitive brush stroke model based on cubic B-splines; a brush stroke is represented by a list of control points $\mathbf{r}_1, ..., \mathbf{r}_N$ and pressure values $p_1, ..., p_N$, which indicate the pressure at which the brush touches the paper. Let $B(\mathbf{r}_1, ..., \mathbf{r}_N)$ indicate a B-spline associated to points $\mathbf{r}_1, ..., \mathbf{r}_N$, which is a piece-wise cubic curve with smooth derivatives until the second order. Let $\mathbf{u}_1, ..., \mathbf{u}_N$ be unit vectors orthogonal to $B(\mathbf{r}_1, ..., \mathbf{r}_N)$ at points $\mathbf{r}_1, ..., \mathbf{r}_N$. Both control points and pressure values are used to compute the trajectory of each bristle of the brush and to draw a colored curve for each

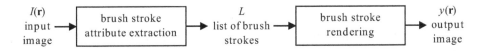

FIGURE 14.1

Schematic representation of brush stroke-based painterly rendering.

FIGURE 14.2

Brush stroke-based painterly rendering using the approach described in Reference [9]: (a) input image and (b) output image.

bristle trajectory. Specifically, the trajectory T of the brush is computed as $B(\mathbf{r}_1, ..., \mathbf{r}_N)$. The trajectory of the kth bristle is computed as $B(\mathbf{r}_1 + \delta_1^{(k)}\mathbf{u}_1, ..., \mathbf{r}_N + \delta_N^{(k)}\mathbf{u}_N)$, where the coefficient $\delta_i^{(k)}$ is associated to the i-th control and the k-th bristle is proportional to the pressure vale p_i. This takes into account the fact that when a brush is pushed against the paper with higher pressure, the bristles will spread out more. Once the trajectory B_k of each bristle is determined, it is rendered by changing the color of those pixels which are crossed by B_k. The color of the trace left out by each bristle can be determined in different ways. The simplest one would be to assign a constant color to all bristles of a given brush stroke. A more complex diffusion scheme can change the color across the brush stroke; the color $c_k(t)$ of the k-th bristle at time t can be computed from the diffusion equation

$$c_k(t + \Delta t) = (1 - \lambda)c_k(t) + \lambda \frac{c_{k-1}(t) + c_{k+1}(t)}{2}, \tag{14.1}$$

where the coefficient $\lambda \in [0, 1]$ is related to the speed of diffusion. Another aspect taken into the account is the amount of ink of each bristle. Specifically, the length of the trace left by each bristle is made proportional to the amount of ink of that bristle. This influences the appearance of the resulting brush stroke, as it will be more "compact" at its starting point and more "hairy" at is end point. In Reference [42], the speed at which bristles become empty is made proportional to the local pressure of the brush.

More sophisticated models perform fluid simulation by means of cellular automata. Cellular automation is described by the following components:

- A lattice $\mathcal{L}$ of cells; here a two-dimensional square lattice is considered, but other geometries, such as hexagonal, could be used as well. Each cell in the lattice is identified by an index $\mathbf{i} \triangleq (i_1, i_2)$, which indicates the position of that cell in the lattice.

- A set $\mathcal{S}$ of states in which each cell can be. The state of the cell in position $\mathbf{i}$ is defined by a set of state variables $a_{\mathbf{i}}$, $b_{\mathbf{i}}$, $c_{\mathbf{i}}$, etc.

- A neighborhood $N_{\mathbf{i}}$, which is a set of cells surrounding the cell in position $\mathbf{i}$. Normally, the neighborhood is translation invariant, that is, $N_{\mathbf{i}+\mathbf{v}} = N_{\mathbf{i}} + \mathbf{v}$, where $\mathbf{v}$ is a 2D vector.

- A state transition function which determines the evolution of the cellular automata over (discretized) time. Specifically, the state $s_{\mathbf{i}}(n+1)$ of the cell $\mathbf{i}$ at time $n+1$ is a function F of the states of all cells $\mathbf{k}$ in the neighborhood $N_{\mathbf{i}}$ of $\mathbf{i}$, that is, $s_{\mathbf{i}}(n+1) = F[s_{\mathbf{k}}(n) : \mathbf{k} \in N_{\mathbf{i}}]$. The transition function is the most important part of a cellular automata, since it determines the state of the entire cellular automata at any time for every initial condition.

The notion of *coupled cellular automata* is also needed; two cellular automata C_1 and C_2 are coupled if the evolution of each one of them is determined by the states of both C_1 and C_2. In other words, for $s_{\mathbf{i}}^{(1)}$ and $s_{\mathbf{i}}^{(2)}$ denoting the states of the $\mathbf{i}$-th cell of C_1 and C_2, respectively, the state transition functions of C_1 and C_2 can be written as follows:

$$s_{\mathbf{i}}^{(u)}(n+1) = F^{(u)}[s_{\mathbf{k}_1}^{(1)}(n), s_{\mathbf{k}_2}^{(2)}(n) : \mathbf{k}_1, \mathbf{k}_2 \in N_{\mathbf{i}}], \tag{14.2}$$

with $u = 1, 2$. The following focuses on cellular automata with a deterministic state transition function, that is, A for which a given configuration of states at time n will always produce the same configuration at time $n + 1$. Stochastic cellular automata could be considered as well, which are well known in the literature as discrete Markov random fields. In general, cellular automata are capable of modeling any phenomenon in which each part interacts only with its neighbors[1]. Moreover, their structure makes them suitable for parallel implementations. A deeper treatment of the subject can be found in Reference [43].

Cellular automata are used in painterly rendering to simulate the diffusion process that occurs when a fluid pigment is transferred from a brush to the paper [44]. A brush stroke simulation system based on cellular automata consists of a cellular automata that models the paper, a cellular automata that models the brush, and a coupling equation. In the case of a cellular automata that models the paper, the state of every cell describes the amount of water $W_{\mathbf{i}}^{(p)}$ and ink $I_{\mathbf{i}}^{(p)}$ that are present in each point of the paper. The state transition function F_p is designed to take into account several phenomena, such as the diffusion of water in paper, the diffusion of ink in water, and water evaporation. The diffusion of water in paper can be isotropic, that is, water flows in all directions at equal speed, or anisotropic. Anisotropy can be due to several factors, such as local directionality in the structure of the paper, or by the effect of gravity, so that water flows downward at higher speed than upward. In the case of a cellular automata that models the brush, the state of every cell describes the amount of water $W_{\mathbf{i}}^{(b)}$ and ink $I_{\mathbf{i}}^{(b)}$ that are present locally at each bristle of the brush. The state transition function F_b regulates the amount of fluid that is transferred to the paper, the flow of fluid from the tip of the brush to the contact point with the paper, the diffusion of fluid between bristle, and the rate at which each bristle gets empty. The last element, a coupling equation, relates the states of the paper cellular automata to the state of the brush A. In the simplest case, this is just a balancing equation between the amount of fluid that leaves the brush and the amount of fluid that comes into the paper at the contact point between paper and brush.

In this framework, a brush stroke is simulated as described below. Let $\mathbf{i}(n)$ be the discretized position of the brush at time n, which is supposed to be given as input to the brush stroke simulation system, and let $\mathbf{0}$ be the index of the cell of the brush cellular automata that touches the paper (as shown in Figure 14.3 for a one-dimensional example). First, all cells of the paper cellular automata are initialized to the value zero (no water or ink) and the cells of the brush cellular automata are initialized to a value that is proportional to the amount of pigment the brush has been filled with. Then, the states of both cellular automata are iteratively updated according to their state transition functions. Specifically, at each iteration, given amounts ΔW and ΔI of water and ink are transferred from the brush to the paper at position $\mathbf{i}(n)$:

$$[W_{\mathbf{i}(n)}^{(p)} + \Delta W, I_{\mathbf{i}(n)}^{(p)} + \Delta I] \Rightarrow [W_{\mathbf{i}(n)}^{(p)}, I_{\mathbf{i}(n)}^{(p)}], [W_{\mathbf{0}}^{(b)} - \Delta W, I_{\mathbf{0}}^{(b)} + \Delta I] \Rightarrow [W_{\mathbf{0}}^{(b)}, I_{\mathbf{0}}^{(b)}]. \quad (14.3)$$

Diffusion is simulated by updating the state of all cells of the paper cellular automata and brush cellular automata according to the respective transition functions F_p and F_b.

---

[1]Due to the maximum speed at which information can be conveyed, namely, the speed of light, one can argue that cellular automata can model every phenomenon in the real word.

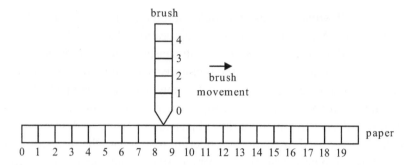

FIGURE 14.3

Brush stroke simulation by means of cellular automata.

Once convergence is reached, the state of the paper cellular automata must be converted into an image. The simplest approach is to modify the color of each pixel in proportion to the amount of ink that is present in each cell of the paper cellular automata. Specifically, let c_b be the color of the ink and $c_p(\mathbf{i})$ the color of the paper at position $\mathbf{i}$ before the application of the brush stroke. Then, the color of the paper at position $\mathbf{i}$ after the application of the brush stroke is given by $I_{\mathbf{i}}c_b + (1 - I_{\mathbf{i}})c_p(\mathbf{i})$. Such a linear combination of c_b and $c_p(\mathbf{i})$ allows the simulation of the transparency of brush strokes which are placed on top of each other. A more sophisticated approach, consists of determining the color of each pixel of the image by simulating the interaction between white light and a layer of dry pigment of a given thickness. Specifically, it can be proved that the reflectance of the pigment layer can be expressed as

$$R_x = 1 + \frac{K}{S} + \sqrt{(\frac{K}{S})^2 + 2\frac{K}{S}}, \tag{14.4}$$

where K and S are the adsorption and scattering coefficients of the pigment [45] and [46].

Reference [21] proposes a more sophisticated fluid simulation model for synthetic watercolor generation. In this approach, the interaction between water, ink, and paper is modeled by means of three layers, each one of which is associated with a cellular automata. Namely, a *shallow water layer* which models the motion of water, a *pigment deposition layer* which models the absorption and desorption of pigment from paper, and a *capillarity layer* which models the penetration of water into the pores of the paper. This model takes into account a large number of factors, such as the fluid velocity in the x and y directions, water pressure, the effect of gravity due to the inclination of the paper with respect to the vertical direction, and physical properties of the solution of ink and water, such as viscosity and viscous drag. As a result, the brush stroke obtained with this system is virtually indistinguishable from true watercolor brush stroke.

14.2.2 Extraction of Brush Stroke Attributes

Size and elongation of the brush strokes are important distinctive features of many painting styles. For example, in Pointillism (Figure 14.4a), artists use very short brush strokes, while in other styles, such as Impressionism (Figure 14.4b), paintings are rendered by means of a few coarse touches and fine details are neglected. Other painters, such as Van

(a)   (b)   (c)

FIGURE 14.4

Example painting styles: (a) Seurat's painting *La Parade de Cirque* from 1889, (b) Monet's painting *Impression, Sunrise* from 1872, and (c) Van Gogh's painting *Road with Cypress and Star* from 1890.

Gogh, used elongated brush strokes to add to a painting a geometric structure which makes it more vibrant (Figure 14.4c). Therefore, length l and width w of brush strokes are usually determined by the desired painting style rather than automatically extracted from the input image. In a number of algorithms, the values of l and w are the same for all brush strokes and are specified by the user depending on the desired effect. In some cases, an interface is provided in which the user selects the painting style and the values of l and w are determined accordingly. Other approaches take into account the fact that artists often render different areas of the paintings at different levels of detail. In particular, artists first make a simplified sketch of their painting by using coarse brush strokes, and then paint on top of it with smaller brushes to render finer details. The so-called coarse-to-fine rendering algorithms [17], [38] produce a final synthetic painting as the superposition of different layers to be rendered on top of each other. On the lowest layers, coarse brush strokes are rendered, while at the highest layers finer brush strokes are present only on those regions which need to be rendered at finer detail. The success of such an approach depends on the strategies deployed to identify which regions of the painting should be present in each layer. In Reference [17], such areas are determined iteratively; specifically, the coarsest layer is initialized by rendering brush strokes on the whole image. Then, in the k-th layer, brush strokes are rendered only in those regions for which the difference in color between the input image and the painting rendered until the $(k-1)$-th layer is below a given threshold. A different approach is presented in Reference [38]; regions to be rendered at a higher level of detail are detected by looking at the frequency content of each edge. In Reference [19], the level of detail in different areas of the painting is determined by a measure of saliency, or visual interest, of the different areas of the painting. Specifically, a saliency map is computed by looking at how frequently the local pattern around each pixel occurs in the image and assigning high saliency to the most rare patterns; afterwards, high saliency areas are rendered with smaller brush strokes.

The simplest method to compute the *position* of each brush stroke is to place them on a square lattice, whose spacing is less than or equal to the width of the brush stroke. This guarantees that every pixel of the output image is covered by at least one brush stroke.

(a)   (b)   (c)

FIGURE 14.5 (See color insert.)

The effect of $\mathbf{v}(\mathbf{r}) \perp \nabla I(\mathbf{r})$ on image quality: (a) input image, (b) synthetic painting obtained by imposing the condition $\mathbf{v}(\mathbf{r}) \perp \nabla I(\mathbf{r})$ on the whole image, and (c) synthetic painting obtained by imposing the condition $\mathbf{v}(\mathbf{r}) \perp \nabla I(\mathbf{r})$ only on high contrast edge points.

The main disadvantage of this approach is that it works well only with brush stroke of fixed size. In more sophisticated approaches [10], the position of each brush stroke is determined by its area. Specifically, an area map $A(\mathbf{r})$ is first extracted from the input image; then, a random point set S is generated, whose local density is a decreasing function of $A(\mathbf{r})$. Each point S is the position of a brush stroke. One possible way to compute the function $A(\mathbf{r})$, proposed in Reference [10], is to analyze the moments up to the second order of regions whose color variation is below a given threshold.

Orientation is another attribute to consider. A vector field $\mathbf{v}(\mathbf{r})$ is introduced such that that a brush stroke placed at point $\mathbf{r}_0$ is locally oriented along $\mathbf{v}(\mathbf{r}_0)$ [9], [17], [38], [47]. The simplest approach to automatic extraction of a suitable vector field from the input image consists of orienting $\mathbf{v}(\mathbf{r})$ orthogonally to the gradient of the input image. This simulates the fact that artists draw brush strokes along the object contours. However, the gradient orientation is a reliable indicator only in the presence of very high contrast edges and tends to be random on textures as well as on regions with slowly varying color. Therefore, the other approach is to impose $\mathbf{v}(\mathbf{r}) \perp \nabla I(\mathbf{r})$ only on points for which the gradient magnitude is sufficiently high, and to compute $\mathbf{v}(\mathbf{r})$ on the other pixels by means of diffusion or interpolation processes [38]. On the one hand, this simple expedient considerably improves the appearance of the final output (Figure 14.5); on the other hand, these gradient-based approaches for extracting $\mathbf{v}(\mathbf{r})$ only look at a small neighborhood of each pixel while neglecting the global geometric structure of the input image. A global method for vector field extraction, inspired on fluid dynamics is proposed in Reference [39]. Specifically, the input image is first partitioned into N regions $\mathcal{R}_k, k = 1, ..., N$ by means of image segmentation. Then, inside each region $\mathcal{R}_k$, the motion of a fluid mass is simulated by solving the Bernoulli equation of fluid dynamics with the constraint that the fluid velocity $\mathbf{v}(\mathbf{r})$ is orthogonal to $\mathcal{R}_k$ on the boundary of $\mathcal{R}_k$. This approach has the advantages of deriving the vector field from general principles, taking into account the global structure of the in-

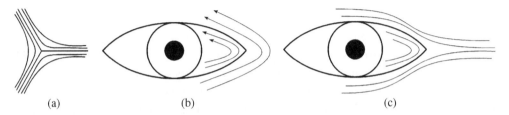

FIGURE 14.6

Illustration of the superiority of tensor hyperstreamlines with respect to vector streamlines: (a) topological configuration of hyperstreamlines, called trisector, which cannot be reproduced by the streamlines of a continuous vector field, (b) streamlines of a vector field, and (c) hyperstreamlines of a tensor field, also oriented along the edges of an eye which form a trisector configuration.

put image, and allowing an easy control of interesting geometric features of the resulting vector field, such as vorticity (the presence of vortices in $\mathbf{v}(\mathbf{r})$ results in whirls in the final output which resemble some Van Gogh paintings). Unfortunately, the fluid dynamic model introduces several input parameters which do not have a clear interpretation in the context of painterly rendering and for which is not obvious how to choose their value (for example, the mixture parameter [39]).

More recently, it has been suggested to locally orient brush strokes along the so-called *hyperstreamlines* of a symmetric tensor field $T(\mathbf{r})$ rather than along a vector field. Hyperstreamlines are defined as lines locally oriented along the eigenvector associated to the largest eigenvalue of $T(\mathbf{r})$. Hyperstreamlines are not defined on points for which the two eigenvalues of $T(\mathbf{r})$ are equal. The main difference between hyperstreamlines of a tensor field and streamlines of a vector field is that the former are *unsigned*. Consequently, tensor fields offer a larger variety of topologies compared to vector fields. An example is given in Figure 14.6. The geometric structure shown in Figure 14.6c, which is more natural for painting an eye, can be reproduced by tensor fields but not by vector fields. On the other hand, both geometric structures in Figures 14.6b and 14.6c can be synthesized by using tensor fields.

The simplest approach to assigning a *shape* to each brush stroke is to fix a constant shape for all brush strokes. The most common choices are rectangles [9], [10], [11], [12]. In the system developed in Reference [38], the user can choose among a larger set of shapes, including irregular rectangloid brush strokes, irregular blob-like shaped brush strokes, and flower shaped brush strokes. While these approaches are simple, they do not take into account the fact that in true paintings artists draw brush strokes of different shapes depending on the image content. A first step toward brush strokes of adaptive shape is made in Reference [17], where curved brush strokes with fixed thickness are generated. Specifically, the shape of each brush stroke is obtained by thickening a cubic B-spline curve with a disk of a given radius. The control points $\mathbf{r}_i$ of the spline are obtained in two steps. First, the starting point $\mathbf{r}_0$ is calculated by means of one of the methods for determining the position. Then, each point is obtained from the previous by moving orthogonally to the gradient of a fixed spacing h as follows:

$$\mathbf{r}_{i+1} = \mathbf{r}_i + h\boldsymbol{\delta}_i, \tag{14.5}$$

where $\boldsymbol{\delta}_i$ is a unit vector oriented orthogonally to the gradient of the input image in $\mathbf{r}_i$. Fully adaptive procedures to determine the brush stroke shape are presented in References [15] and [18]. The idea is to presegment the input image into a large number of components, and to generate brush strokes by merging segments of similar colors. In Reference [15], morphological techniques are used to prevent brush strokes with holes.

In all approaches discussed so far, the algorithms deployed to extract the brush stroke attributes from the input image are not derived from general principles. As a result, all such algorithms can imitate a very limited number of painting styles and need the introduction of many additional parameters whose values are not theoretically justified. To overcome these limitations, the task of computing the brush stroke attributes can be formulated as an optimization problem [19], [48]. Specifically, let $\mathbf{b}$ be a vector whose components are the attributes of all brush strokes in the painting, and let P_b be the image obtained by rendering the brush stroke of B with one of the techniques reviewed in Section 14.2.1. Let also $E(\mathbf{b})$ be an energy function which measures the dissimilarity between P_b a painting in a given style. Then, the brush stroke attributes are computed in order to minimize $E(\mathbf{b})$.

The most challenging aspects of this approach are the definition of a suitable energy function and the development of search algorithms able to minimize an energy in the extremely high dimensional space in which $\mathbf{b}$ is set. Reference [48] proposes an energy function which is a weighted sum of four terms:

$$E(\mathbf{b}) \triangleq E_{app} + w_{area}E_{area} + w_{nstr}E_{nstr} + w_{cov}E_{cov}, \qquad (14.6)$$

where the coefficients w_{area}, w_{nstr}, and w_{cov} are different for each painting style. The term $E_{app} \triangleq \int w(app(\mathbf{r})[I(\mathbf{r}) - p_b(\mathbf{r})]^2 d^2\mathbf{r}$ measures the dissimilarity between the input image $I(\mathbf{r})$ and the resulting painting $p_b(\mathbf{r})$. The term E_{area} is proportional the total area of all brush strokes and reflects the total amount of paint that is used by the artist. Since brush strokes overlap on top of each other, this term could be much higher than the total area of the painting. Therefore, minimizing E_{area} corresponds to minimizing the brush stroke overlap and reflects the principle of economy that is followed in art. The term E_{nstr} is proportional to the number of brush strokes in the painting. A high number of brush strokes will result in a very detailed image, while a low number will give rise to very coarse and sketchy representation. Therefore, the weight assigned to E_{nstr} influences the painting style in terms of the level of detail at which the image is rendered. The term E_{cov} is proportional to the number of pixels that are not covered by any brush stroke. Assigning a high weight to this term will prevent areas of the canvas to remain uncovered by brush strokes. As can be seen, the minimization of $E(\mathbf{b})$ results in a trade off between competing goals, such as the amount of details in the rendered image and the number of brush strokes. If, following Reference [48], one assumes that painting styles differ for the importance that is given to each subgoal, it results that different styles can be imitated by simply changing the weights of each one of the four terms in $E(\mathbf{b})$. An important advantage of energy minimization-driven painterly rendering algorithms is that new painting styles can be included by simply modifying the cost function.

Once a suitable energy function $E(\mathbf{b})$ is found, there is still the problem of developing procedures to minimize it. At the current state of the art, this problem is very far from being solved, especially due to the extremely high dimensionality of the search space in which $\mathbf{b}$ is set and the huge number of local minima in which a search algorithm can be trapped. Reference [48] addresses this problem by means of a relaxation algorithm in combination with heuristic search. Specifically, the algorithm is initialized by an empty painting. Then, a trial-and-error iterative procedure is followed. At each iteration, several modifications of the painting are attempted (such as adding or removing a brush stroke, or modifying the attributes of a given brush stroke), and the new painting energy is computed. Changes which result in an energy decrease are adopted and the entire process is reiterated until convergence. Unfortunately, the algorithm is tremendously slow and, in general, converges neither to a global optimum nor even to a local minimum. A more principled approach to minimize $E(\mathbf{b})$ can be found in Reference [19], where genetic algorithms are deployed with some strategies to avoid undesired local minima.

14.3 Filtering-Based Painterly Rendering

The previous section described techniques which simulate the process performed by artists and the classical tools used in art. This section will focus on a different class of algorithms, which only look at the visual properties that distinguish a painting from a photographic image, irrespective of the tools and the techniques used to generate them. This class of algorithms can be categorized into approaches based on edge preserving and edge enhancing smoothing, techniques which add a synthetic texture to the input image to simulate the geometric pattern induced by brush strokes, and algorithms based on image analogies.

14.3.1 Edge Preserving and Enhancing Smoothing

Two important visual properties of many painting styles are the absence of texture details and the increased sharpness of edges as compared to photographic images. This suggests that painting-like artistic effects can be achieved from photographic images by filters that smooth out texture details, while preserving or enhancing edges and corners. However, practice shows that not all the existing algorithms for edge preserving smoothing are suitable for this purpose. The following describes some simple edge preserving smoothers which are useful in painterly rendering.

Let $f(x,y)$ be a function defined on $\mathbb{R}^2$ which, for the moment, is interpreted here as an altitude profile. Let us now imagine filling every valley with water until the area of water surface on each cavity is not greater than a given value A. Informally speaking, a new altitude profile $f_C(x,y;A)$ is considered which is equal to the original profile $f(x,y)$ on points (x,y) that are not reached by water, and equal to the water level on the other points. Using the *area closing* operator $\mathscr{C}_A$ produces $f_C(x,y;A)$ as output by taking in input $f(x,y)$ and A, that is, $f_C(x,y;A) \triangleq \mathscr{C}_A\{f(x,y)\}$. Dually, *area opening* $\mathscr{O}_A$ can be defined as the negative of the result of an area closing applied to the function $-f(x,y)$, that is, $\mathscr{O}_A\{f(x,y)\} \triangleq -\mathscr{C}_A\{-f(x,y)\}$. Two other operators, *area open-closing* and *area close-opening*, can be introduced as the results, respectively, of the application of an area opening followed by an area closing, and an area closing followed by an area opening. These operators are illustrated in Figure 14.7 for a one-dimensional case. As can be seen, area

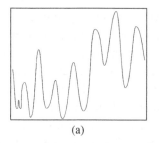 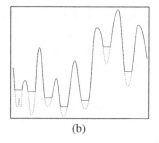 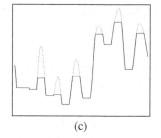

(a)   (b)   (c)

FIGURE 14.7

Morphological operators applied to a one-dimensional signal: (a) input $f(x)$, (b) output $f_C(x)$ of area opening, and (c) output $f_{CO}(x)$ of area close-opening.

(a)   (b)

FIGURE 14.8 (See color insert.)

Morphological color image processing: (a) input image and (b) output image obtained via close-opening applied separately to each RGB component.

closing flattens local minima to a certain area, while area opening flattens local maxima. The combined application of a closing and an opening reduces the amount of texture in a signal while preserving edges. Figure 14.8 shows the result of an area open-closing applied independently to each RGB component of a photographic image, this operator effectively adds an artistic effect to the input image. The rationale beyond it is that area open-closing induces flat zones with irregular shapes which simulate irregular brush strokes. A more formal introduction to these operators, as well as efficient algorithms for their computation, can be found in Reference [49].

The following describes the Kuwahara filter and its variants. Let us consider a gray level image $I(x,y)$ and a square of length $2a$ centered around a point (x,y) which is partitioned

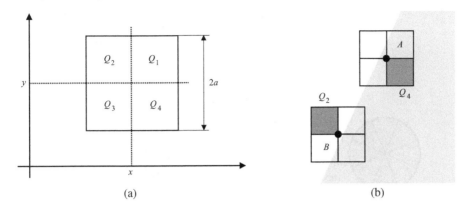

(a)   (b)

FIGURE 14.9

Kuwahara filtering: (a) regions Q_i on which local averages and standard deviations are computed, and (b) the square with the smallest standard deviation, delineated by a thick line, determines the output of the filter. © 2007 IEEE

into four identical squares $Q_1, ..., Q_4$ (Figure 14.9a). Let $s_i(x,y)$ and $m_i(x,y)$ be the local average and the local standard deviation, respectively, computed on each square $Q_i(x,y)$, for $i = 1, ..., 4$. For a given point (x,y), the output $\Phi(x,y)$ of the Kuwahara filter is given by the value of $m_i(x,y)$ that corresponds to the i-th square providing the minimum value of $m_i(x,y)$ [50]. Figure 14.9b shows the behavior of the Kuwahara operator in the proximity of an edge. When the central point (x,y) is on the dark side of the edge (point A), the chosen value of m_i corresponds to the square that completely lies on the dark side (Q_4 here), as this is the most homogeneous area corresponding to minimum s_i. On the other hand, as soon as the point (x,y) moves to the bright side (point B), the output is determined by the square that lies completely in the bright area (Q_2 here), since now it corresponds to the minimum standard deviation s_i. This flipping mechanism guarantees the preservation of edges and corners, while the local averaging smooths out texture and noise.

One limitation of Kuwahara filtering is the block structure of the output, particularly evident on textured areas (Figure 14.9), that is due to the square shape of the regions Q_1 to Q_4 and to the Gibbs phenomenon [51]. This problem can be avoided by using different shapes for the regions Q_i and by replacing the local averages with weighted local averages. For example, the squares Q_i can be replaced by pentagons and hexagons [52]. Reference [53] uses circular regions whereas References [54] and [55] take into account a larger set of overlapping windows. Namely, Reference [55] avoids the Gibbs phenomenon by using Gaussian weighted local averages (Gaussian Kuwahara filtering) instead of the local averages. Other solutions are based on smoothed circular [56] and elliptical [57] sectors. Specifically, in Reference [56] local averages and standard deviations are computed as follows:

$$m_i(x,y) = w_i(x,y) \star I(x,y), \qquad s_i^2(x,y) = w_i(x,y) \star I^2(x,y) - m_i^2(x,y), \qquad (14.7)$$

where the weighting functions $w_i(x,y)$ are given by the product of a two-dimensional isotropic Gaussian function with an angular sector.

A more serious problem is that the Kuwahara filter is not a mathematically well defined operator; every time the minimum value of s_i is reached by two or more squares, the output

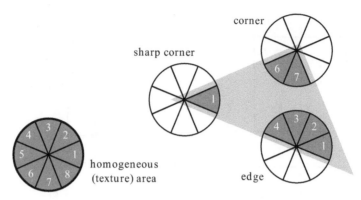

FIGURE 14.10

Sector selection in various situations. The sectors selected to determine the output are delineated by a thick line. © 2007 IEEE

cannot be uniquely determined because it is unclear which subregion should be chosen. To solve this problem, Reference [56] replaces the minimum standard deviation criterion with a weighted sum of the values m_i, where the weights are decreasing functions of the values s_i. Specifically, the output is expressed as

$$\Phi_q(x,y) = \frac{\sum_i m_i(x,y) s_i^{-q}(x,y)}{\sum_i s_i^{-q}(x,y)},\qquad(14.8)$$

where $q > 0$ is an input parameter which controls the sharpness of the edges. For $q = 0$ this reduces to linear filtering which has high stability to noise but poor edge preservation; conversely, for $q \to \infty$, only the term in the sum with minimum s_i survives, thus reducing to the minimum standard deviation criterion of Kuwahara filtering which has high edge preserving performance but poor noise stability. This operator is thus an intermediate case between linear filtering and Kuwahara-like filtering, taking on the advantages of both.

Another advantage of this combination criterion is that it automatically selects the most interesting subregions. This is illustrated in Figure 14.10 for the circular sectors deployed in Reference [56]. On areas that contain no edges (case a), the s_i values are very similar to each other, therefore the output q is close to the average of the m_i values. The operator behaves very similarly to a Gaussian filter; texture and noise are averaged out and the Gibbs phenomenon is avoided. On the other hand, in presence of an edge (case b), the sectors placed across it give higher s_i values with respect to the other sectors. If q is sufficiently large (for instance, $q = 4$) the sectors intersected by the edge (S_5 - S_8) give a negligible contribution to the value of q. Similarly, in presence of corners (case c) and sharp corners (case d), only those sectors which are placed inside the corner (S_6, S_7) for case c and S_1 for case d give an appreciable contribution to value of q whereas the others are negligible.

In all methods described so far, the values of m_i and s_i are computed over regions with fixed shapes. In Reference [57], adaptive anisotropic subregions are deployed. Specifically, an ellipse of a given size is placed on each point of the input image, where orientation and ellipticity of each ellipse are determined by means of the structure tensor of the input image [58]. Then, the concerned subregions are sectors of such ellipses. Deploying sectors

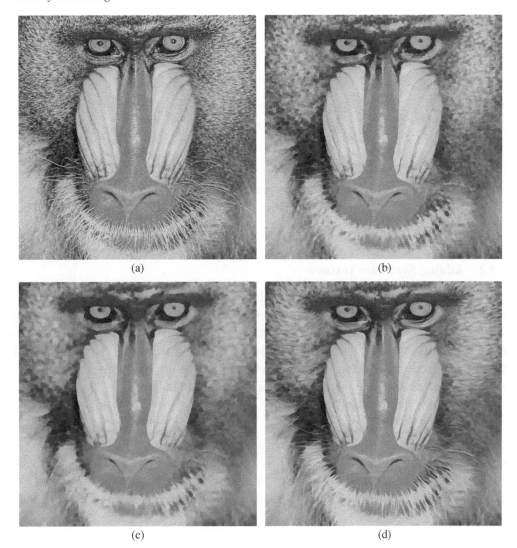

(a)

(b)

(c)

(d)

FIGURE 14.11

Painterly rendering by edge preserving smoothing: (a) input image, (b) output of the Kuwahara filter [50], (c) output of the isotropic operator proposed in Reference [56], and (d) output of the anisotropic approach deployed in Reference [57].

of adaptable ellipses instead of circles results in better behavior in the presence of elongated structures and in a smaller error in the presence of edges, since it does not deal with a small number of primary orientations. On the other hand, this makes the algorithm much slower, since spatially variant linear filters cannot be evaluated in the Fourier domain.

A comparison is shown in Figure 14.11. The blocky structure and the Gibbs phenomenon, well visible in the Kuwahara output (Figure 14.11b), are avoided by the approaches proposed in Reference [56] (Figure 14.11c) and Reference [57] (Figure 14.11d). Moreover, the use of adaptable windows better preserves elongated structures, such as the whiskers of the baboon.

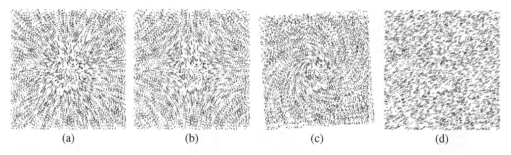

(a)   (b)   (c)   (d)

FIGURE 14.12

Glass patterns obtained by several geometric transformations: (a) isotropic scaling, (b) expansion and compression in the horizontal and vertical directions, respectively, (c) combination of rotation and isotropic scaling, and (d) translation. © 2009 IEEE

14.3.2   Adding Synthetic Texture

The methods described in the previous subsection produce artistic images by removing natural texture from photographic images. A method, based on the theory of glass patterns, which modifies the texture of the input image in order to imitate various painting styles, is presented below. After the introduction of the glass pattern theory, the formalism will be extended to the continuous case, and finally a way to produce artistic images by transferring the microstructure of a glass pattern to the input image will be described.

14.3.2.1   Glass Patterns

Glass patterns [59], [60], [61] are defined as the superposition of two random point sets, one of which is obtained from the other by means of a small geometric transformation (Figure 14.12). More formally, let $\mathbf{v}(\mathbf{r})$ be a vector field defined on $\mathbb{R}^2$ and $d\mathbf{r}/dt = \mathbf{v}(\mathbf{r})$ be the corresponding differential equation. The solution of this equation can be expressed as follows:

$$\mathbf{r}(t) = \Phi_{\mathbf{v}}(\mathbf{r}_0, t), \tag{14.9}$$

with the initial condition $\mathbf{r}(0) = \mathbf{r}_0$. For a fixed value of t, the term $\Phi_{\mathbf{v}}$ denotes a map from $\mathbb{R}^2$ to $\mathbb{R}^2$, which satisfies the condition $\Phi_{\mathbf{v}}(\mathbf{r}, 0) = \mathbf{r}$. Let $S = \{\mathbf{r}_1, ..., \mathbf{r}_N\}$ be a random point set and $\Phi_{\mathbf{v}}(S, t) \triangleq \{\Phi_{\mathbf{v}}(\mathbf{r}, t) | \mathbf{r} \in S\}$. Using this notation, the glass pattern $G_{\mathbf{v}, t}(S)$ associated with S, $\mathbf{v}$, and t is defined as $G_{\mathbf{v}, t}(S) \triangleq S \bigcup \Phi_{\mathbf{v}}(S, t)$. In general, the geometrical structure exhibited by a glass pattern is related to the streamlines of $\mathbf{v}(\mathbf{r})$.

Due to their randomness and geometric structure, glass patterns capture the essence of the motives induced by brush strokes in several impressionistic paintings, and provide corresponding mathematical models (Figure 14.13). The following will show that transferring the microstructure of a glass pattern to an input image results in outputs perceptually similar to paintings.

14.3.2.2   Continuous Glass Patterns

The first step toward painterly rendering based on the theory of glass patterns is the extension of the related formalism to the continuous case [47]. A binary field $b_S(\mathbf{r})$ associated with a point set S can be defined as a function which takes the value 1 for points of S and

(a)   (b)

FIGURE 14.13

Comparison of the real and synthesized painting styles: (a) Vincent Van Gogh's painting *Road with Cypress and Star*, and (b) manually generated glass pattern. The two images exhibit similar geometric structures.

the value 0 for other points. Clearly, the binary field associated with the superposition of two point sets S_1 and S_2 is equal to

$$b_{S_1 \cup S_2}(\mathbf{r}) = \max[b_{S_1}(\mathbf{r}), b_{S_2}(\mathbf{r})], \tag{14.10}$$

the binary field associated with a glass pattern is thus equal to $b_{G_{\mathbf{v},t}(S)}(\mathbf{r}) = \max\{b_S(\mathbf{r}), b_S[\Phi_{\mathbf{v}}(\mathbf{r},t)]\}$. With this notation, generalizing glass patterns to the continuous case is straightforward. Namely, a continuous set of patterns $b_S[\Phi(\mathbf{r},\tau)]$ with $\tau \in [0,1]$ is considered instead of only two patterns, and any real valued random image $z(\mathbf{r})$ can be used instead of $b_S(\mathbf{r})$. Specifically, a *continuous glass pattern* $\mathfrak{G}_{\mathbf{v}}(\mathbf{r})$ is defined as follows:

$$\mathfrak{G}_{\mathbf{v}}(\mathbf{r}) \triangleq \max_{\tau \in [0,1]} \{z[\Phi_{\mathbf{v}}(\mathbf{r},\tau)]\}. \tag{14.11}$$

This formula can be rewritten more compactly as

$$\mathfrak{G}_{\mathbf{v}}(\mathbf{r}) = \max_{\boldsymbol{\rho} \in A_{\mathbf{v}}(\mathbf{r})} \{z(\boldsymbol{\rho})\}, \tag{14.12}$$

where $A_{\mathbf{v}}(\mathbf{r})$ is the arc of streamline $\mathbf{r}(t) = \Phi_{\mathbf{v}}(\mathbf{r},t)$ with $t \in [0,1]$, that is, $A_{\mathbf{v}}(\mathbf{r}) \triangleq \{\Phi_{\mathbf{v}}(\mathbf{r},t)|t \in [0,1]\}$.

Given a vector field $\mathbf{v}(\mathbf{r})$ and a random image $z(\mathbf{r})$, the glass pattern can easily be computed by integrating the differential equation $d\mathbf{r}/dt = \mathbf{v}(\mathbf{r})$ with the Euler algorithm [62] and by taking the maximum of $z(\mathbf{r})$ over an arc of the solving trajectory $\Phi_{\mathbf{v}}(\mathbf{r},t)$. Examples

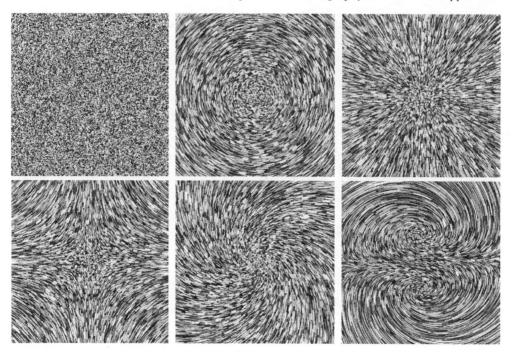

FIGURE 14.14

Random image $z(\mathbf{r})$ and examples of continuous glass patterns obtained from it. Their geometrical structure is analogous to the discrete case. © 2009 IEEE

of continuous glass patterns are shown in Figure 14.14, with the histograms of all images being equalized for visualization purposes. As can be seen, these patterns exhibit similar geometric structures to the corresponding discrete patterns.

A simple way to obtain painterly images from continuous glass patterns is depicted in Figures 14.15 and 14.16. The first step is edge preserving smoothing with the output $I_{EPS}(\mathbf{r})$, which can be obtained using the techniques described above. The second step is the generation of *synthetic painterly texture* $U_{SPT}(\mathbf{r})$ which simulates oriented brush strokes (Figure 14.16c). This is simply a continuous glass pattern associated with a vector field which forms a constant angle ϕ with the color gradient of the input image. An example of such a texture is shown in Figure 14.16c for $\theta_0 = \pi/2$ and $a = 18$, for an image of size 320×480 pixels. It can be seen that the geometric structure of $U_{SPT}(\mathbf{r})$ is similar to the elongated brush strokes that artists use in paintings. For $\theta_0 = \pi/2$, such strokes are oriented orthogonally to $\nabla_\sigma I_{EPS}(\mathbf{r})$. This mimics the fact that artists usually tend to draw

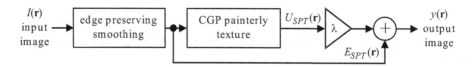

FIGURE 14.15

Schematic representation of artistic image generation by means of continuous glass patterns.

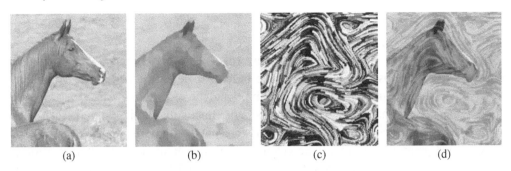

(a)   (b)   (c)   (d)

FIGURE 14.16 (See color insert.)

Artistic image generation for the example of Figure 14.15: (a) input image $I(\mathbf{r})$, (b) edge preserving smoothing output $I_{EPS}(\mathbf{r})$, (c) associated synthetic painterly texture, and (d) final output $y(\mathbf{r})$. © 2009 IEEE

brush strokes along object contours. Moreover, it is easy to prove that for $\theta_0 = \pi/2$ the streamlines of $V(\mathbf{r},t)$ are closed curves [47]. Thus, the brush strokes tend to form whirls which are typical of some impressionist paintings.

Finally, the artistic effect is achieved by adding the synthetic texture to the smoothed image, thus obtaining the final output $y(\mathbf{r}) \triangleq I_{EPS}(\mathbf{r}) + \lambda U_{SPT}(\mathbf{r})$ (Figure 14.16d). The parameter λ controls the strength of the synthetic texture. Comparing Figures 14.16a and 14.16d

(a)

(b)

(c)

(d)

FIGURE 14.17 (See color insert.)

Comparison of various artistic effects: (a) input image, (b) glass pattern algorithm, (c) artistic vision [15], and (d) impressionistic rendering [9]. © 2009 IEEE

reveals that natural texture of the input image is replaced by $U_{SPT}(\mathbf{r})$. Such a simple texture manipulation produces images which look like paintings.

An example of how well this approach performs is shown in Figure 14.17, in comparison with two of the most popular brush stroke-based artistic operators described previously: namely, the *impressionist rendering* algorithm [9] and the so-called *artistic vision* [15] technique. It can be seen that the glass pattern operator effectively mimics curved brush strokes oriented along object contours while the whirls present in contourless areas resemble some impressionist paintings. As to artistic vision, though simulation of curved brush strokes is attempted, several artifacts are clearly visible. Impressionistic rendering does not produce artifacts, but it tends to render blurry contours and small object details are lost. Moreover, impressionistic rendering is less effective in rendering impressionist whirls.

14.3.2.3 Pattern Transfer

An important limitation of the approach described above is a possible mismatch between the streamlines of $\mathbf{v}(\mathbf{r})$ and the actual object contours, which can result in unrealistic effects (Figure 14.18). This can be avoided using an operator called cross continuous glass patterns [47], [63], [64]. The idea is to generate a colored continuous glass pattern which already contains the color profile of the input image, rather than superimposing a graylevel texture to $I(\mathbf{r})$. In order to do this, a cross continuous glass pattern can be defined as follows [47]:

$$C_{\mathbf{v}}\{z(\mathbf{r}), \mathbf{I}(\mathbf{r})\} \triangleq \mathbf{I}\{\boldsymbol{\rho}_0(\mathbf{r})\}, \qquad (14.13)$$

with $\boldsymbol{\rho}_0(\mathbf{r}) \triangleq \arg\max_{\boldsymbol{\rho} \in A_{\mathbf{v}}(\mathbf{r})}\{z(\boldsymbol{\rho})\}$. In other words, instead of directly considering the maximum of $z(\mathbf{r})$ over $A_{\mathbf{v}}(\mathbf{r})$, the point $\boldsymbol{\rho}_0(\mathbf{r})$ which maximizes $z(\boldsymbol{\rho})$ is first identified, and the value of $\mathbf{I}(\mathbf{r})$ at that point is taken. It is easy to see that if the input image $\mathbf{I}(\mathbf{r})$ coincides with $z(\mathbf{r})$, then the output coincides with a continuous glass pattern $C_{\mathbf{v}}\{z(\mathbf{r}), z(\mathbf{r})\} = \mathfrak{G}_{\mathbf{v}}(\mathbf{r})$ defined above. An efficient implementation of cross continuous glass patterns can be found in References [47] and [63].

Examples of cross continuous glass patterns are given in Figure 14.19, which are respectively related to the vector fields $\mathbf{v}(\mathbf{r}) = [x, y]/\sqrt{x^2 + y^2}$, $\mathbf{v}(\mathbf{r}) = [-y, x]/\sqrt{x^2 + y^2}$, and $\mathbf{v}(\mathbf{r}) \perp \nabla I(\mathbf{r})$. Figures 14.19b and 14.19c show similar microstructure whereas Fig-

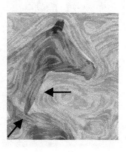

FIGURE 14.18

Artifact that arises with the approach proposed in Reference [47] when the streamlines of $\mathbf{v}(\mathbf{r})$ do not match the object contours of the input image (marked by arrows). This happens especially in presence of sharp corners.
© 2009 IEEE

(a)

(b)

(c)

(d)

FIGURE 14.19 (See color insert.)

Examples of cross continuous glass patterns: (a) input image, (b) output corresponding to $\mathbf{v}(\mathbf{r}) = [x,y]/\sqrt{x^2+y^2}$, (c) output corresponding to $\mathbf{v}(\mathbf{r}) = [-y,x]/\sqrt{x^2+y^2}$, and (d) output corresponding to $\mathbf{v}(\mathbf{r}) \perp \nabla I(\mathbf{r})$. The last image is perceptually similar to a painting.

ure 14.19d is perceptually similar to a painting. Other examples are shown in Figure 14.18. In Figure 14.20b the vector field $\mathbf{v}(\mathbf{r})$ is orthogonal to the color gradient of the input image, while in Figure 14.20c it forms an angle of 45° with ∇I. As can be seen, though the streamlines of $\mathbf{v}(\mathbf{r})$ strongly mismatch the object contours, no artifacts similar to those shown in Figure 14.18 are present (see also Figures 14.19b and 14.19c). It can also be observed that this approach, as well as all techniques based on vector fields, is very versatile since substantially different artistic images can be achieved by varying a few input parameters. Specifically, when the angle between $\mathbf{v}(\mathbf{r})$ and the gradient direction is equal to $\theta = \pi/2$, the strokes follow the object contours and form whirls in flat areas, while for $\theta = \pi/4$, the strokes are orthogonal to the contours and build star-like formations in flat regions.

14.4 Conclusions

The classical approach to generate synthetic paintings consists in rendering an ordered list of brush strokes on a white or canvas textured background. Alternatively, the possi-

FIGURE 14.20

Cross continuous glass patterns generated using different values of θ: (a) input image, (b) output for $\theta = \pi/4$, and (c) output for $\theta = \pi/2$. © 2009 IEEE

bility to render strokes on top of an image instead of a white background (underpainting) is explored in Reference [15]. The advantages of this class of algorithms are high realism and the possibility to imitate different painting styles by changing the values of some input parameters [17], [38], [48]. Moreover, a brush stroke-based representation of a synthetic painterly image results in a (lossless) compression ratio with respect to well established general purpose (lossy) image compression methods [12]. Another advantage of representing a synthetic painting as a list of brush strokes is that true brush strokes can be rendered on a true canvas by means of a robot [65].

These systems for painterly rendering consist of two major components; namely, techniques to render a digital brush stroke with given descriptive attributes and algorithms which automatically extract a list of brush stroke attributes from an input image. As to the former, the physical simulation of artistic media [21], [66] gives rise to excellent results in terms of realism; however, it comes at the cost of high computational complexity because many iterations of a complex cellular automaton need to be made for each brush stroke. Physical simulation is perfectly suitable for interactive applications in which the user draws each brush stroke by hand on an electric canvas; in such applications, computation time is not a pressing issue while high realism is imperative. Conversely, for automatic painterly rendering applications, where hundreds or thousands of brush strokes need to be

(a) (b)

FIGURE 14.21

Painterly rendering using simpler brush strokes models: (a) input image and (b) output of the algorithm presented in Reference [15] which does not perform physical brush stroke simulation. Small brush strokes are rendered properly, such as on the trees. On the other hand, visible undesired artifacts are present on large brush strokes, such as on the sky.

rendered for each image, simpler brush stroke models based on predefined intensity and texture profiles are usually deployed [9], [38]. Such simpler models still give acceptable results for small brush strokes but might produce visible artifacts for larger brush strokes (Figure 14.21). Extracting brush stroke attributes from an input image is a much harder task, because it involves some knowledge about how artists see the world. The basic idea behind this approach is to look for regions where the color profile of the input image is constant or slowly varying, and then to extract position, shape, and orientation of each brush stroke. There exist a large variety of complex algorithms, with many predetermined parameters and without a general guiding principle. To overcome this problem, the extraction of brush stroke attributes can be formulated as an optimization problem. However, this leads to cost functions defined on an extremely high dimensional space, thus making such an optimization extremely difficult carry out in practice. Moreover, it is not obvious that every artistic effect can be achieved by the mere modification of a cost function.

Due to these intrinsic difficulties, some authors propose to focus on the visual properties which distinguish a painting from a photographic image while abstracting from the process deployed by an artist to generate them. Examples of such properties are sharp edges, absence of natural texture, or presence of motives induced by brush strokes in several impressionistic paintings (see, for instance, Figure 14.13a). An effective and efficient approach to painterly rendering is smoothing the input image while preserving or sharpening edges. It should be noted, however, that not all existing smoothing operators which preserve and enhance edges produce satisfactory artistic images. Examples are bilateral filtering, median filtering, and structural opening and closing [67]. Area open-closing can produce artistic effects, but only on images that are rich in texture and with sharp edges. In presence of blurry edges and relatively flat areas, area opening does not modify the input image substantially. In contrast, the approach proposed by Kuwahara and the subsequent extensions are much more effective for every input image. Another interesting way to produce artistic images is based on glass patterns. The theory of glass patterns naturally combines three essential aspects of painterly artwork: perception, randomness, and geometric structure. Therefore, it constitutes a suitable framework for the development of mathematical models of the visual properties that distinguish paintings from photographic images. Transferring the microstructure of a glass pattern to an input image produces outputs that are perceptually similar to a painting. The related algorithms have low computational complexity and do not require predetermined parameters. Usually, filtering-driven painterly rendering leads to conceptually and computationally simpler algorithms with respect to approaches based on brush stroke simulation. Algorithms are more efficient and are expressed in a more compact mathematical form, which makes them suitable for further theoretical analysis. Moreover, the input parameters are typically easy to interpret. However, it is difficult to imitate specific painting styles with these algorithms.

In general, at the current state of the art, it can be said that the output of painterly rendering algorithms will differ from the input image only in the textural details. Moreover, such an output does not give the sense of three-dimensionality that is observed in true paintings, for instance, due to the variable thickness of the paint or the non-flatness of the paper in watercolor paintings. It can be concluded that a larger step in painterly rendering can be achieved by introducing operators which perform a more radical modification of the input image, such as some form of exaggeration seen in caricatures.

Acknowledgment

Figures 14.9 and 14.10 are reprinted from Reference [67] and Figures 14.12, 14.14, 14.16, 14.17, 14.18, and 14.20 are reprinted from Reference [47], with the permission of IEEE.

References

[1] R. Arnheim, *Art and Visual Perception: A Psychology of the Creative Eye*. Berkeley, CA: University of California Press, September 1974.

[2] R. Arnheim, *Toward a Psychology of Art*. Berkeley, CA: University of California Press, May 1972.

[3] R. Arnheim, *Visual Thinking*. Berkeley, CA: University of California Press, 1969.

[4] E. Loran, *Cézanne's Composition: Analysis of His Form with Diagrams And Photographs of His Motifs*. Berkeley, CA: University of California Press, 1963.

[5] S. Zeki, *Inner Vision*. New York: Oxford University Press, February 2000.

[6] S. Zeki, "Trying to make sense of art," *Nature*, vol. 418, pp. 918–919, August 2002.

[7] V. Ramachandran and W. Hirstein, "The science of art," *Journal of Consciousness Studies*, vol. 6, no. 6-7, pp. 15–51(37), 1999.

[8] B. Pinna, *Art and Perception Towards a Visual Science of Art*. Koninklijke Brill NV, Leiden, The Netherlands, November 2008.

[9] P. Litwinowicz, "Processing images and video for an impressionist effect," in *Proceedings of the 24th Annual Conference on Computer Graphics and Interactive Techniques*, Los Angeles, CA, USA, August 1997, pp. 407–414.

[10] M. Shiraishi and Y. Yamaguchi, "An algorithm for automatic painterly rendering based on local source image approximation," in *Proceedings of the 1st International Symposium on Non-Photorealistic Animation and Rendering*, Annecy, France, June 2000, pp. 53–58.

[11] N. Li and Z. Huang, "Feature-guided painterly image rendering," in *Proceedings of the IEEE International Conference on Image Processing*, Rochester, New York, USA, September 2002, pp. 653–656.

[12] L. Kovács and T. Szirányi, "Painterly rendering controlled by multiscale image features," in *Proceedings of Spring Conference on Computer Graphics*, Budmerice, Slovakia, April 2004, pp. 177–184.

[13] P. Haeberli, "Paint by numbers: Abstract image representations," *ACM SIGGRAPH Computer Graphics*, vol. 24, no. 4, pp. 207–214, August 1990.

[14] D. De Carlo and A. Santella, "Stylization and abstraction of photographs," *ACM Transactions on Graphics*, vol. 21, no. 3, pp. 769–776, July 2002.

[15] B. Gooch, G. Coombe, and P. Shirley, "Artistic vision: Painterly rendering using computer vision techniques," in *Proceedings of the 2nd International Symposium on Non-Photorealistic Animation and Rendering*, Annecy, France, June 2002, pp. 83–90.

[16] M. Schwar, T. Isenberg, K. Mason, and S. Carpendale, "Modeling with rendering primitives: An interactive non-photorealistic canvas," in *Proceedings of the 5th International Symposium on Non-Photorealistic Animation and Rendering*, San Diego, CA, USA, August 2007, pp. 15–22.

[17] A. Hertzmann, "Painterly rendering with curved brush strokes of multiple sizes," in *Proceedings of the 25th Annual Conference on Computer Graphics and Interactive Techniques*, Orlando, FL, USA, July 1998, pp. 453–460.

[18] A. Kasao and K. Miyata, "Algorithmic painter: A NPR method to generate various styles of painting," *The Visual Computer*, vol. 22, no. 1, pp. 14–27, January 2006.

[19] J. Collomosse and P. Hall, "Salience-adaptive painterly rendering using genetic search," *International Journal on Artificial Intelligence Tools*, vol. 15, no. 4, pp. 551–575, August 2006.

[20] A. Orzan, A. Bousseau, P. Barla, and J. Thollot, "Structure-preserving manipulation of photographs," in *Proceedings of the 5th International Symposium on Non-Photorealistic Animation and Rendering*, San Diego, CA, USA, August 2007, pp. 103–110.

[21] C. Curtis, S. Anderson, J. Seims, K. Fleischer, and D. Salesin, "Computer-generated watercolor," in *Proceedings of the 24th Annual Conference on Computer Graphics and Interactive Techniques*, Los Angeles, CA, USA, August 1997, pp. 421–430.

[22] E. Lum and K. Ma, "Non-photorealistic rendering using watercolor inspired textures and illumination," in *Proceedings of Pacific Conference on Computer Graphics and Applications*, Tokyo, Japan, October 2001, pp. 322–331.

[23] T. Van Laerhoven, J. Liesenborgs, and F. Van Reeth, "Real-time watercolor painting on a distributed paper model," in *Proceedings of Computer Graphics International*, Crete, Greece, June 2004, pp. 640–643.

[24] E. Lei and C. Chang, "Real-time rendering of watercolor effects for virtual environments," in *Proceedings of IEEE Pacific-Rim Conference on Multimedia*, Tokyo, Japan, December 2004, pp. 474–481.

[25] H. Johan, R. Hashimota, and T. Nishita, "Creating watercolor style images taking into account painting techniques," *Journal of the Society for Art and Science*, vol. 3, no. 4, pp. 207–215, 2005.

[26] A. Bousseau, M. Kaplan, J. Thollot, and F. Sillion, "Interactive watercolor rendering with temporal coherence and abstraction," in *Proceedings of the 4th International Symposium on Non-Photorealistic Animation and Rendering*, Annecy, France, June 2006, pp. 141–149.

[27] D. Small, "Simulating watercolor by modeling diffusion, pigment, and paper fibers," in *Proceedings of SPIE*, vol. 1460, p. 140, February 1991.

[28] E. Lum and K. Ma, "Non-photorealistic rendering using watercolor inspired textures and illumination," in *Proceedings of Pacific Conference on Computer Graphics and Applications*, Tokyo, Japan, October 2001, pp. 322–331.

[29] E. Lei and C. Chang, "Real-time rendering of watercolor effects for virtual environments," in *Proceedings of IEEE Pacific-Rim Conference on Multimedia*, Tokyo, Japan, December 2004, pp. 474–481.

[30] T. Van Laerhoven, J. Liesenborgs, and F. Van Reeth, "Real-time watercolor painting on a distributed paper model," in *Proceedings of Computer Graphics International*, Crete, Greece, June 2004, pp. 640–643.

[31] H. Johan, R. Hashimota, and T. Nishita, "Creating watercolor style images taking into account painting techniques," *Journal of the Society for Art and Science*, vol. 3, no. 4, pp. 207–215, 2005.

[32] A. Bousseau, M. Kaplan, J. Thollot, and F. Sillion, "Interactive watercolor rendering with temporal coherence and abstraction," in *Proceedings of the 4th International Symposium on Non-Photorealistic Animation and Rendering*, Annecy, France, June 2006, pp. 141–149.

[33] T. Luft and O. Deussen, "Real-time watercolor illustrations of plants using a blurred depth test," in *Proceedings of the 4th International Symposium on Non-Photorealistic Animation and Rendering*, Annecy, France, June 2006, pp. 11–20.

[34] A. Bousseau, F. Neyret, J. Thollot, and D. Salesin, "Video watercolorization using bidirectional texture advection," *Transactions on Graphics*, vol. 26, no. 3, July 2007.

[35] A. Santella and D. DeCarlo, "Abstracted painterly renderings using eye-tracking data," in *Proceedings of the Second International Symposium on Non-photorealistic Animation and Rendering* Annecy, France, June 2002, pp. 75–82.

[36] S. Nunes, D. Almeida, V. Brito, J. Carvalho, J. Rodrigues, and J. du Buf, "Perception-based painterly rendering: functionality and interface design," *Ibero-American Symposium on Computer Graphics*, Santiago de Compostela, Spain, July 2006, pp. 53–60.

[37] A. Hertzmann, "A survey of stroke-based rendering," *IEEE Computer Graphics and Applications*, vol. 23, no. 4, pp. 70–81, July/August 2003.

[38] J. Hays and I. Essa, "Image and video based painterly animation," *Proceedings of the 3rd International Symposium on Non-Photorealistic Animation and Rendering*, Annecy, France, June 2004, pp. 113–120.

[39] S. Olsen, B. Maxwell, and B. Gooch, "Interactive vector fields for painterly rendering," *Proceedings of Canadian Annual Conference on Graphics Interface*, Victoria, British Columbia, May 2005, pp. 241–247.

[40] E. Stavrakis and M. Gelautz, "Stereo painting: Pleasing the third eye," *Journal of 3D Imaging*, vol. 168, pp. 20–23, Spring 2005.

[41] S. Stavrakis and M. Gelautz, "Computer generated stereoscopic artwork," in *Proceedings of Workshop on Computational Aesthetics in Graphics, Visualization and Imaging*, Girona, Spain, May 2005, pp. 143–149.

[42] S. Strassmann, "Hairy brushes," *ACM SIGGRAPH Computer Graphics*, vol. 20, no. 4, pp. 225–232, August 1986.

[43] J. Hopcroft, R. Motwani, and J. Ullman, *Introduction to Automata Theory, Languages, and Computation.* Addison-Wesley, July 2006.

[44] Q. Zhang, Y. Sato, T. Jy, and N. Chiba, "Simple cellular automaton-based simulation of ink behaviour and its application to suibokuga-like 3D rendering of trees," *The Journal of Visualization and Computer Animation*, vol. 10, no. 1, pp. 27–37, April 1999.

[45] C. Haase and G. Meyer, "Modeling pigmented materials for realistic image synthesis," *ACM Transactions on Graphics*, vol. 11, no. 4, pp. 305–335, October 1992.

[46] G. Kortüm, *Reflectance Spectroscopy: Principles, Methods, Applications.* New York: Springer-Verlag, January 1969.

[47] G. Papari and N. Petkov, "Continuous glass patterns for painterly rendering," *IEEE Transactions on Image Processing*, vol. 18, no. 3, pp. 652–664, March 2009.

[48] A. Hertzmann, "Paint by relaxation," in *Proceedings of Computer Graphics International*, Hong Kong, July 2001, pp. 47–54.

[49] M. Wilkinson, H. Gao, W. Hesselink, J. Jonker, and A. Meijster, "Concurrent computation of attribute filters on shared memory parallel machines," *IEEE Transactions on Pattern Analysis and Machine Intelligence*, vol. 30, no. 10, pp. 1800–1813, October 2008.

[50] M. Kuwahara, K. Hachimura, S. Eiho, and M. Kinoshita, "Processing of ri-angiocardiographic images," *Digital Processing of Biomedical Images*, pp. 187–202, 1976.

[51] A. Oppenheim, R. Schafer, and J. Buck, *Discrete-Time Signal Processing.* Englewood Cliffs, NJ: Prentice Hall, 1989.

[52] M. Nagao and T. Matsuyama, "Edge preserving smoothing," *Computer Graphics and Image Processing*, vol. 9, no. 4, pp. 394–407, 1979.

[53] P. Bakker, L. Van Vliet, and P. Verbeek, "Edge preserving orientation adaptive filtering," in *Proceedings of the IEEE Conference on Computer Vision and Pattern Recognition*, Ft. Collins, CO, USA, June 1999, pp. 535–540.

[54] M.A. Schulze and J.A. Pearce, "A morphology-based filter structure for edge-enhancing smoothing," in *Proceedings of the IEEE International Conference on Image Processing*, Austin, Texas, USA, November 1994, pp. 530–534.

[55] R. van den Boomgaard, "Decomposition of the Kuwahara-Nagao operator in terms of a linear smoothing and a morphological sharpening," in *Proceedings of International Symposium on Mathematical Morphology*, Sydney, NSW, Australia, April 2002, p. 283.

[56] G. Papari, N. Petkov, and P. Campisi, "Edge and corner preserving smoothing for artistic imaging," *Proceedings of SPIE*, vol. 6497, pp. 649701–64970J, 2007.

[57] J. Kyprianidis, H. Kang, and J. Döllner, "Image and video abstraction by anisotropic Kuwahara filtering," *Computer Graphics Forum*, vol. 28, no. 7, 2009.

[58] T. Brox, J. Weickert, B. Burgeth, and P. Mrázek, "Nonlinear structure tensors," *Image and Vision Computing*, vol. 24, no. 1, pp. 41–55, January 2006.

[59] L. Glass, "Moiré effect from random dots," *Nature*, vol. 223, pp. 578–580, 1969.

[60] L. Glass and R. Perez, "Perception of random dot interference patterns," *Nature*, vol. 246, pp. 360–362, 1973.

[61] L. Glass and E. Switkes, "Pattern recognition in humans: Correlations which cannot be perceived," *Perception*, vol. 5, no. 1, pp. 67–72, 1976.

[62] G. Hall and J. Watt, *Modern Numerical Methods for Ordinary Differential Equations.* Oxford, UK: Clarendon Press, October 1976.

[63] G. Papari and N. Petkov, "Spatially variant dilation for unsupervised painterly rendering," in *Abstract Book of the 9th International Symposium on Mathematical Morphology*, Groningen, Netherlands, August 2009, pp. 56–58.

[64] G. Papari and N. Petkov, "Reduced inverse distance weighting interpolation for painterly rendering," in *Proceedings of International Conference on Computer Analysis of Images and Patterns*, Mnster, Germany, September 2009, pp. 509–516.

[65] C. Aguilar and H. Lipson, "A robotic system for interpreting images into painted artwork," in *Proceedings of Generative Art Conference*, Milano, Italy, December 2008, pp. 372–387.

[66] W. Baxter, J. Wendt, and M. Lin, "Impasto: A realistic, interactive model for paint," in *Proceedings of the 3rd International Symposium on Non-Photorealistic Animation and Rendering*, Annecy, France, June 2004, pp. 45–148.

[67] G. Papari, N. Petkov, and P. Campisi, "Artistic edge and corner enhancing smoothing," *IEEE Transactions on Image Processing*, vol. 29, no. 10, pp. 2449–2462, October 2007.

15

Machine Learning Methods for Automatic Image Colorization

Guillaume Charpiat, Ilja Bezrukov, Matthias Hofmann, Yasemin Altun, and Bernhard Schölkopf

15.1   Introduction

Automatic image colorization is the task of adding colors to a grayscale image without any user intervention. This problem is ill-posed in the sense that there is not a unique colorization of a grayscale image without any *prior* knowledge. Indeed, many objects can have different colors. This is not only true for artificial objects, such as plastic objects

FIGURE 15.1 (See color insert.)

Failure of standard colorization algorithms in the presence of texture: (left) manual initialization, (right) result of Reference [1] with the code available at http://www.cs.huji.ac.il/~yweiss/Colorization/. Despite the general efficiency of this simple method, based on the mean and the standard deviation of local intensity neighborhoods, the texture remains difficult to deal with. Hence texture descriptors and learning edges from color examples are required.

which can have random colors, but also for natural objects such as tree leaves which can have various nuances of green and brown in different seasons, without significant change of shape.

The most common color *prior* in the literature is the user. Most image colorization methods allow the user to determine the color of some areas and extend this information to the whole image, either by presegmenting the image into (preferably) homogeneous color regions or by spreading color flows from the user-defined color points. The latter approach involves defining a color flow function on neighboring pixels and typically estimates this as a simple function of local grayscale intensity variations [1], [2], [3], or as a predefined threshold such that color edges are detected [4]. However, this simple and efficient framework cannot deal with texture examples of Figure 15.1, whereas simple oriented texture features such as Gabor filters can easily overcome these limitations. Hence, an image colorization method should incorporate texture descriptors for satisfactory results. More generally, the manually set criteria for the edge estimation are problematic, since they can be limited to certain scenarios. The goal of this chapter is to *learn* the variables of image colorization modeling in order to overcome the limitations of manual assignments.

User-based approaches have the advantage that the user has an interactive role, for example, by adding more color points until a satisfactory result is obtained or by placing color points strategically in order to give indirect information on the location of color boundaries. The methods proposed in this chapter can easily be adapted to incorporate such user-provided color information. *Predicting* the colors, that is, providing an initial fully automatic colorization of the image prior to any possible user intervention, is a much harder but arguably more useful task. Recent literature investigating this task [5], [6], [7], [8] yields mixed conclusions. An important limitation of these methods is their use of local predictors. Color prediction involves many ambiguities that can only be resolved at the global level. In general, local predictions based on texture are most often very noisy and not reliable. Hence, the information needs to be integrated over large regions in order to provide a significant signal. Extensions of local predictors to include global information has been limited to using automatic tools (such as automatic texture segmentation [7]) which can introduce errors due to the cascaded nature of the process or incorporating small neighborhood information, such as a one-pixel-radius filter [7]. Hence, an important design criterion in learning to predict colors is to develop global methods that do not rely on limited neighborhood texture-based classification.

The color assignment ambiguity also occurs when the shape of an object is relevant for determining the color of the whole object. More generally, it appears that the boundaries of the object contain useful information, such as the presence of edges in the color space, and significant details which can help to identify the whole object. This scenario again states the importance of global methods for image colorization so that the colorization problem cannot be solved at the local level of pixels. Another source of prior information is the motion and time coherency as in the case of the video sequences to be colored [1]. Hence, a successful automatic color predictor should be general enough to incorporate various sources of information in a global manner. Section 15.2 briefly discusses related work.

Machine learning methods, in particular *nonparametric* methods such as Parzen window estimators and support vector machines (SVMs), provide a natural and efficient way of incorporating information from various sources. This chapter reformulates the problem of automatic image colorization as a prediction problem and investigates applications of machine learning techniques for it. Although colors are continuous variables, considering color prediction as a regression problem is problematic due to the multimodal nature of the problem. In order to cope with the multimodality, both the color space discretization and multiclass machine learning methods are investigated in this chapter. Section 15.3 outlines the limitations of the regression approach and describes representation of the color space as well as the local grayscale texture space.

Three machine learning methods are proposed for learning local color predictions and spatial coherence functions. Spatial coherency criteria are modeled by the likelihood of color variations which is estimated from training data. The Parzen window method is a probabilistic, nonparametric, scalable, and easy to implement machine learning algorithm. Section 15.4 describes an image colorization method which uses Parzen windows to learn local color predictors and color variations given a set of colored images. Section 15.5 outlines the second approach for automatic image colorization. This approach is based on SVMs, which constitute a more sophisticated machine learning method that can learn a more general class of predictive functions and has stronger theoretical guarantees.

Once the local color prediction functions along with spatial coherency criteria are learned, they can be employed in *graph-cut algorithms*. Graph-cut algorithms are optimization techniques commonly used in computer vision in order achieve optimal predictions on complete images. They combine local predictions with spatial coherency functions across neighboring pixels. This results in global interaction across pixel colorings and yields the best coloring for a grayscale image with respect to both predictors. The details of using graph-cuts for image colorization are given in Section 15.6.

One shortcoming of the approaches outlined above is the independent training of the two components; the local color predictor and the spatial coherency functions. It can be argued that a joint optimization of these models can find the optimal parameters, whereas independent training may yield suboptimal models. The third proposed approach investigates this issue and uses *structured output* prediction techniques where the two models are trained jointly. Section 15.7 provides the details of applying structured SVMs to automatic image colorization.

Section 15.8 focuses on an experimental analysis of the proposed machine learning methods on datasets of various sizes. All proposed approaches perform well with a large number of colors and outperform existing methods. It will be shown that the Parzen window approach provides natural colorization, especially when trained on small datasets, and performs reasonably well on big datasets. On large training data, SVMs and structured SVMs leverage the information more efficiently and yield more natural colorization, with more color details, at the expense of longer training times. Although experiments presented in this chapter focus on colorization of still images, the proposed framework can be readily extended to movies. It is believed that the framework has the potential to enrich existing movie colorization methods that are suboptimal in the sense that they heavily rely on user input. Further discussion on the future work and conclusions are offered in Section 15.9.

15.2 Related Work

Colorization based on examples of color images is also known as *color transfer* in the literature. A survey of this field can be found in Reference [9]. The first results [5], [6] in the field of fully automatic colorization, though promising, seem to deal with only a few colors and many small artifacts can be observed. These artifacts can be attributed to the lack of a suitable spatial coherency criterion. Indeed, References [5] and [6] deal with the colorization process which is iterative and consists of searching for each pixel, in scan-line order, as the best match in the training set. These approaches are thus not expressed mathematically; in particular, it is not clear whether an energy function is minimized.

Reference [7] proposes finding *landmark* points in the image where a color prediction algorithm reaches the highest confidence and applying the method presented in Reference [1] as if these points were given by the user. This approach assumes the existence of a training set of colored images, that is segmented by the user into regions. The new image is automatically segmented into locally homogeneous regions whose texture is similar to one of the colored regions in the training data, and the colors are transferred. The limitations of this approach relate to preprocessing and spatial coherency. The preprocessing step involves segmentation of images into regions of homogeneous texture either by the user or by automatic segmentation tools. Given that fully automatic segmentation is known to be a difficult problem, an automatic image colorization method that does not rely on automatic segmentation, such as the approaches described in this chapter, can be more robust. Reference [7] incorporates spatial coherency at a local level via a one-pixel-radius filter and automatic segments. The proposed approach can capture global spatial coherency via the graph-cut algorithm which assigns the best coloring to the global image.

15.3 Model for Colors and Grayscale Texture

In the image colorization problem, two important quantities to be modeled are i) the output space, that is, the color space, and ii) the input space, that is, the feature representation of the grayscale images. Let I denote a grayscale image to be colored, $\mathbf{p}$ the location of one particular pixel, and C a colorization of image I. Hence, I and C are images of the same size, and the color of the pixel $\mathbf{p}$, denoted by $C(\mathbf{p})$, is in the standard RGB color space. Since the grayscale information is already given by $I(\mathbf{p})$, the term $C(\mathbf{p})$ is restricted such that computing the grayscale intensity of $C(\mathbf{p})$ yields $I(\mathbf{p})$. Thus, the dimension of the color space to be explored is intrinsically two rather than three.

This section presents the model chosen for the color space, the limitations of a regression approach for color prediction, and the proposed color space discretization method. It also discusses how to express probability distributions of continuous valued colors given a discretization and describes the feature space used for the description of grayscale patches.

15.3.1 Lab Color Space

In order to measure the similarity of two colors, a metric on the space of colors is needed. This metric is also employed to associate a saturated color to its corresponding gray level, that is, the closest unsaturated color. It is also at the core of the color coherency problem. An object with uniform reflectance shows different colors in its illuminated and shadowed parts since they have different gray levels. This behavior creates the need of a definition that is robust against changes of lightness. More precisely, the modeling of the color space should specify how colors are expected to vary as a function of the gray level and how a dark color is projected onto the subset of all colors that share a specific brighter gray level.

There are various color models, such as RGB, CMYK, XYZ, and *Lab*. Among these, the *Lab* space is chosen here because its underlying metric has been designed to express color coherency. Based on psychophysical experiments, this color space was designed such that the Euclidean distance between the coordinates of any colors in this space approximates the human perception of distances between colors as accurately as possible. The L component expresses the luminance or lightness and consequently denotes the grayscale axis. The two other components, a and b, stand for the two orthogonal color axes. The transformation from standard RGB colors to *Lab* is achieved by applying first the gamma correction, then a linear function in order to obtain the XYZ color space, and finally a highly nonlinear function which is basically a linear combination of the cubic roots of the coordinates in XYZ. More details on color spaces can be found in Reference [10]. In the following, L and (a, b) are referred to as *gray level* and *two-dimensional (2D) color*, respectively. Since the gray level $I(\mathbf{p})$ of the color $C(\mathbf{p})$ at pixel $\mathbf{p}$ is given, the search can be done only for the remaining 2D color, denoted by $ab(\mathbf{p})$.

15.3.2 Need for Multimodality

In automatic image colorization using machine learning methods, the goal is to learn a function that associates the right color for a pixel $\mathbf{p}$ given a local description of grayscale patches centered at $\mathbf{p}$. Since colors are continuous variables, regression tools such as support vector regression or Gaussian process regression [11] can be employed for this task. Unfortunately, a regression approach performs poorly and there is an intuitive explanation for this performance. Namely, many objects with the same or similar local descriptors can have different colors. For instance, balloons at a fair could be green, red, blue, etc. Even if the task of recognizing a balloon was easy and it is known that that the observed balloon colors should be used to predict the color of a new balloon, a regression approach would recommend using the average color of the observed balloons. This problem is not specific to objects of the same class, but also extends to objects with similar local descriptors. For example, the local descriptions of grayscale patches of skin and sky are very similar. Hence, a method trained on images including both objects would recommend purple for skin and sky, without considering the fact that this average value is never probable. Therefore, an image colorization method requires multimodality, that is, the ability to predict different colors if needed, or more precisely, the ability to predict scores or probability values of *every* possible color at each pixel.

15.3.3 Discretization of the Color Space

Due to the multimodal nature of the color prediction problem, the machine learning methods proposed in this chapter first infer distributions for discrete colors given a pixel and then project the predicted colors to the continuous color space. The following discusses a discretization of the 2D color space and a projection method for continuous valued colors.

There are numerous ways of discretization, for instance via K-means. Instead of setting a regular grid in the color space, a discretization can be defined which adapts to the colors in the training dataset such that each color bin contains approximately the same number of pixels. Indeed, some zones of the color space are useless for many real image datasets. Allocating more color bins to zones with higher density allows the models to have more nuances where it makes statistical sense. Figure 15.2 shows the densities of colors corresponding to some images, as well as the discretization of the color space into 73 bins resulting from these densities. This discretization is obtained by using a polar coordinate system in *ab*, cutting color bins recursively with highest numbers of points at their average color into four parts, and assigning the average color to each bin.

Given the densities in the discrete color space, the densities for continuous colors on the whole *ab* plane can be expressed via interpolation. In order to interpolate the information given by each color bin i continuously, Gaussian functions are placed on the average color μ_i, with standard deviation proportional to the empirical standard deviation σ_i (see last column of Figure 15.2). The interpolation of the densities $d(i)$ in the discrete color space to any point x in the *ab* plane is given by

$$d_G(x) = \sum_i \frac{1}{\pi(\alpha\sigma_i)^2} e^{-\frac{\|x-\mu_i\|^2}{2(\kappa\sigma_i)^2}} d(i).$$

It is observed that $\kappa \approx 2$ yields successful experimental results. For better performance, it is possible to employ cross-validation for the optimal κ value for a given training set.

15.3.4 Grayscale Patches and Features

As discussed in Section 15.1, the gray level of one pixel is not informative for color prediction. Additional information such as texture and local context is necessary. In order to extract as much information as possible to describe local neighborhoods of pixels in the grayscale image, SURF descriptors [12] are computed at three different scales for each

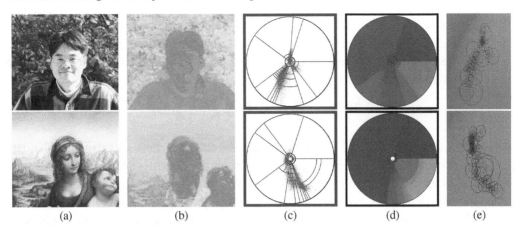

FIGURE 15.2 (See color insert.)

Examples of color spectra and associated discretizations: (a) color image, (b) corresponding 2D colors, (c) the location of the observed 2D colors in the *ab*-plane (a red dot for each pixel) and the computed discretization in color bins, (d) color bins filled with their average color, and (e) continuous extrapolation with influence zones of each color bin in the *ab*-plane (each bin is replaced by a Gaussian, whose center is represented by a black dot; red circles indicate the standard deviation of colors within the color bin, blue ones are three times larger).

pixel. This leads to a vector of 192 features per pixel. Using principal component analysis (PCA), only the first 27 eigenvectors are kept, in order to reduce the number of features and to condense the relevant information. Furthermore, as supplementary components, the pixel gray level as well as two biologically inspired features are included. Namely, these feature are a weighted standard deviation of the intensity in a 5×5 neighborhood (whose meaning is close to the norm of the gradient), and a smooth version of its Laplacian. This 30-dimensional vector, computed at each pixel $\mathbf{q}$, is referred to as *local description*. It is denoted by $\mathbf{v}(\mathbf{q})$ or $\mathbf{v}$, when the text uniquely identifies $\mathbf{q}$.

15.4 Parzen Windows for Color Prediction

Given a set of colored images and a new grayscale image I to be colored, the color prediction task is to extract knowledge from the training set to predict colors C for the new image. This knowledge is represented in two models, namely a local color predictor and a spatial coherency function. This section outlines how to use the Parzen window method in order to learn these models, based on the representation described in Section 15.3.

15.4.1 Learning Local Color Prediction

Multimodality of the color prediction problem creates the need of predicting scores or probability values for *all* possible colors at each pixel. This can be accomplished by modeling the conditional probability distribution of colors *knowing* the local description of the grayscale patch around the considered pixel. The conditional probability of the color c_i at pixel $\mathbf{p}$ given the local description $\mathbf{v}$ of its grayscale neighborhood can be expressed as the fraction, amongst colored examples $e_j = (\mathbf{w}_j, c(j))$ whose local description $\mathbf{w}_j$ is similar to $\mathbf{v}$, of those whose observed color $c(j)$ is in the same color bin B_i. This can be estimated with a Gaussian Parzen window model

$$p(c_i|\mathbf{v}) = \left(\sum_{\{j:c(j)\in B_i\}} k(\mathbf{w}_j, \mathbf{v}) \right) \Big/ \sum_j k(\mathbf{w}_j, \mathbf{v}), \qquad (15.1)$$

where $k(\mathbf{w}_j, \mathbf{v}) = e^{-\|\mathbf{w}_j - \mathbf{v}\|^2/2\sigma^2}$ is the Gaussian kernel. The best value for the standard deviation σ can be estimated by cross-validation on the densities. Parzen windows also allow one to express how reliable the probability estimation is; its confidence depends directly on the density of examples around $\mathbf{v}$, since an estimation far from the clouds of observed points loses significance. Thus, the confidence on a probability estimate is given by the density in the feature space as follows:

$$p(\mathbf{v}) \propto \sum_j k(\mathbf{w}_j, \mathbf{v}).$$

Note that both distributions, $p(c_i|\mathbf{v})$ and $p(\mathbf{v})$, require computing the similarities $k(\mathbf{v}, \mathbf{w}_j)$ of all pixel pairs, which can be expensive during both training and prediction. For computational efficiency, these can be approximated by restricting the sums to K-nearest neighbors of $\mathbf{v}$ in the training set with a sufficiently large K chosen as a function of the σ and the Parzen densities can be estimated based on these K points. In practice, $K = 500$ is chosen. Using fast nearest neighbor search techniques, such as kD-tree in the TSTOOL package available at http://www.physik3.gwdg.de/tstool/ without particular optimization, the time needed to compute the predictions for all pixels of a 50×50 image is only 10 seconds (for a training set of hundreds of thousands of patches) and this scales linearly with the number of test pixels.

15.4.2 Local Color Variation Prediction

Instead of choosing a prior for spatial coherence, based either on detection of edges, the Laplacian of the intensity, or pre-estimated complete segmentation, it is possible to directly *learn* how likely it is to observe a color variation at a pixel knowing the local description of its grayscale neighborhood, based on a training set of real color images. The technique is similar to the one detailed in the previous section. For each example $\mathbf{w}_j$ of a colored patch, the norm g_j of the gradient of the 2D color (in the *Lab* space) is computed at the center of the patch. The expected color variation $g(\mathbf{v})$ at the center of a new grayscale patch $\mathbf{v}$ is then given by

$$g(\mathbf{v}) = \frac{\sum_j k(\mathbf{w}_j, \mathbf{v}) g_j}{\sum_j k(\mathbf{w}_j, \mathbf{v})}.$$

15.5 Support Vector Machines for Color Prediction

The method proposed in Section 15.4 is an improvement of existing image colorization approaches by learning color variations and local color predictors using the Parzen window method. Section 15.6 outlines how to use these estimators in a graph-cut algorithm in order to get spatially coherent color predictions. Before describing the details of this technique, further improvements over the Parzen window approach are proposed, by employing support vector machines (SVMs) [13] to learn the local color prediction function.

Equation 15.1 describes the Parzen window estimator for the conditional probability of the colors given a local grayscale description $\mathbf{v}$. A more general expression for the color prediction function is given by

$$s(c_i|\mathbf{v};\alpha_i) = \sum_j \alpha_i(j)k(\mathbf{w}_j,\mathbf{v}), \qquad (15.2)$$

where the kernel k satisfies $k(\mathbf{v},\mathbf{v}') = \langle \mathbf{f}(\mathbf{v}),\mathbf{f}(\mathbf{v}') \rangle$ for all $\mathbf{v}$ and $\mathbf{v}'$ in a certain space of features $\mathbf{f}(\mathbf{v})$, embedded with an inner product $\langle \cdot,\cdot \rangle$ between feature vectors (more details can be found in Reference [11]). In Equations 15.1 and 15.2, the expansions for each color c_i are linear in the feature space. The decision boundary between different colors, which tells which color is the most probable, is consequently an hyperplane. The α_i can be considered as a dual representation of the normal vector λ_i of the hyperplane separating the color c_i from other colors. The estimator in this *primal* space can then be represented as

$$s(c_i|\mathbf{v};\lambda_i) = \langle \lambda_i, \mathbf{f}(\mathbf{v}) \rangle. \qquad (15.3)$$

In the Parzen window estimator, all α values are nonzero constants. In order to overcome computational problems, Section 15.4 proposes a restriction of α parameters of pixels $\mathbf{p}_j$ that are not in the neighborhood of $\mathbf{v}$ to be zero. A more sophisticated classification approach is obtained using SVMs which differ from Parzen window estimators in terms of patterns whose α values are active (i.e., nonzero) and in terms of finding the optimal values for these parameters. In particular, SVMs remove the influence of correctly classified training points that are far from the decision boundary, since they generally do not improve the performance of the estimator and removing such instances (setting their corresponding α values to 0) reduces the computational cost during prediction. Hence, the goal in SVMs is to identify the instances that are close to the boundaries, commonly referred as *support vectors*, for each class c_i and find the optimal α_i. More precisely, the goal is to discriminate the observed color $c(j)$ for each colored pixel $e_j = (\mathbf{w}_j, c(j))$ from the other colors as much as possible while keeping a sparse representation in the dual space. This can be achieved by imposing the *margin* constraints

$$s(c(j)|\mathbf{w}_j;\lambda_{c(j)}) - s(c_i|\mathbf{w}_j;\lambda_i) \geq 1, \quad \forall j, \forall c_i \neq c(j), \qquad (15.4)$$

where the decision function is given in Equation 15.3. If these constraints are satisfiable, one can find multiple solutions by simply scaling the parameters. In order to overcome this problem, it is common to search for parameters that satisfy the constraints with minimal complexity. This can be accomplished by minimizing the norm of the solution λ. In cases where the constraints cannot be satisfied, one can allow violations of the constraints by adding *slack variables* ξ_j for each colored pixel e_j and penalize the violations in the optimization, where K denotes the trade-off between the loss term and the regularization term [14]:

$$\frac{1}{2}\sum_i ||\lambda_i||^2 + K\sum_j \xi_j, \quad \text{subject to} \tag{15.5}$$
$$s(c(j)|\mathbf{w}_j; \lambda_{c(j)}) - s(c|\mathbf{w}_j; \lambda_c) \geq 1 - \xi_j, \quad \forall j, \forall c \neq c(j)$$
$$\xi_j \geq 0, \quad \forall j.$$

If the constraint is satisfied for a pixel e_j and a color c_i, SVM yields 0 for $\alpha_i(j)$. The pixel-color pairs with nonzero $\alpha_i(j)$ are the pixels that are difficult (and hence critical) for the color prediction task. These pairs are the support vectors and these are the only training data points that appear in Equation 15.2.

The constraint optimization problem of Equation 15.5 can be rewritten as a quadratic program (QP) in terms of the dual parameters α_i for all colors $c(i)$. Minimizing this function yields sparse α_i, which can be used in the local color predictor function (Equation 15.2). While training SVMs is more expensive than training Parzen window estimators, SVMs yield often better prediction performance. More details on SVMs can be found in Reference [11]. Note that in the experiments, an SVM library publicly available at http://www.csie.ntu.edu.tw/~cjlin/libsvm/ was used. A Gaussian kernel was used in both Parzen windows and SVMs.

15.6 Global Coherency via Graph Cuts

For each pixel of a new grayscale image, it is possible now to estimate scores of all possible colors (within a large finite set of colors due to the discretization of the color space into bins) using the techniques outlined in Section 15.4 and in Section 15.5. Similarly, it is possible to estimate the probability of a color variation for each pixel. If the spatial coherency criterion given by the color variation function is incorporated into the color predictor, the choice of the best color for a pixel is affected by the probability distributions in the neighborhood. Since all pixels are connected through neighborhoods, it results in a global interaction across all pixels. Hence, in order to get spatially coherent colorization the solution should be computed globally, since any local search can yield suboptimal results. Indeed it may happen that, in some regions that are supposed to be homogeneous, a few different colors may seem to be the most probable ones at a local level, but that the winning color at the scale of the region is different, because in spite of its only second rank probability at the local level, it ensures a good probability everywhere in the whole region. On the opposite end of this spectrum are the cases where a color is selected in a whole homogeneous region because of its very high probability at a few points with high confidence. The problem is consequently not trivial, and the issue is to find a global solution. It is proposed here to use local predictors and color variation models in graph cuts in order to find spatially coherent colorization.

15.6.1 Energy Minimized by Graph Cuts

The graph cut or max flow algorithm is an optimization technique widely used in computer vision [15], [16] because it is fast, suitable for many image processing problems, and guarantees to find a good local optimum. In the multilabel case with α-expansion [17], it can be applied to all energies of the form $\sum_i V_i(x_i) + \sum_{i \sim j} D_{i,j}(x_i, x_j)$ where x_i are the unknown variables that take values in a finite set $\mathscr{L}$ of labels, V_i are any functions, and $D_{i,j}$ are any pairwise interaction terms with the restriction that each $D_{i,j}(\cdot, \cdot)$ should be a metric on $\mathscr{L}$. For the swap-move case, the constraints are weaker [18]:

$$D_{i,j}(\alpha, \alpha) + D_{i,j}(\beta, \beta) \leq D_{i,j}(\alpha, \beta) + D_{i,j}(\beta, \alpha) \tag{15.6}$$

for a pair of labels α and β.

The image colorization problem can be formulated as an optimization problem

$$\sum_{\mathbf{p}} V_{\mathbf{p}}(c(\mathbf{p})) + \rho \sum_{\mathbf{p} \sim \mathbf{q}} \frac{|c(\mathbf{p}) - c(\mathbf{q})|_{Lab}}{g_{\mathbf{p},\mathbf{q}}}, \tag{15.7}$$

where $V_{\mathbf{p}}(c(\mathbf{p}))$ is the cost of choosing color $c(\mathbf{p})$ locally for pixel $\mathbf{p}$ (whose neighboring texture is described by $\mathbf{v}(\mathbf{p})$) and where $g_{\mathbf{p},\mathbf{q}} = 2\left(g(\mathbf{v}(\mathbf{p}))^{-1} + g(\mathbf{v}(\mathbf{q}))^{-1}\right)^{-1}$ is the harmonic mean of the estimated color variation at pixels $\mathbf{p}$ and and $\mathbf{q}$. An eight-neighborhood is considered for the interaction term, and $\mathbf{p} \sim \mathbf{q}$ denotes that $\mathbf{p}$ and $\mathbf{q}$ are neighbors.

The interaction term between pixels penalizes color variation where it is not expected, according to the variations predicted in the previous paragraph. The hyper-parameter ρ enables a trade-off between local color scores and spatial coherence score. It can be estimated using cross validation.

Two methods that yield scores to local color prediction were described earlier in this chapter. These can be used to define $V_{\mathbf{p}}(c(\mathbf{p}))$. When using the Parzen window estimator, the local color cost $V_{\mathbf{p}}(c(\mathbf{p}))$ can be defined as follows:

$$V_{\mathbf{p}}(c(\mathbf{p})) = -\log\left(p(\mathbf{v}(\mathbf{p}))\right) p(c(\mathbf{p})|\mathbf{v}(\mathbf{p})). \tag{15.8}$$

Then, $V_{\mathbf{p}}$ penalizes colors which are not probable at the local level according to the probability distributions obtained in Section 15.4.1, with respect to the confidence in the predictions.

When using SVMs, there exist two options to define $V_{\mathbf{p}}(c(\mathbf{p}))$. Even though SVMs are not probabilistic, methods exist to convert SVM decision scores to probabilities [19]. Hence, the $p(c(\mathbf{p})|\mathbf{v}(\mathbf{p}))$ term in Equation 15.8 can be replaced with the probabilistic SVM scores and the graph cut algorithm can be used to find spatially coherent colorization. However, since V is not restricted to be a probabilistic function, $V_{\mathbf{p}}(c(\mathbf{p}))$ can be directly used as $-s(c(\mathbf{p})|\mathbf{v}(\mathbf{p}))$. This way does not require to get the additional $p(\mathbf{v}(\mathbf{p}))$ estimate in order to model the confidence of the local predictor; $s(c(\mathbf{p})|\mathbf{v}(\mathbf{p}))$ already captures the confidence via the margin concept and renders the additional (possibly noisy) estimation unnecessary.

The graph cut package [18] available at http://vision.middlebury.edu/MRF/code/ was used in the experiments. The solution for a 50×50 image and 73 possible colors is obtained by graph cuts in a fraction of second and is generally satisfactory. The computation time scales approximately quadratically with the size of the image, which is still fast, and the algorithm performs well even on significantly downscaled versions of the image so that a good initial colorization can still be given quickly for very large images as well. The computational costs compete with those of the fastest colorization techniques [20] while achieving more spatial coherency.

15.6.2　Refinement in the Continuous Color Space

The proposed method so far makes color predictions in the discrete space. In order to refine the predictors in the continuous color space, some smoothing should be performed. This can be achieved naturally for the Parzen window approach. Once the density estimation is achieved in the discrete space, probability distributions $p(c_i|\mathbf{v}(\mathbf{p}))$ estimated at each pixel $\mathbf{p}$ for each color bin i are interpolated to the whole space of colors with the technique described in Section 15.3. This renders $V_{\mathbf{p}}(c)$ well defined for continuous color values as well. The energy function given in Equation 15.7 can consequently be minimized in the continuous space of colors. In order to do so, the solution obtained by graph cuts is refined with a gradient descent. This refinement step generally does not introduce large changes such as changing the color of whole regions, but introduces more nuances.

15.7　Structured Support Vector Machines for Color Prediction

The methods described above improve existing image colorization approaches by learning color variations and the local color predictors separately and combining them via graphcut algorithm. It is now proposed to learn the local color predictor and spatial coherence jointly, as opposed to learning them independently as described in Sections 15.4 and 15.5. This can be accomplished by *structured prediction* methods. In particular, this section describes the application of structured support vector machines (SVMstruct) [21] for automatic image colorization. SVMstruct is a machine learning method designed to predict *structured objects*, such as images where color prediction for each pixel is influenced by the prediction of neighboring pixels as well as the local input descriptors.

15.7.1　Joint Feature Functions and Joint Estimator

The decision function of SVMstruct is computed with respect to feature functions that are defined over the joint input-output variables. The feature functions should capture the dependency of a color to the local characteristics of the gray scale image as well as the dependency of a color on the colors of neighboring pixels. The feature functions were already defined for local dependencies in Section 15.3.4; these features were denoted by $\mathbf{v}$. Furthermore, Section 15.6 outlined an effective way of capturing color dependencies across neighboring pixels,

$$\bar{f}(c,c') = |c - c'|_{Lab},\tag{15.9}$$

which are later scaled with respect to the color variations g. It is conceivable that this dependency is more pronounced for some color pairs than others. In order to allow the model to learn such distinctions in case of their existence, feature functions given in Equation 15.9 will be defined and a parameter $\bar{\lambda}_{cc'}$ will be learned for each color pair c, c'.

The decision function of SVMstruct can now be defined with respect to $\mathbf{v}$ and the $\bar{f}$ function as follows:

$$s(C|I) = \sum_{\mathbf{p}} \langle \lambda_{C(p)}, \mathbf{v}(\mathbf{p}) \rangle + \sum_{\mathbf{p}\sim\mathbf{q}} \bar{\lambda}_{C(\mathbf{p})C(\mathbf{q})} \bar{f}(C(\mathbf{p}), C(\mathbf{q})),\tag{15.10}$$

where C refers to a color assignment for image I and $C(p)$ denotes its restriction to pixel $\mathbf{p}$, hence the color assigned to the pixel. As in the case of standard SVMs, there is a kernel expansion of the joint predictor given by

$$s(C|I) = \sum_{\mathbf{p}} \sum_j \alpha_{C(\mathbf{p})}(j)k(\mathbf{w}_j, \mathbf{v}(\mathbf{p})) + \sum_{\mathbf{p} \sim \mathbf{q}} \bar{\lambda}_{C(\mathbf{p})C(\mathbf{q})} \bar{f}(C(\mathbf{p}), C(\mathbf{q})). \quad (15.11)$$

The following discusses this estimator with respect to the previously considered functions. Compared to the SVM-based local prediction function given in Equation 15.2, this estimator is defined over a full grayscale image I and its possible colorings C as opposed to the SVM case which is defined over an individual grayscale pixel $\mathbf{p}$ and its colorings c. Furthermore, the spatial coherence criteria (the second term in Equation 15.11) are incorporated directly rather than by two-step approaches used in the Parzen window and SVM-based methods. It can be also observed that the proposed joint predictor, Equation 15.11, is simply a variation of the energy function used in the graph-cut algorithm given in Equation 15.7, where different parameters for spatial coherence can now be estimated by a joint learning process as opposed to learning color variation and finding λ in the energy via cross-validation. With the additional symmetry constraint $\bar{\lambda}_{cc'} = \bar{\lambda}_{c'c}$ for each color pair c, c', the energy function can be optimized using the graph cuts swap move algorithm. Hence, SVMstruct provides a more unified approach for learning parameters and removes the necessity of the hyper-parameter ρ.

15.7.2 Training Structured SVMs

Given the joint estimator in Equation 15.11, the learning procedure is now defined to estimate optimal parameters α_i for all colors c_i and $\bar{\lambda}_{cc'}$ for all color pairs c, c'. The training procedure is similar to SVMs where the norm of the parameters is minimized with respect to margin constraints. Note that the margin constraints are now defined on colored images (I_j, C_j) for all images j in the training data,

$$s(C_j|I_j) - s(C|I_j) \geq 1 - \xi_j, \quad \forall j, \forall C \neq C_j.$$

As in SVMs, the goal is to separate the observed coloring C_j of an image I_j from all possible colorings C of I. This formulation can be extended by quantifying the quality of a particular coloring with respect to the observed coloring of an image. If a coloring C is similar to the observed coloring C_j for the training image I_j, the model should relax the margin constraints for C and j. In order to employ this idea in the proposed joint optimization framework, a *cost function* $\Delta(C, C')$, that measures the distance between C and C' and imposes margin constraints with respect to this cost function, is defined as:

$$s(C_j|I_j) - s(C|I_j) \geq \Delta(C_j, C) - \xi_j, \forall j, \forall C \neq C_j.$$

The incorporation of Δ renders the constraints of colorings similar to C_j essentially ineffective and leads to more reliable predictions. This cost function can be defined as the average perceived difference between two colors across all pixels:

$$\Delta(C, \bar{C}) = \sum_{\mathbf{p}} \frac{||C(\mathbf{p}) - \bar{C}(\mathbf{p})||}{\max_{c,c'} ||c - c'||}. \quad (15.12)$$

The normalization term ensures that the local color differences are between 0 and 1.

Coloring a painting given another painting by the same painter: (a) training image, (b) test image, (c) image colored using Parzen windows — the border is not colored because of the window size needed for SURF descriptors, (d) color variation predicted — white stands for homogeneity and black for color edge, (e) most probable color at the local level, and (f) 2D color chosen by graph cuts.

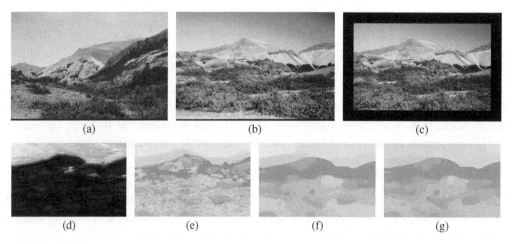

Landscape example with Parzen windows: (a) training image, (b) test image, (c) output image, (d) predicted color variation, (e) most probable color locally, (f) 2D color chosen by graph cuts, (g) colors obtained after refinement step.

There are efficient training algorithms for this optimization [21]. In this chapter, the experiments were done using the SVMstruct implementation available at http://svmlight.joachims.org/ with a Gaussian kernel.

15.8 Experiments

This section presents experimental results achieved using the proposed automatic colorization methods on different datasets.

15.8.1 Colorization Based on One Example

Figure 15.3 shows a painting colored using the Parzen window method given another painting by the same painter. The two paintings are significantly different and textures are relatively dissimilar. The prediction of color variation performs well and helps significantly to determine the boundaries of homogeneous color regions. The multimodality framework proves extremely useful in areas such as Mona Lisa's forehead or neck where the texture of skin can be easily mistaken for the texture of sky at the local level. Without the proposed global optimization framework, several entire skin regions would be colored in blue, disregarding the fact that skin color is the second probable colorization for these areas. This makes sense at the global level since they are surrounded by skin-colored areas, with low probability of edges. Note that the input of previous texture-based approaches is very similar to the "most probable color" prediction, whereas the proposed framework considers the probabilities of all possible colors at all pixels. This means that given a certain quality of texture descriptors, the proposed framework handles much more information.

Figure 15.4 shows the outcome of similar experiments with photographs of landscapes. The effect of the refinement step can be observed in the sky where nuances of blue vary more smoothly. Both SVMs and SVMstruct produce only slightly different results, hence the colorization results are not presented.

The proposed Parzen window method is now compared with the method of Reference [7], on their own example. Figure 15.5 shows the results; the task is easier and therefore results are similar. The method of Reference [7] colored a several pixel wide band of grass around the zebra's legs and abdomen as if it were part of the zebra. The boundaries of color regions produced using the proposed method fit better to the zebra contour, as seen in Figure 15.6. However, grass areas near the zebra are colored according to the grass observed at similar locations around the zebra in the training image, thus creating color halos which are visually not completely satisfactory. It is expected that this bias will disappear with larger training sets since the color of the background becomes independent of zebra's presence.

15.8.2 Colorization Based on a Small Set of Different Images

In the following, a very difficult task is considered. Namely, an image to be colored is from a Charlie Chaplin movie, with many different objects and textures, such as a brick wall, a door, a dog, a head, hands, and a loose suit. Because of the number of objects and because of their particular arrangement, it is unlikely to find a single color image with a

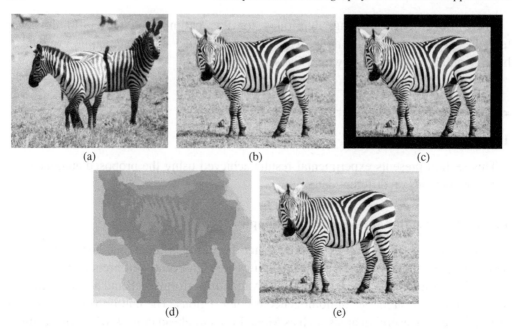

FIGURE 15.5

Comparison with the method of Reference [7]: (a) color *zebra* example, (b) test image, (c) proposed method output, (d) 2D colors predicted using the proposed method, and (e) output using Reference [7] with the assumption that this is a binary classification problem. © Eurographics Association 2005

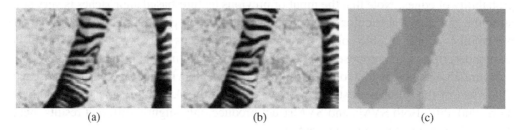

FIGURE 15.6

Zoomed portion of images in Figure 15.5: (a) method of Reference [7], (b) proposed method, and (c) colors predicted using the proposed method.

similar scene that can be used as a training image. Therefore, a small set of three different images is considered; each image from this set shares a partial similarity with the Charlie Chaplin image. The underlying difficulty is that each training image also contains parts which should not be reused in this target image. Figure 15.7 shows the results obtained using Parzen windows. The result is promising considering the training set. In spite of the difficulty of the task, the prediction of color edges and of homogeneous regions remains significant. The brick wall, the door, the head, and the hands are globally well colored. The large trousers are not in the training set; the mistakes in the colors of Charlie Chaplin's dog are probably due to the blue reflections on the dog in the training image and to the light brown of its head. Dealing with larger training datasets increases the computation time only logarithmically during the kD-tree search.

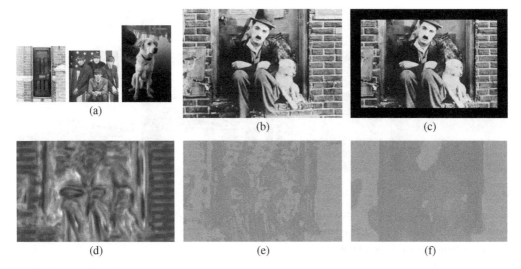

FIGURE 15.7 (See color insert.)

Image colorization using Parzen windows: (a) three training images, (b) test image, (c) colored image, (d) prediction of color variations, (e) most probable colors at the local level, and (f) final colors.

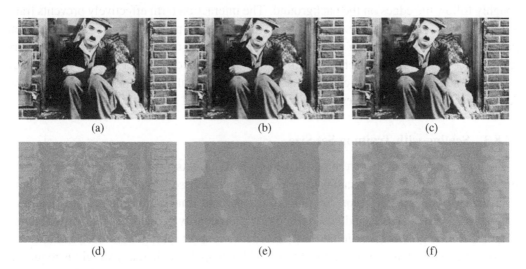

FIGURE 15.8 (See color insert.)

SVM-driven colorization of Charlie Chaplin frame using the training set of Figure 15.7: (a,d) SVM, (b,e) SVM with spatial regularization — Equation 15.7, and (c,f) SVMstruct.

Figure 15.8 shows the results for SVM-based prediction. In the case of SVM colorization, the contours of Charlie Chaplin and the dog are well recognizable in the color-only image. Face and hands do not contain nonskin colors, but the abrupt transitions from very pale to warm tones are visually not satisfying. The dog contains a large fraction of skin colors; this could be attributed to a high textural similarity between human skin and dog body regions. The large red patch on the door frame is probably caused by a high local similarity with the background flag in the first training image. Application of the spatial coherency criterion from Equation 15.7 yields a homogeneous coloring, with skin areas being well

FIGURE 15.9

Eight of 20 training images from the Caltech Pasadena Houses 2000 collection available at http://www.vision.caltech.edu/archive.html. The whole set can be seen at http://www-sop.inria.fr/members/ Guillaume.Charpiat/color/houses.html.

represented. The coloring of the dog does not contain the mistakes from Figure 15.7, and appears consistent. The door is colored with regions of multiple, but similar colors, which roughly follow the edges on the background. The interaction term effectively prevents transitions between colors that are too different, when the local probabilities are similar. When colorizing with the interaction weights learned using SVMstruct, the overall result appears less balanced; the dog has patches of skin color and the face consists of two not very similar color regions. Given the difficulty of the task and the small amount of training data, the result is presentable, as all interaction weights were learned automatically.

15.8.3 Scaling with Training Set Size

Larger-scale experiments were performed on two datasets. The first dataset is based on the first 20 images of the Pasadena houses Caltech database (Figure 15.9). This training data was used to predict colors for the 21st and 22nd images. Figure 15.10 illustrates colorization using Parzen windows. The colorization quality is relatively similar to the one obtained for the previous small datasets. In order to remove texture noise (for example the unexpected red dots in trees in Figure 15.10), a higher spatial coherency weight is required, which explains the lack of nuances. The use of more discriminative texture features can improve texture identification, reduce texture noise and consequently allow more nuances.

Figure 15.11 shows images colored using SVMs and SVMstruct. In order to reduce the training time, every 3rd pixel on a regular grid was used for all training images.

The shapes of the windows, doors, and stairs can be identified from the color-only image, which is not the case for the Parzen window method, demonstrating the ability of SVM to discriminate between fine details on local level. The colorization of the lawn is irregular, the local predictor tends to switch between green and gray, which would correspond to a decision between grass or asphalt. This effect does not occur in the Parzen window classifier and can be attributed to the subsampling of the training data, leading to a comparatively low number of training points for lawn areas. Application of spatial regularization indeed leads to the effect of the lawn becoming completely gray. When the spatial coherency criterion from Equation 15.7 is applied, the coloring becomes homogeneous and is comparable

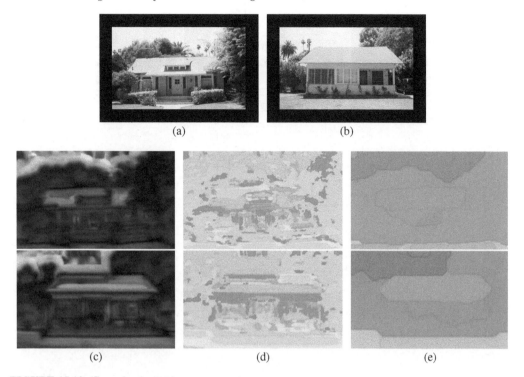

FIGURE 15.10 (See color insert.)

Colorization of the 21st and 22nd images from the Pasadena houses Caltech dataset: (a) 21st image colored, (b) 22nd image colored, (c) predicted edges, (d) most probable color at the pixel level, and (e) colors chosen.

to the results from the Parzen window method, but with a higher number of different color regions and thus a more realistic appearance. SVMstruct-based colorization preserves a much higher number of different color regions while removing most of the inconsistent color patches and keeping finer details. It is able to realistically color the small bushes in front of the second house. The spatial coherency weights for transitions between different colors were learned from training data without the need for cross validation for the adjustment of the hyper-parameter λ. These improvements come at the expense of longer training time, which scales quadratically with the training data.

Finally, an ambitious experiment was performed to evaluate how the proposed approach deals with quantities of different textures on similar objects; that is, how it scales with the number of textures observed. A portrait database of 53 paintings with very different styles (Figure 15.12) was built. Five other portraits were colored using Parzen windows (Figure 15.13) and the same parameters. Although given that red color is indeed the dominant color in an important proportion of the training set, the colorizations sometimes appears rather reddish. The surprising part is the good quality of the prediction of the colored edges, which yields a segmentation of the test images into homogeneous color regions. The boundaries of skin areas in particular are very well estimated, even in images which are very heavily textured. The good estimation of color edges helped the colorization process to find suitable colors inside the supposedly-homogeneous areas, despite locally noisy color predictions. Note that neither SVM nor SVMstruct were evaluated in this experiment due to their expensive training.

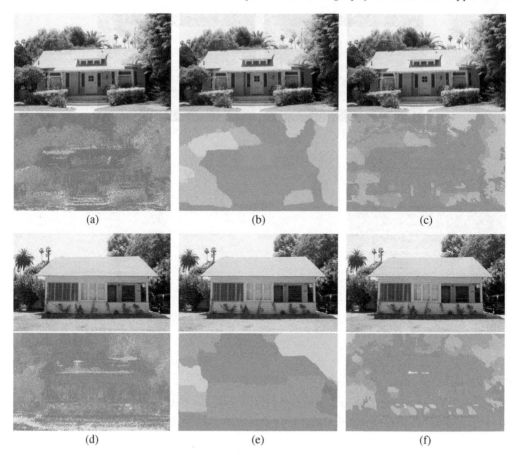

FIGURE 15.11 (See color insert.)

SVM-driven colorization of the 21st and 22nd images from the Pasadena houses Caltech dataset; results and colors chosen are displayed: (a-c) 21st image, (d-f) 22nd image; (a,d) SVM, (b,e) SVM with spatial regularization, and (c,f) SVMstruct.

15.9 Conclusion

This chapter presented three machine learning methods for automatic image colorization. These methods do not require any intervention by the user other than the choice of relatively similar training data. The color prediction task was formally stated as an optimization problem with respect to an energy function. Since the proposed approaches retain the multimodality until the prediction step, they extract information from training data effectively using different machine learning methods. The fact that the problem is solved directly at the global level with the help of graph cuts makes the proposed framework more robust to noise and local prediction errors. It also allows resolving large scale ambiguities as opposed to previous approaches. The multimodality framework is not specific to image colorization and could be used in any prediction task on images. For example, Reference [22] outlines a similar approach for medical imaging to predict computed tomography scans for patients whose magnetic resonance scans are known.

FIGURE 15.12

Some of 53 portraits used as a training set. Styles of paintings vary significantly, with different kinds of textures and different ways of representing edges. The full training set is available at http://www-sop.inria.fr/members/Guillaume.Charpiat/color/.

The proposed framework exploits features derived from various sources of information. It provides a principal way of learning local color predictors along with spatial coherence criteria as opposed to the previous methods which chose the spatial coherence criteria manually. Experimental results on small and large scale experiments demonstrate the validity of the proposed approach which produces significant improvements over the methods in References [5] and [6], in terms of the spatial coherency formulation and the large number of possible colors. It requires less or similar user-intervention than the method in Reference [7], and can handle cases which are more ambiguous or have more texture noise.

Currently, the proposed automatic colorization framework does not employ decisive information which is commonly used in user-interactive approaches. However, the proposed framework can easily incorporate user-provided information such as the color c at pixel $\mathbf{p}$ in order to modify a colorization that has been obtained automatically. This can be achieved by *clamping* the local prediction to the color provided by the user with high confidence. For example, in the Parzen window method, $p(c|\mathbf{v}(\mathbf{p})) = 1$ and the confidence $p(\mathbf{v}(\mathbf{p}))$ is set to a very large value. Similar clamping assignments are possible for SVM-based approaches. Consequently, the proposed optimization framework is usable for further interactive colorization. A recolorization with user-provided color landmarks does not require the re-estimation of color probabilities, and therefore requires only a fraction of second. This interactive setting will be addressed in future work.

Acknowledgment

The authors would like to thank Jason Farquhar, Peter Gehler, Matthew Blaschko, and Christoph Lampert for very fruitful discussions.

FIGURE 15.13

Portrait colorization: (top) result, (middle) colors chosen without grayscale intensity, and (bottom) predicted color edges. Predicted color variations are particularly meaningful and correspond precisely to the boundaries of the principal regions. Thus, the color edge estimator can be seen as a segmentation tool. The background colors cannot be expected to be correct since the database focuses on faces. The same parameters were used for all portraits.

Figures 15.1 to 15.5, and 15.7 are reprinted with permission from Reference [8]. Figure 15.5 is reprinted from Reference [7], with the permission of Eurographics Association. Figure 15.9 contains photos from the Caltech Pasadena Houses 2000 collection (http://www.vision.caltech.edu/archive.html), reproduced with permission.

References

[1] A. Levin, D. Lischinski, and Y. Weiss, "Colorization using optimization," *ACM Transactions on Graphics*, vol. 23, no. 3, pp. 689–694, August 2004.

[2] L. Yatziv and G. Sapiro, "Fast image and video colorization using chrominance blending," *IEEE Transactions on Image Processing*, vol. 15, no. 5, pp. 1120–1129, May 2006.

[3] T. Horiuchi, "Colorization algorithm using probabilistic relaxation," *Image Vision Computing*, vol. 22, no. 3, pp. 197–202, March 2004.

[4] T. Takahama, T. Horiuchi, and H. Kotera, "Improvement on colorization accuracy by partitioning algorithm in CIELAB color space," *Lecture Notes in Computer Science*, vol. 3332, pp. 794–801, November 2004.

[5] T. Welsh, M. Ashikhmin, and K. Mueller, "Transferring color to greyscale images," *ACM Transactions on Graphics*, vol. 21, no. 3, pp. 277–280, July 2002.

[6] A. Hertzmann, C.E. Jacobs, N. Oliver, B. Curless, and D.H. Salesin, "Image analogies," in *Proceedings of the 28th Annual Conference on Computer Graphics and Interactive Techniques*, Los Angeles, CA, USA, August 2001, pp. 327–340.

[7] R. Irony, D. Cohen-Or, and D. Lischinski, "Colorization by example," in *Proceedings of Eurographics Symposium on Rendering*, Konstanz, Germany, June 2005, pp. 201–210.

[8] G. Charpiat, M. Hofmann and B. Schölkopf, "Automatic image colorization via multimodal predictions," *Lecture Notes on Computer Science*, vol. 5304, pp. 126–139, October 2008.

[9] F. Pitie, A. Kokaram, and R. Dahyot, *Single-Sensor Imaging: Methods and Applications for Digital Cameras*, ch. Enhancement of digital photographs using color transfer techniques, R. Lukac (ed.), Boca Raton, FL: CRC Press / Taylor & Francis, September 2008, pp. 295–321.

[10] R.W.G. Hunt, *The Reproduction of Colour*, Chichester, England: John Wiley, November 2004.

[11] B. Schölkopf and A.J. Smola, *Learning with Kernels: Support Vector Machines, Regularization, Optimization, and Beyond*. Cambridge, MA: MIT Press, December 2001.

[12] H. Bay, A. Ess, T. Tuytelaars, and L. Van Gool, "SURF: Speeded up robust features," *Computer Vision and Image Understanding* vol. 110, no. 3, pp. 346–359, June 2008.

[13] V. Vapnik, *Statistical Learning Theory*, New York, NY, USA: Wiley-Interscience, September 1998.

[14] J. Weston and C. Watkins, "Support vector machines for multi-class pattern recognition," in *Proceedings of the European Symposium on Artificial Neural Networks*, Bruges, Belgium, April 1999.

[15] Y. Boykov and V. Kolmogorov, "An experimental comparison of min-cut/max-flow algorithms for energy minimization in vision," in *Proceedings of the Third International Workshop on Energy Minimization Methods in Computer Vision and Pattern Recognition*, London, UK, September 2001, pp. 359–374.

[16] V. Kolmogorov and R. Zabih, "What energy functions can be minimized via graph cuts?," in *Proceedings of the European Conference on Computer Vision*, Copenhagen, Denmark, May 2002, pp. 65–81.

[17] Y. Boykov, O. Veksler, and R. Zabih, "Fast approximate energy minimization via graph cuts," in *Proceedings of the International Conference on Computer Vision*, Kerkyra, Greece, September 1999, pp. 377–384.

[18] R. Szeliski, R. Zabih, D. Scharstein, O. Veksler, V. Kolmogorov, A. Agarwala, M.F. Tappen, and C. Rother, "A comparative study of energy minimization methods for markov random fields," *Lecture Notes in Computer Science*, vol. 3952, pp. 16–29, May 2006.

[19] J. Platt, *Advances in Large Margin Classifiers*, ch. Probabilistic outputs for support vector machines and comparisons to regularized likelihood methods, P.B. Alexander and J. Smola (eds.), Cambridge, MA: MIT Press, October 2000, pp. 61–74.

[20] G. Blasi and D.R. Recupero, "Fast Colorization of Gray Images," in *Proceedings of Eurographics Italian Chapter*, Milano, Italy, September 2003, pp. 1120–1129.

[21] I. Tsochantaridis, T. Hofmann, T. Joachims, and Y. Altun, "Support vector machine learning for interdependent and structured output spaces," in *Proceedings of International Conference on Machine Learning*, Banf, AB, Canada, July 2004, pp. 823–830.

[22] M. Hofmann, F. Steinke, V. Scheel, G. Charpiat, J. Farquhar, P. Aschoff, M. Brady, B. Schölkopf, and B.J. Pichler, "MR-based attenuation correction for PET/MR: A novel approach combining pattern recognition and atlas registration," *Journal of Nuclear Medicine*, vol. 49, no. 11, pp. 1875–1883, November 2008.

16

Machine Learning for Digital Face Beautification

Gideon Dror

16.1 Introduction

Beauty, particularly of the human face, has fascinated human beings from the very dawn of mankind, inspiring countless artists, poets, and philosophers. Numerous psychological studies find high cross-cultural agreement in facial attractiveness ratings among raters from different ethnicities, socioeconomic classes, ages, and gender [1], [2], [3], [4], indicating that facial beauty is a universal notion, transcending the boundaries between different cultures. These studies suggest that the perception of facial attractiveness is *data-driven*; the properties of a particular set of facial features are the same irrespective of the perceiver.

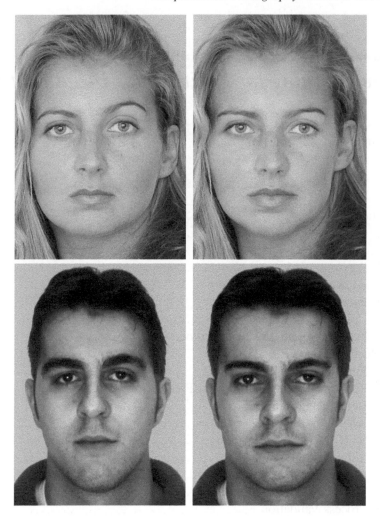

FIGURE 16.1

Digital face beautification: (left) input facial images, and (right) the modified images generated using the proposed method. The changes are subtle, yet their impact is significant. Notice that different modifications are applied to men and women, according to preferences learned from human raters. © 2008 ACM

The universality of the notion of facial attractiveness along with the ability to reliably and automatically rate the facial beauty from a facial image [5], [6] has motivated this work. Specifically, this chapter presents a novel tool capable of automatically enhancing the attractiveness of a face in a given frontal portrait. It aims at introducing only subtle modifications to the original image, such that the resulting beautified face maintains a strong, unmistakable similarity to the original, as demonstrated in Figure 16.1 by the pairs of female and male faces. This is a highly nontrivial task, since the relationship between the ensemble of facial features and the degree of facial attractiveness is anything but simple.

Professional photographers have been retouching and deblemishing their subjects ever since the invention of photography. It may be safely assumed that any model present on a magazine cover today has been digitally manipulated by a skilled, talented retouching

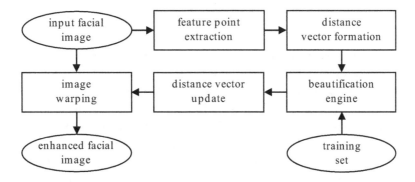

FIGURE 16.2

The proposed digital face beautification process.

artist. Since the human face is arguably the most frequently photographed object on earth, a tool such as the one described in this chapter would be a useful and welcome addition to the ever-growing arsenal of image enhancement and retouching tools available in today's digital image editing packages. The potential of such a tool for motion picture special effects, advertising, and dating services, is also quite obvious.

Given a face, a variety of predetermined facial locations are identified to compute a set of distances between them. These distances define a point in a high-dimensional face space. This space is searched for a nearby point that corresponds to a more attractive face. The key component in this search is an automatic facial beauty rating machine, that is, two support vector regressors trained separately on a database of female and male faces with accompanying facial attractiveness ratings collected from a group of human raters. Once such a point is found, the corresponding facial distances are embedded in the plane and serve as a target to define a two-dimensional (2D) warp field which maps the original facial features to their beautified locations. The process is schematically depicted in Figure 16.2.

Experimental results indicate that the proposed method is capable of effectively improving the facial attractiveness of most images of female faces used in experiments. In particular, its effectiveness was experimentally validated by a group of test subjects who consistently found the beautified faces to be more attractive than the original ones.

The proposed beauty regressor was trained using frontal portraits of young Caucasian males and females with neutral expression and roughly uniform lighting. Thus, it currently can only be expected to perform well on facial images with similar characteristics. However, it may be directly extended to handle additional ethnic groups, simply by using it with beauty regressors trained on suitable collections of portraits.

16.1.1 Overview

Section 16.2 describes psychological findings related to perception of human facial beauty, that accumulated in the past three decades. These notions are important for understanding the reasoning behind the proposed methods. Section 16.3 discusses how to construct a model of facial beauty using supervised learning methods. The models are based on sets of face images of females and males, rated by human raters. This section also describes the features used to represent facial geometry, that are based on psychological research as well as detailed empirical tests. Section 16.4 presents two alternative methods of face beautification. The first uses direct optimization of a beauty function and the other is based on heuristics motivated by well-known psychological effects of beauty perception.

The next two sections describe techniques and methods related to the process of face beautification. Namely, Section 16.5 presents the methods used to locate facial features and identify 84 canonical points on a face, including points at specific locations on the mouth, nose, eyes, brows, and the contour of the face, which provide a succinct description of the geometry of a face. These points are the anchors for warping a faces to a beautified version thereof. Section 16.6 describes distance embedding required to carry out the warping, and the warping process, modified for the specific task of warping human faces.

Section 16.7 presents examples of beautified faces of both females and males. An empirical validation based on a large set of faces is described, showing that face images produced by the process are indeed significantly more pleasing than the original images. This section concludes by pointing out some applications of face beautification. Finally, conclusions are offered in Section 16.8 which also discusses some ideas for extending the proposed method to handle nonfrontal portraits and nonneutral expressions.

16.2 Background

Philosophers, artists and scientists have been trying to capture the nature of beauty since the early days of philosophy. Although in modern days a common laymans notion is that judgments of beauty are a matter of subjective opinion alone, recent findings suggest that people share a common taste for facial attractiveness and that their preferences may be an innate part of our primary constitution. Indeed, several rating studies have shown high cross-cultural agreement in attractiveness rating of faces of different ethnicities [1], [2], [3].

Other experimental studies demonstrated consistent relations between attractiveness and various facial features, which were categorized as neonate (features such as small nose and high forehead), mature (e.g., prominent cheekbones) and expressive (e.g., arched eyebrows). They concluded that beauty is not an inexplicable quality which lies only in the eye of the beholder [7], [8].

Further experiments have found that infants ranging from two to six months of age prefer to look longer at faces rated as attractive by adults than at faces rated as unattractive [9], [10]. They also found that twelve month old infants prefer to play with a stranger with an attractive face compared with a stranger with an unattractive face.

Such findings give rise to the quest for common factors which determine human facial attractiveness. Accordingly, various hypotheses, from cognitive, evolutional, and social perspectives, have been put forward to describe and interpret the common preferences for facial beauty. Inspired by the photographic method of composing faces presented in Reference [11], Reference [12] proposed to create averaged faces (Figure 16.3) by morphing multiple images together. Human judges found these averaged faces to be attractive and rated them with attractiveness ratings higher than the mean rating of the component faces composing them, proposing that averageness is the answer for facial attractiveness [12],

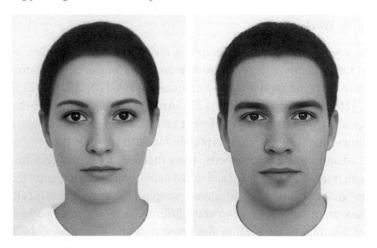

FIGURE 16.3 (See color insert.)

Female and male face-composites, each averaging sixteen faces. It has been empirically shown that average faces tend to be considerably more attractive than the constituent faces.

[13]. Investigating symmetry and averageness of faces, it was found that symmetry is more important than averageness in facial attractiveness [14]. Other studies have agreed that average faces are attractive but claim that faces with certain extreme features, such as extreme sexually dimorphic traits, may be more attractive than average faces [4].

Many contributors refer to the evolutionary origins of attractiveness preferences [15]. According to this view, facial traits signal mate quality and imply chances for reproductive success and parasite resistance. Some evolutionary theorists suggest that preferred features might not signal mate quality but that the "good taste" by itself is an evolutionary adaptation (individuals with a preference for attractiveness will have attractive offspring that will be favored as mates) [15]. Another mechanism explains attractiveness preferences through a cognitive theory — a preference for attractive faces might be induced as a by-product of general perception or recognition mechanisms [16]. Attractive faces might be pleasant to look at since they are closer to the cognitive representation of the face category in the mind. It was further demonstrated that not just average faces are attractive but also birds, fish, and automobiles become more attractive after being averaged with computer manipulation [17]. Such findings led researchers to propose that as perceivers can process an object more fluently, aesthetic response becomes more positive [18]. A third view suggests that facial attractiveness originates in a social mechanism, where preferences may be dependent on the learning history of the individual and even on his social goals [16]. Other studies have used computational methods to analyze facial attractiveness. In several cases faces were averaged using morphing tools [3]. Laser scans of faces were put into complete correspondence with the average face in order to examine the relationship between facial attractiveness, age, and averageness [19]. Machine learning methods have been used recently to investigate whether a machine can predict attractiveness ratings by learning a mapping from facial images to their attractiveness scores [5], [6]. The predictor presented by the latter achieved a correlation of 0.72 with average human ratings, demonstrating that facial beauty can be learned by a machine with human-level accuracy.

16.2.1 Previous Work

Much of the research in computer graphics and computer vision has concentrated on techniques and tools specifically geared to human faces. In particular, there is an extensive body of literature on facial modeling and animation [19], [20], [21], [22], [23], [24], face detection [25], [26], and face recognition [27]. Among these previous works, the most relevant to the proposed approach are the different methods for 2D facial image morphing [28], [29] and the three-dimensional (3D) morphable facial models [30].

Similarly to image morphing methods, the proposed approach also makes use of 2D image warping to transform the input face into a beautified one. However, the goals of these two approaches are very different. In image morphing, the goal is typically to produce a continuous transformation between two very different faces (or other pairs of objects). The challenge there lies mainly in finding the corresponding features of the two faces, and defining an appropriate warp. In the proposed approach, the challenge lies in finding the target shape into which the source image is to be warped, such that the changes are subtle yet result in a noticeable enhancement of facial attractiveness.

Perceptual psychologists also often use image compositing, morphing, and warping to gain a better understanding of how humans perceive various facial attributes. For example, warping towards, and away from average faces has been used to study facial attractiveness and aging [31], [32]. Again, in this case the target shape for the morph, or the direction of the warp, is predefined.

Reference [30] presents a 3D morphable face model suitable to manipulate a number of facial attributes such as masculinity or fullness, or even to generate new facial expressions. This morphable model is formed by a linear combination of a set of prototype faces. Its underlying working assumption is that the markedness of the attribute of interest is a linear function. Consequently, increasing or decreasing the markedness is achieved by moving along a single optimal direction in the space of faces. At first glance, it may appear that the beautification task could be carried out using such a method (indeed, such an attempt was made in Reference [33]). However, as demonstrated in this chapter, facial attractiveness is a highly nonlinear attribute.

The proposed approach does not require fitting a 3D model to a facial image in order to beautify it; rather, it operates directly on the 2D image data. The approach relies on the availability of experimental data correlating facial attractiveness with 2D distances in a facial image, while no equivalent data exists yet for distances between landmarks on a 3D facial mesh. The beautification process could, however, assist in obtaining a beautified 3D model, by applying the proposed beautification technique to an input image in a preprocess, followed by fitting a 3D morphable model to the beautified result.

Reference [34] focuses on performance-driven animation. This approach transfers expressions, visemes, and head motions from a recorded performance of one individual to animate another, and is not concerned with synthesizing new faces or with modifying the appearance of a face. A genetic algorithm-driven approach, guided by interactive user selections to evolve a "most beautiful" female face, is presented in Reference [35]. However, although there are several insights about the constituents of facial beauty (e.g., golden ratios and averageness hypothesis), the first successful attempt at an automatic, software guided beautification of faces was carried out in Reference [36].

16.3 Machine Learning of Facial Beauty

16.3.1 Face Datasets

As shown in Figure 16.2, the beautification engine is based on a beauty predictor, which is trained on a set of rated human faces. Since the characteristics of facial beauty differ in males and females, a separate dataset for each gender should be used.

The female dataset was composed of 91 facial images of American females. All 91 samples were frontal color photographs of young Caucasian females with a neutral expression. The subjects' portraits had no accessories or other distracting items such as jewelry, and minimal makeup was used.

The male dataset consisted of 32 facial images [37]. All images were frontal color photographs of young Caucasian males with a neutral expression. All samples were of similar age, skin color and gender. Subjects' facial hair was restricted to allow precise determination of facial features such as lips, lower jaw, and nostrils.

16.3.2 Collection of Ratings

The facial images in the datasets were rated for attractiveness by 28 human raters (15 males, 13 females) on a 7-point Likert scale (1 = very unattractive, 7 = very attractive) [6]. Ratings were collected with a specifically designed html interface. Each rater was asked to view the entire set before rating in order to acquire a notion of attractiveness scale. There was no time limit for judging the attractiveness of each sample and raters could go back and adjust the ratings of previously rated samples. The images were presented to each rater in a random order and each image was presented on a separate page. The final attractiveness rating of each sample (male or female) was its mean rating across all raters.

The following tests were made to validate that the number of ratings collected adequately represented the collective attractiveness rating:

- For the females dataset, an independent average rating from Reference [38] was obtained. The Pearson correlation between the two sets of ratings was 0.92.

- Increasing the number of ratings to 60, with approximately the same ratio of males to female raters, introduced no significant modification of the average ratings.

- The raters were divided into two disjoint groups of equal size. The mean rating for each facial image in each group was calculated to determine the Pearson correlation between the mean ratings of the two groups. This process was repeated 1000 times. This procedure was taken separately for the male and female datasets. The mean correlation between two groups for both datasets was higher than was 0.9, with a standard deviation of $\sigma < 0.03$. It should be noted that the split-half correlations reported were high in all 1000 trials (as evident from the low standard deviation) and not only over the average. Experimental results show that there is a greater agreement on human ratings of female faces while male face preferences are more variable, in accordance with Reference [39]. This correlation corresponds well to the known level of consistency among groups of raters reported in the literature [1].

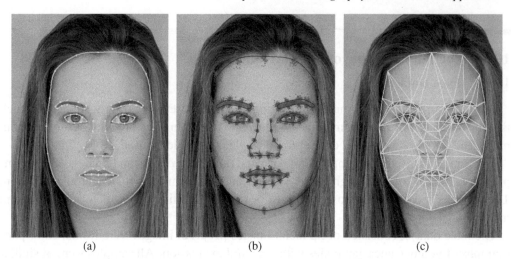

(a)	(b)	(c)

FIGURE 16.4 (See color insert.)

An example of the eight facial features (two eyebrows, two eyes, the inner and outer boundaries of the lips, the nose, and the boundary of the face), composed of a total of 84 feature points, used in the proposed algorithm: (a) output feature points from the active shape model search, (b) scatter of the aligned 84 landmark points of 92 sample training data and their average, and (c) 234 distances between these points. © 2008 ACM

Hence, the mean ratings collected are stable indicators of attractiveness that can be used for the learning task. The facial set contained faces in all ranges of attractiveness. Final attractiveness ratings range from 1.42 to 5.75, with the mean rating equal to 3.33 and $\sigma = 0.94$.

16.3.3 Data Representation and Preprocessing

Experimentations with various ways of representing a facial image [5], [6] have systematically shown that features based on measured proportions, distances, and angles of faces are most effective in capturing the notion of facial attractiveness. Other representation of faces, such as the eigenface decomposition [40], were found significantly inferior. The representation adopted here is specifically designed to capture the facial geometry.

To extract facial features, an automatic engine was developed to identify 84 feature points located on the outlines of eight different facial features: two eyebrows, two eyes, the inner and outer boundaries of the lips, the nose, and the boundary of the face; as shown in Figure 16.4a. Feature points were selected in accordance with facial attributes shown to be strongly related with facial attractiveness for both males and females: size of eyes and lips, distance between eyes, height of forehead, width of lower jaw, distance between the lower lip and the chin etc.

Several regions are automatically suggested for sampling nongeometric characteristics of faces that are known to be related with facial beauty: mean hair color, mean skin color, and skin texture. These features are not manipulated by the beautification process, but are used to adjust average beauty scores (see Section 16.4). The feature extraction process was basically automatic but some coordinates needed to be manually adjusted in some of the images.

The mean (normalized) positions of the extracted feature points, see Figure 16.4b, are used to construct a Delaunay triangulation. The triangulation consists of 234 edges, and the lengths of these edges in each face form its 234-dimensional distance vector. Figure 16.4c is an example for face triangulation and the associated distance vector. The distances are normalized by the square root of the face area to make them invariant of scale. The proposed method works with distances between feature points, rather than with their spatial coordinates, as such distances are more directly correlated with the perceived attractiveness of a face. Furthermore, working with a facial mesh, rather than some other planar graph, imposes some rigidity on the beautification process, preventing it from generating distances which may possess a high score but do not correspond to a valid face.

16.4 Face Beautification

16.4.1 Support Vector Regression

Support vector regression (SVR) [41] is an induction algorithm for fitting multidimensional data (see Appendix). Being based on the ideas of structural risk minimization, it has excellent generalization performance. By using various kernels, SVR can fit highly nonlinear functions. An SVR model is constructed by training it with a sparse set of samples $(\vec{x}_i, y_i)$, where $\vec{x}_i \in R^d$ and $y_i \in R$.

In the proposed method, $\vec{x}_i$ is made of features that represent the geometry of the ith face and y_i are the corresponding averaged beauty scores. Specifically, the mean (normalized) positions of the extracted feature points (Figure 16.4b) were used to construct a Delaunay triangulation. The triangulation consists of 234 edges, and the lengths of these edges in each face form its 234-dimensional *distance vector*, $\vec{x}$ (Figure 16.4c). The distances are normalized by the square root of the face area to make them invariant of scale. The resulting regressor defines a smooth function $f_b : R^d \to R$ used to estimate the beauty scores of distance vectors corresponding to faces outside the training set.

Following extensive preliminary experimentation, a radial basis function kernel was chosen to model the nonlinear behavior expected for such a problem. Further model selection was performed by a grid search using ten-fold cross-validation over the width of the kernel σ, the slack parameter C, and the the tube width parameter ε (see Appendix for more details). A soft margin SVR implemented in SVM^{light} [42] is used throughout this work.

Notice that in contrast to the regressors described in References [5] and [6] which attempted to use all relevant features, the proposed regressor makes no use of nongeometric features, such as hair color, and skin texture. The proposed beautification engine is thus designed to modify only the geometry of the face, thereby making nongeometric features irrelevant to the process.

It was therefore necessary to adjust the beauty scores so as to discount the effect of the nongeometric features. Specifically, linear regression was used to model the effect of the nongeometric features. To this end, the nongeometric features, $\vec{z}_i$, whose Pearson correlation with the beauty score is significant to a level of 0.01 were selected to calculate the regression line $y_i^{lin} = a + \vec{b} \cdot \vec{z}$. The proposed regressor was then trained on the difference

$y = y^{orig} - y^{lin}$, where y^{orig} are the original beauty scores and y^{lin} is the regression hyperplane, based on the nongeometrical features above.

16.4.2 Performance of Facial Beauty Predictors

Machine attractiveness ratings of female images obtained a high Pearson correlation of 0.56 (Pvalue $= 3.2 \times 10^{-9}$) with the mean ratings of human raters (the learning targets). For the male dataset, the correlation was much lower, 0.34 (Pvalue $= 0.0197$). The reason for the inferior performance of the male predictor is twofold: i) the small size of the male dataset is in particularly unfavorable due to the large number of features, and ii) it is well known that the notion of male beauty is not as well defined as that of the female, and probably learning it by a machine might be a considerably harder task.

16.4.3 Beautification Process

Let $\vec{x}$ denote the normalized distance vector extracted from an input facial image. The goal of the beautification process is to generate a nearby vector $\vec{x}'$ with a higher beauty score $f_b(\vec{x}') > f_b(\vec{x})$. Since many points in the proposed 234-dimensional feature space do not correspond to distance vectors of faces at all, the main challenge is to keep the synthesized vectors $\vec{x}'$ inside the subspace of valid faces. Many vectors in this space could possess a higher score, but any such vector must be projected back into the subspace of valid faces, and the score might be reduced in the process. The assumption here is that f_b is smooth enough to allow climbing it incrementally using local optimization techniques.

Two complementary techniques based on weighted K-nearest neighbors (KNN) search (Section 16.4.4) and SVR-driven optimization (Section 16.4.5) were used to achieve this objective. Assuming that face space is locally convex, the KNN-based method guarantees that the resulting beautified faces lie within this space. The SVR-based method optimizes a given face according to the beauty function, f_b. Since the latter method does not assume local convexity, it has a more fundamental flavor. However, since the problem is very sparse, and since the SVR is trained on a rather small set of samples, the regression function could exhibit strong artifacts away from the regions populated by the training samples. Therefore, the search is constrained to a compact region in face space by applying regularization.

16.4.4 KNN-Based Beautification

Psychological experiments demonstrate that average faces are generally considered attractive [9], [14]. While this may be true as a general rule of thumb, this was shown to be an oversimplification in later studies. For example, it was also found that composites of beautiful faces were rated as more attractive than an average composite face [3], [43].

Experiments with the proposed SVR regressor also showed that there are many more local maxima for which the beauty score is higher than the global average. It was further found that an effective way of beautifying a face while maintaining a close resemblance to the original is to modify the distance vector of the face in the direction of the beauty-weighted average of the K nearest neighbors of that face. In agreement with the literature, the beauty scores of faces modified in this manner are typically higher than those resulting from moving towards the global unweighted average.

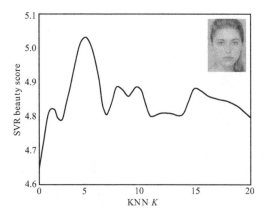

FIGURE 16.5

The beauty score plotted as a function of K in the proposed KNN-based technique applied to one of the faces in the test database. The optimal value of K is 5 with an associated SVR beauty score of 5.03. The initial beauty score for this face is 4.38, and the simple average score, $K \rightarrow \infty$, is 4.51. The proposed SVR-based beautifier succeeds in finding a distance vector with a higher score of 5.20. © 2008 ACM

More specifically, let $\vec{x}_i$ and y_i denote the set of distance vectors corresponding to the training set samples and their associated beauty scores, respectively. Now, given a distance vector $\vec{x}$, the beauty-weighted distances w_i can be defined as follows:

$$w_i = \frac{y_i}{\|\vec{x} - \vec{x}_i\|}.$$

Notice that y_i gives more weight to the more beautiful samples, in the neighborhood of $\vec{x}$. The best results are obtained by first sorting $\{\vec{x}_i\}$ in descending order, such that $w_i \geq w_{i+1}$, and then searching for the value of K maximizing the SVR beauty score f_b of the weighted sum

$$\vec{x}' = \frac{\sum_{i=1}^{K} w_i \vec{x}_i}{\sum_{i=1}^{K} w_i}. \tag{16.1}$$

The chart in Figure 16.5 shows how the beauty score changes for different values of K. Note that the behavior of the beauty score is nontrivial. However, generally speaking, small values of K tend to produce higher beauty scores than that of the average face. Some examples of KNN-beautified faces with different choices of K are shown in Figure 16.6.

Rather than simply replacing the original distances $\vec{x}$ with the beautified ones $\vec{x}'$, more subtle beautification effects can be produced. This can be achieved through trading-off the degree of the beautification for resemblance to the original face, by linearly interpolating between $\vec{x}$ and $\vec{x}'$ before performing the distance embedding described in Section 16.6.

16.4.5 SVR-Based Beautification

The SVR-based beautification is a numerical optimization treating the SVR beauty function as a *potential field* over the feature space constructed by distance vectors. Thus, f_b is used directly to seek beautified feature distance vectors. Whereas the KNN-based approach only produces convex combinations of the training set samples, SVR-based optimization

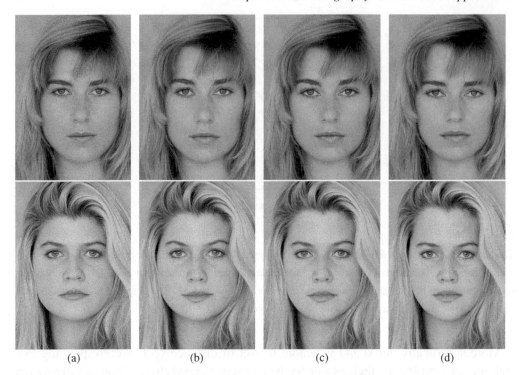

| (a) | (b) | (c) | (d) |

FIGURE 16.6

Face beautification using KNN and SVR: (a) original faces, (b) KNN-beautified images with $K = 3$, (c) KNN-beautified images with optimal K, and (d) SVR-beautified images. © 2008 ACM

is limited by no such constraint. Figure 16.6 demonstrates the differences between KNN-based and SVR-based beautification.

Formally, the beautified distance vector $\vec{x}'$ is defined as follows:

$$\vec{x}' = \underset{\vec{x}}{\operatorname{argmin}}\ E(\vec{x}) = \underset{\vec{x}}{\operatorname{argmin}}\ (-f_b(\vec{x})). \tag{16.2}$$

Here, the standard no-derivatives direction set method [44] is used to numerically perform this minimization. To accelerate the optimization, principal component analysis (PCA) is performed on the feature space to reduce its dimensionality from 234 to 35. Thus, the minimization process can be applied in the low dimensional space, with $\vec{u}$ denoting the projection of $\vec{x}$ on this lower dimensional space.

For the majority of the facial images in the test database using only the beauty function as a guide produces results with higher beauty score than the KNN-based approach. However, for some samples, the SVR-based optimization yields distance vectors which do not correspond to valid human face distances. To constrain the search space, the energy functional, Equation 16.2, can be regularized by adding a log-likelihood term (*LP*):

$$E(\vec{u}) = (\alpha - 1)f_b(\vec{u}) - \alpha LP(\vec{u}),$$

where α controls the importance of the log-likelihood term, with $\alpha = 0.3$ being sufficient to enforce probable distance vectors. This technique is similar to the one used in Reference [24].

The likelihood function P is approximated by modeling face space as a multivariate Gaussian distribution. When projected onto PCA subspace, P may be expressed as follows:

$$P(\vec{u}) = \frac{1}{(2\pi)^{N/2}|\Sigma|^{1/2}} \exp\left(-\frac{1}{2}(\vec{u}-\vec{\mu})^\top \Sigma^{-1}(\vec{u}-\vec{\mu})\right),$$

where $\vec{\mu}$ is the expectation value of $\vec{u}$ and Σ is its covariance matrix.

Since $\vec{u}$ is already a projection on PCA space, the covariance matrix Σ, is already diagonal and the log-likelihood term becomes

$$LP(\vec{u}) = -\sum_{j=1}^{d'} \frac{(\vec{u}-\vec{\mu})^2}{2\Sigma_{jj}} + \text{const},$$

where the constant term is independent of $\vec{u}$ and d' denotes the dimensionality of $\vec{u}$.

16.5 Facial Feature Extraction

The extraction of the distance vector from a facial image involves the nontrivial task of automatically identifying the facial feature points. As shown in Figure 16.4, the feature points are located on the prominent facial features. Each of these features is approximated by a spline. There is extensive literature that deals with the task of snapping such splines to their corresponding facial features. The reader is referred to Reference [27] for a survey of these techniques.

The proposed method uses the Bayesian tangent shape model (BTSM) [45], a technique that improves the well-known active shape model (ASM) [46]. ASM consists of a point distribution model capturing shape variations of valid object instances, and a set of gray gradient distribution models which describe local texture of each landmark point. The model is constructed using a training set and its parameters are actively updated as new examples are added. This bootstrapping process is semiautomatic. At the early stages of the training, considerable user intervention is necessary, but as the training set increases, user assistance is only required in rare cases. The major advantage of ASM is that the model can only deform in the ways learnt. That is, it can accommodate considerable variability and it is still specific to the class of objects it intends to represent.

The facial analysis process requires aligning and normalizing the $2N$-dimensional space of feature data, where N is the number of landmarks. This process takes advantage of the correlated nature of the landmarks based on PCA (in ASM) or through a Bayesian framework (in BTSM).

Given a new facial image, the ASM algorithm requires an initial guess for the locations of the landmarks. The average shape is a good choice, yet finding the initial scale and orientation greatly improves the accuracy of the detected locations and reduces the need for manual adjustments. For this purpose, the OpenCV Haar classifier cascade [47] is used.

In the proposed method, the ASM training set consists of 92 samples, each containing 84 landmarks. The distribution of these landmarks is illustrated in Figure 16.4b over one of the facial images in the training set. To beautify a new facial image, this new image is first analyzed and its feature landmarks are extracted in the same way as was done for the training images. In most cases, the input image analysis is fully automatic. In rare cases some user intervention is required, typically, when large parts of the face are occluded by hair.

16.6 Distance Embedding and Warping

The beautification engine yields a beautified distance vector $\vec{v}'$. These distances have to be now converted to a set of new facial landmarks. Since $\vec{v}'$ is not guaranteed to correspond to distances of edges in a planar facial mesh, the goal is to find the target landmark positions $q_i = (x_i, y_i)$ that provide the best fit, in the least squares sense, for the distances in $\vec{v}'$. Formally, it is possible to define

$$E(q_1, \ldots, q_N) = \sum_{e_{ij}} \alpha_{ij} \left(\|q_i - q_j\|^2 - d_{ij}^2 \right)^2, \tag{16.3}$$

where e_{ij} denotes the facial mesh connectivity matrix. To reduce nonrigid distortion of facial features, α_{ij} is set to 1 for intra-feature edges (edges that connect two feature points from different facial features), and to 10 for inter-feature edges. The target distance term d_{ij} is the entry in $\vec{v}'$ corresponding to the edge e_{ij}.

The target landmark positions q_i are obtained by minimizing E. This kind of optimization has been recently studied in the context of graph drawing [48]. It is referred to as a *stress minimization* problem, originally developed for multidimensional scaling [49]. Here, the Levenberg-Marquardt (LM) algorithm is used to efficiently perform this minimization [50], [51], [52]. The LM algorithm is an iterative nonlinear minimization algorithm which requires reasonable initial positions. However, in this case, the original geometry provides a good initial guess, since the beautification always modifies the geometry only a little.

The embedding process has no knowledge of the semantics of facial features. However, human perception of faces is extremely sensitive to the shape of the eyes. Specifically, even a slight distortion of the pupil or the iris into a noncircular shape significantly detracts from the realistic appearance of the face. Therefore, a postprocess that enforces a similarity transform on the landmarks of the eyes, independently for each eye, should be performed. A linear least squares problem in the four free variables of the similarity transform

$$S = \begin{bmatrix} a & b & t_x \\ -b & a & t_y \\ 0 & 0 & 1 \end{bmatrix},$$

can be solved by minimizing $\sum \|Sp_i - q_i\|^2$ for all feature points of the eyes, where p_i are original landmark locations and q_i are their corresponding embedded positions (from Equation 16.3). Then Sp_i replaces q_i to preserve the shape of the original eyes. The embedding works almost perfectly with an average beauty score drop of only 0.005, before applying the above similarity transform correction to the eyes. However, there is an additional small loss of 0.232 on average in beauty score after the similarity correction.

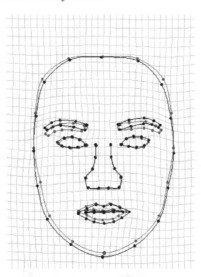

FIGURE 16.7 (See color insert.)

The warp field is defined by the correspondence between the source feature points (in dark gray) and the beautified geometry (in black). © 2008 ACM

The distance embedding process maps the set of feature points $\{p_i\}$ from the source image to the corresponding set $\{q_i\}$ in the beautified image. Next, a warp field that maps the source image into a beautified one according to this set of correspondences is computed. For this purpose, the multilevel free-form deformation (MFFD) technique [53] is adapted. The warp field is illustrated in Figure 16.7, where the source feature points are shown using dark gray and the corresponding beautified positions are indicated using black.

The MFFD consists of a hierarchical set of free-form deformations of the image plane where, at each level, the warp function is a free-form deformation defined by B-spline tensor products. The advantage of the MFFD technique is that it guarantees a one-to-one mapping (no foldovers). However, this comes at the expense of a series of hierarchical warps [53]. To accelerate the warping of high resolution images, the explicit hierarchical composition of transformations is first unfold into a flat one by evaluating the MFFD on the vertices of a fine lattice.

16.7 Results

To demonstrate performance of the proposed digital beautification technique, a simple interactive application, which was used to generate all of the examples in this chapter, has been implemented. After loading a portrait, the application automatically detects facial features, as described in Section 16.5. The user is able to examine the detected features, and adjust them if necessary. Next, the user specifies the desired degree of beautification, as well as the beautification function used, f_b (males or females) and the application computes and displays the result.

FIGURE 16.8 (See color insert.)

Beautification examples: (top) input portraits and (bottom) their beautified versions. © 2008 ACM

Figure 16.8 shows a number of input faces and their corresponding beautified versions. The degree of beautification in all these examples is 100 percent, and the proposed beautification process increases the SVR beauty score by roughly 30 percent. Note that in each of these examples, the differences between the original face and the beautified one are quite subtle, and thus the resemblance between the two faces is unmistakable. Yet the subtle changes clearly succeed in enhancing the attractiveness of each of these faces.

The faces shown in Figure 16.8 are part of the set of 92 faces, which were photographed by a professional photographer, and used to train the SVR, as described in Section 16.4. However, the resulting beautification engine is fully effective on faces outside that set. This is demonstrated by the examples in Figure 16.9 for females and Figure 16.10 for males. All images are part of the AR face database [54]. Note that the photographs of this open repository appear to have been taken under insufficient illumination.

In some cases, it is desirable to let the beautification process modify only some parts of the face, while keeping the remaining parts intact. This mode is referred to as *beautification by parts*. For example, the operator of the proposed application may request that only the eyes should be subject to beautification, as shown in Figure 16.11. These results demonstrate that sometimes a small local adjustment may result in an appreciable improvement in the facial attractiveness. Figure 16.12 is another example of beautification by parts, where all of the features except the rather unique lips of this individual were subject to adjustment.

Performing beautification by parts requires only those distances where at least one endpoint is located on a feature designated for beautification. This reduces the dimensionality of the feature space and enables the algorithm to search only among the beautified features. This technique implicitly assumes that features that are part of a beautiful face are beautiful on their own.

As mentioned earlier, it is possible to specify the desired degree of beautification, with 0 percent corresponding to the original face and 100 percent corresponding to the face

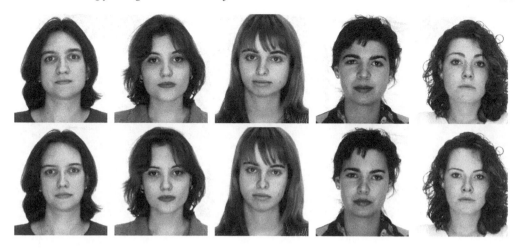

FIGURE 16.9

Beautification of female faces that were not part of the 92 training faces set for which facial attractiveness ratings were collected: (top) input portraits, and (bottom) their beautified versions. © 2008 ACM

FIGURE 16.10

Beautification of male faces that were not part of the male training faces: (top) input portraits, and (bottom) their beautified versions.

defined by the beautification process. Degrees between 0 and 100 are useful in cases where the fully beautified face is too different from the original, as demonstrated in Figure 16.13.

16.7.1 Process Validation

The proposed techniques can improve the facial attractiveness of a majority of the faces used in experiments. However, there are also cases where the beautifier does not introduce any appreciable change. To objectively validate the proposed beautification procedure, an experiment was conducted in which human raters were presented with 93 pairs of faces (original and beautified), of males (45 faces) and females (48 faces). The raters were asked to indicate the more attractive face in each pair. The positions of the faces in each pair (left

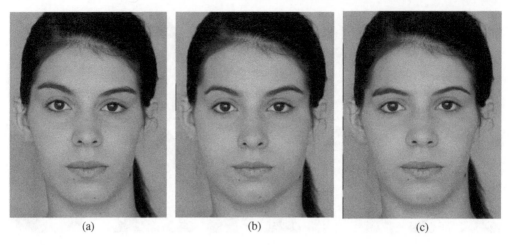

FIGURE 16.11

Beautification by parts: (a) original image, (b) full beautification, and (c) only the eyes are designated for beautification. © 2008 ACM

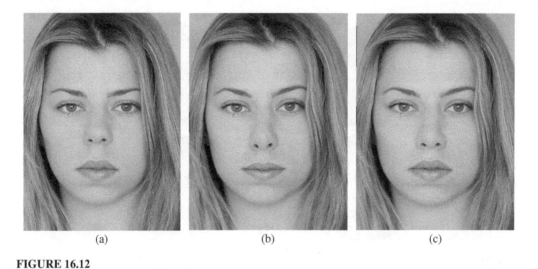

FIGURE 16.12

Beautification by parts: (a) original image, (b) full beautification, and (c) the mouth region is excluded from beautification. © 2008 ACM

or right) were determined randomly, and the 93 pairs were shown in random order. All of the 93 original faces were obtained from the AR face database [54]. In total, 37 raters, both males and females aged between 25 and 40, participated in the experiment.

As could be expected, the agreement between raters is not uniform for all portraits. Still, for all 48 female portraits, the beautified faces were chosen as more attractive by most raters, and in half of the portraits the beautified versions were preferred by more than 80 percent of the raters. Finally, on average, in 79 percent of the cases the raters chose the beautified version as the more attractive face. This result is very significant statistically (P-value $= 7.1 \times 10^{-13}$), proving that on average the proposed tool succeeds in significantly improving the attractiveness of female portraits.

(a) (b) (c)

FIGURE 16.13

Varying the degree of beautification: (a) original image, (b) 50 percent, and (c) 100 percent, where the differences with respect to the original image may be too conspicuous. © 2008 ACM

As for the male portraits, 69 percent of the beautified versions were chosen as more attractive. Notice that this result, although not as striking as that for females, is still statistically significant (P-value = 0.006). The possible reasons for the different strengths of the female and male validation are twofold: i) the male training set was considerably smaller than that for females; and ii) the notion of male attractiveness is not as well defined as that of females, so the consensus in both training set ratings and ratings of beautified images versus original images is not as uniform. Both issues are also reflected in the lower performance of the males' beauty regressor.

16.7.2 Applications

Professional photographers have been retouching and deblemishing their subjects ever since the invention of photography. It may be safely assumed that any model that appears on a magazine cover today has been digitally manipulated by a skilled, talented retouching artist. It should be noted that such retouching is not limited to manipulating color and texture, but also to wrinkle removal and changes in the geometry of the facial features. Since the human face is arguably the most frequently photographed object on earth, face beautification would be a useful and welcome addition to the ever-growing arsenal of image enhancement and retouching tools available in today's digital image editing packages. The potential of such a tool for motion picture special effects, advertising, and dating services, is also quite obvious.

Another interesting application of the proposed technique is the construction of facial collages when designing a new face for an avatar or a synthetic actor. Suppose a collection of facial features (eyes, nose, mouth, etc.) originating from different faces to synthesize a new face with these features. The features may be assembled together seamlessly using Poisson blending [55], but the resulting face is not very likely to look attractive, or even natural, as demonstrated in Figure 16.14. Applying the proposed tool to the collage results in a new face that is more likely to be perceived as natural and attractive.

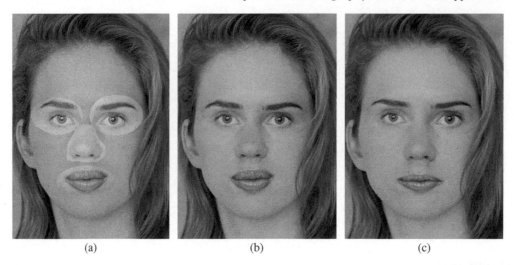

(a)   (b)   (c)

FIGURE 16.14

Beautification using a collection of facial features: (a) a collage with facial features taken from a catalog, (b) result of Poisson-blending the features together, and (c) result after applying the proposed technique to the middle image. © 2008 ACM

16.8   Conclusions

A face beautification method based on an optimization of a beauty function modeled by a support vector regressor has been developed. The challenge was twofold: first, the modeling of a high dimensional nonlinear beauty function, and second, climbing that function while remaining within the subspace of valid faces. It should be emphasized that the synthesis of a novel valid face is a particularly challenging task, since humans are extremely sensitive to every single nuance in a face. Thus, the smallest artifact may be all that is needed for a human observer to realize that the face he is looking at is a fake. Currently, the proposed method is limited to beautifying faces in frontal views and with a neutral expression only. Extending this technique to handle general views and other expressions is a challenging direction for further research.

In the proposed method, beautification is obtained by manipulating only the geometry of the face. However, as was mentioned earlier, there are also important nongeometric attributes that have a significant impact on the perceived attractiveness of a face. These factors include color and texture of hair and skin, and it would be interesting to investigate how changes in these attributes might be incorporated in the proposed digital beautification framework.

Finally, it should be noted that the goal of this research was not to gain a deeper understanding of how humans perceive facial attractiveness. Thus, no specific explicit beautification guidelines, such as making the lips fuller or the eyes larger, were proposed. Instead, this work aimed at developing a more general methodology that is based on raw beauty ratings data. It is hoped, however, that perceptual psychologists will find the proposed technique useful in their quest to better understanding of the perception of beauty.

Appendix

Suppose l examples $\{\vec{x}_i, y_i\}$, with $\vec{x}_i \in R^d$ and $y_i \in R$ for all $i = 1, 2 \ldots l$. Let us also assume that ε-support vector regression (SVR) [41] finds a smooth function $f(\vec{x})$ that has at most ε deviation from the actual values of the target data y_i, and at the same time is as flat as possible. In other words, errors are ignored as long as they are less than ε. In the simplest case, f is a linear function, taking the form $f(\vec{x}) = \vec{w} \cdot \vec{x}_i + b$ where $\vec{w} \in R^d$ and $b \in R$. In the linear case flatness simply means small $\|\vec{w}\|$.

Formally one can write this as constraint optimization problem [41] requiring

$$
\begin{aligned}
\text{minimize } & \frac{1}{2}\|\vec{w}\|^2 + C\sum_{i=1}^{l}(\xi_i + \xi_i^*), \\
\text{subject to } & y_i - \vec{w} \cdot \vec{x}_i - b \le \varepsilon + \xi_i, \\
& \vec{w} \cdot \vec{x}_i + b - y_i \le \varepsilon + \xi_i^*, \\
& \xi_i, \xi_i^* \ge 0.
\end{aligned}
\tag{16.4}
$$

This formulation, referred to as soft margin SVR, introduced the slack variables ξ_i and ξ_i^* in order to allow for some outliers. Figure 16.15 illustrates the ε-insensitive band as well as the meaning of the slack variables for a one-dimensional regression problem. The positive constant C determines the trade off between the flatness of f and the tolerance to outliers.

Nonlinear functions f can be elegantly incorporated into the SVR formalism by using a mapping $\vec{\Phi}$ from the space of input examples R^n into some feature feature space and then applying the standard SVR formulation. The transformation $\vec{\Phi}(\vec{x})$ need not be carried out explicitly, due to the fact that the SVR algorithm depends only on inner products between various examples. Therefore, it suffices to replace all inner products in the original formulation by a kernel function $k(x, x') = \vec{\Phi}(\vec{x}) \cdot \vec{\Phi}(\vec{x}')$. This forms a quadratic optimization problem and therefore allows efficient numeric solutions. Not surprisingly, there are constraints on kernel functions, which are known as Mercer's conditions [41], [56].

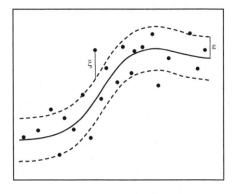

FIGURE 16.15

ε-insensitive regression. The ξ and ξ_i^* variables are nonzero only for examples that reside outside the region bounded between the dashed lines.

The solution of Equation 16.4 usually proceeds via the dual formulation. Each constraint is associated by a Lagrange multiplier, in terms of which one obtains a quadratic optimization problem, which is easy to solve [57].

The radial basis function kernel $k(x, x') = \exp(-\|\vec{x} - \vec{x'}\|^2 / (2\sigma^2))$ is especially useful since it smoothly interpolates between linear models obtained with $\sigma \to \infty$ and highly nonlinear models obtained for small values of σ.

Acknowledgment

Figures 16.1, 16.4 to 16.9, and 16.11 to 16.14 are reprinted from Reference [36], with the permission of ACM.

References

[1] M.R. Cunningham, A.R. Roberts, C.H. Wu, A.P. Barbee, and P.B. Druen, "Their ideas of beauty are, on the whole, the same as ours: Consistency and variability in the cross-cultural perception of female attractiveness," *Journal of Personality and Social Psychology*, vol. 68, no. 2, pp. 261–279, February 1995.

[2] D. Jones, *Physical Attractiveness and the Theory of Sexual Selection: Results from Five Populations*, Ann Arbor: University of Michigan Press, June 1996.

[3] D.I. Perrett, K.A. May, and S. Yoshikawa, "Facial shape and judgements of female attractiveness," *Nature*, vol. 368, pp. 239–242, March 1994.

[4] D.I. Perrett, K.J. Lee, I.S. Penton-Voak, D.A. Rowland, S. Yoshikawa, M.D. Burt, P. Henzi, D.L. Castles, and S. Akamatsu, "Effects of sexual dimorphism on facial attractiveness," *Nature*, vol. 394, pp. 884–887, August 1998.

[5] Y. Eisenthal, G. Dror, and E. Ruppin, "Facial attractiveness: Beauty and the machine," *Neural Computer*, vol. 18, no. 1, pp. 119–142, January 2006.

[6] A. Kagian, G. Dror, T. Leyvand, I. Meilijson, D. Cohen-Or, and E. Ruppin, "A machine learning predictor of facial attractiveness revealing human-like psychophysical biases," *Vision Research*, vol. 48, no. 2, pp. 235–243, January 2008.

[7] M.R. Cunningham, "Measuring the physical in physical attractivenes: Quasi experiments on the sociobiology of female facial beauty," *Journal of Personality and Social Psychology*, vol. 50, no. 5, pp. 925–935, May 1986.

[8] M.R. Cunningham, A.P. Barbee, and C.L. Philhower, *Facial Attractiveness: Evolutionary, Cognitive, and Social Perspectives*, ch. Dimensions of facial physical attractiveness: The intersection of biology and culture, G. Rhodes and L.A. Zebrowitz (eds.), Westport, CT: Ablex Publishing, October 2001, pp. 193–238.

[9] J.H. Langlois, L.A. Roggman, R.J. Casey, J.M. Ritter, L.A. Rieser-Danner, and V.Y. Jenkins, "Infant preferences for attractive faces: Rudiments of a stereotype?," *Developmental Psychology*, vol. 23, no. 5, pp. 363–369, May 1987.

[10] A. Slater, C.V. der Schulenberg, E. Brown, G. Butterworth, M. Badenoch, and S. Parsons, "Newborn infants prefer attractive faces," *Infant Behavior and Development*, vol. 21, no. 2, pp. 345–354, 1998.

[11] F. Galton, "Composite portraits," *Journal of the Anthropological Institute of Great Britain and Ireland*, vol. 8, pp. 132–142, 1878.

[12] J.H. Langlois and L.A. Roggman, "Attractive faces are only average," *Psychological Science*, vol. 1, no. 2, pp. 115–121, March 1990.

[13] A. Rubenstein, J. Langlois, and L. Roggman, *Facial Attractiveness: Evolutionary, Cognitive, and Social Perspectives*, ch. What makes a face attractive and why: The role of averageness in defining facial beauty, G. Rhodes and L.A. Zebrowitz (eds.), Westport, CT: Ablex Publishing, October 2001, pp. 1–33.

[14] K. Grammer and R. Thornhill, "Human facial attractiveness and sexual selection: The role of symmetry and averageness," *Journal of Comparative Psychology*, vol. 108, no. 3, pp. 233–242, September 1994.

[15] R. Thornhill and S.W. Gangsted, "Facial attractiveness," *Trends in Cognitive Sciences*, vol. 3, no. 12, pp. 452–460, December 1999.

[16] L.A. Zebrowitz and G. Rhodes, *Facial Attractiveness: Evolutionary, Cognitive, and Social Perspectives*, ch. Nature let a hundred flowers bloom: The multiple ways and wherefores of attractiveness, G. Rhodes and L.A. Zebrowitz (eds.), Westport, CT: Ablex Publishing, October 2001, pp. 261–293.

[17] J.B. Halberstadt and G. Rhodes, "It's not just average faces that are attractive: Computer-manipulated averageness makes birds, fish, and automobiles attractive," *Psychonomic Bulletin and Review*, vol. 10, no. 1, pp. 149–156, March 2003.

[18] R. Reber, N. Schwarz, and P. Winkielman, "Processing fluency and aesthetic pleasure: Is beauty in the perceiver's processing experience?," *Personality and Social Psychology Review*, vol. 8, no. 4, pp. 364–382, 2004.

[19] A.J. O'Toole, T. Price, T. Vetter, J.C. Bartlett, and V. Blanz, "3D shape and 2D surface textures of human faces: The role of "averages" in attractiveness and age," *Image Vision Computing*, vol. 18, no. 1, pp. 9–19, December 1999.

[20] F.I. Parke and K. Waters, *Computer Facial Animation*, Wellesley, MA: A K Peters, September 1996.

[21] Y. Lee, D. Terzopoulos, and K. Waters, "Realistic modeling for facial animation," in *Proceedings of the 22nd Annual Conference on Computer Graphics and Interactive Techniques*, New York, USA, August 1995, pp. 55–62.

[22] B. Guenter, C. Grimm, D. Wood, H. Malvar, and F. Pighin, "Making faces," in *Proceedings of the 25th Annual Conference on Computer Graphics and Interactive Techniques*, New York, USA, July 1998.

[23] F. Pighin, J. Hecker, D. Lischinski, R. Szeliski, and D. H. Salesin, "Synthesizing realistic facial expressions from photographs," in *Proceedings of the 25th Annual Conference on Computer Graphics and Interactive Techniques*, New York, USA, July 1998, pp. 75–84.

[24] V. Blanz and T. Vetter, "A morphable model for the synthesis of 3d faces," in *Proceedings of the 26th Annual Conference on Computer Graphics and Interactive Techniques*, New York, USA, August 1999, pp. 187–194.

[25] P.A. Viola and M.J. Jones, "Robust real-time face detection," *International Journal of Computer Vision*, vol. 57, no. 2, pp. 137–154, May 2004.

[26] M.H. Yang, D. Kriegman, and N. Ahuja, "Detecting faces in images: A survey," *IEEE Transactions on Pattern Analysis and Machine Intelligence*, vol. 24, no. 1, pp. 34–58, January 2002.

[27] W. Zhao, R. Chellappa, P.J. Phillips, and A. Rosenfeld, "Face recognition: A literature survey," *ACM Computing Surveys*, vol. 35, no. 4, pp. 399–458, December 2003.

[28] T. Beier and S. Neely, "Feature-based image metamorphosis," in *Proceedings of the 19th Annual Conference on Computer Graphics and Interactive Techniques Conference*, New York, USA, July 1992, pp. 35–42.

[29] S.Y. Lee, G. Wolberg, and S.Y. Shin, "Scattered data interpolation with multilevel B-splines," *IEEE Transactions on Visualization and Computer Graphics*, vol. 3, no. 3, pp. 228–244, July-September 1997.

[30] V. Blanz and T. Vetter, "A morphable model for the synthesis of 3D faces," in *Proceedings of the 26th Annual Conference on Computer Graphics and Interactive Techniques*, New York, USA, August 1999, pp. 187–194.

[31] D.I. Perrett, D.M. Burt, I.S. Penton-Voak, K.J. Lee, D.A. Rowland, and R. Edwards, "Symmetry and human facial attractiveness," *Evolution and Human Behavior*, vol. 20, no. 5, pp. 295–307, September 1999.

[32] A. Lanitis, C.J. Taylor, and T.F. Cootes, "Toward automatic simulation of aging effects on face images," *IEEE Transactions on Pattern Analysis and Machine Intelligence*, vol. 24, no. 4, pp. 442–455, April 2002.

[33] V. Blanz, "Manipulation of facial attractiveness." Available nnline http://www.mpi-inf.mpg.de/ blanz/data/attractiveness/, 2003.

[34] D. Vlasic, M. Brand, H. Pfister, and J. Popovic, "Face transfer with multilinear models," *ACM Transactions on Graphics*, vol. 24, no. 3, pp. 426–433, July 2005.

[35] V.S. Johnston and M. Franklin, "Is beauty in the eye of the beholder?," *Ethology and Sociobiology*, vol. 14, pp. 183–199, May 1993.

[36] T. Leyvand, D. Cohen-Or, G. Dror, and D. Lischinski, "Data-driven enhancement of facial attractiveness," *ACM Transactions on Graphics*, vol. 27, no. 3, pp. 1–9, August 2008.

[37] C. Braun, M. Gruendl, C. Marberger, and C. Scherber, "Beautycheck - ursachen und folgen von attraktivitaet." Available online: http://www.uni-regensburg. de/Fakultaeten/phil_Fak_II/Psychologie/Psy_II/beautycheck/english/2001.

[38] B. Fink, N. Neave, J. Manning, and K. Grammer, "Facial symmetry and judgements of attractiveness, health and personality," *Personality and Individual Differences*, vol. 41, no. 3, pp. 491–499, August 2006.

[39] A.C. Little, I.S. Penton-Voak, D.M. Burt, and D.I. Perrett, *Facial Attractiveness: Evolutionary, Cognitive, and Social Perspectives*, ch. Evolution and individual differences in the perception of attractiveness: How cyclic hormonal changes and self-perceived attractiveness influence female preferences for male faces, G. Rhodes and L.A. Zebrowitz (eds.), Westport, CT: Ablex Publishing, October 2001, pp. 68–72.

[40] M.A. Turk and A.P. Pentland, "Face recognition using eigenfaces," in *Proceedings of the IEEE International Conference on Computer Vision and Pattern Recognition*, Maui, HI, USA, June 1991, pp. 586–591.

[41] V. Vapnik, *The Nature of Statistical Learning Theory*, New York: Springer, 1995.

[42] T. Joachims, *Advances in Kernel Methods: Support Vector Learning*, ch. Making large-scale SVM learning practical, Cambridge, MA: MIT Press, 1999, pp. 169–184.

[43] T.R. Alley and M.R. Cunningham, "Averaged faces are attractive, but very attractive faces are not average," *Psychological Science*, vol. 2, no. 2, pp. 123–125, March 1991.

[44] W.H. Press, B.P. Flannery, S.A. Teukolsky, and W.T. Vetterling, *Numerical Recipes: The Art of Scientific Computing*, Cambridge, UK: Cambridge University Press, 2nd Edition, 1992.

[45] Y. Zhou, L. Gu, and H.J. Zhang, "Bayesian tangent shape model: Estimating shape and pose parameters via Bayesian inference," in *Proceedings of the IEEE Computer Society Conference on Computer Vision and Pattern Recognition 2003*, Los Alamitos, CA, USA, June 2003, pp. 109–118.

[46] T.F. Cootes, C.J. Taylor, D.H. Cooper, and J. Graham, "Active shape models - Their training and their applications," *Computer Vision and Image Understanding*, vol. 61, no. 1, pp. 38–59, January 1995.

[47] G. Bradski, "The OpenCV library," *Dr. Dobb's Journal*, vol. 25, no. 11, pp. 120–125, November 2000.

[48] J.D. Cohen, "Drawing graphs to convey proximity: An incremental arrangement method," *ACM Transactions on Computer-Human Interaction*, vol. 4, no. 3, pp. 197–229, September 1997.

[49] J.W. Sammon, "A nonlinear mapping for data structure analysis," *IEEE Transactions on Computers*, vol. C-18, no. 5, pp. 401–409, May 1969.

[50] K. Levenberg, "A method for the solution of certain problems in least squares," *The Quarterly of Applied Mathematics*, vol. 2, pp. 164–168, 1944.

[51] D. Marquardt, "An algorithm for least-squares estimation of nonlinear parameters," *SIAM Journal of Applied Mathematics*, vol. 11, no. 2, pp. 431–441, June 1963.

[52] M. Lourakis, "Levmar: Levenberg-Marquardt non-linear least squares algorithms in C/C++." Available online http://www.ics.forth.gr/ lourakis/levmar, 2004.

[53] S.Y. Lee, G. Wolberg, K.Y. Chwa, and S.Y. Shin, "Image metamorphosis with scattered feature constraints," *IEEE Transactions on Visualization and Computer Graphics*, vol. 2, no. 4, pp. 337–354, December 1996.

[54] A.M. Martinez and R. Benavente, "The AR face database," Tech. Rep. 24, CVC, 1998.

[55] P. Pérez, M. Gangnet, and A. Blake, "Poisson image editing," *ACM Transactions on Graphics*, vol. 22, no. 3, pp. 313–318, July 2003.

[56] R. Courant and D. Hilbert, *Methods of Mathematical Physics*. Interscience, 1953.

[57] A.J. Smola and B. Schölkopf, "A tutorial on support vector regression," *Statistics and Computing*, vol. 14, no. 3, pp. 199–222, August 2004.

17

High-Quality Light Field Acquisition and Processing

Chia-Kai Liang and Homer H. Chen

17.1 Introduction

Computational photography is changing the way of capturing images. While traditional photography simply captures two-dimensional (2D) projection of three-dimensional (3D) world, computational photography captures additional information by using generalized optics. The captured image may not be visually attractive, but together with the additional information, it enables novel postprocessing that can deliver quality images and, more importantly, generate data such as scene geometry that were unobtainable in the past. These new techniques overwrite the concept of traditional photography and transform a normal camera into a powerful device.

Light field acquisition is of fundamental importance among all aspects of computational photography. A complete four-dimensional (4D) light field contains most visual information of a scene and allows various photographic effects to be generated in a physically correct way. However, existing light field cameras [1], [2], [3], [4] manipulate the light rays by means of lens arrays or attenuating masks that trade the spatial resolution for the angular resolution. Even with the latest sensor technology, one can hardly generate a light field with mega-pixel spatial resolution. Moreover, the light field captured by light field cameras appears to have a common photometric distortion and aliasing that, if not properly managed, may render the data useless.

This chapter presents the hardware and software approaches to address those issues. First, a new device is described, called *programmable aperture* [5], [6], for high resolution light field acquisition. It exploits the fast multiple-exposure feature of digital sensors without trading off sensor resolution to capture light field sequentially, which, in turn, enables the multiplexing of light rays. In summary, the programmable aperture has several advantages over the previous devices:

- It can capture a light field at full spatial resolution, that is, the same as the sensor resolution of the camera.

- It has better acquisition efficiency due to the multiplexing technique.

- It has adjustable angular resolution and prefilter kernel. When the angular resolution is set to one, the light field camera becomes a conventional camera.

- The device is compact and economic. The programmable aperture can be placed in, and nicely integrated with, a conventional camera.

Second, two algorithms are presented to enhance the captured light field. The first is a calibration algorithm to remove the photometric distortion unique to a light field without using any reference object. The distortion is directly estimated from the captured light field. The other is a depth estimation algorithm utilizing the multi-view property of the light field and visibility reasoning to generate view-dependent depth maps for view interpolation.

This chapter also presents a simple light transport analysis of the light field cameras. The device and algorithms constitute a complete system for high quality light field acquisition. In comparison with other light field cameras, the spatial resolution of the proposed camera is increased by orders of magnitude, and the angular resolution can be easily adjusted during operation or postprocessing. The photometric calibration enables more consistent rendering and more accurate depth estimation. The multi-view depth estimation effectively increases the angular resolution for smoother transitions between views and makes depth-aware image editing possible.

In the remaining part of this chapter, Section 17.2 presents the relevant work in light field rendering and computational photography. Section 17.3 shortly discusses the light transport process in image acquisition and Section 17.4 describes the programmable aperture, including the concepts and the implementations. Section 17.5 presents the novel postprocessing algorithms for the light field data. Section 17.6 presents experimental results and Section 17.7 discusses the limitations and future work. Finally, the conclusions are drawn in Section 17.8.

17.2 Related Work

This chapter is inspired by previous research in light field acquisition, computational photography, and illumination multiplexing. This section reviews the remarkable progress in these areas. The work related to the postprocessing is briefly reviewed in Section 17.5.

17.2.1 Light Field Acquisition

4D light field representation of the ray space was first proposed for image-based rendering and then applied to various fields [7], [8]. There are several ways to capture the light field. The simplest method uses a single moving camera whose position for each exposure is located by a camera gantry [7] or estimated by a structure-from-motion algorithm [8]. This method is slow and only works in a controlled environment. Another method simultaneously captures the full 4D dataset by using a camera array [9], [10], which is cumbersome and expensive.

The third method, which is most related to the proposed approach, inserts additional optical elements or masks in the camera to avoid the angular integration of the light field. The idea dates back to nearly a century ago, called *integral photography* or *parallax panoramagrams* and realized using a fly-eye lens (i.e., lenslet) array or a slit plate [11], [12], [13]. Compact implementations and theoretical analysis of this method have been recently developed. In a *plenoptic camera*, for example, a microlens array is placed at the original image plane inside the camera [1], [14]. The resulting image behind each microlens records the angular distribution of the light rays. Alternatively, one can place a positive lens array in front of the camera [2]. Along the same line, the slit plate can be replaced with a cosine mask to improve efficiency [4]. To this end, these devices manipulate the 4D light field spectrum by modulation or reparameterization to fit in a 2D sensor slice [3].

All these devices have the following common drawbacks. First, the spatial resolution, or spectrum bandwidth, is traded for the angular resolution. Although high resolution sensors can be made, capturing a light field with high spatial and high angular resolutions is still difficult. Second, inserting masks or optical elements in a camera automatically imposes a fixed sampling pattern. These components are usually permanently installed and cannot be easily removed from the camera to capture regular pictures.

Recently, a *focused* plenoptic camera was presented [15], in which each lenslet is focused on the virtual image plane. This design provides a better tradeoff between the angular and spatial resolution, as discussed in the next section. However, the other drawbacks of the plenoptic camera remain. The proposed device can be configured in that way to achieve the similar result.

17.2.2 Computational Photography

Two popular techniques, coded aperture and multiple capturing, are closely related to the proposed work. The former treats the aperture (or shutter) as an optical modulator to preserve the high-frequency components of motion-blurred images [16] to provide high-dynamic-range or multispectral imaging [17], [18], to split the field of view [19], or to capture stereoscopic images [20]. Reference [21] uses a coded aperture to estimate the depth of a near-Lambertian scene from a single image. Similar results can be obtained by placing a color-filter array at the aperture [22]. In contrast to those methods, the proposed method directly captures the 4D light field and estimates the depth from it when possible.

The multiple capturing technique captures the scene many times sequentially, or simultaneously by using beam splitters and camera arrays. At each exposure the imaging parameters, such as lighting [23], exposure time, focus, viewpoints [24], or spectral sensitivity [19], are made different. Then a quality image or additional information, for instance, alpha matte, is obtained by computation. This technique can be easily implemented in digital cameras since the integration duration of the sensor can be electronically controlled. For example, Reference [25] splits a given exposure time into a number of time steps and samples one image in each time step. The resulting images are then registered for correcting hand-shaking.

17.2.3 Illumination Multiplexing

Capturing the appearances of an object under different lightings is critical for image-based relighting and object recognition. Since the dimensionality of the signal (a 4D incident light field) is higher than that of the sensor (a 2D photon sensor array), the signal must be captured sequentially, one subset of the signal at a time. Multiplexing can be used to reduce the acquisition time and improve the signal-to-noise ratio by turning on multiple light sources at each exposure and recovering the signal corresponding to a single light source [26], [27], [28].

Both the coded aperture and multiple capturing techniques are exploited in the proposed system. More specifically, multiple exposures are used to avoid the loss of spatial resolution whereas coded aperture are employed to perform multiplexing for quality improvement. Although the proposed method requires sequential multiple exposures, capturing a clear light field dataset takes the same amount of time as capturing a clear image with a conventional camera.

17.3 Light Transport in Photography

This section gives a brief review of the light field representation and the light transport theory of the photography process. For simplicity, only 2D geometry is considered here, although the result can be easily extended to 3D. A similar analysis focusing on the defocus effect can be found in Reference [29].

A light field can be represented as a function that maps the geometric entities of a light ray in free space to the radiance along the light ray. Each light ray is specified by the intersections of two planes with the light ray [7], [8]. There are several ways to define the two planes, see Figure 17.1. For example, one plane can be located on the object surface of interest, and the other at unit distance from, and parallel to, the first one. The coordinates of the intersection of a light ray with the second plane are defined with respect to the intersection of the light ray with the first plane (x and u on the left in Figure 17.1) [30].

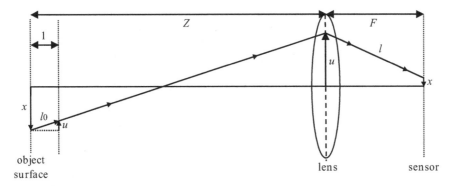

FIGURE 17.1

Light field and light transport. A light ray emitting from a point on an object surface at Z can be represented by $l_0([x\,u]^T)$ or $l([x\,u]^T)$ after refraction by lens. These two representations only differ by a linear transformation (Equation 17.1).

Another common representation places the two planes at the lens and the film (sensor) of a camera and defines independent coordinate systems for these two planes [31] (x and u on the right in Figure 17.1).

Suppose there is a light ray emitting from an object surface point and denote its radiance by the light field $l_0([x\,u]^T)$, where x and u are the intersections of the light ray with the two coordinate planes. The light ray first traverses the space to the lens of the camera at distance Z from the emitting point, as illustrated in Figure 17.1. According to the light transport theory, this causes a shearing to the light field [30]. Next, the light ray changes its direction after it leaves the lens. According to the matrix optics, this makes another shearing to the light field [3]. As the light ray traverses to the image plane at distance F from the lens plane, one more shearing is produced. Finally, the light field is reparameterized into the coordinate system used in the camera. Since the shearings and the reparameterization are all linear transformations, they can be concatenated into a single linear transformation. Hence, the transformed light field $l([x\,u]^T)$ can be represented by

$$l([x\,u]^T) = l_0(M[x\,u]^T) = l_0\left(\begin{bmatrix} -\frac{Z}{F} & Z\Delta \\ \frac{1}{F} & \frac{1}{f} - \frac{1}{F} \end{bmatrix}\begin{bmatrix} x \\ u \end{bmatrix}\right), \tag{17.1}$$

where f is the focal length of the lens and $\Delta = 1/Z + 1/F - 1/f$. This transformation, plus modulation due to the blocking of the aperture [4], describes various photographic effects such as focusing [3].

In traditional photography, a sensor integrates the radiances along rays from all directions into an irradiance sample and thus loses all angular information of the light rays. The goal of this work is to capture the transformed light field $l([x\,u]^T)$ that contains both the spatial and the angular information. In other words, the goal is to avoid the integration step in the traditional photography.

In capturing the light field, although the sampling grids on the lens plane and the sensor plane are usually fixed, the camera parameters, F and f, can be adjusted to modify the transformation in Equation 17.1, thus changing the actual sampling grid for the original light field l_0. For example, it is well-known that in natural Lambertian scenes, while the

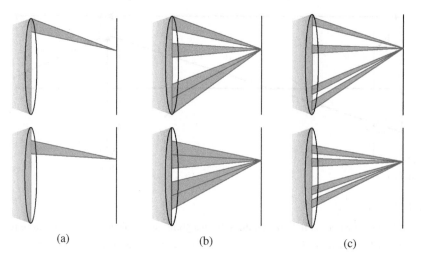

FIGURE 17.2

Configurations of the programmable aperture: (a) capturing one single sample at a time, (b) aggregating several samples at each exposure for quality improvement, and (c) adjusting the prefilter kernel without affecting the sampling rate.

spatial information is rich (i.e, complex structures, texture, shadow, etc.), the angular information is usually of low-frequency. Therefore, setting a high sampling rate along the angular axis u is wasteful. By properly adjusting F, the spatial information can be moved to the angular domain to better utilize the sample budget, as shown in Reference [15].

17.4 Programmable Aperture Camera

This section describes how a programmable aperture camera captures the light field and how multiplexing improves the acquisition efficiency. The prototypes of the camera are described, as well.

17.4.1 Sequential Light Field Acquisition

In a traditional camera, only the size of the aperture can be adjusted to change the depth of field. The captured image is always a 2D projection of the 3D scene, and the angular information is unrecoverably lost. However, if the *shape* of the aperture is modified so that only the light rays arriving in a small specified region of the aperture can pass through the aperture, the angular integration can be avoided. More specifically, if the aperture blocks all light rays but those around u, the resulting image is a subset of the light field. Denote such a *light field image* by I_u:

$$I_u(x) = l([x\,u]^T). \tag{17.2}$$

By capturing images with different aperture shapes (Figure 17.2a), a complete light field can be constructed. However, unlike previous devices that manipulate the light rays after they enter the camera [1], [3], [4], the proposed method blocks the undesirable light rays and captures one subset of the data at a time. The spatial resolution of the light field is thus the same as the sensor resolution. For the method to take effect, a *programmable aperture* is needed. Its transmittance has to be spatially variant and controllable.

An intuitive approach to such a programmable aperture is to replace the lens module with a volumetric light attenuator [19]. However, according to the following frequency analysis of light transport, it can be found that the lens module should be preserved for efficient sampling. Let $L_0([f_x\ f_u]^T)$ and $L([f_x\ f_u]^T)$ denote the Fourier transform of $l_0([x\ u]^T)$ and $l([x\ u]^T)$, respectively. By Equation 17.1 and the Fourier linear transformation theory, L_0 and L are related as follows:

$$L([f_x\ f_u]^T) = |\det(M)|^{-1} L_0(M^{-T}[f_x\ f_u]^T)$$

$$= \frac{1}{|\det(M)|} L_0 \left(\begin{bmatrix} 1 - \frac{F}{f} & 1 \\ FZ\Delta & Z \end{bmatrix} \begin{bmatrix} f_x \\ f_u \end{bmatrix} \right). \tag{17.3}$$

Consider the case where the scene is a Lambertian plane perpendicular to the optical path at $Z = 3010$, $f = 50$, and the camera is focused at $Z = 3000$ (so $F = 50.8475$). If the lens module is removed, $f \to \infty$ and the sampling rate along the f_u axis has to be increased by a factor of 18059 to capture the same signal content. As a result, millions of images need to be captured for a single dataset, which is practically infeasible. Therefore, the lens module must be preserved. The light rays are bent inwards at the lens due to refraction and consequently the spectrum of the transformed light field is compressed. With the lens module and by carefully selecting the in-focus plane, the spectrum can be properly reshaped to reduce aliasing. A similar analysis is developed for multi-view displays [32].

17.4.2 Light Field Multiplexing

A light field with angular resolution N requires N exposures, one for each angular co-ordinate u. Compared with traditional photography, the light collection efficiency of this straightforward acquisition is decreased because only a small aperture is open at each exposure and each exposure time is only $1/N$ of the total acquisition time. As a result, given the same acquisition time, the captured images are noisier than those captured by conventional cameras. To solve this problem, the light field images should be *multiplexed* at each exposure. Specifically, because the radiances of the light rays are additive, multiple light field samples should be aggregated at each exposure by opening multiple regions of the aperture and individual signals should be recovered afterwards.

By opening multiple regions of the aperture, each captured image M_u is a linear combination of N light field images (Figure 17.2b):

$$M_u(x) = \sum_{k=0}^{N-1} w_{uk} I_k(x). \tag{17.4}$$

The weights $w_{uk} \in [0, 1]$ of the light field images can be represented by a vector $\mathbf{w}_u = [w_{u0}, w_{u1}, \dots, w_{u(N-1)}]$ and is referred to as a *multiplexing pattern* since $\mathbf{w}_u$ is physically realized as a spatial-variant mask on the aperture. After N captures with N different multiplexing patterns, the light field images can be recovered by *demultiplexing* the captured images, if the chosen multiplexing patterns form an invertible linear system.

FIGURE 17.3 (See color insert.)

Performance improvement by multiplexing: (a) light field image captured without multiplexing, (b) demultiplexed light field image, (c) image captured with multiplexing, that is, $M_u(x)$ in Equation 17.4, and (d) enlarged portions of (a) and (b). The insets in (a-c) show the corresponding multiplexing patterns.

Intuitively, one should open as many regions as possible, that is, maximize $\|\mathbf{w}_u\|$, to allow the sensor to gather as much light as possible. In practice, however, noise is always involved in the acquisition process and complicates the design of the multiplexing patterns. In the case where the noise is independent and identically-distributed (i.i.d.), Hadamard code-based patterns are best in terms of the quality of the demultiplexed data [5], [26], [33]. However, noise in a digital sensor is often correlated with the input signal [34], [35]. For example, the variance of the shot noise grows linearly with the number of incoming photons. In this case, using the Hadamard code-based patterns actually degrades the data quality [27]. Another drawback of the Hadamard code-based patterns is that they only exist for certain sizes.

Instead, multiplexing patterns can be obtained through optimization. Given the noise characteristics of the device and the true signal value, the mean square error of the demultiplexed signal is proportional to a function $E(\mathbf{W})$:

$$E(\mathbf{W}) = \mathrm{Trace}((\mathbf{W}^T \mathbf{W})^{-1}), \tag{17.5}$$

where $\mathbf{W}$ is an $N \times N$ matrix and each row of $\mathbf{W}$ is a multiplexing pattern $\mathbf{w}_u$. Finding a matrix $\mathbf{W}^\star$ that minimizes $E(\mathbf{W})$ can be formulated as a constrained convex optimization problem and solved by the projected gradient method [36] or brute-force search when the size of $\mathbf{W}$ is small. Because most entities (w_{uk}) of the $\mathbf{W}^\star$ thus obtained are either ones or zeros and because binary masks can be made more accurately in practice, all the entities of $\mathbf{W}^\star$ can be enforced to be binary. This only slightly affects the performance. A result of multiplexing is given in Figure 17.3 which shows that the demultiplexed image is much clearer than the one captured without multiplexing.

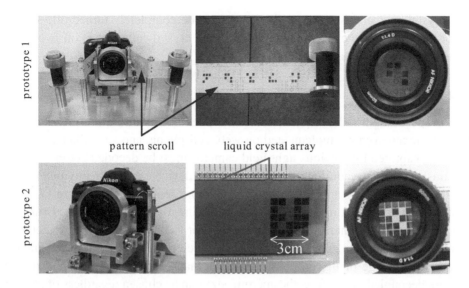

FIGURE 17.4

Prototypes of the programmable aperture cameras: (top) with aperture patterns on an opaque slip of paper and (bottom) on an electronically controlled liquid crystal array.

17.4.3 Prototypes

Two prototypes of the programmable aperture camera shown in Figure 17.4 were implemented using a regular Nikon D70 digital single-lens reflex (DSLR) camera and a 50mm f/1.4D lens module. For simplicity, the lens module was dismounted from the Nikon camera to insert the programmable aperture between this module and the camera. Hence the distance (F in Figure 17.1) between the lens and the sensor is lengthened and the focus range is shortened as compared to the original camera.

The optimization of the multiplexing patterns requires information of the noise characteristics of the camera and the scene intensity. The former is obtained by calibration and the latter is assumed to be one half of the saturation level. Both prototypes can capture the light field with or without multiplexing. The maximal spatial resolution of the light field is 3039×2014 and the angular resolution is adjustable.

In the first prototype, the programmable aperture is made up of a *pattern scroll*, which is an opaqued slit of paper used for film protection. The aperture patterns are manually cut and scrolled across the optical path. The pattern scroll is long enough to include tens of multiplexing patterns and the traditional aperture shapes. This quick and dirty method is simple and performs well except one minor issue; the blocking cell ($w_{uk} = 0$) cannot stay on the pattern scroll if it loses support. This is solved by leaving a gap between cells. Since the pattern scroll is movable, its position may drift out of place. However, this can be simply solved in the industrial level manufacture.

In the second prototype, the programmable aperture is made up of a liquid crystal array (LCA) controlled by a Holtek HT49R30A-1 micro control unit that supports C language. Two different resolutions, 5×5 and 7×7, of the LCA are made. The LCA is easier to program and mount than the pattern scroll, and the multiplexing pattern is no longer limited to binary. However, the light rays can leak from the gaps (used for routing) in between the liquid crystal cells and from the cells that cannot be completely turned off. To compensate for the leakage, an extra image with all liquid crystal cells turned off is captured and subtracted from other images.

17.4.4 Summary

The proposed light field acquisition scheme does not require a high resolution sensor. Therefore, it can even be implemented on web, cell-phone, and surveillance cameras, etc. The image captured by previous light field cameras must be decoded before visualization. In contrast, the image captured using the proposed device can be directly displayed. Even when the multiplexing is applied, the in-focus regions remain sharp (Figure 17.3c).

It should be noted that multiplexing cannot be directly applied to the existing methods like a single moving camera, a plenoptic camera, or a camera array. These methods use a permanent optical design and cannot dynamically select the light rays for integration.

Another advantage of the proposed device is that the sampling grid and the prefilter kernel are decoupled. Therefore, the aperture size can be chosen regardless of the sampling rate (Figure 17.2c). A small prefilter is chosen to preserve the details and remove aliasing by view interpolation. Also the sampling lattice on the lens plane in the proposed device is not restricted to rectangular grids. These parameters, including the number of samples, the sampling grid, and the size of the prefilter kernel, can all be adjusted dynamically.

17.5 Postprocessing Algorithms

The photometric distortion and aliasing due to undersampling have to be addressed before the captured light field can be applied.

17.5.1 Photometric Calibration

The light fields captured by either the proposed programmable aperture or other light field cameras have a noticeable photometric distortion. The light field images corresponding to the boundary of the aperture would appear very different from that corresponding to the center of the aperture, as shown in Figure 17.5. While being termed as vignetting collectively, this photometric distortion is attributed to several sources, such as the cosine fall-off [37], the blocking of the lens diaphragm [38], and pupil aberrations [39]. Because this distortion breaks the common photometric consistency assumption, it must be removed or it can obstruct view interpolation, depth estimation, and many other applications.

The exact physical model of the vignetting effect is difficult to construct. In general, a simplified model that describes the ratio between the distorted light field image $I_u^d(x)$ and the clean image $I_u(x)$ by a $2(D-1)$-degree polynomial function $f_u(x)$ is adopted:

$$I_u^d(x) = f_u(x)I_u(x) = \left(\sum_{i=0}^{D-1} a_{ui} \|x - c_u\|_2^{2i} \right) I_u(x), \quad (17.6)$$

where $\{a_{ui}\}$ are the polynomial coefficients, c_u is the vignetting center, and $\| \cdot \|_2$ is the

FIGURE 17.5 (See color insert.)

The effect of the photometric distortion. The images shown here are two of the nine light field images of a static scene. The insets show the corresponding aperture shape.

Euclidean distance (the coordinates are normalized to $(0, 1)$). The function f_u, called *vignetting field*, is a smooth field across the image. It is large when the distance between x and c_u is small and gradually decreases as the distance increases.

Since the number of unknown variables in Equation 17.6 is larger than the number of observations, the estimation problem is inherently ill-posed. A straightforward method is to capture a uniformly lit object so the distortion-free image $I_u(x)$ becomes *a priori*. However, the vignetting field changes when the camera parameters, including the focus, aperture size, and lens module, are adjusted. It is impractical to perform the calibration whenever a single parameter is changed.

Existing photometric calibration methods that require no specific reference object generally use two assumptions to make the problem tractable [38], [40]. First, the scene points have multiple registered observations with different levels of distortions. This assumption is usually valid in panoramic imaging where the panorama is stitched from many images of the same view point. Second, the vignetting center c_u. This assumption is valid in most traditional cameras, where the optics and the sensors are symmetric along the optical path. Some recent methods remove the first assumption by exploiting the edge and gradient priors in natural images [41], but the second assumption is still needed.

However, both assumptions are inappropriate for the light field images for two reasons. First, the registration of the light field images taken from different view points requires an accurate per-pixel disparity map that is difficult to obtain from the distorted inputs. Second, in each light field image, the parameters, $\{a_{ui}\}$ and c_u, of the vignetting function, are image-dependent and coupled. Therefore, simultaneously estimating the parameters and the clean image is an under-determined nonlinear problem. Another challenge specific to the proposed camera is that the vignetting function changes with the lens and the aperture settings (such as the size of the prefilter kernel) and hence is impossible to tabulate.

An algorithm is proposed here to automatically calibrate the photometric distortion of the light field images. The key idea is that the light field images closer to the center of the optical path have less distortion. Therefore, it can be assumed that $I_0^d \approx I_0$ and then other I_u's can be approximated by properly transforming I_0 to estimate the vignetting field. This way, the problem is greatly simplified. The approach can also be generalized to handle the distortions of other computational cameras, particularly previous light field cameras.

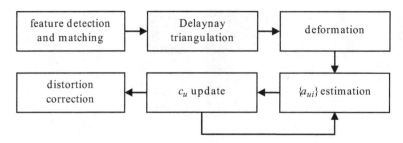

FIGURE 17.6

The photometric calibration flow.

Figure 17.6 depicts the flowchart of the proposed algorithm, with example images shown in Figure 17.7. First, the scale-invariant feature transform (SIFT) method, which is well immune to local photometric distortions [42], is used to detect the feature points in an input I_u^d and find their valid matches in I_0 (Figures 17.7a and 17.7b). Next, the Delaunay triangulation is applied to the matched points in I_u^d to construct a mesh (Figure 17.7c). For each triangle A of the mesh, the displacement vectors of its three vertices are used to determine an affine transform. By affinely warping all triangles, an image I_u^w from I_0 is obtained (Figure 17.7d).

The warped image I_u^w is close enough to the clean image I_u unless there are triangles including objects of different depths or incorrect feature matchings. Such erroneous cases can be effectively detected and removed by measuring the variance of the associated displacement vectors. By dividing the distorted image I_u^d with the warped image I_u^w and excluding the outliers, an estimated vignetting field is obtained (Figure 17.7e). Comparing this image with the vignetting field estimated from the image without warping (Figure 17.7f) reveals that the warping operation effectively finds a rough approximation of the smooth vignetting field and the outliers around the depth discontinuities are successfully excluded.

After the approximation of the vignetting field is obtained, the parametric vignetting function (Equation 17.6) is estimated by minimizing an objective function $E(\{a_{ui}\}, \mathbf{c}_u)$:

$$E(\{a_{ui}\}, \mathbf{c}_u) = \sum_x \left(\frac{I_u^d(x)}{I_u^w(x)} - f_u(x) \right)^2. \tag{17.7}$$

Since $\{a_{ui}\}$ and $\mathbf{c}_u$ are coupled, this objective function is nonlinear and can be minimized iteratively. Given an initial estimate, the vignetting center $\mathbf{c}_u$ is fixed first, as this makes the Equation 17.7 linear in $\{a_{ui}\}$, which can be easily solved by a least square estimation. Then, $\{a_{ui}\}$ can be fixed and $\mathbf{c}_u$ can be updated. This is done by a gradient descent method. The goal is to find a displacement $\mathbf{d}_u$ such that $E(\{a_{ui}\}, \mathbf{c}_u + \mathbf{d}_u)$ is minimal.

Specifically, let r_i denote the distance between the i-th pixel at (x_i, y_i) and $\mathbf{c}_u = (c_{u,x}, c_{u,y})$, the N-D vector $\mathbf{r} = [r_1, r_1, ..., r_N]^T$ denote the distances between all points and $\mathbf{c}_u$, and $\mathbf{I}_u$ denote the estimated vignetting field, that is, the ratio I_u^d / I_u^w. Since $\mathbf{c}_u$ is the only variable, the vignetting function $f_u(x)$ can be redefined as a vector function $\mathbf{f}(\mathbf{c}_u) = [f_u(x_1), f_u(x_2), ..., f_u(x_N)]^T$. Equation 17.7 is then equivalent to the l_2 norm of the error vector ε, that is, $\|\varepsilon\| = \|\mathbf{I}_u - \mathbf{f}(\mathbf{c}_u)\|$. The optimal displacement $\mathbf{d}_u$ at iteration t can then be obtained by solving the normal equation:

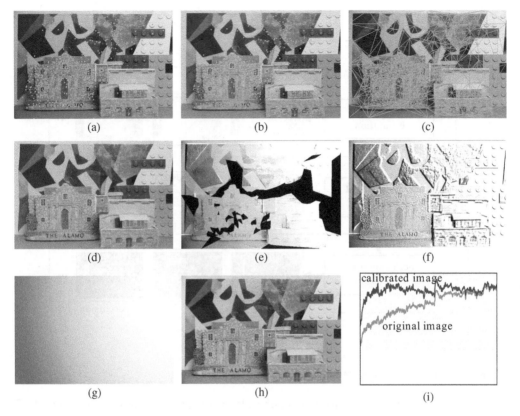

FIGURE 17.7 (See color insert.)

Photometric calibration results: (a) input image I_u^d to be corrected — note the darker left side, (b) reference image I_0, (c) triangulation of the matched features marked in the previous two images, (d) image I_u^w warped using the reference image based on the triangular mesh, (e) approximated vignetting field with suspicious areas removed, (f) vignetting field approximated without warping, (g) estimated parametric vignetting field, (h) calibrated image I_u, and (i) intensity profile of the 420th scanline before and after the calibration.

$$\mathbf{J}^T \mathbf{J} \mathbf{d}_u = -\mathbf{J}^T \varepsilon_{t-1}, \tag{17.8}$$

where $\mathbf{J}$ is the Jacobian matrix ($\mathbf{J} = \frac{d\mathbf{f}}{d\mathbf{c}_u}$) and ε_{t-1} is the error vector of the previous iteration. By setting $D = 4$, the Jacobian is defined as:

$$\frac{d\mathbf{f}}{d\mathbf{c}_v} = \begin{bmatrix} -(x_0 - c_{u,x})(2a_1 + 4a_2 r_0^2 + 6a_3 r_0^4) & -(y_0 - c_{u,y})(2a_1 + 4a_2 r_0^2 + 6a_3 r_0^4) \\ -(x_1 - c_{u,x})(2a_1 + 4a_2 r_1^2 + 6a_3 r_1^4) & -(y_1 - c_{u,y})(2a_1 + 4a_2 r_1^2 + 6a_3 r_1^4) \\ \vdots & \vdots \\ -(x_N - c_{u,x})(2a_1 + 4a_2 r_N^2 + 6a_3 r_N^4) & -(y_N - c_{u,y})(2a_N + 4a_2 r_N^2 + 6a_3 r_N^4) \end{bmatrix}. \tag{17.9}$$

Note that this Jacobian is evaluated using the vignetting center obtained in the previous iteration and the coefficients estimated in this iteration. In this way the convergence speed is increased. Since the number of parameters is small, the image can be subsampled to reduce the computation. Usually 1000 to 2000 inlier samples are sufficient. For each image, 50 iterations are performed and the parameters with minimal objective value are chosen here as the result. One obtained vignetting field is shown in Figure 17.7g.

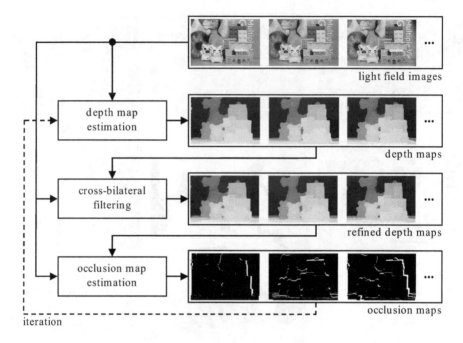

FIGURE 17.8

Overview of the proposed multi-view depth estimation algorithm.

Finally, I_u^d is divided by f_u to recover the clean image I_u, as shown in Figure 17.7h. One scanline profile is also shown in Figure 17.7i for comparison. It can be seen that the recovered image has much less distortion.

17.5.2 Multi-View Depth Estimation

Images corresponding to new viewpoints or focus settings can be rendered from the captured light field by resampling. However, the quality of the rendered image is dictated by the bandwidth of the light field, which strongly depends on the scene geometry [43], [44]. Generally speaking, a scene with larger depth range requires a higher angular resolution for aliasing-free rendering. When the sampling rate is not enough, aliasing effect is observed in rendered images (see Section 17.6).

Traditionally, the aliasing is removed by prefiltering [7] or postfiltering [45]. In this way the out-of-focus object are blurred. These methods implicitly require the depth range of the scene but do not fully utilize this information. The image can be segmented into blocks, each assigned an optimal depth value [43]. If the user wants the best visual quality, this method would require the per-pixel depth value. Finally, although one can adjust the angular resolution of the programmable aperture camera, a high angular sampling rate requires a long capture duration and a large storage, which may not be always affordable.

To solve this problem, a multi-view depth estimation algorithm (Figure 17.8) is proposed here to generate view-dependent depth maps for view interpolation. By depth-dependent view interpolation, the angular sampling rate can be greatly reduced for the near-Lambertian scene. The estimated depth maps can also benefit other applications, such as z-keying, matting [22], and robot vision.

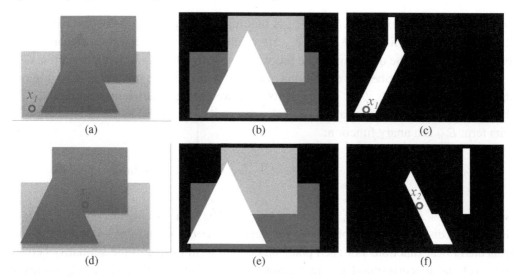

FIGURE 17.9

(a,d) Two images of a simple scene with different viewpoints, (b,e) the corresponding depth maps, and (c,f) the occlusion maps. The black region is unoccluded and white region is occluded.

The multi-view depth estimation problem is similar to the traditional stereo correspondence problem [46]. However, the visibility reasoning is extremely important for multi-view depth estimation since the occluded views should be excluded from the depth estimation. Previous methods that determine the visibility by hard constraint [47] or greedy progressive masking [48] can easily be trapped in local minima because they cannot recover from incorrect occlusion guess. Inspired by the symmetric stereo matching algorithm [49], this problem can be alleviated by iteratively optimizing a view-dependent *depth map* D_u for each image I_u and an *occlusion map* O_{uv} for each pair of neighboring images I_u and I_v. If a scene point projected onto a point x in I_u is occluded in I_v, it does not have a valid correspondence. When this happens, $O_{uv}(x)$ is set to one to exclude it from the matching process (x_1 in Figure 17.9). On the other hand, if the estimated correspondence x' of x_u in I_v is marked as invisible, that is, $O_{vu}(x') = 1$, the estimate is unreliable (x_2 in Figure 17.9).

The depth and occlusion estimation are now reformulated as a discrete labeling problem. For each pixel x_u, a discrete depth value $D_u(x) \in \{0, 1, ..., d_{max}\}$ and a binary occlusion value $O_{uv}(x) \in \{0, 1\}$ need to be determined. More specifically, given a set of light field images $\mathcal{I} = \{I_u\}$, the goal is to find a set of depth maps $\mathcal{D} = \{D_u\}$ and a set of occlusion maps $\mathcal{O} = \{O_{uv}\}$ to minimize the energy functional defined as follows:

$$E(\mathcal{D}, \mathcal{O}|\mathcal{I}) = \sum_u \left\{ E_{dd}(D_u|\mathcal{O}, \mathcal{I}) + E_{ds}(D_u|\mathcal{O}, \mathcal{I}) \right\}$$
$$+ \sum_u \sum_{v \in \mathcal{N}(u)} \left\{ E_{od}(O_{uv}|D_u, \mathcal{I}) + E_{os}(O_{uv}) \right\}, \qquad (17.10)$$

where E_{dd} and E_{ds} are, respectively, the data term and the smoothness (or regularization) term of the depth map, and E_{od} and E_{os} denote, respectively, the data term and the smoothness term of the occlusion map. The term $\mathcal{N}(u)$ denotes the set of eight viewpoints that are closest to u. The energy minimization is performed iteratively. In each iteration, first the occlusion maps are fixed and $E_{dd} + E_{ds}$ is minimized by updating the depth maps. Then, the depth maps are fixed and $E_{od} + E_{os}$ are minimized by updating the occlusion maps. Figure 17.8 shows the overview of the proposed algorithm for minimizing Equation 17.10.

The following describes the definitions of energy terms and the method used to minimize these terms. Let α, β, γ, ζ, and η denote the weighting coefficients and K and T the thresholds. These parameters are empirically determined and fixed in the experiments. The data term E_{dd} is a unary function:

$$E_{dd}(D_u|\mathcal{O},\mathcal{I}) = \sum_x \left\{ \sum_{v \in \mathcal{N}(u)} \left(\bar{O}_{uv}(x) C(I_u(x) - I_v(\rho)) + \alpha O_{vu}(\rho) \right) \right\}, \quad (17.11)$$

where $\rho = x + D_{uv}(x)$ and $\bar{O}_{uv}(x) = 1 - O_{uv}(x)$. The term $\bar{O}_{uv}(x) = 1 - O_{uv}(x)$, $D_{uv}(x)$ denotes the disparity corresponding to the depth value $D_u(x)$ and $C(k) = \min(|k|,K)$ is a truncated linear function. For each pixel x_u, the first term measures the similarity between the pixel and its correspondence in I_v, and the second term adds a penalty to an invalid correspondence.

The pairwise smoothness term E_{ds} is based on a generalized Potts model:

$$E_{ds}(D_u|\mathcal{O},\mathcal{I}) = \sum_{\substack{(x,y) \in \mathscr{P}, \\ O_u(x) = O_u(y)}} \beta \min(|D_u(x) - D_u(y)|, T), \quad (17.12)$$

where $\mathscr{P}$ is the set of all pairs of neighboring pixels and $O_u = \bigcap O_{uv}$, which is true only when x_u is occluded in all other images. This term encourages the depth map to be piecewise smooth.

Since the depth maps $\mathscr{D}$ are fixed in the second step of each iteration, the prior of an occlusion map can be obtained by warping the depth map. Specifically, let W_{uv} denote a binary map. The value of $W_{uv}(x)$ is one when the depth map D_v warped to the viewpoint u is null at x_u, and zero otherwise. If $W_{uv}(x) = 1$, x_u might be occluded in I_v. With this prior and for $\rho = x + D_{uv}(x)$, the data term E_{od} is formulated as:

$$E_{od}(O_{uv}|D_u,\mathcal{I}) = \sum_x \left(\bar{O}_{uv}(x) C(I_u(x) - I_v(\rho)) + \gamma O_{uv}(x) + \zeta |O_{uv}(x) - W_{uv}(x)| \right). \quad (17.13)$$

The first term above biases a pixel to be non-occluded if it is similar to its correspondence. The second term penalizes the occlusion ($O = 1$) to prevent the whole image from being marked as occluded, and the third term favors the occlusion when the prior W_{uv} is true. Finally, the smoothness term E_{os} is based on the Potts model:

$$E_{os}(O_{uv}) = \sum_{(x,y) \in \mathscr{P}} \eta |O_{uv}(x) - O_{uv}(y)|. \quad (17.14)$$

The solution of the energy minimization problem is a maximum *a posteriori* estimate of a Markov random field (MRF), for which high-performance algorithms have been recently developed. The MRF optimization library [50] is used here to address this problem. Three leading algorithms in the library, the alpha-expansion graph cut [51], the asynchronous belief propagation [52], and the tree-reweighted message passing [53] perform well. The latter two methods give slightly better results at the cost of execution time. Finally, a

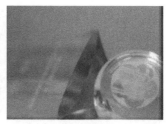

FIGURE 17.10

Digital refocusing without depth information. The angular resolution is 4×4.

modified cross bilateral filtering is applied to the depth maps at the end of each iteration to improve their quality and make the iteration converge faster [54]. When the resolution of the light field is too large to fit in the memory, the tile-based belief propagation [55], which requires much smaller memory and bandwidth than the previous methods, is used.

17.5.3 Summary

The light field images captured using the proposed programmable aperture camera have several advantages for depth estimation. First, the viewpoints of the light field images are well aligned with the 2D grid on the aperture, and thus the depth estimation can be performed without camera calibration. Second, the disparity corresponding to a depth value can be adjusted by changing the camera parameters without any additional rectification as required in camera array systems. Finally, unlike depth-from-defocus methods [21], [56], there is no ambiguity in the scene points behind and in front of the in-focus object.

Finally, when the scene is out-of-focus, only the disparity cue between the light field images are used for depth estimation. It is possible to combine the defocus cue and further remove the defocus blur, as in Reference [57]. Also, it is possible to iteratively estimate the vignetting fields and depth maps to obtain better results.

17.6 Results

All data in the experiments are captured indoors. Images shown in Figures 17.10 and 17.11 are captured using the first prototype and the rest are captured using the second one. The shutter speed of each exposure is set to 10ms for images shown in Figures 17.3, 17.11, and 17.14, and 20ms for the rest. These settings are chosen for the purpose of fair comparison. For example, it takes 160ms with an aperture setting of f/8 to capture a clean and large depth of field image for the scene in Figure 17.11; therefore 10ms is chosen for the proposed device.

All the computations are performed on a Pentium IV 3.2GHz computer with 2GB memory. Demultiplexing one light field dataset takes three to five seconds. To save the computational cost, the light field images are optionally downsampled to 640×426 after demultiplexing. The photometric calibration takes 30 seconds per image, and the multi-view depth

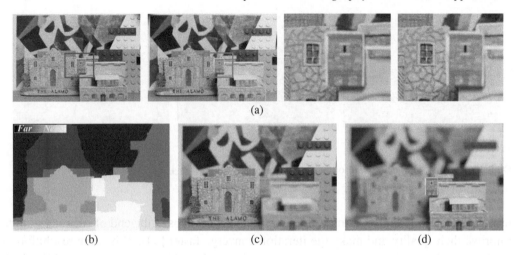

(a)

(b)   (c)   (d)

FIGURE 17.11 (See color insert.)

(a) Two demultiplexed light field images generated by the proposed system; the full 4D resolution is $4 \times 4 \times 3039 \times 2014$. (b) The estimated depth map of the left image of (a). (c,d) Postexposure refocused images generated from the light field and the depth maps.

estimation takes around 30 minutes. The following demonstrates still images with various effects generated from the captured light field and the associated depth maps. The video results are available on the project website http://mpac.ee.ntu.edu.tw/ chiakai/pap.

Figure 17.10 shows a scene containing a transparent object in front of a nearly uniform background. The geometry of this scene is difficult to estimate. However, since the proposed acquisition method does not impose any restriction on the scene, the light field can be captured with 4×4 angular resolution and faithful refocused images can be generated through dynamic reparameterization [44].

The dataset shown in Figure 17.11 is used to evaluate the performance of the proposed postprocessing algorithms. Here a well-known graph cut stereo matching algorithm without occlusion reasoning is implemented for comparison [51]. The photo-consistency assumption is violated in the presence of the photometric distortion, and thus poor result is obtained (Figure 17.12a). With the photometric calibration, the graph cut algorithm generates a good depth map but errors can be observed at the depth discontinuities (Figure 17.12b). On the contrary, the proposed depth estimation algorithm can successfully identify these discontinuities and generate a more accurate result (Figure 17.11b).

Both the light field data and the postprocessing algorithms are indispensable for generating plausible photographic effects. To illustrate this, a single light field image and its associated depth map are used as the input of the Photoshop Lens Blur tool to generate a defocused image. The result shown in Figure 17.12c contains many errors, particularly at the depth discontinuities (Figure 17.12d). In contrast, the results of the proposed algorithm (Figures 17.11c and 17.11d) are more natural. The boundaries of the defocused objects are semitransparent and thus the objects behind can be partially seen.

Figure 17.13 shows the results of view interpolation. The raw angular resolution is 3×3. If a simple bilinear interpolation is used, ghosting effect due to aliasing is observed (Figure 17.13b). While previous methods use filtering to remove the aliasing [7], [44], a mod-

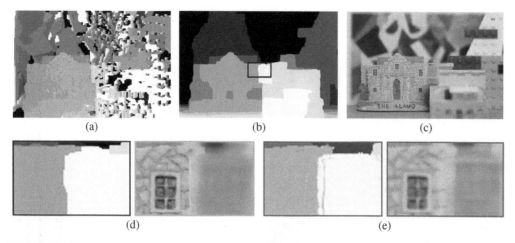

FIGURE 17.12

(a) Depth map estimated *without* photometric calibration and occlusion reasoning, (b) depth map estimated *without* occlusion reasoning, and (c) defocusing by the Photoshop Lens Blur tool. (d) Close-up of (b) and (c). (e) Corresponding close-up of Figures 17.11b and 17.11c.

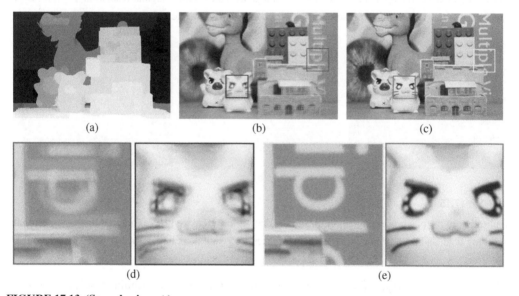

FIGURE 17.13 (See color insert.)

(a) An estimated depth map. (b) Image interpolated without depth information. (c) Image interpolated with depth information. (d-e) Close-up of (b-c). The angular resolution of the light field is 3×3.

ified projective texture mapping [58] is used here instead. Given a viewpoint, three closest images are warped according to their associated depth maps. The warped images are then blended; the weight of each image is inversely proportional to the distance between its viewpoint and the given viewpoint. This method greatly suppresses the ghosting effect without blurring (Figure 17.13c). Note that unlike the single-image view morphing method [21], hole-filling is not performed here due to the multi-view nature of the light field. In most cases, the region occluded in one view is observed in others.

FIGURE 17.14 (See color insert.)

(a) An estimated depth map. (b) Digital refocused image with the original angular resolution 4×4. (c) Digital refocused image with the angular resolution 25×25 boosted by view interpolation. (d-e) Close-up of (b-c).

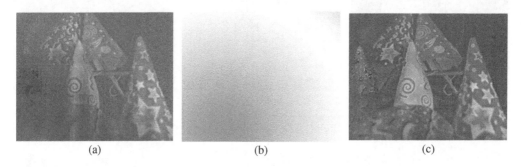

FIGURE 17.15

Application of the postprocessing algorithms to the dataset in Reference [4]: (a) original image, (b) estimated vignetting field, and (c) processed image. Image courtesy of Ashok Veeraraghavan.

Figure 17.14 shows another digital refocusing result. The raw angular resolution is 4×4. Though the in-focus objects are sharp, the out-of-focus objects are subject to the ghost effect due to aliasing (Figure 17.14b). With the estimated depth maps, the angular resolution is first increased to 25×25 by view interpolation described above and then digital refocusing is performed. As can be seen in Figure 17.14c, the out-of-focus objects are blurry while the in-focus objects are unaffected.

Finally, to illustrate the robustness of the proposed algorithms, these algorithms are applied to the noisy and photometrically distorted data captured by the heterodyned light field camera [4]. Four clear images are selected from the data to perform photometric calibration and multi-view depth estimation and synthesize the whole light field by view interpolation. As seen in Figure 17.15, the interpolated image is much cleaner than the original one.

17.7 Discussion

This section discusses the performance and limitations of the proposed camera and the directions of future research.

17.7.1 Performance Comparisons

Three different devices, a conventional camera, a plenoptic camera [1], [14], and a programmable aperture camera, are compared. Because no light ray is blocked or attenuated in the plenoptic camera, it is superior to other mask-based light field cameras [4], [12]. Without loss of generality, it is assumed that the default number of sensors in these devices is M^2, and the angular resolution of the two light field cameras is N^2. The total exposure duration for capturing a single dataset is fixed. Therefore, each exposure in the proposed device is $1/N^2$ of the total exposure.

A signal-to-noise ratio (SNR) analysis of these devices is performed using a simple noise model. There are typically two zero-mean noise sources in the imaging process; one with a constant variance σ_c^2 and another with a variance σ_p^2 proportional to the received irradiance P of the sensor. The results of the SNR analysis are listed in Table 17.1. The image captured by a conventional camera with a large aperture has the best quality, but it has a shallow depth of field. A light field image is equivalent to the image captured by a conventional camera with a small aperture and thus its quality is lower. However, this can be improved by digital refocusing. Light rays emitted from an in-focus scene point are recorded by N^2 light field samples. The refocusing operation averages these samples and thus increases the SNR by N.

TABLE 17.1

Performance comparison between the conventional camera, the plenoptic camera, and the programmable aperture camera.[†]

Device	AS	#shot	SED	SNR_{LFS}	SNR_{RI}	A× SR
CCSA	A	1	T	$P/\sqrt{\sigma_p^2+\sigma_c^2}$	—	$1 \times M^2$
CCLA	N^2A	1	T	$N^2P/\sqrt{N^2\sigma_p^2+\sigma_c^2}$	—	$1 \times M^2$
PCAM	N^2A	1	T	$N^2P/\sqrt{N^2\sigma_p^2+\sigma_c^2}$	$N^3P/\sqrt{N^2\sigma_p^2+\sigma_c^2}$	$N^2 \times M^2/N^2$
PCAMS	N^2A	1	T	$P/\sqrt{\sigma_p^2+\sigma_c^2}$	$NP/\sqrt{\sigma_p^2+\sigma_c^2}$	$N^2 \times M^2$
PAC	A	N^2	T/N^2	$N^{-2}P/\sqrt{\sigma_p^2/N^2+\sigma_c^2}$	$N^{-1}P/\sqrt{\sigma_p^2/N^2+\sigma_c^2}$	$N^2 \times M^2$
PACM	$\approx N^2A/2$	N^2	T/N^2	$\approx NS_1/2$	$\approx N^2S_1/2$	$N^2 \times M^2$

† CCSA – conventional camera with small aperture, CCLA – conventional camera with large aperture, PCAM – plenoptic camera, PCAMS – plenoptic camera with N^2M^2 sensors, PAC – programable aperture camera, PACM – programable aperture camera with multiplexing, AS – aperture size, SED – single exposure duration, SNR_{LFS} – SNR of the light field samples, SNR_{RI} – SNR of the refocused image, A×SR – angular × spatial resolution.

Though the light field data is noisier then the normal picture, it enables better postprocessing abilities. Because the image from the light field can be simply refocused, there is no longer need for setting up the focus and aperture size. In traditional photography much longer time is usually spent on these settings than on the exposure.

The plenoptic camera is slightly better than the programmable aperture camera at the same angular and spatial resolutions. Nevertheless, it requires N^2M^2 sensors. To capture a light field of the same resolution as the dataset shown in Figure 17.11, the plenoptic camera requires an array of nearly 100 million sensors, which is expensive, if not difficult, to make.

17.7.2 Limitation and Future Direction

The proposed device has great performance and flexibility, but it requires that the scene and the camera be static because the data are captured sequentially. However, as mentioned in Section 17.4.4, the sharpness of in-focus objects is unaffected by multiplexing. Hence the proposed system can capture a moving in-focus object amid static out-of-focus objects and then recover the light field and scene geometry of the static objects.

On the other hand, other devices capture the light field in one exposure at the expense of spatial resolution. However, it should be pointed out that the proposed method is complementary to the existing ones. A cosine mask or a microlens array can be placed near the image plane to capture a coarse angular resolution light field and the programmable aperture can be used to provide the fine angular resolution needed.

Multiplexing a light field is equivalent to transforming the light field to another representation by basis projection. While the goal here is to obtain a reconstruction with minimal error from a fixed number of projected images ($M_u(x)$ in Equation 17.4), an interesting research direction is to reduce the number of images required for reconstruction. The compressive sensing theory states that if a signal of dimension n has a sparse representation, less than n projected measurements can be used to recover the full signal [59]. Finding a proper set of bases to perform compressive sensing is worth pursuing in the future.

The light field captured by light field cameras can be applied to many different applications, with only a few basic ones demonstrated here. For example, using the 4D frequency analysis to the light transport process given in Section 17.3, it is possible to detect the major objects of the scene at different depths without estimating the per-pixel depth map [60]. Given this information, it is possible to perform refocusing to the major objects or all-focused rendering without any user intervention. It is also possible to combine the multi-view nature of the light field and the active light sources for photometric stereo. Moreover, since the light field images of different viewpoints capture the sub-pixel movement of the scene, this information can be utilized for creating super-resolution images.

17.8 Conclusion

This chapter described a system for capturing light field using a programmable aperture with an optimal multiplexing scheme. Along with the programmable aperture, two postprocessing algorithms for photometric calibration and multi-view depth estimation were developed. This system is probably the first single-camera system that generates light field at the same spatial resolution as that of the sensor, has adjustable angular resolution, and is free of photometric distortion. In addition, the programmable aperture is fully backward compatible with conventional apertures.

While this work focused on the light field acquisition, the programmable aperture camera can be further exploited for other applications. For example, it can be used to realize a computational camera with a fixed mask. It is believed that by replacing the traditional aperture with the proposed programmable aperture, a camera will become much more versatile than before.

References

[1] R. Ng, M. Levoy, M. Brédif, G. Duval, M. Horowitz, and P. Hanrahan, "Light field photography with a hand-held plenoptic camera," CSTR 2005-02, Stanford University, April 2005.

[2] T. Georgiev, K.C. Zheng, B. Curless, D. Salesin, S. Nayar, and C. Intwala, "Spatio-angular resolution tradeoff in integral photography," in *Proceedings of the 17th Eurographics Workshop on Rendering*, Nicosia, Cyprus, June 2006, pp. 263–272.

[3] T. Georgiev, C. Intwala, and D. Babacan, "Light-field capture by multiplexing in the frequency domain," Technical Report, Adobe Systems Incorporated, 2007.

[4] A. Veeraraghavan, R. Raskar, A. Agrawal, A. Mohan, and J. Tumblin, "Dappled photography: Mask enhanced cameras for heterodyned light fields and coded aperture refocusing," *ACM Transactions on Graphics*, vol. 26, no. 3, pp. 69:1–69:10, July 2007.

[5] C.K. Liang, G. Liu, and H.H. Chen, "Light field acquisition using programmable aperture camera," in *Proceedings of the IEEE International Conference on Image Processing*, San Antonio, TX, USA, September 2007, pp. 233–236.

[6] C.K. Liang, T.H. Lin, B.Y. Wong, C. Liu, and H.H. Chen, "Programmable aperture photography: Multiplexed light field acquisition," *ACM Transactions on Graphics* , vol. 27, no. 3, pp. 55:1–10, August 2008.

[7] M. Levoy and P. Hanrahan, "Light field rendering," in *Proceedings of the 23rd Annual Conference on Computer Graphics and Interactive Techniques*, New York, NY, USA, August 1996, pp. 31–42.

[8] S.J. Gortler, R. Grzeszczuk, R. Szeliski, and M.F. Cohen, "The lumigraph," in *Proceedings of the 23rd Annual Conference on Computer Graphics and Interactive Techniques*, New York, NY, USA, August 1996, pp. 43–54.

[9] J.C. Yang, M. Everett, C. Buehler, and L. McMillan, "A real-time distributed light field camera," in *Proceedings of the 13th Eurographics Workshop on Rendering*, Pisa, Italy, June 2002, pp. 77–85.

[10] B. Wilburn, N. Joshi, V. Vaish, E.V. Talvala, E. Antunez, A. Barth, A. Adams, M. Horowitz, and M. Levoy, "High performance imaging using large camera arrays," *ACM Transactions on Graphics*, vol. 24, no. 3, pp. 765–776, July 2005.

[11] M.G. Lippmann, "Epreuves reversible donnant la sensation du relief," *Journal de Physics*, vol. 7, pp. 821–825, 1908.

[12] H.E. Ive, "Parallax panoramagrams made with a large diameter lens," *Journal of the Optical Society of America*, vol. 20, no. 6, pp. 332–342, June 1930.

[13] T. Okoshi, *Three-dimensional imaging techniques*, New York: Academic Press New York, 1976.

[14] E.H. Adelson and J.Y.A. Wang, "Single lens stereo with a plenoptic camera," *IEEE Transactions on Pattern Analysis and Machine Intelligence*, vol. 14, no. 2, pp. 99–106, February 1992.

[15] A. Lumsdaine and T. Georgiev, "The focused plenoptic camera," in *Proceedings of the First IEEE International Conference on Computational Photography*, San Francisco, CA, USA, April 2009.

[16] R. Raskar, A. Agrawal, and J. Tumblin, "Coded exposure photography: Motion deblurring using fluttered shutter," *ACM Transactions on Graphics*, vol. 25, no. 3, pp. 795–804, July 2006.

[17] S.K. Nayar and V. Branzoi, "Adaptive dynamic range imaging: Optical control of pixel exposures over space and time," in *Proceedings of the 9th IEEE International Conference on Computer Vision*, Nice, France, October 2003, pp. 1168–1175.

[18] Y.Y. Schechner and S.K. Nayar, "Uncontrolled modulation imaging," in *Proceedings of the IEEE Conference on Computer Vision and Pattern Recognition*, Washington, DC, USA, June 2004, pp. 197–204.

[19] A. Zomet and S.K. Nayar, "Lensless imaging with a controllable aperture," *Proceedings of the IEEE Conference on Computer Vision and Pattern Recognition*, New York, USA, June 2006, pp. 339–346.

[20] H. Farid and E.P. Simoncelli, "Range estimation by optical differentiation," *Journal of the Optical Society of America A*, vol. 15, no. 7, pp. 1777–1786, July 1998.

[21] A. Levin, R. Fergus, F. Durand, and W.T. Freeman, "Image and depth from a conventional camera with a coded aperture," *ACM Transactions on Graphics*, vol. 26, no. 3, p. 70, July 2007.

[22] Y. Bando, B.Y. Chen, and T. Nishita, "Extracting depth and matte using a color-filtered aperture," *ACM Transactions on Graphics*, vol. 27, no. 5, pp. 134:1–9, December 2008.

[23] R. Raskar, K.H. Tan, R. Feris, J. Yu, and M. Turk, "Non-photorealistic camera: Depth edge detection and stylized rendering using multi-flash imaging," *ACM Transactions on Graphics*, vol. 23, no. 3, pp. 679–688, August 2004.

[24] N. Joshi, W. Matusik, and S. Avidan, "Natural video matting using camera arrays," *ACM Transactions on Graphics*, vol. 25, no. 3, pp. 779–786, July 2006.

[25] C. Senkichi, M. Toshio, H. Toshinori, M. Yuichi, and K. Hidetoshi, "Device and method for correcting camera-shake and device for detecting camera shake," JP Patent 2003-138436, 2003.

[26] Y.Y. Schechner, S.K. Nayar, and P.N. Belhumeur, "A theory of multiplexed illumination," in *Proceedings of the 9th IEEE International Conference on Computer Vision*, Nice, France, October 2003, pp. 808–815.

[27] A. Wenger, A. Gardner, C. Tchou, J. Unger, T. Hawkins, and P. Debevec, "Performance relighting and reflectance transformation with time-multiplexed illumination," *ACM Transactions on Graphics*, vol. 24, no. 3, pp. 756–764, July 2005.

[28] N. Ratner, Y.Y. Schechner, and F. Goldberg, "Optimal multiplexed sensing: Bounds, conditions and a graph theory link," *Optics Express*, vol. 15, no. 25, pp. 17072–17092, December 2007.

[29] A. Levin, S.W. Hasinoff, P. Green, F. Durand, and W.T. Freeman, "4D frequency analysis of computational cameras for depth of field extension," *ACM Transactions on Graphics*, vol. 28, no. 3, pp. 97:1–14, July 2009.

[30] F. Durand, N. Holzschuch, C. Soler, E. Chan, and F.X. Sillion, "A frequency analysis of light transport," *ACM Transactions on Graphics*, vol. 24, no. 3, pp. 1115–1126, July 2005.

[31] R. Ng, "Fourier slice photography," *ACM Transactions on Graphics*, vol. 24, no. 3, pp. 735–744, July 2005.

[32] M. Zwicker, W. Matusik, F. Durand, and H. Pfister, "Antialiasing for automultiscopic 3D displays," in *Proceedings of the 17th Eurographics Symposium on Rendering*, Nicosia, Cyprus, June 2006, pp. 73–82.

[33] M. Harwit and N.J. Sloane, *Hadamard Transform Optics*. New York: Academic Press, July 1979.

[34] HP components group, "Noise sources in CMOS image sensors," Technical Report, Hewlett-Packard Company, 1998.

[35] Y. Tsin, V. Ramesh, and T. Kanade, "Statistical calibration of CCD imaging process," in *Proceedings of the 8th IEEE International Conference on Computer Vision*, Vancouver, BC, Canada, July 2001, pp. 480–487.

[36] N. Ratner and Y.Y. Schechner, "Illumination multiplexing within fundamental limits," in *Proceedings of the IEEE Conference on Computer Vision and Pattern Recognition*, Minneapolis, MN, USA, June 2007, pp. 1–8.

[37] B.K. Horn, *Robot Vision*. MIT Press, March 1986.

[38] D.B. Goldman and J.H. Chen, "Vignette and exposure calibration and compensation," in *Proceedings of the 10th IEEE International Conference on Computer Vision*, Beijing, China, October 2005, pp. 899–906.

[39] M. Aggarwal, H. Hua, and N. Ahuja, "On cosine-fourth and vignetting effects in real lenses," in *Proceedings of the 8th IEEE International Conference on Computer Vision*, Vancouver, BC, Canada, July 2001, pp. 472–479.

[40] A. Litvinov and Y. Y. Schechner, "Addressing radiometric nonidealities: A unified framework," in *Proceedings of the IEEE Conference on Computer Vision and Pattern Recognition*, San Diego, CA, USA, June 2005, pp. 52–59.

[41] Y. Zheng, J. Yu, S.B. Kang, S. Lin, and C. Kambhamettu, "Single-image vignetting correction using radial gradient symmetry," in *Proceedings of the IEEE Conference on Computer Vision and Pattern Recognition*, Anchorage, AK, USA, June 2008, pp. 1–8.

[42] D.G. Lowe, "Distinctive image features from scale-invariant keypoints," *International Journal of Computer Vision*, vol. 60, no. 2, pp. 91–110, November 2004.

[43] J.X. Chai, X. Tong, S.C. Chan, and H.Y. Shum, "Plenoptic sampling," in *Proceedings of the 27th Annual Conference on Computer Graphics and Interactive Techniques*, New York, NY, USA, July 2000, pp. 307–318.

[44] A. Isaksen, L. McMillan, and S.J. Gortler, "Dynamically reparameterized light fields," in *Proceedings of the 27th Annual Conference on Computer Graphics and Interactive Techniques*, New York, NY, USA, July 2000, pp. 297–306.

[45] J. Stewart, J. Yu, S.J. Gortler, and L. McMillan, "A new reconstruction filter for undersampled light fields," in *Proceedings of the 14th Eurographics Workshop on Rendering*, Leuven, Belgium, June 2003, pp. 150–156.

[46] D. Scharstein and R. Szeliski, "A taxonomy and evaluation of dense two-frame stereo correspondence algorithms," *International Journal of Computer Vision*, vol. 47, no. 1-3, pp. 7–42, April 2002.

[47] V. Kolmogorov and R. Zabih, "Multi-camera scene reconstruction via graph cuts," *Proceedings of the European Conference on Computer Vision*, Copenhagen, Denmark, May 2002, pp. 82–96.

[48] S.B. Kang and R. Szeliski, "Extracting view-dependent depth maps from a collection of images," *International Journal of Computer Vision*, vol. 58, no. 2, pp. 139–163, July 2004.

[49] J. Sun, Y. Li, S.B. Kang, and H.Y. Shum, "Symmetric stereo matching for occlusion handling," in *Proceedings of the IEEE Conference on Computer Vision and Pattern Recognition*, San Diego, CA, USA, June 2005, pp. 399–406.

[50] R. Szeliski, R. Zabih, D. Scharstein, O. Veksler, V. Kolmogorov, A. Agarwala, M. Tappen, and C. Rother, "A comparative study of energy minimization methods for markov random fields," *IEEE Transactions on Pattern Analysis and Machine Intelligence*, vol. 30, no. 6, pp. 1068–1080, June 2008.

[51] Y. Boykov, O. Veksler, and R. Zabih, "Fast approximate energy minimization via graph cuts," *IEEE Transactions on Pattern Analysis and Machine Intelligence*, vol. 23, no. 11, pp. 1222–1239, November 2001.

[52] M. Tappen and W.T. Freeman, "Comparison of graph cuts with belief propagation for stereo, using identical MRF parameters," in *Proceedings of the 9th IEEE International Conference on Computer Vision*, Nice, France, October 2003, pp. 900–907.

[53] V. Kolmogorov, "Convergent tree-reweighted message passing for energy minimization," *IEEE Transactions on Pattern Analysis and Machine Intelligence*, vol. 28, no. 10, pp. 1568–1583, October 2006.

[54] Q. Yang, R. Yang, J. Davis, and D. Nister, "Spatial-depth super resolution for range images," in *Proceedings of the IEEE Conference on Computer Vision and Pattern Recognition*, Minneapolis, MN, USA, June 2007, pp. 1–8.

[55] C.K. Liang, C.C. Cheng, Y.C. Lai, L.G. Chen, and H.H. Chen, "Hardware-efficient belief propagation," in *Proceedings of the IEEE Conference on Computer Vision and Pattern Recognition*, Miami, FL, US), June 2009, pp. 80–87.

[56] P. Green, W. Sun, W. Matusik, and F. Durand, "Multi-aperture photography," *ACM Transactions on Graphics*, vol. 26, no. 3, pp. 68:1–68:10, July 2007.

[57] T. Bishop, S. Zanetti, and P. Favaro, "Light field superresolution," in *Proceedings of the 1st IEEE International Conference on Computational Photography*, San Francisco, CA, USA, April 2009.

[58] P.E. Debevec, C.J. Taylor, and J. Malik, "Modeling and rendering architecture from photographs: A hybrid geometry- and image-based approach," in *Proceedings of the 23rd Annual Conference on Computer Graphics and Interactive Techniques*, New York, NY, USA, August 1996, pp. 11–20.

[59] D. Donoho, "Compressed Sensing," *IEEE Transactions on Information Theory*, vol. 52, no. 4, pp. 1289–1306, April 2006.

[60] Y.H. Kao, C.K. Liang, L.W. Chang, and H.H. Chen, "Depth detection of light field," in *Proceedings of the International Conference on Acoustics, Speech, and Signal Processing*, Honolulu, HI, USA), April 2007, pp. 893–897.

18

Dynamic View Synthesis with an Array of Cameras

Ruigang Yang, Huaming Wang, and Cha Zhang

18.1   Introduction

Being able to look freely through a scene has long been an active research topic in the computer graphics community. Historically, computer graphics research has been focused on *rendering*. That is, given a three-dimensional (3D) model, how to generate new images faster, better, and more realistically. *View synthesis* addresses a typically more challenging problem. It is aimed to generate new images using only a set of two-dimensional (2D) images, instead of 3D models.

There are many different ways to categorize and introduce the vast variety of existing view synthesis methods. For example, these methods can be categorized based on the type of scenes they can handle, the number of input images required, or the level of automation. In this chapter, existing methods are categorized using their internal representations of the scene. Based on this criterion, there is a continuum of view synthesis methods shown in Figure 18.1. They vary on the dependency of images samples vs. geometric primitives.

Approaches on the left side of the continuum are categorized as *geometry-based*. Given a set of input images, a 3D model is extracted, either manually or algorithmically, and can then be rendered from novel viewing angles using computer graphics rendering techniques. In this category, the primary challenge is in the creation of the 3D model. Automatic extraction of 3D models from images has been one of the central research topics in the field of *computer vision* for decades. Although many algorithms and techniques exist, such as the extensively studied stereo vision techniques, they are relatively fragile and prone to error in practice. For instance, most 3D reconstruction algorithms assume a Lambertian (diffuse) scene, which is only a rough approximation of real-world surfaces.

By contrast, approaches on the right side of the continuum are categorized as *image-based modeling and rendering* (IBMR) — a popular alternative for view synthesis in recent years. The basic idea is to synthesize new images directly from input images, partly or completely bypassing the intermediate 3D model. In other words, IBMR methods typically represent the scene as a collection of images, optionally augmented with additional information for view synthesis. Light field rendering (LFR) [1], [2] represents one extreme of such techniques; it uses many images (hundreds or even thousands) to construct a light field function that completely characterizes the flow of light through unobstructed space in a scene. Synthesizing different views becomes a simple lookup of the light field function. This method works for any scene and any surface; the synthesized images are usually so realistic that they are barely distinguishable from real photos. But the success of this method ultimately depends on having a very high sampling rate, and the process of capturing, storing, and retrieving many samples from a real environment can be difficult or even impossible.

In the middle of the continuum are some hybrid methods that represent the scene as a combination of images samples and geometrical information. Typically these methods require a few input images as well as some additional information about the scene, usually in the form of approximate geometric knowledge or correspondence information. By using this information to set constraints, the input images can be correctly warped to generate novel views. To avoid the difficult shape recovery problem, successful techniques usually

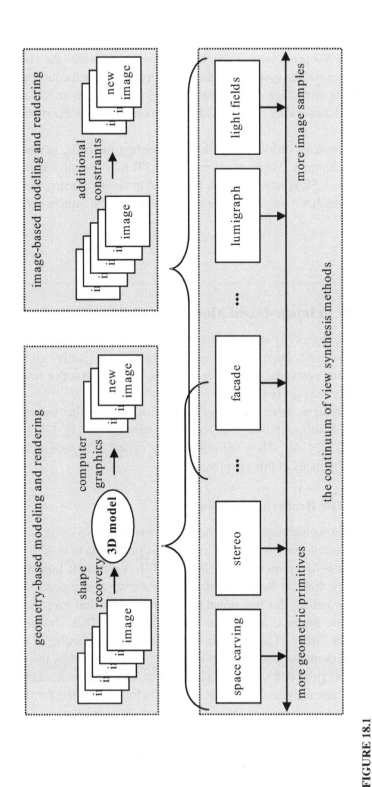

FIGURE 18.1

The continuum for view synthesis methods. Relative positions of some well-known methods are indicated. The two subfigures show two main subgroups for view synthesis. Note that the boundary between these two groups is blurry.

require a human operator to be involved in the process and use *a priori* domain knowledge to constrain the problem. Because of the required user interaction, these techniques are typically categorized under the IBMR paradigm. For example, in the successful Façade system [3] designed to model and render architecture from photographs, an operator first manually places simple 3D primitives in rough positions and specifies the corresponding features in the input images. The system automatically optimizes the location and shape of the 3D primitives, taking advantage of the regularity and symmetry in architectures. Once a 3D model is generated, new views can be synthesized using traditional computer graphics rendering techniques.

This chapter briefly reviews various image-based modeling methods, in particular light field-style rendering techniques. Then, an extension of LFR for dynamic scenes, for which the notion of space-time LFR is introduced, is discussed in detail. Finally, *reconfigurable* LFR is introduced, in which not only the scene content, but also the camera configurations, can be dynamic.

18.2   Related Work in Image-Based Modeling and Rendering

As pioneered in References [4] and [5], the basic idea of *image-based modeling and rendering* (IBMR) is to use a large number of input images to (partly) circumvent the difficult 3D reconstruction problems. IBMR has been one of the most active research topics in computer graphics for the past few years.

Within the IBMR paradigm, there are a large number of methods that use many images to generate novel views without relying on geometric information [1], [2], [6], [7], [8], [9], [10], [11], [12], [13], [14]. They are collectively called *light field*-style rendering techniques, which are the focus of this chapter.

18.2.1   Light Field-Style Rendering Techniques

Light field style rendering techniques are formulated around the *plenoptic function* (from the Latin root *plenus*, meaning complete or full, and *optic*, pertaining to vision). The notion of plenoptic functions was first proposed in Reference [15]. A plenoptic function describes all of the radiant energy that can be perceived by an observer at any point in space and time. Reference [15] formalizes this functional description by providing a parameter space over which the plenoptic function is valid, as shown in Figure 18.2. Imagine that one wants to look at a scene freely. The location of an idealized eye can be at any point in space (V_x, V_y, V_z). From there one can perceive a bundle of rays defined by the range of the azimuth and elevation angles (θ, ϕ), as well as a band of wavelengths λ. The time t can also be selected if the scene is dynamic. This results in the following form for the plenoptic function:

$$\rho = P(\theta, \phi, \lambda, V_x, V_y, V_z, t). \tag{18.1}$$

This seven-dimensional function can be used to develop a taxonomy for evaluating models of low-level vision [15]. By introducing the plenoptic function to the computer graphics

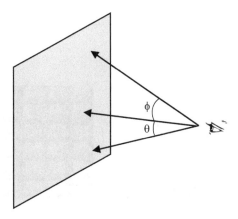

FIGURE 18.2

The plenoptic function describes all of the image information visible from any given viewing position.

community, all image-based modeling and rendering approaches can be cast as attempts to reconstruct the plenoptic function from a sample set of that function [4].

From a computer graphics standpoint, one can consider a plenoptic function as a scene representation that describes the flow of light in all directions from any point at any time. In order to generate a view from a given point in a particular direction, one would merely need to plug in appropriate values for (V_x, V_y, V_z) and select from a range of (θ, ϕ, λ) for some constant t.

This plenoptic function framework provides many venues for exploration, such as the representation, optimal sampling, and reconstruction of the plenoptic function. The following sections discuss several popular parameterizations of the plenoptic function under varying degrees of simplification.

18.2.1.1 5D Light Field

Assuming that there is a static scene and that the effects of different wavelengths can be ignored, the representation of the reduced five-dimensional (5D) plenoptic function can be seen as a set of panoramic images at different 3D locations [4], [5]. In this implementation, computer vision techniques are employed to compute stereo disparities between panoramic images to avoid a dense sampling of the scene.[1] After the disparity images are computed, the input images can be interactively warped to new viewing positions. Visibility is resolved by forward-warping [16] the panoramic images in a back-to-front order. This hidden-surface algorithm is a generalization of a visible line method [17] to arbitrary projected grid surfaces.

18.2.1.2 4D Light Field

Reference [1] points out that the 5D representation offered in References [4] and [5] can be reduced to the four-dimensional (4D) representation in free space (regions free of

---

[1]References [4], [5] could be included in the second category, in which geometry information is used. However, they are presented here, since the plenoptic function theory presented in these references paved the way for subsequent light field rendering techniques.

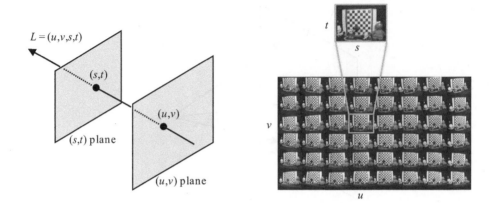

FIGURE 18.3

Light field rendering: the light slab representation and its construction.

occluders[2]) [1]. This is based on the observation that the radiance along a line does not change unless blocked. In Reference [1], the 4D plenoptic function is called the *light field* function which may be interpreted as a functions on the space of oriented light rays. Such reduction in dimensions has been used to simplify the representation of radiance emitted by luminaries [18], [19].

Reference [1] parameterizes light rays based on their intersections with two planes (see Figure 18.3). The coordinate system is (u,v) on the first plane, and (s,t) on the second plane. An oriented light ray is defined by connecting a point on the uv plane to a point on the st plane; this representation is called a light slab. Intuitively, a light slab represents the beam of light entering one quadrilateral and exiting another quadrilateral. To construct a light slab, one can simply take a 2D array of images. Each image can be considered a slice of the light slab with a fixed (u,v) coordinate and a range of (s,t) coordinates.

Generating a new image from a light field is quite different than previous view interpolation approaches. First, the new image is generally formed from many different pieces of the original input images, and does not need to look like any of them. Second, no model information, such as depth values or image correspondences, is needed to extract the image values. The second property is particularly attractive since automatically extracting depth information from images is a very challenging task. However, many image samples are required to completely reconstruct the 4D light field functions. For example, to completely capture a small Buddha as shown in Figure 18.3, hundreds or even thousands of images are required. Obtaining so many images samples from a real scene may be difficult or even impossible. Reference [2] augments the two-plane light slab representation with a rough 3D geometric model that allows better quality reconstructions using fewer images. However, recovering even a rough geometric model raises the difficult 3D reconstruction problem.

---

[2]Such reduction can be used to represent scenes and objects as long as there is no occluder between the desired viewpoint and the scene. In other words, the effective viewing volume must be *outside* the convex hull of the scene. A 4D representation cannot thus be used, for example, in architecture workthroughs to explore from one room to another.

Reference [8] presents a more flexible parameterization of the light field function. In essence, one of the two planes of the light slab is allowed to move. Because of this additional degree of freedom, it is possible to simulate a number of dynamic photograph effects, such as depth of field and apparent focus [8]. Furthermore, this reparameterization technique makes it possible to create integral photography-based [20], auto-stereoscopic displays for direct viewing of light fields.

Besides the popular two-plane parameterization of the plenoptic function, there is the spherical representation introduced in Reference [6]. The object-space algorithm presented in the same work can easily be embedded into the traditional polygonal rendering system and accelerated by 3D graphics boards.

Light field rendering is the first image-based rendering method that does not require *any* geometric information about the scene. However, this advantage is acquired at the cost of many image samples. The determination of the minimum number of samples required for light field rendering involves complex relationships among various factors, such as the depth and texture variation of the scene, the input image resolutions, and the desired rendering resolutions. Details can be found in Reference [21].

18.2.1.3 3D Plenoptic Function: Line Light Field and Concentric Mosaic

If camera (viewing) motion is further constrained to a continuous surface or curved manifold, the plenoptic function can be reduced to a 3D function. Reference [9] presents a 3D parameterization of the plenoptic function. By constraining camera motion along a line, the 4D light field parameterization can be reduced to a 3D function. Such reduction is necessary for time-critical rendering given limited hardware resources. A (u,s,t) representation is used where u parameterizes the camera motion, and (s,t) parameterizes the other plane. Moving along the line provides parallax in the motion direction. To achieve complete coverage of the object, the camera can move along four connected perpendicular lines, for instance, a square.

Reference [10] presents an alternative parameterization for a 3D plenoptic function called *concentric mosaics*. This is achieved by constraining camera motion to planar concentric circles and creating concentric mosaics through the composition of slit images taken at different locations along each circle. Concentric mosaics index all input image rays naturally according to three parameters: radius, rotation angle, and vertical elevation. Compared to a 4D light field, concentric mosaics have much smaller file sizes because only a 3D plenoptic function is constructed. Concentric mosaics allow a user to move freely in a circular region and observe significant parallax without recovering the geometric and photometric scene models. Compared to the method in Reference [9], concentric mosaics offer uniform and continuous sampling of the scene. However, rendering with concentric mosaics could produce some distortions in the rendered images, such as bending of straight lines and distorted aspect ratios. Detailed discussion about the causes of and possible corrections for these problems can be found in Reference [10].

18.2.1.4 2D Plenoptic Function: Environment Map (Panoramas)

An environment map records the incident light arriving from all directions at a point. Under the plenoptic function framework, a 2D plenoptic function in which only the gaze direction (θ, ϕ) varies can be reconstructed from a single environment map. While the original use of environment maps is to efficiently approximate reflections of the environment on a surface [22], [23], environment maps can be used to quickly display any outward-looking view of the environment from a fixed location but at a variable orientation — this is the basis of the Apple QuickTimeVR system [24]. In this system, environment maps are created at key locations in the scene. A user is able to navigate discretely from one location to another and, while at each location, continuously change the viewing direction.

While it is relatively easy to generate computer-generated environmental maps [23], it is more difficult to capture panoramic images from real scenes. A number of techniques have been developed. Some use special hardware [25], [26], [27], such as panoramic cameras or cameras with parabolic mirrors; others use regular cameras to capture many images that cover the whole viewing space, then stitch them into a complete panoramic image [28], [29], [30], [31], [32].

18.3   Space-Time Light Field Rendering for Dynamic Scenes

With technological advancement, it is now possible to build camera arrays to capture a dynamic light field for moving scenes [33], [34], [35], [36]. However, most existing camera systems rely on synchronized input, thus treating each set of images taken at the same instant as a separate light field. The only exception is the high-performance light field array presented in Reference [37]. It uses customized cameras that can be triggered in a staggered pattern to increase the temporal sampling rate. Nevertheless, even with an increased temporal sampling rate, a viewer can explore the temporal domain in a discrete way.

Reference [38] presents the notion of *space-time light field rendering* (ST-LFR) that allows *continuous* exploration of a dynamic light field in both the spatial and temporal domain. A space-time light field is defined as a collection of video sequences captured by a set of tightly packed cameras that may or may not be synchronized. The basic goal of ST-LFR is to synthesize novel images from an arbitrary viewpoint at an arbitrary time instant t. Traditional LFR is therefore a special case of ST-LFR at some fixed-time instant.

As shown in Figure 18.4, the ST-LFR approach has the following major steps: image registration, temporal offset estimation (optional), synchronized image generation (temporal interpolation), and light field rendering (spatial interpolation). In the first step (Section 18.3.1), spatial-temporal optical flow algorithm is developed to establish feature correspondences among successive frames for each camera's video sequence. In cases when video sequences are not synchronized or relative temporal offsets are unknown, the relative time stamp for each frame needs be estimated (Section 18.5). Once feature correspondences are established, new globally synchronized images are synthesized using an edge-guided image morphing method (Section 18.3.3). Synchronized video sequences are eventually used as inputs to synthesize images in the spatial domain using traditional LFR techniques (Section 18.3.4).

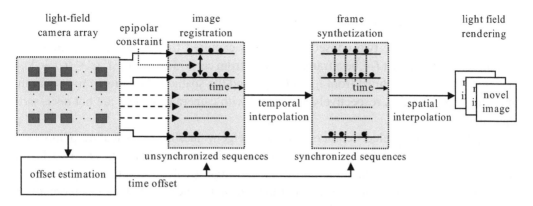

FIGURE 18.4

The structure of the proposed space-time light field rendering algorithm. © 2007 IEEE

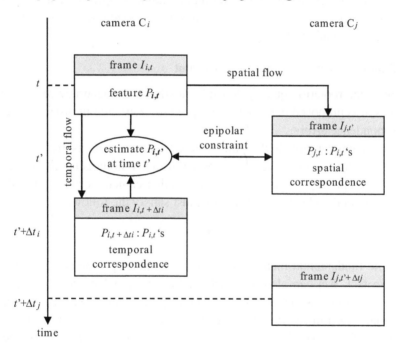

FIGURE 18.5

The registration process of the two-camera case. © 2007 IEEE

18.3.1 Image Registration

The ST-LFR framework starts with computing the temporal optical flow between two successive frames for each camera. The spatial optical flow between different camera sequences is also computed to remove outliers, assuming the relative temporal offset between them is known. The following presents the algorithm with a two-camera setup, and then shows how this algorithm can be extended to handle multiple camera inputs. Figure 18.5 illustrates the two-camera registration process.

18.3.1.1 Spatial-Temporal Flow Computation

Let $I_{i,t}$ and $I_{i,t+\Delta t_i}$ be two consecutive frames captured from camera C_i at time t and $t+\Delta t_i$, respectively. Let $I_{j,t'}$ be a frame captured from camera C_j at time t' ($t' \in [t, t+\Delta t_i]$), the temporal flow be defined from $I_{i,t}$ to $I_{i,t+\Delta t_i}$ and the spatial flow from $I_{i,t}$ to $I_{j,t'}$. Frame time steps, Δt_i and Δt_j, respectively associated with camera C_i and C_j are not necessarily equivalent.

The procedure first selects corners with large eigenvalues on $I_{i,t}$ as feature points using Harris corner detector [39]. A minimum distance is enforced between any two feature points to prevent them from gathering in a small high gradient region. Then, the sparse optical flow are calculated using tracking functions in OpenCV [40], which is based on the classic Kanade-Lucas-Tomasi (KLT) feature tracker [41], [42].

It chosen here not to calculate a dense, per-pixel flow since it is difficult to calculate in certain areas such as occlusion boundaries and textureless regions. A sparse flow formulation, which includes salient features important for view synthesis, is not only more robust, but also more computationally efficient.

18.3.1.2 Flow Correction by Epipolar Constraint

Since image interpolation quality depends mostly on the accuracy of feature point correspondences, it is proposed here to use the *epipolar constraint* in order to detect temporal point correspondence errors.

Given two images captured at the same instant from two cameras, the epipolar constraint states that if a point $\vec{p}_i = [u_i, v_i, 1]^T$ (expressed in image homogeneous coordinates) from one camera and a point $\vec{p}_j = [u_j, v_j, 1]^T$ from another camera correspond to the same stationary 3D point in the physical world, they must satisfy the following:

$$\vec{p}_j^T \mathbf{F} \vec{p}_i = 0, \tag{18.2}$$

where $\mathbf{F}$ is the fundamental matrix encoding the epipolar geometry between the two images [43]. In fact, $\mathbf{F}\vec{p}_i$ can also be considered as an epipolar line in the second image, Equation 18.2 thus means that $\vec{p}_j$ must lie on the epipolar line $\mathbf{F}\vec{p}_i$, and vice versa.

The epipolar constraint is incorporated to verify the temporal flow from $I_{i,t}$ to $I_{i,t+\Delta t_i}$, given the help of the spatial flow from $I_{i,t}$ to $I_{j,t'}$. Let $\vec{p}_{i,t}$ and $\vec{p}_{i,t+\Delta t_i}$ be the projection of a moving 3D point on camera C_i at time t and $\Delta t + t_i$, respectively; the projection of this 3D point forms a trajectory connecting between $\vec{p}_{i,t}$ and $\vec{p}_{i,t+\Delta t_i}$. Given the spatial correspondence $\vec{p}_{j,t'}$ of $\vec{p}_{i,t}$ from camera C_j at time t', the epipolar constraint is described as follows:

$$\vec{p}_{j,t'}^T \mathbf{F_{ij}} \vec{p}_{i,t'} = 0, \tag{18.3}$$

where $\mathbf{F_{ij}}$ is the fundamental matrix between camera C_i and C_j. Since $\vec{p}_{i,t'}$ is not actually observed by camera C_i at time t', it can be estimated assuming that the 3D point moves locally linear in the image space, as follows:

$$\vec{p}_{i,t'} = \frac{t+\Delta t_i - t'}{\Delta t_i} \vec{p}_{i,t} + \frac{t'-t}{\Delta t_i} \vec{p}_{i,t+\Delta t_i}. \tag{18.4}$$

Therefore, in order to evaluate the correctness of $\vec{p}_{i,t}$'s temporal correspondence $\vec{p}_{i,t+\Delta t_i}$, the procedure first estimates $\vec{p}_{i,t'}$ using the linear motion assumption and then verifies $\vec{p}_{i,t'}$

(a)   (b)   (c)

FIGURE 18.6

Feature points and epipolar line constraints: (a) $\vec{p}_{i,t}$ on image $I_{i,t}$, (b) $\vec{p}_{i,t+\Delta t_i}$ on image $I_{i,t+\Delta t_i}$, and (c) $\vec{p}_{j,t'}$ and $\vec{p}_{i,t'}$'s epipolar lines on image $I_{j,t'}$. © 2007 IEEE

by the epipolar constraint using the spatial correspondence $\vec{p}_{j,t'}$. If both spatial and temporal correspondences are correct, the epipolar constraint should be closely satisfied. This is the criterion used to validate the spatial-temporal flow computation. Figure 18.6 shows that a correct correspondence satisfies the epipolar constraint, while a wrong temporal correspondences causes an error in $\vec{p}_{i,t'}$, which leads to a wrong epipolar line that $\vec{p}_{j,t'}$ fails to meet with.

The fundamental matrix is directly computed from cameras' world position and projection matrices. Due to various error sources such as camera noise, inaccuracy in camera calibration or feature localization, a band of certainty is defined along the epipolar line. For each feature point $\vec{p}_{i,t}$, if the distance from $\vec{p}_{j,t'}$ to $\vec{p}_{i,t'}$'s epipolar line is greater than a certain tolerance threshold, either the temporal or the spatial flow is considered to be wrong. This feature will then be discarded. In the experiment, three pixels are used as the distance threshold.

It should be noted that the proposed correction scheme for unsynchronized input is only reasonable when the motion is roughly linear in the projective space. Many real world movements, such as rotation, do not satisfy this requirement. Fortunately, when cameras have a sufficiently high rate with respect to the 3D motion, such a locally temporal linearization is generally acceptable [44]. Achieved experimental results also support this assumption. In Figure 18.6, correct feature correspondences satisfy the epipolar constraint well even though the magazine was rotating fast. This figure also demonstrates the amount of pixel offset casual motion could introduce: in two successive frames captured at 30fps, many feature points moved more than 20 pixels — a substantial amount that cannot be ignored in view synthesis.

18.3.2 Multi-Camera Flow Correction

In a multi-camera system, more than one reference camera or multiple frames can be used to justify a particular temporal flow using the epipolar constraint. Since the spatial flow itself can contain errors, there exists a trade-off between accurate temporal correspondences and the number of false-negatives, that is, correct temporal correspondences are removed because of erroneous spatial correspondences. Therefore, a selection scheme is needed to choose the best frame(s) to compare. Intuitively, closer cameras are relatively good candidates because of fewer occlusions.

Another factor to consider is the ambiguity along the epipolar line, that is, the correspondence error along the direction of the epipolar line cannot be detected. Many light field acquisition systems put cameras on a regular grid, in which the epipolar lines for cameras in the same row or column are almost aligned with each other. In this case, using more reference cameras does not necessarily improve error detection.

Given a reference camera, one can prefer to choose reference frames captured close to time $t + \Delta t_i$. Those frames can fully reveal the temporal flow error since the error only exists in the temporal correspondence $\vec{p}_{i,t+\Delta t_i}$ associated with $\vec{p}_{i,t}$. Note that $\vec{p}_{i,t}$ is always assumed to be the correct image location of some 3D point $\vec{m}_t$.

A selection function W_j is employed here to determine whether camera C_j should be used to test C_i's temporal flow from $I_{i,t}$ to $I_{i,t+\Delta t_i}$:

$$W_j = Close(i,j) + d \min_{\substack{t':I_{j,t'} \in \\ C_j \text{ sequence}}} \left| t' - (t + \Delta t_i) \right| \tag{18.5}$$

and the reference frame $I_{j,t'}$ from C_j is selected as follows:

$$t' = \operatorname*{arg\,min}_{t':I_{j,t'} \in C_j \text{ sequence}} \left| t' - (t + \Delta t_i) \right|, \tag{18.6}$$

where d is a constant to balance the influence from the camera spatial closeness and the capture time difference. If the multi-camera system is constructed regularly as a camera array, the closeness $Close(i,j)$ can be simply evaluated according to the array indices. A single best camera is chosen along the row and the column respectively to provide both horizontal and vertical epipolar constraints. If all cameras have an identical frame rate, the same camera will always be selected using Equation 18.5.

18.3.3 Temporal Interpolation with Edge Guidance

Once the temporal correspondence between two successive frames from one camera is obtained, intermediate frames need to be generated at any time for this camera so that synchronized video sequences can be produced for traditional light field rendering. One possible approach is to simply triangulate the whole image and blend two images by texture mapping. However, the resulting quality is quite unpredictable since it depends heavily on the local triangulation of feature points. Even a single feature mismatch can affect a large neighboring region on the image. A better interpolation scheme is to use the image morphing method [45], which incorporates varying weights using the image gradient. In essence, line segments are formed automatically from point correspondences to cover the entire image. Segments aligned with image edges will gain extra weights to preserve edge straightness — an important visual cue. Therefore, this robust view synthesis method can synthesize smooth and visually appealing images even in the presence of missing or wrong feature correspondence.

FIGURE 18.7

Real feature edges: (left) the gradient magnitude map and (right) feature edges detected by testing feature points pairwise. © 2007 IEEE

18.3.3.1 Extracting Edge Correspondences

Segments can be constructed by extracting edge correspondences. Given a segment e_t connecting two feature points on image I_t and a segment $e_{t+\Delta t}$ connecting the corresponding feature points on $I_{t+\Delta t}$ from the same camera, it is desirable to know whether this segment pair $(e_t, e_{t+\Delta t})$ are both aligned with image edges, that is, forming an *edge correspondence*. The decision is based on two main factors: edge strength and edge length.

Regarding edge strength, the algorithm examines whether a potential edge matches with both image gradient magnitude and orientation, similar to the Canny edge detection method [46]. On a given segment, a set of samples $s_1, s_2, ..., s_n$ are uniformly selected to calculate the fitness function for this segment as follows:

$$f(e) = \min_{s_k \in e}(|\nabla I(s_k)| + \beta \frac{\nabla I(s_k)}{|\nabla I(s_k)|} \bullet N(e)). \tag{18.7}$$

The first term in Equation 18.7 gives the gradient magnitude and the second term indicates how well the gradient orientation matches with the edge normal direction $N(e)$. The parameter β is used to balance the influence between two terms. The fitness values are calculated for both e_t and $e_{t+\Delta t}$. If both values are greater than a threshold, it means that both e_t and $e_{t+\Delta t}$ are in strong gradient regions and the gradient direction matches the edge normal. Therefore, e_t and $e_{t+\Delta t}$ may represent a real surface or texture boundary in the scene. Figure 18.7 shows a gradient magnitude map and detected feature edges.

The second factor is the edge length. It can be assumed that the edge length is nearly a constant from frame to frame, provided that the object distortion is relatively slow compared with the frame rate. This assumption is used to discard edge correspondences if $|e_{t+\Delta t}|$ changes too much from $|e_t|$. Most changes in edge length are caused by wrong temporal correspondences. Nevertheless, the edge length change may be caused by reasonable point correspondences as well. For instance, in Figure 18.8, the dashed arrows show the intersection boundaries between the magazine and the wall. They are correctly tracked, however, since they do not represent real 3D points, they do not follow the magazine motion at all. Edges ended with those points can be detected from length changes, and be removed to avoid distortions during temporal interpolation. Similarly, segments connect-

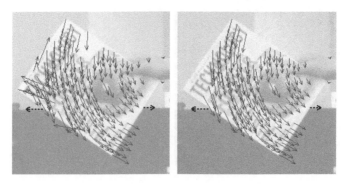

FIGURE 18.8

The optical flow: (left) before applying the epipolar constraint and (right) after applying the epipolar constraint. Correspondence errors on the top of the magazine are detected and removed. © 2007 IEEE

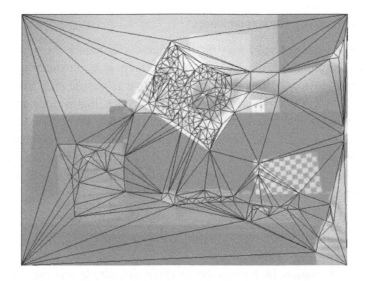

FIGURE 18.9

The edge map constructed after constraining Delaunay triangulation. © 2007 IEEE

ing one static point (that is, a point that does not move in two temporal frames) with one dynamic point should not be considered.

18.3.3.2 Forming the Edge Set

Since the real edge correspondences generated as described in the above section may not be able to cover the whole region sometimes, virtual edge correspondences to the edge set are added to avoid distortions caused by sparse edge sets. Virtual edges are created by triangulating the whole image region using *constrained* Delaunay triangulation [47]. Existing edge correspondences are treated as edge constraints, and four image corners are added. Figure 18.9 shows a sample triangulation result.

FIGURE 18.10

Interpolation results: (left) without virtual feature edges and (right) with virtual feature edges. © 2007 IEEE

18.3.3.3 Morphing

The image morphing method [45] is extended to interpolate two frames temporally. Real edges are allowed to have more influence weights than virtual edges since real edges are believed to have physical meanings in the real world. In Reference [45], the edge weight is calculated using the edge length L and the point-edge distance D as follows:

$$\text{weight}_0 = \left(\frac{L^\rho}{(a+D)} \right)^b, \tag{18.8}$$

where a, b, and ρ are constants to affect the line influence.

The weight for both real and virtual feature edge is calculated here using the formula above. The weight for real edge is then further boosted as follows:

$$\text{weight} = \text{weight}_0 \cdot (1 + \frac{f(e) - f_{\min}(e)}{f_{\max}(e) - f_{\min}(e)})^\tau, \tag{18.9}$$

where $f(e)$ is the edge samples' average fitness value from Equation 18.7. The terms $f_{\min}(e)$ and $f_{\max}(e)$ denote the minimum and maximum fitness value among all real feature edges, respectively. The parameter τ is used to scale the boosting effect exponentially.

Two frames are temporally interpolated using both the forward temporal flow from $I_{i,t}$ to $I_{i,t+\Delta t_i}$ and the backward temporal flow from $I_{i,t+\Delta t_i}$ to $I_{i,t}$. The final pixel intensity is calculated by linear interpolation as follows:

$$P_{t'} = \frac{t + \Delta t_i - t'}{\Delta t_i} \cdot P_{forward} + \frac{t' - t}{\Delta t_i} \cdot P_{backward}, \tag{18.10}$$

where $P_{forward}$ is the pixel color calculated only from frame $I_{i,t}$ and $P_{backward}$ is only from frame $I_{i,t+\Delta t_i}$. This reflects more confidence in features associated with $I_{i,t}$ in the forward flow and features associated with $I_{i,t+\Delta t_i}$ in the backward flow. These features are selected directly on images.

Figure 18.10 shows the results without and with virtual edges. Figure 18.11 shows the results using different interpolation schemes. Since some features are missing on the top of the magazine, the interpolation quality improves when real feature edges get extra weights according to Equation 18.9. Even without additional weights, image morphing generates more visually pleasing images in the presence of bad feature correspondences.

FIGURE 18.11

Interpolated results using different schemes: (a) image morphing without epipolar test, (b) direct image triangulation with epipolar test, (c) image morphing with epipolar test, but without extra edge weights, and (d) image morphing with both epipolar test and extra edge weights. © 2007 IEEE

18.3.4 View Synthesis (Spatial Interpolation)

Once synchronized images for a given time instant are generated, traditional light field rendering techniques can ne used to synthesize views from novel viewpoints. To this end, the unstructured lumigraph rendering technique [48], which can utilize graphics texture hardware to blend appropriate pixels together from nearest cameras in order to compose the desired image, is adopted here as discussed below.

This technique requires an approximation of the scene (a geometric *proxy*) as an input. Two options are implemented. The first is a 3D planar proxy in which the plane's depth can be interactively controlled by the user. Its placement determines which region of the scene

FIGURE 18.12

Reconstructed 3D mesh proxy: (left) oblique view and (right) front view. © 2007 IEEE

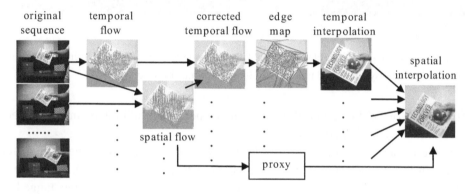

FIGURE 18.13

The space-time light field rendering pipeline. © 2007 IEEE

is in focus. The second option is a reconstructed proxy using the spatial correspondences. Every pair of spatial correspondences defines a 3D point. All the 3D points, plus the four corners of a background plane, are triangulated to create a 3D mesh. Figure 18.12 shows an example mesh.

Whether to use a plane or a mesh proxy in rendering depends mainly on the camera configuration and scene depth variation. The planar proxy works well for scenes with small depth variations. On the other hand, using a reconstructed mesh proxy improves both the depth of field and the 3D effect for oblique-angle viewing.

18.3.5 Results

This section presents some results from the space-time light field rendering framework. Figure 18.13 shows a graphical representation of the entire processing pipeline. To facilitate data capturing, two multi-camera systems were developed. The first system uses eight Point Grey *dragonfly* cameras [49]. This system can provide a global time stamp for each frame in hardware. The second system includes eight SONY color digital fire-wire cameras, but the global time stamp is not available. In both systems, the cameras are approximately 60 mm apart, limited by their form factor. Based on the analysis from Reference [21], the

FIGURE 18.14

Synthesized results using a uniform blending weight, that is, pixels from all frames are averaged together: (left) traditional LFR with unsynchronized frames and (right) space-time LFR. © 2007 IEEE

effective depth of field is about 400 mm. The cameras are calibrated, and all images are rectified to remove lens distortions.

Since the cameras are arranged in a horizontal linear array, only one camera for the epipolar constraint is selected according to the discussion in Section 18.3.2. The closeness is just the camera position distance. The results are demonstrated using data tagged with a global time stamp, that is, the temporal offsets are known. The final images are synthesized using either a planar proxy or a reconstructed proxy from spatial correspondences.

18.3.5.1 View Synthesized with a Planar Proxy

To better demonstrate the advantage of ST-LFR over traditional LFR, this section first shows some results using a planar proxy for spatial interpolation. The datasets used in this section are captured by the Point Grey camera system with known global time stamps. The first dataset includes a person waving the hand, as shown in Figure 18.10. Since the depth variation from the hands to the head is slightly beyond the focus range (400 mm), some tiny horizontal ghosting effects in the synthesized image can be observed, which are entirely different from vertical mismatches caused by the hand's vertical motion. It will be shown later that these artifacts can be reduced by using a reconstructed proxy.

The importance of virtual synchronization is emphasized in Figure 18.14, where the image is synthesized with a constant blending weight $(1/8)$ for all eight cameras. The scene contains a moving magazine. Without virtual synchronization, the text on the magazine cover is illegible. This problem is rectified by registering feature points and generating virtually synchronized images.

The next dataset is a moving open book. In Figure 18.15, several views are synthesized using traditional LFR and ST-LFR. Results from ST-LFR remain sharp from different viewpoints. The noticeable change of intensity is due to the mixed use of color and grayscale cameras.

FIGURE 18.15

Synthesized results of the book image: (top) traditional LFR with unsynchronized frames and (bottom) space-time LFR. © 2007 IEEE

FIGURE 18.16

Synthesized results of the book sequence: (left) plane proxy adopted and (right) 3D reconstructed mesh proxy adopted. © 2007 IEEE

18.3.5.2 View Synthesized with a Reconstructed Proxy

As mentioned in Section 18.3.4, building a 3D mesh geometry proxy improves the synthesis results for scenes with large depth range. This method is adopted in the following two datasets and the results are compared with plane proxy. The first dataset is the open book as shown in Figure 18.16. It is obvious that the back cover of the book in this example is better synthesized with the 3D mesh proxy for its improved depth information. The second example shown in Figure 18.17 is a human face viewed from two different angles. The face rendered with the planar proxy shows some noticeable distortions in side views.

FIGURE 18.17

Synthesized results of a human face from two different angles: (top) plane proxy adopted and (bottom) 3D mesh proxy adopted. © 2007 IEEE

18.4 Reconfigurable Array

This section presents a self-reconfigurable camera array system which captures video sequences from an array of *mobile* cameras. In other words, not only the scene is dynamic, the cameras are moving around, too. The benefit of such a mobile array is that it can reconfigure the camera positions on the fly in order to achieve better rendering quality. An overview of the camera array is presented below, together with a multi-resolution mesh-based algorithm for view synthesis and a related algorithm for camera self-reconfiguration.

18.4.1 System Overview

As shown in Figure 18.18, the self-reconfigurable camera array system is composed of 48 (8 × 6) Axis 205 network cameras placed on six linear guides. The linear guides are 1600 mm in length, thus the average distance between cameras is about 200 mm. Vertically the

FIGURE 18.18

A self-reconfigurable camera array system with 48 cameras. © 2007 IEEE

FIGURE 18.19

The mobile camera unit. © 2007 IEEE

cameras are 150 mm apart. They can capture up to 640×480 pixel^2 images at maximally 30 fps. The cameras have built-in HTTP servers, which respond to HTTP requests and send out motion JPEG sequences. The JPEG image quality is controllable. The cameras are connected to a central computer through 100 Mbps Ethernet cables.

The cameras are mounted on a mobile platform, as shown in Figure 18.19. Each camera is attached to a standard pan servo capable of rotating for about 90 degrees. They are mounted on a platform which is equipped with another sidestep servo. The sidestep servo is a hacked one, and can rotate continuously. A gear wheel is attached to the sidestep servo which allows the platform to move horizontally with respect to the linear guide. The gear rack is added to avoid slippery during the motion. The two servos on each camera

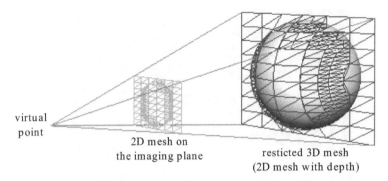

virtual
point

2D mesh on
the imaging plane

resticted 3D mesh
(2D mesh with depth)

FIGURE 18.20

The multi-resolution 2D mesh with depth information on its vertices.

unit allow the camera to have two degrees of freedom: pan and sidestep. However, the 12
cameras at the leftmost and rightmost columns have fixed positions and can only pan.

The servos are controlled by the Mini SSC II servo controller [50]. Each controller is
in charge of no more than eight servos (either standard servos or hacked ones). Multiple
controllers can be chained, thus up to 255 servos can be controlled simultaneously through
a single serial connection to a computer. The current system uses altogether 11 Mini SSC
II controllers to control 84 servos (48 pan servos, 36 sidestep servos).

The system is controlled by a single computer with an Intel Xeon 2.4 GHz dual processor, 1 GB of memory and a 32 MB NVIDIA Quadro2 EX graphics card. As will be
detailed later, the proposed rendering algorithm is so efficient that the region of interest
(ROI) identification, JPEG (Joint Photographic Experts Group) image decompression, and
camera lens distortion correction, which were usually performed with dedicated computers
in previous systems, can all be conducted during the rendering process for a camera array
in the considered system. On the other hand, it is not difficult to modify the system and
attribute ROI identification and image decoding to dedicated computers, as is done in the
distributed light field camera described in Reference [51].

The system software runs as two processes, one for capturing and the other for rendering. The capturing process is responsible for sending requests to and receiving data from
the cameras. The received images (in JPEG compressed format) are directly copied to
some shared memory that both processes can access. The capturing process is often lightly
loaded, consuming about 20% of one of the processors in the computer. When the cameras start to move, their external calibration parameters need to be calculated in real-time.
Since the internal parameters of the cameras do not change during their motion, they are
calibrated offline. To calibrate the external parameters on the fly, a large planar calibration
pattern is placed in the scene and the algorithm presented in Reference [52] is used for the
calibrating the external parameters. The calibration process runs very fast on an employed
processor (150 to 180 fps at full speed).

18.4.2 Real-Time Rendering

The real-time rendering algorithm reconstructs the geometry of the scene as a 2D multiresolution mesh (MRM) with depths on its vertices, as shown in Figure 18.20. The 2D mesh

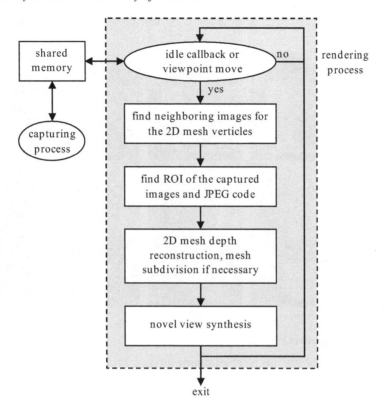

FIGURE 18.21

The flow chart of the rendering algorithm.

is positioned on the imaging plane of the virtual view, thus the geometry is view-dependent (similar to that in References [53], [54], and [55]). The MRM solution significantly reduces the amount of computation spent on depth reconstruction, making it possible to be implemented efficiently in software.

The flow chart of the rendering algorithm is shown in Figure 18.21. A novel view is rendered when there is an idle callback or the user moves the viewpoint. An initial sparse and regular 2D mesh on the imaging plane of the virtual view are constructed first. For each vertex of the initial 2D mesh, the procedure looks for a subset of images that will be used to interpolate its intensity during the rendering. Once such information has been collected, it is easy to identify the ROIs of the captured images and decode them when necessary. The depths of the vertices in the 2D mesh are then reconstructed through a plane sweeping algorithm. During plane sweeping, a set of depth values are hypothesized for a given vertex, and the color consistency verification (CCV) score for the projections on the nearby images is computed based on the mean-removed correlation coefficient as follows:

$$r_{ij} = \frac{\sum_k (I_{ik} - \bar{I}_i)(I_{jk} - \bar{I}_j)}{\sqrt{\left[\sum_k (I_{ik} - \bar{I}_i)^2\right]\left[\sum_k (I_{jk} - \bar{I}_j)^2\right]}}, \tag{18.11}$$

where I_{ik} and I_{jk} are the kth pixel intensity in the projected small patches of nearby image

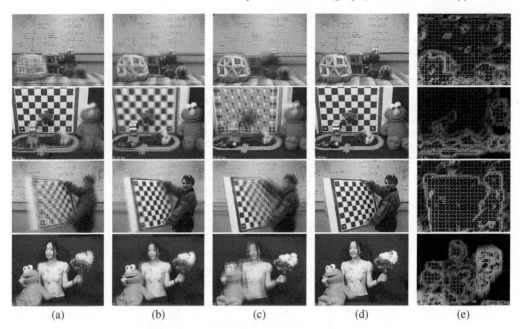

(a)   (b)   (c)   (d)   (e)

FIGURE 18.22 (See color insert.)

Scenes captured and rendered with the proposed camera array: (a) rendering with a constant depth at the background, (b) rendering with a constant depth at the middle object, (c) rendering with a constant depth at the closest object, (d) rendering with the proposed method, and (e) multi-resolution 2D mesh with depth reconstructed on-the-fly, brighter intensity means smaller depth. Captured scenes from top to bottom: *toys*, *train*, *girl and checkerboard*, and *girl and flowers*.

#i and #j, respectively. The terms $\bar{I}_i$ and $\bar{I}_j$ denote the mean of pixel intensities in the two patches. Equation 18.11 is widely used in traditional stereo matching algorithms [56]. The overall CCV score of the nearby input images is one minus the average correlation coefficient of all the image pairs. The depth plane resulting in the lowest CCV score will be selected as the scene depth.

If a certain triangle in the mesh bears large depth variation, subdivision is performed to obtain more detailed depth information. After the depth reconstruction, the novel view can be synthesized through multi-texture blending, similar to that in the unstructured lumigraph rendering (ULR) [48]. Lens distortion is corrected in the last stage, although the procedure also compensates distortion during the depth reconstruction stage.

The proposed camera array system is used to capture a variety of scenes, both static and dynamic. The speed of rendering process is about four to ten fps. The rendering results of some static scenes are shown in Figure 18.22. Note that the cameras are evenly spaced on the linear guide. The rendering positions are roughly on the camera plane but not too close to any of the capturing cameras. Figures 18.22a to 18.22c show results rendered with the constant depth assumption. The ghosting artifacts are very severe due to the large spacing between the cameras. Figure 18.22d shows the result from the proposed algorithm. The improvement is significant. Figure 18.22e shows the reconstructed 2D mesh with depth information on its vertices. The grayscale intensity represents the depth; the brighter the intensity, the closer the vertex. Like many other geometry reconstruction algorithms, the

(a)   (b)   (c)   (d)

FIGURE 18.23 (See color insert.)

Real-world scenes rendered with the proposed algorithm: (top) scene *train*, (bottom) scene *toys*; (a,c) used per-pixel depth map to render, and (b,d) the proposed adaptive mesh to render.

geometry obtained using the proposed method contains some errors. For example, in the background region of scene *toys*, the depth should be flat and far, but the achieved results have many small "bumps." This is because part of the background region has no texture, which is prone to error for depth recovery. However, the rendered results are not affected by these errors because view-dependent geometry is used and the local color consistency always holds at the viewpoint.

Figure 18.23 gives the comparison of the rendering results using a dense depth map and the proposed adaptive mesh. Using adaptive mesh produces rendering images at almost the same quality as using dense depth map, but with a much smaller computational cost.

18.4.3 Self-Reconfiguration of the Cameras

Intuitively, the CCV score mentioned above quantitatively measures the potential rendering error around a vertex in the 2D mesh. When the score is high, the reconstructed depth tends to be wrong (usually in occluded or non-Lambertian regions), and the rendered scene tends to have low quality, as shown in the first and third line of Figure 18.24c. Therefore, to improve the rendering quality, one shall move the cameras closer to these regions. In theory, such a camera self-reconfiguration problem can be formulated and solved with a recursive weighted vector quantization scheme [57]. The following briefly presents an ad-hoc algorithm for the mobile array in Figure 18.18, where the cameras are are constrained on the linear guides.

1. *Locate the camera plane and the linear guides* (as line segments on the camera plane). The camera positions in the world coordinate are obtained through the calibration process. Although they are not strictly on the same plane, an approximated one is used here, which is parallel to the checkerboard pattern in the scene. The linear guides are located by averaging the vertical positions of each row of cameras on the camera plane. As shown in Figure 18.25, the vertical coordinates of the linear guides on the camera plane are denoted as Y_j, for $j = 1, 2, ..., 6$.

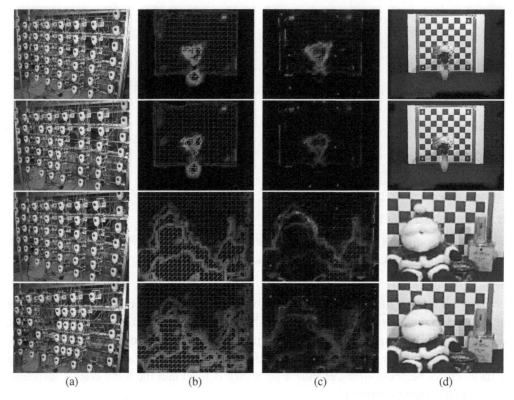

(a)   (b)   (c)   (d)

FIGURE 18.24 (See color insert.)

Scenes rendered by reconfiguring the proposed camera array: (a) the camera arrangement, (b) reconstructed depth map, brighter intensity means smaller depth, (c) CCV score of the mesh vertices and the projection of the camera positions to the virtual imaging plane denoted by dots — darker intensity means better consistency, and (d) rendered image. Captured scenes from top to bottom: *flower* with cameras evenly spaced, *flower* with cameras self-reconfigured (6 epochs), *Santa* with cameras are evenly spaced, and *Santa* with cameras self-reconfigured (20 epochs). © 2007 IEEE

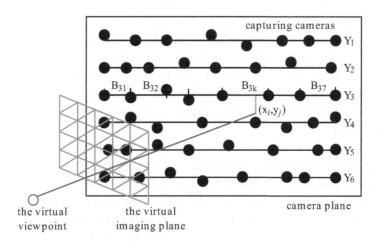

FIGURE 18.25

Self-reconfiguration of the cameras. © 2007 IEEE

2. *Back-project the vertices of the mesh model to the camera plane.* In Figure 18.25, one mesh vertex is back-projected as (x_i, y_i) on the camera plane. Note that such back-projection can be performed even if there are multiple virtual views to be rendered, thus the proposed algorithm is applicable to situations where there exist multiple virtual viewpoints.

3. *Collect CCV score for each pair of neighboring cameras on the linear guides.* The capturing cameras on each linear guide naturally divide the guide into seven segments. Let these segments be B_{jk}, where j is the row index of the linear guide, k is the index of bins on that guide, $1 \le j \le 6$, $1 \le k \le 7$. If a back-projected vertex (x_i, y_i) satisfies

$$Y_{j-1} < y_i < Y_{j+1} \quad and \quad x_i \in B_{jk}, \tag{18.12}$$

the CCV score of the vertex is added to the bin B_{jk}. After all the vertices have been back-projected, the procedure obtains a set of accumulated CCV scores for each linear guide, denoted as S_{jk}, where j is the row index of the linear guide and k is the index of bins on that guide.

4. *Determine which camera to move on each linear guide.* Given a linear guide j, the procedure looks for the largest S_{jk}, for $1 \le k \le 7$. Let it be denoted as S_{jK}. If the two cameras forming the corresponding bin B_{jK} are not too close to each other, one of them will be moved towards the other (thus reducing their distance). Note that each camera is associated with two bins. To determine which one of the two cameras should move, the procedure checks their other associated bin and moves the camera with a smaller accumulated CCV score in its other associated bin.

5. *Move the cameras.* Once the moving cameras are decided, the procedure issues them commands such as "move left" or "move right." Once the cameras are moved, the process waits until it is confirmed that the movement is finished and the cameras are re-calibrated. Then it jumps back to Step 1 for the next epoch of movement.

Some results of the proposed self-reconfiguration algorithm are shown in Figure 18.24. In the first and third line of this figure, the capturing cameras are evenly spaced on the linear guide; note that scene *flower* is rendered behind the camera plane whereas *Santa* is rendered in front of the camera plane. Due to depth discontinuities, some artifacts can be observed in the corresponding rendered images shown in Figure 18.24d along the object boundaries. Figure 18.24b shows the reconstructed depth of the scene at the virtual viewpoint. Figure 18.24c depicts the CCV score obtained during the depth reconstruction. It is obvious that the CCV score is high along the object boundaries, which usually means wrong or uncertain reconstructed depth, or bad rendering quality. The dots in Figure 18.24c are the projections of the capturing camera positions to the virtual imaging plane.

The second and fourth line in Figure 18.24 show the rendering result after reconfiguration; note that the result for scene *flower* is achieved using six epochs of camera movement, whereas the results for scene *Santa* is after twenty epochs. It can be seen from the CCV score map (Figure 18.24c) that the consistency generally gets better after the camera movement (indicated by the dots). The cameras move towards the regions where the CCV score is high, which effectively increases the sampling rate for the rendering of those regions. Figure 18.24d shows that the rendering results after self-reconfiguration are much better than those obtained using evenly spaced cameras.

The major limitation of the self-reconfigurable camera array is that the motion of the cameras is generally slow. During the self-reconfiguration of the cameras, it is necessary to

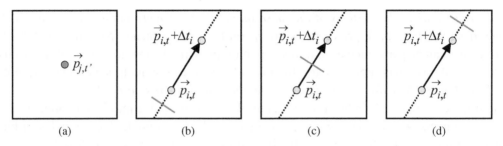

FIGURE 18.26

An illustration of various cases of the temporal offset $\tilde{t}$ between two frames: (a) a feature point in camera C_j at time t', (b-d) its epipolar line in C_i can intersect the feature trajectory, from t to $t + \Delta t_i$, in C_i in three ways.

assume that the scene is either static or moving very slowly, and the viewer is not changing his/her viewpoint all the time. A more practical system might be to have a much denser set of cameras, and limit the total number of cameras actually used for rendering the scene. The camera selection problem can be solved in a similar fashion as the recursive weighted vector quantization scheme in Reference [57].

18.5   Estimation of Temporal Offset

Low-cost commodity cameras, such as these used in the system presented in the previous section, usually provide no means of intra-camera synchronization or time-code. With unsynchronized input, it is necessary to estimate the temporal offset from the video sequences. A robust technique to estimate the temporal offset in software is presented below. Based on the epipolar constraint, this technique is similar to the temporal flow correction method described in Section 18.3.1.2, except that now the temporal offset is treated as an unknown.

Given $\vec{p}_{i,t}$'s temporal correspondence $\vec{p}_{i,t+\Delta t_i}$ from camera C_i and spatial correspondence $\vec{p}_{j,t'}$ from camera C_j, the temporal offset $\tilde{t} = t' - t$ satisfies the following:

$$\vec{p}_{i,t'} \cdot \mathbf{F_{ij}} \cdot \vec{p}_{j,t'} = 0, \quad \vec{p}_{i,t'} = \frac{\Delta t_i - \tilde{t}}{\Delta t_i}\vec{p}_{i,t} + \frac{\tilde{t}}{\Delta t_i}\vec{p}_{i,t+\Delta t_i}, \tag{18.13}$$

where $\mathbf{F_{ij}}$ is the fundamental matrix between C_i and C_j and $\vec{p}_{i,t'}$ is the estimated location at time t', assuming temporary linear motion. If all feature correspondences are correct, the equation system above can be organized as a single linear equation of one unknown $\tilde{t}$. Geometrically, this equation finds the intersection between $\vec{p}_{j,t'}$'s epipolar line and the straight line defined by $\vec{p}_{i,t}$ and $\vec{p}_{i,t+\Delta t_i}$. As shown in Figure 18.26, if the intersection happens between $\vec{p}_{i,t}$ and $\vec{p}_{i,t+\Delta t_i}$, then $0 < \tilde{t} < \Delta t$; if the intersection happens before $\vec{p}_{i,t}$, then $\tilde{t} < 0$; and if beyond $p_{i,t+\Delta t_i}$, then $\tilde{t} > \Delta t$.

Ideally given frames $I_{i,t}$, $I_{i,t+\Delta t_i}$ and $I_{j,t'}$, the offset $\tilde{t}$ can be calculated using just a single feature by Equation 18.13. Unfortunately, the flow computation is not always correct in practice. To provide a more robust and accurate estimation of the temporal offset, the following procedure is used:

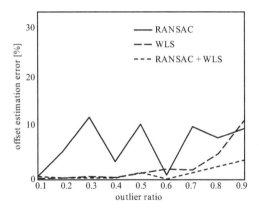

FIGURE 18.27

Temporal offset estimation error using synthetic tracked features. The ratio of outliers varies from 10% to 90%. Using three variations of the proposed techniques, the relative errors to the ground truth are plotted. The error is no greater than 15% in any test case.

1. Select salient features on image $I_{i,t}$, calculate the temporal flow from $I_{i,t}$ to $I_{i,t+\Delta t_i}$ and the spatial flow from $I_{i,t}$ to $I_{j,t'}$ using the technique described in Section 18.3.1.1.

2. Calculate the time offset $\tilde{t}_{[1]}$, $\tilde{t}_{[2]}$, ... $\tilde{t}_{[N]}$ for each feature $P_{[1]}$, $P_{[2]}$, ... $P_{[N]}$ using Equation 18.13.

3. Detect and remove time offset outliers using the random sample consensus (RANSAC) algorithm [58], assuming that outliers are primarily caused by random optical flow errors.

4. Calculate the final time offset using a weighted least squares (WLS) method, given the remaining inliers from RANSAC. The cost function is defined as follows:

$$C = \sum_{k=1}^{M} w_k(\tilde{t} - \tilde{t}_k)^2, \tag{18.14}$$

where M is the total number of remaining inliers and the weight factor w_k is defined as:

$$w_k = e^{\gamma|\tilde{t} - \tilde{t}_k|}, \quad \gamma \geq 0. \tag{18.15}$$

The last step (WLS) is repeated several times. During each regression step, the weight factor is recalculated for each inlier in order to recalculate the weighted average offset $\tilde{t}$. Since RANSAC has already removed most outliers, the weighted least squares fitting can converge fast.

18.5.1 Experiments on Temporal Offset Estimation

This section presents some experimental evaluation of the estimation procedure. The procedure was first tested with a simulated dataset in which feature points are moving

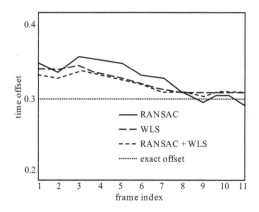

FIGURE 18.28

Estimated time offset using real image sequences (containing 11 frames). Exact time offset (0.3 frame time, or 10 ms) is also shown.

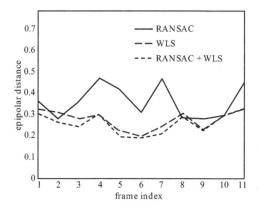

FIGURE 18.29

Weighted average epipolar distance (in pixels) from a pair of real sequences 11 frames long. The distance of each feature $\vec{p}$ is from $\vec{p}_{j,t'}$'s epipolar line on camera C_j to $\vec{p}_{i,t'}$ that is calculated using the estimated offset.

randomly. Random correspondences (outliers) are added and the ground-truth feature correspondences (inliers) are perturbed by adding a small amount of Gaussian noise. Figure 18.27 plots the accuracy with various outlier ratios. It shows that the technique is extremely robust even when 90% of the offsets are outliers.

The second dataset is a pair of video sequences with the ground-truth offset. Each sequence, captured at 30 fps, contains 11 frames. The temporal offset between the two is 10 ms, that is, 0.3 frame time. Figure 18.28 shows the estimated temporal offset, the error is typically with in 0.05 frame time. Figure 18.29 plots the weighted average epipolar distance using the estimated time offset. Ideally, the epipolar line should go right through the linearly interpolated feature point (as in Equation 18.13). Figure 18.29 shows that subpixel accuracy can be achieved.

From the above experiments with known ground truth, it can be seen that the proposed approach can produce very accurate and robust estimation of the temporal offset. In ad-

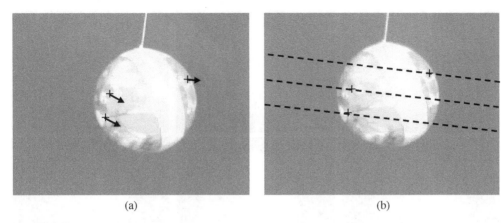

(a) (b)

FIGURE 18.30

Verifying estimated time offset using epipolar lines: (a) temporal optical flow on camera C_i and (b) the epipolar line of $\vec{p}_{i,t'}$, which is linearly interpolated based on the temporal offset, and $\vec{p}_{j,t'}$ on camera C_j.

dition, combining RANSAC and weighted least squares fitting yields better results than either of these techniques alone.

18.5.1.1 View Synthesis Results

The last experiment uses a dataset *without* a global time stamp. It contains a swinging toy ball. All cameras are started simultaneously to capture a set of roughly aligned sequences. The temporal offset for each sequence pair is estimated, using the sequence from the middle camera as the reference. Since the ground truth offset is unknown, one can only verify the offset accuracy by looking at the epipolar distances for temporally interpolated feature points. Figure 18.30a shows the temporal optical flow seen on a frame $I_{i,t}$ and Figure 18.30b shows the $\vec{p}_{i,t'}$'s epipolar lines created using the estimated time offset on the frame $I_{j,t'}$. Note that the spatial correspondences $\vec{p}_{j,t'}$ are right on the epipolar lines, which verifies the correctness of the estimated temporal offset. Using a planar proxy, the synthesized results are shown in Figure 18.31. A visualization of the motion trajectory of the ball can be found in the last row of Figure 18.32.

18.6 Conclusion

This chapter provided a brief overview of an important branch for view synthesis, namely methods based on the concept of light field rendering (LFR). The technical discussions were focused on extending traditional LFR to the temporal domain to accommodate dynamic scenes. Instead of capturing the dynamic scene in strict synchronization and treating each image set as an independent static light field, the notion of a *space-time light field* simply assumes a collection of video sequences. These sequences may or may not be synchronized and they can have different capture rates.

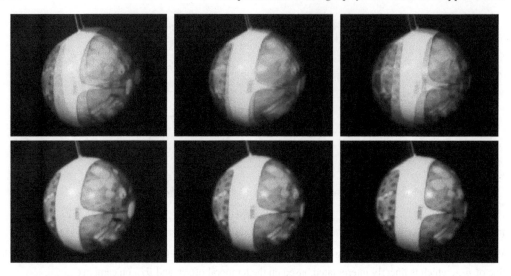

FIGURE 18.31 (See color insert.)

Synthesized results of the toy ball sequence: (top) traditional LFR with unsynchronized frames and (bottom) space-time LFR using estimated temporal offsets among input sequences. Note that the input dataset does not contain global time stamps.

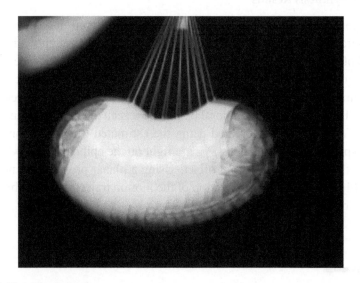

FIGURE 18.32 (See color insert.)

Visualization of the trajectory.

In order to be able to synthesize novel views from any viewpoint at any time instant, feature correspondences are robustly identified across frames. They are used as land markers to digitally synchronize the input frames and improve view synthesis quality. Furthermore, this chapter presented a reconfigurable camera array in which the cameras' placement can be automatically adjusted to achieve optimal view synthesis results for different scene contents. With the ever-decreasing cost of web cameras and the increased computational and communication capability of modern hardware, it is believed that light field rendering techniques can be adopted in many interesting applications such as 3D video teleconferencing, remote surveillance, and tele-medicine.

Acknowledgment

Figures 18.4 to 18.17 are reprinted from Reference [38] and Figures 18.18, 18.19, 18.24, and 18.25 are reprinted from Reference [57], with the permission of IEEE.

References

[1] M. Levoy and P. Hanrahan, "Light field rendering," in *Proceedings of the 23rd International Conference on Computer Graphics and Interactive Techniques*, New Orleans, LA, USA, August 1996, pp. 31–42.

[2] S.J. Gortler, R. Grzeszczuk, R. Szeliski, and M.F. Cohen, "The lumigraph," in *Proceedings of the 23rd International Conference on Computer Graphics and Interactive Techniques*, New Orleans, LA, USA, August 1996, pp. 43–54.

[3] P.E. Debevec, C.J. Taylor, and J. Malik, "Modeling and rendering architecture from photographs: A hybrid geometry-and image-based approach," in *Proceedings of the 23rd International Conference on Computer Graphics and Interactive Techniques*, New Orleans, LA, United States, August 1996, pp. 11–20.

[4] L. McMillan and G. Bishop, "Plenoptic modeling: An image-based rendering system.," in *Proceedings of the 22nd International Conference on Computer Graphics and Interactive Techniques*, Los Angeles, CA, USA, August 1995, pp. 39–46.

[5] L. McMillan, *An Image-Based Approach to Three-Dimensional Computer Graphics*, PhD thesis, University of North Carolina at Chapel Hill, 1997.

[6] I. Ihm, S. Park, and R.K. Lee, "Rendering of spherical light fields," in *Proceedings of Pacific Conference on Computer Graphics and Applications*, Seoul, Korea, October 1997, p. 59.

[7] E. Camahort, A. Lerios, and D. Fussell, "Uniformly sampled light fields," in *Proceedings of Eurographics Rendering Workshop*, Vienna, Austria, July 1998, pp. 117–130.

[8] A. Isaksen, L. McMillan, and S.J. Gortler, "Dynamically reparameterized light fields," in *Proceedings of the 27th International Conference on Computer Graphics and Interactive Techniques*, New Orleans, LA, USA, August 2000, pp. 297–306.

[9] P.P. Sloan, M.F. Cohen, and S.J. Gortler, "Time critical lumigraph rendering," in *Proceedings of Symposium on Interactive 3D Graphics*, Providence, RI, USA, April 1997, pp.17–23.

[10] H.Y. Shum and L.W. He, "Rendering with Concentric Mosaics," in *Proceedings of the 24th International Conference on Computer Graphics and Interactive Techniques*, Los Angeles, CA, USA, July 1997, pp. 299–306.

[11] W. Li, Q. Ke, X. Huang, and N. Zheng, "Light field rendering of dynamic scenes," *Machine Graphics and Vision*, vol. 7, no. 3, pp. 551–563, 1998.

[12] H. Schirmacher, C. Vogelgsang, H.P. Seidel, and G. Greiner, "Efficient free form light field rendering," in *Proceedings of Vision, Modeling, and Visualization*, Stuttgart, Germany, November 2001, pp. 249–256.

[13] P.P. Sloan and C. Hansen, "Parallel lumigraph reconstruction," in *Proceedings of Symposium on Parallel Visualization and Graphics*, San Francisco, CA, USA, October 1999, pp. 7–15.

[14] H. Schirmacher, W. Heidrich, and H.P. Seidel, "High-quality interactive lumigraph rendering through warping," in *Proceedings of Graphics Interface*, Montreal, Canada, May 2000, pp. 87–94.

[15] E.H. Adelson and J. Bergen, *Computational Models of Visual Processing*, ch. "The plenoptic function and the elements of early vision," M.S. Landy and J.A. Movshon (eds.), Cambridge, MA: MIT Press, August 1991, pp. 3–20.

[16] L.A. Westover, "Footprint Evaluation for Volume Rendering," in *Proceedings of the 17th International Conference on Computer Graphics and Interactive Techniques*, Dallas, TX, USA, August 1990, pp. 367–376.

[17] D. Anderson, "Hidden Line Elimination in Projected Grid Surfaces," *ACM Transactions on Graphics*, vol. 4, no. 1, pp. 274–288, October 1982.

[18] R. Levin, "Photometric characteristics of light controlling apparatus," *Illuminating Engineering*, vol. 66, no. 4, pp. 205–215, 1971.

[19] I. Ashdown, "Near-field photometry: A new approach," *Journal of the Illuminating Engineering Society*, vol. 22, no. 1, pp. 163–180, Winter 1993.

[20] T. Okoshi, *Three-Dimensional Imaging Techniques*. New York: Academic Press, Inc., February 1977.

[21] J.X. Chai, X. Tong, S.C. Chan, and H.Y. Shum, "Plenoptic Sampling," in *Proceedings of the 27th International Conference on Computer Graphics and Interactive Techniques*, New Orleans, Louisiana, USA, July 2000, pp. 307–318.

[22] J. Blinn and M. Newell, "Texture and reflection in computer generated images," *Communications of the ACM*, vol. 19, no. 10, pp. 542–547, October 1976.

[23] N. Greenem, "Environment mapping and other applications of world projections," *IEEE Computer Graphics and Applications*, vol. 6, no. 11, pp. 21–29, November 1986.

[24] S.E. Chen, "Quicktime VR: An image-based approach to virtual environment navigation," in *Proceedings of SIGGRAPH 1995*, Los Angeles, LA, USA, August 1995, pp. 29–38.

[25] J. Meehan, *Panoramic Photography*. New York: Amphoto, March 1996.

[26] B. Technology. http://www.behere.com.

[27] S.K. Nayar, "Catadioptric omnidirectional camera," in *Proceedings of Conference on Computer Vision and Pattern Recognition*, San Juan, Puerto Rico, June 1997, p. 482.

[28] R. Szeliski and H.Y. Shum, "Creating full view panoramic image mosaics and environment maps," in *Proceedings of the 24th International Conference on Computer Graphics and Interactive Techniques*, Los Angeles, CA, USA, August 1997, pp. 251–258.

[29] M. Irani, P. Anandan, and S. Hsu, "Mosaic based representations of video sequences and their applications," in *Proceedings of International Conference on Computer Vision*, Cambridge, MA, USA, June 1995, p. 605.

[30] S. Mann and R.W. Picard, "Virtual bellows: Constructing high-quality images from video.," in *Proceedings of International Conference on Image Processing*, Austin, TX, USA, November 1994, pp. 363–367.

[31] R. Szeliski, "Image mosaicing for tele-reality applications," in *Proceedings of IEEE Workshop on Applications of Computer Vision*, Sarasota, FL, USA, December 1994, pp. 44–53.

[32] R. Szeliski, "Video mosaics for virtual environments," *IEEE Computer Graphics and Applications*, vol. 16, no. 2, pp. 22–30, March 1996.

[33] T. Naemura, J. Tago, and H. Harashima, "Realtime video-based modeling and rendering of 3D scenes," *IEEE Computer Graphics and Applications*, vol. 22, no. 2, pp. 66–73, March/April 2002.

[34] J.C. Yang, M. Everett, C. Buehler, and L. McMillan, "A real-time distributed light field camera," in *Proceedings of the 13th Eurographics Workshop on Rendering*, Pisa, Italy, June 2002, pp. 77–86.

[35] B. Wilburn, M. Smulski, H. Lee, and M. Horowitz, "The light field video camera," in *Proceedings of SPIE Electronic Imaging Conference*, San Jose, CA, USA, January 2002.

[36] W. Matusik and H. Pfister, "3D TV: A scalable system for real-time acquisition, transmission, and autostereoscopic display of dynamic scenes," *ACM Transactions on Graphics*, vol. 23, no. 3, pp. 814–824, August 2004.

[37] B. Wilburn, N. Joshi, V. Vaish, E.V. Talvala, E. Antunez, A. Barth, A. Adams, M. Horowitz, and M. Levoy, "High performance imaging using large camera arrays," *ACM Transactions on Graphics*, vol. 24, no. 3, pp. 765–776, July 2005.

[38] H. Wang, M. Sun, and R. Yang, "Space-time light field rendering," *IEEE Transactions on Visualization and Computer Graphics*, vol. 13, no. 4, pp. 697–710, July/August 2007.

[39] C.J. Harris and M. Stephens, "A combined corner and edge detector," in *Proceedings of 4th Alvey Vision Conference*, Manchester, UK, August 1988, pp. 147–151.

[40] J.Y. Bouguet, "Pyramidal implementation of the Lucas Kanade feature tracker description of the algorithm," Technical Report, 1999.

[41] B.D. Lucas and T. Kanade, "An Iterative Image Registration Technique with an Application to Stereo Vision," in *Proceedings of International Joint Conference on Artificial Intelligence*, Vancouver, BC, Canada, August 1981, pp. 674–679.

[42] C. Tomasi and T. Kanade, "Detection and tracking of point features," Technical Report CMU-CS-91-132, Carnegie Mellon University, 1991.

[43] O. Faugeras, *Three-Dimensional Computer Vision: A Geometric Viewpoint*. Cambridge, MA: MIT Press, November 1993.

[44] L. Zhang, B. Curless, and S. Seitz, "Spacetime stereo: Shape recovery for dynamic scenes," in *Proceedings of Conference on Computer Vision and Pattern Recognition*, Madison, WI, USA, June 2003, pp. 367–374.

[45] T. Beier and S. Neely, "Feature based image metamorphosis," *SIGGRAPH Computer Graphics*, vol. 26, no. 2, pp. 35–42, July 1992.

[46] J. Canny, "A computational approach to edge detection," *IEEE Transactions on Pattern Analysis and Machine Intelligence*, vol. 8, no. 6, pp. 679–698, November 1986.

[47] J.R. Shewchuk, "Triangle: Engineering a 2D quality mesh generator and Delaunay triangulato," *Lecture Notes in Computer Science*, vol. 1148, pp. 203–222, August 1996.

[48] C. Buehler, M. Bosse, L. McMillan, S. Gortler, and M. Cohen, "Unstructured lumigraph rendering," in *Proceedings of the 28th International Conference on Computer Graphics and Interactive Techniques*, Los Angeles, CA, USA, August 2001, pp. 405–432.

[49] Point Grey Research Inc., Available online: http://www.ptgrey.com.

[50] MiniSSC-II, "Scott Edwards Electronics Inc., Available online: http://www.seetron.com/ssc.htm,"

[51] J.C. Yang, M. Everett, C. Buehler, and L. McMillan, "A real-time distributed light field camera," in *Proceedings of Eurographics Workshop on Rendering*, Pisa, Italy, June 2002.

[52] Z. Zhang, "A flexible new technique for camera calibration," Technical Report, MSR-TR-98-71, 1998.

[53] R. Yang, G. Welch, and G. Bishop, "Real-time consensus-based scene reconstruction using commodity graphics hardware," in *Proceedings of Pacific Conference on Computer Graphics and Applications*, Beijing, China, October 2002.

[54] G. G. Slabaugh, R. W. Schafer, and M. C. Hans, "Image-based photo hulls," Technical Report HPL-2002-28, HP Labs, 2002.

[55] W. Matusik, C. Buehler, and L. McMillan, "Polyhedral visual hulls for real-time rendering," in *Proceedings of Eurographics Workshop on Rendering*, London, UK, June 2001.

[56] O. Faugeras, B. Hotz, H. Mathieu, T. Viéville, Z. Zhang, P. Fua, E. Théron, L. Moll, G. Berry, J. Vuillemin, P. Bertin, and C. Proy, "Real time correlation-based stereo: Algorithm, implementations and applications," Technical Report 2013, INRIA, 1993.

[57] C. Zhang and T. Chen, "Active rearranged capturing of image-based rendering scenes - Theory and practice," *IEEE Transactions on Multimedia*, vol. 9, no. 3, pp. 520–531, April 2007.

[58] M.A. Fischler and R.C. Bolles, "Random sample consensus: A paradigm for model fitting with applications to image analysis and automated cartography," *Communication of the ACM*, vol. 24, no. 6, pp. 381–395, June 1981.

Index

T - #0046 - 101024 - C32 - 254/178/32 [34] - CB - 9781439817490 - Gloss Lamination